LONDON MATHEMATICAL SOCIETY LECTURE NOTE SERIES

Managing Editor: Professor M. Reid, Mathematics Institute,
University of Warwick, Coventry CV4 7AL, United Kingdom

The titles below are available from booksellers, or from Cambridge University Press at
www.cambridge.org/mathematics

247 Analytic number theory, Y. MOTOHASHI (ed)
248 Tame topology and O-minimal structures, L. VAN DEN DRIES
249 The atlas of finite groups - ten years on, R.T. CURTIS & R.A. WILSON (eds)
250 Characters and blocks of finite groups, G. NAVARRO
251 Gröbner bases and applications, B. BUCHBERGER & F. WINKLER (eds)
252 Geometry and cohomology in group theory, P.H. KROPHOLLER, G.A. NIBLO & R. STÖHR (eds)
253 The q-Schur algebra, S. DONKIN
254 Galois representations in arithmetic algebraic geometry, A.J. SCHOLL & R.L. TAYLOR (eds)
255 Symmetries and integrability of difference equations, P.A. CLARKSON & F.W. NIJHOFF (eds)
256 Aspects of Galois theory, H. VÖLKLEIN, J.G. THOMPSON, D. HARBATER & P. MÜLLER (eds)
257 An introduction to noncommutative differential geometry and its physical applications (2nd Edition), J. MADORE
258 Sets and proofs, S.B. COOPER & J.K. TRUSS (eds)
259 Models and computability, S.B. COOPER & J. TRUSS (eds)
260 Groups St Andrews 1997 in Bath I, C.M. CAMPBELL et al (eds)
261 Groups St Andrews 1997 in Bath II, C.M. CAMPBELL et al (eds)
262 Analysis and logic, C.W. HENSON, J. IOVINO, A.S. KECHRIS & E. ODELL
263 Singularity theory, W. BRUCE & D. MOND (eds)
264 New trends in algebraic geometry, K. HULEK, F. CATANESE, C. PETERS & M. REID (eds)
265 Elliptic curves in cryptography, I. BLAKE, G. SEROUSSI & N. SMART
267 Surveys in combinatorics, 1999, J.D. LAMB & D.A. PREECE (eds)
268 Spectral asymptotics in the semi-classical limit, M. DIMASSI & J. SJÖSTRAND
269 Ergodic theory and topological dynamics of group actions on homogeneous spaces, M.B. BEKKA & M. MAYER
271 Singular perturbations of differential operators, S. ALBEVERIO & P. KURASOV
272 Character theory for the odd order theorem, T. PETERFALVI. Translated by R. SANDLING
273 Spectral theory and geometry, E.B. DAVIES & Y. SAFAROV (eds)
274 The Mandelbrot set, theme and variations, T. LEI (ed)
275 Descriptive set theory and dynamical systems, M. FOREMAN, A.S. KECHRIS, A. LOUVEAU & B. WEISS (eds)
276 Singularities of plane curves, E. CASAS-ALVERO
277 Computational and geometric aspects of modern algebra, M. ATKINSON et al (eds)
278 Global attractors in abstract parabolic problems, J.W. CHOLEWA & T. DLOTKO
279 Topics in symbolic dynamics and applications, F. BLANCHARD, A. MAASS & A. NOGUEIRA (eds)
280 Characters and automorphism groups of compact Riemann surfaces, T. BREUER
281 Explicit birational geometry of 3-folds, A. CORTI & M. REID (eds)
282 Auslander-Buchweitz approximations of equivariant modules, M. HASHIMOTO
283 Nonlinear elasticity, Y.B. FU & R.W. OGDEN (eds)
284 Foundations of computational mathematics, R. DEVORE, A. ISERLES & E. SÜLI (eds)
285 Rational points on curves over finite fields, H. NIEDERREITER & C. XING
286 Clifford algebras and spinors (2nd Edition), P. LOUNESTO
287 Topics on Riemann surfaces and Fuchsian groups, E. BUJALANCE, A.F. COSTA & E. MARTÍNEZ (eds)
288 Surveys in combinatorics, 2001, J.W.P. HIRSCHFELD (ed)
289 Aspects of Sobolev-type inequalities, L. SALOFF-COSTE
290 Quantum groups and Lie theory, A. PRESSLEY (ed)
291 Tits buildings and the model theory of groups, K. TENT (ed)
292 A quantum groups primer, S. MAJID
293 Second order partial differential equations in Hilbert spaces, G. DA PRATO & J. ZABCZYK
294 Introduction to operator space theory, G. PISIER
295 Geometry and integrability, L. MASON & Y. NUTKU (eds)
296 Lectures on invariant theory, I. DOLGACHEV
297 The homotopy category of simply connected 4-manifolds, H.-J. BAUES
298 Higher operads, higher categories, T. LEINSTER (ed)
299 Kleinian groups and hyperbolic 3-manifolds, Y. KOMORI, V. MARKOVIC & C. SERIES (eds)
300 Introduction to Möbius differential geometry, U. HERTRICH-JEROMIN
301 Stable modules and the D(2)-problem, F.E.A. JOHNSON
302 Discrete and continuous nonlinear Schrödinger systems, M.J. ABLOWITZ, B. PRINARI & A.D. TRUBATCH
303 Number theory and algebraic geometry, M. REID & A. SKOROBOGATOV (eds)
304 Groups St Andrews 2001 in Oxford I, C.M. CAMPBELL, E.F. ROBERTSON & G.C. SMITH (eds)
305 Groups St Andrews 2001 in Oxford II, C.M. CAMPBELL, E.F. ROBERTSON & G.C. SMITH (eds)

306 Geometric mechanics and symmetry, J. MONTALDI & T. RATIU (eds)
307 Surveys in combinatorics 2003, C.D. WENSLEY (ed.)
308 Topology, geometry and quantum field theory, U.L. TILLMANN (ed)
309 Corings and comodules, T. BRZEZINSKI & R. WISBAUER
310 Topics in dynamics and ergodic theory, S. BEZUGLYI & S. KOLYADA (eds)
311 Groups: topological, combinatorial and arithmetic aspects, T.W. MÜLLER (ed)
312 Foundations of computational mathematics, Minneapolis 2002, F. CUCKER *et al* (eds)
313 Transcendental aspects of algebraic cycles, S. MÜLLER-STACH & C. PETERS (eds)
314 Spectral generalizations of line graphs, D. CVETKOVIĆ, P. ROWLINSON & S. SIMIĆ
315 Structured ring spectra, A. BAKER & B. RICHTER (eds)
316 Linear logic in computer science, T. EHRHARD, P. RUET, J.-Y. GIRARD & P. SCOTT
     (eds)
317 Advances in elliptic curve cryptography, I.F. BLAKE, G. SEROUSSI & N.P. SMART
     (eds)
318 Perturbation of the boundary in boundary-value problems of partial differential equations,
     D. HENRY
319 Double affine Hecke algebras, I. CHEREDNIK
320 L-functions and Galois representations, D. BURNS, K. BUZZARD & J. NEKOVÁŘ (eds)
321 Surveys in modern mathematics, V. PRASOLOV & Y. ILYASHENKO (eds)
322 Recent perspectives in random matrix theory and number theory, F. MEZZADRI & N.C.
     SNAITH (eds)
323 Poisson geometry, deformation quantisation and group representations, S. GUTT *et al*
     (eds)
324 Singularities and computer algebra, C. LOSSEN & G. PFISTER (eds)
325 Lectures on the Ricci flow, P. TOPPING
326 Modular representations of finite groups of Lie type, J.E. HUMPHREYS
327 Surveys in combinatorics 2005, B.S. WEBB (ed)
328 Fundamentals of hyperbolic manifolds, R. CANARY, D. EPSTEIN & A. MARDEN (eds)
329 Spaces of Kleinian groups, Y. MINSKY, M. SAKUMA & C. SERIES (eds)
330 Noncommutative localization in algebra and topology, A. RANICKI (ed)
331 Foundations of computational mathematics, Santander 2005, L.M PARDO, A. PINKUS,
     E. SÜLI & M.J. TODD (eds)
332 Handbook of tilting theory, L. ANGELERI HÜGEL, D. HAPPEL & H. KRAUSE (eds)
333 Synthetic differential geometry (2nd Edition), A. KOCK
334 The Navier–Stokes equations, N. RILEY & P. DRAZIN
335 Lectures on the combinatorics of free probability, A. NICA & R. SPEICHER
336 Integral closure of ideals, rings, and modules, I. SWANSON & C. HUNEKE
337 Methods in Banach space theory, J. M. F. CASTILLO & W. B. JOHNSON (eds)
338 Surveys in geometry and number theory, N. YOUNG (ed)
339 Groups St Andrews 2005 I, C.M. CAMPBELL, M.R. QUICK, E.F. ROBERTSON & G.C.
     SMITH (eds)
340 Groups St Andrews 2005 II, C.M. CAMPBELL, M.R. QUICK, E.F. ROBERTSON &
     G.C. SMITH (eds)
341 Ranks of elliptic curves and random matrix theory, J.B. CONREY, D.W. FARMER, F.
     MEZZADRI & N.C. SNAITH (eds)
342 Elliptic cohomology, H.R. MILLER & D.C. RAVENEL (eds)
343 Algebraic cycles and motives I, J. NAGEL & C. PETERS (eds)
344 Algebraic cycles and motives II, J. NAGEL & C. PETERS (eds)
345 Algebraic and analytic geometry, A. NEEMAN
346 Surveys in combinatorics 2007, A. HILTON & J. TALBOT (eds)
347 Surveys in contemporary mathematics, N. YOUNG & Y. CHOI (eds)
348 Transcendental dynamics and complex analysis, P.J. RIPPON & G.M. STALLARD (eds)
349 Model theory with applications to algebra and analysis I, Z. CHATZIDAKIS, D.
     MACPHERSON, A. PILLAY & A. WILKIE (eds)
350 Model theory with applications to algebra and analysis II, Z. CHATZIDAKIS, D.
     MACPHERSON, A. PILLAY & A. WILKIE (eds)
351 Finite von Neumann algebras and masas, A.M. SINCLAIR & R.R. SMITH
352 Number theory and polynomials, J. MCKEE & C. SMYTH (eds)
353 Trends in stochastic analysis, J. BLATH, P. MÖRTERS & M. SCHEUTZOW (eds)
354 Groups and analysis, K. TENT (ed)
355 Non-equilibrium statistical mechanics and turbulence, J. CARDY, G. FALKOVICH & K.
     GAWEDZKI
356 Elliptic curves and big Galois representations, D. DELBOURGO
357 Algebraic theory of differential equations, M.A.H. MACCALLUM & A.V. MIKHAILOV
     (eds)
358 Geometric and cohomological methods in group theory, M.R. BRIDSON, P.H.
     KROPHOLLER & I.J. LEARY (eds)
359 Moduli spaces and vector bundles, L. BRAMBILA-PAZ, S.B. BRADLOW, O.
     GARCÍA-PRADA & S. RAMANAN (eds)
360 Zariski geometries, B. ZILBER
361 Words: Notes on verbal width in groups, D. SEGAL
362 Differential tensor algebras and their module categories, R. BAUTISTA, L. SALMERÓN
     & R. ZUAZUA
363 Foundations of computational mathematics, Hong Kong 2008, F. CUCKER, A. PINKUS
     & M.J. TODD (eds)
364 Partial differential equations and fluid mechanics, J.C. ROBINSON & J.L. RODRIGO
     (eds)
365 Surveys in combinatorics 2009, S. HUCZYNSKA, J.D. MITCHELL & C.M.
     RONEY-DOUGAL (eds)
366 Highly oscillatory problems, B. ENGQUIST, A. FOKAS, E. HAIRER & A. ISERLES
     (eds)

London Mathematical Society Lecture Note Series: 367

Random Matrices: High Dimensional Phenomena

GORDON BLOWER
Lancaster University

CAMBRIDGE
UNIVERSITY PRESS

CAMBRIDGE
UNIVERSITY PRESS

University Printing House, Cambridge CB2 8BS, United Kingdom

One Liberty Plaza, 20th Floor, New York, NY 10006, USA

477 Williamstown Road, Port Melbourne, VIC 3207, Australia

314-321, 3rd Floor, Plot 3, Splendor Forum, Jasola District Centre, New Delhi - 110025, India

103 Penang Road, #05-06/07, Visioncrest Commercial, Singapore 238467

Cambridge University Press is part of the University of Cambridge.

It furthers the University's mission by disseminating knowledge in the pursuit of education, learning and research at the highest international levels of excellence.

www.cambridge.org
Information on this title: www.cambridge.org/9780521133128

© G. Blower 2009

First published 2009

A catalogue record for this publication is available from the British Library

ISBN 978-0-521-13312-8 Paperback

To the memory of my father
Ronald Frederick Blower

Contents

Introduction   *page* 1

1 **Metric measure spaces**   4
   1.1 Weak convergence on compact metric spaces   4
   1.2 Invariant measure on a compact metric group   10
   1.3 Measures on non-compact Polish spaces   16
   1.4 The Brunn–Minkowski inequality   22
   1.5 Gaussian measures   25
   1.6 Surface area measure on the spheres   27
   1.7 Lipschitz functions and the Hausdorff metric   31
   1.8 Characteristic functions and Cauchy transforms   33

2 **Lie groups and matrix ensembles**   42
   2.1 The classical groups, their eigenvalues and norms   42
   2.2 Determinants and functional calculus   49
   2.3 Linear Lie groups   56
   2.4 Connections and curvature   63
   2.5 Generalized ensembles   66
   2.6 The Weyl integration formula   72
   2.7 Dyson's circular ensembles   78
   2.8 Circular orthogonal ensemble   81
   2.9 Circular symplectic ensemble   83

3 **Entropy and concentration of measure**   84
   3.1 Relative entropy   84
   3.2 Concentration of measure   93
   3.3 Transportation   99
   3.4 Transportation inequalities   103
   3.5 Transportation inequalities for uniformly
        convex potentials   106
   3.6 Concentration of measure in matrix ensembles   109

3.7	Concentration for rectangular Gaussian matrices	114
3.8	Concentration on the sphere	123
3.9	Concentration for compact Lie groups	126

4   Free entropy and equilibrium   **132**

4.1	Logarithmic energy and equilibrium measure	132
4.2	Energy spaces on the disc	134
4.3	Free versus classical entropy on the spheres	142
4.4	Equilibrium measures for potentials on the real line	147
4.5	Equilibrium densities for convex potentials	154
4.6	The quartic model with positive leading term	159
4.7	Quartic models with negative leading term	164
4.8	Displacement convexity and relative free entropy	169
4.9	Toeplitz determinants	172

5   Convergence to equilibrium   **177**

5.1	Convergence to arclength	177
5.2	Convergence of ensembles	179
5.3	Mean field convergence	183
5.4	Almost sure weak convergence for uniformly convex potentials	189
5.5	Convergence for the singular numbers from the Wishart distribution	193

6   Gradient flows and functional inequalities   **196**

6.1	Variation of functionals and gradient flows	196
6.2	Logarithmic Sobolev inequalities	203
6.3	Logarithmic Sobolev inequalities for uniformly convex potentials	206
6.4	Fisher's information and Shannon's entropy	210
6.5	Free information and entropy	213
6.6	Free logarithmic Sobolev inequality	218
6.7	Logarithmic Sobolev and spectral gap inequalities	221
6.8	Inequalities for Gibbs measures on Riemannian manifolds	223

7   Young tableaux   **227**

7.1	Group representations	227
7.2	Young diagrams	229
7.3	The Vershik Ω distribution	237
7.4	Distribution of the longest increasing subsequence	243
7.5	Inclusion-exclusion principle	250

8   Random point fields and random matrices   253
   8.1   Determinantal random point fields   253
   8.2   Determinantal random point fields on the real line   261
   8.3   Determinantal random point fields and orthogonal
         polynomials   270
   8.4   De Branges's spaces   274
   8.5   Limits of kernels   278

9   Integrable operators and differential equations   281
   9.1   Integrable operators and Hankel integral operators   281
   9.2   Hankel integral operators that commute with second
         order differential operators   289
   9.3   Spectral bulk and the sine kernel   293
   9.4   Soft edges and the Airy kernel   299
   9.5   Hard edges and the Bessel kernel   304
   9.6   The spectra of Hankel operators and rational
         approximation   310
   9.7   The Tracy–Widom distribution   315

10   Fluctuations and the Tracy–Widom distribution   321
   10.1   The Costin–Lebowitz central limit theorem   321
   10.2   Discrete Tracy–Widom systems   327
   10.3   The discrete Bessel kernel   328
   10.4   Plancherel measure on the partitions   334
   10.5   Fluctuations of the longest increasing subsequence   343
   10.6   Fluctuations of linear statistics over unitary
          ensembles   345

11   Limit groups and Gaussian measures   352
   11.1   Some inductive limit groups   352
   11.2   Hua–Pickrell measure on the infinite unitary group   357
   11.3   Gaussian Hilbert space   365
   11.4   Gaussian measures and fluctuations   369

12   Hermite polynomials   373
   12.1   Tensor products of Hilbert space   373
   12.2   Hermite polynomials and Mehler's formula   375
   12.3   The Ornstein–Uhlenbeck semigroup   381
   12.4   Hermite polynomials in higher dimensions   384

13   From the Ornstein–Uhlenbeck process to the
**   Burgers equation**   392
   13.1   The Ornstein–Uhlenbeck process   392

13.2   The logarithmic Sobolev inequality for the
       Ornstein–Uhlenbeck generator   396
13.3   The matrix Ornstein–Uhlenbeck process   398
13.4   Solutions for matrix stochastic differential equations   401
13.5   The Burgers equation   408

14   Noncommutative probability spaces   411
14.1   Noncommutative probability spaces   411
14.2   Tracial probability spaces   414
14.3   The semicircular distribution   418
References   424
Index   433

Introduction

The purpose of this book is to introduce readers to certain topics in random matrix theory that specifically involve the phenomenon of concentration of measure in high dimension. Partly this work was motivated by researches in the EC network *Phenomena in High Dimension*, which applied results from functional analysis to problems in statistical physics. Pisier described this as the transfer of technology, and this book develops this philosophy by discussing applications to random matrix theory of:

  (i) optimal transportation theory;
 (ii) logarithmic Sobolev inequalities;
(iii) exponential concentration inequalities;
(iv) Hankel operators.

Recently some approaches to functional inequalities have emerged that make a unified treatment possible; in particular, optimal transportation links together seemingly disparate ideas about convergence to equilibrium. Furthermore, optimal transportation connects familiar results from the calculus of variations with the modern theory of diffusions and gradient flows.

I hope that postgraduate students will find this book useful and, with them in mind, have selected topics with potential for further development. Prerequisites for this book are linear algebra, calculus, complex analysis, Lebesgue integration, metric spaces and basic Hilbert space theory. The book does not use stochastic calculus or the theory of integrable systems, so as to widen the possible readership.

In their survey of random matrices and Banach spaces, Davidson and Szarek present results on Gaussian random matrices and then indicate that some of the results should extend to a wider context by the

1

theory of concentration of measure [152]. This book follows this programme in the context of generalized orthogonal ensembles and compact Lie groups. While the Gaussian unitary ensemble and Wishart
ensembles have special properties, they provide a helpful model for
other cases. The book covers the main examples of the subject, such
as Gaussian random matrices, within the general context of invariant
ensembles.

The coverage of material is deliberately uneven, in that some topics
are treated more thoroughly than others and some results from other areas of analysis are recalled with minimal discussion. There are detailed
accounts of familiar topics such as the equilibrium measure of the quartic potential, since these illustrate techniques that are useful in many
problems. The book develops classical and free probability in parallel,
in the hope that the analogy makes free probability more accessible.

The presentation is mainly rigorous, although some important proofs
are omitted. In order to understand the standard ensembles of random
matrix theory, the reader must have some knowledge of Lie groups, so
the book contains an abbreviated treatment which covers the main cases
that are required and emphasizes the classical compact linear groups.
Likewise, the presentations of Gaussian measures in Chapter 11 and the
Ornstein–Uhlenbeck process in Chapters 12 and 13 are self-contained,
but do not give a complete perspective on the theory. Similarly, the
treatment of free probability describes only one aspect of the topic.

Some of the results and proofs are new, although the lack of a specific
reference does not imply originality. In preparing the Sections 2.3, 2.4
and 2.6 on Lie groups, I have used unpublished notes from lectures given
by Brian Steer in Oxford between 1987 and 1991. Chapter 5 features results originally published by the author in [17] and [19], with technical
improvements due to ideas from Bolley's thesis [29]. The material in
Chapter 6 on gradient flows was originally written for an instructional
lecture to postgraduate students attending the North British Functional Analysis Seminar at Lancaster in 2006; likewise, Sections 8.1, 8.2,
and 7.3 are drawn from postgraduate lectures at Lancaster. Conversely,
Sections 7.2, 12.2 and 2.5 are based upon dissertations written by my
former students Katherine Peet, Stefan Olphert and James Groves. Sub-
stantial portions of Chapter 9 and Section 10.3 are taken from Andrew
McCafferty's PhD thesis [113], which the author supervised.

In his authoritative guide to lakeland hillwalking [168], Wainwright
offers the general advice that one should keep moving, and he discusses
6 possible ascents of Scafell Pike, the optimal route depending upon the

starting point, the time available and so on. Similarly, the author proposes 6 routes through the book, in addition to the obvious progression 1.1–14.3 which goes over the local maxima.

(1) Compact groups feature in the first half of the book, especially in Sections 1.2, 2.3–2.9, 3.8, 3.9, 5.1, 7.4, 10.5.
(2) Generalized orthogonal ensembles feature in the middle of the book, particularly in 1.5, 2.5, 3.4–3.7, 4.4–4.7, 6.3.
(3) Convergence to equilibrium distributions is the topic in 1.1, 3.4–3.9, 5.2–5.5, 10.6, 11.4.
(4) Free probability features in 4.3, 4.5, 4.8, 6.5, 6.6, 13.5, 14.1–3.
(5) Semicircular and similar special distributions appear in 4.4–4.7, 5.5, 7.3, 13.5, 14.3.
(6) Integrable operators appear in 9.1–9.7 and 11.2.

To summarize the contents of sections or the conclusions of examples, we sometimes give lists of results or definitions with bullet points. These should be considered in context, as they generally require elaboration. There are exercises that the reader should be able to solve in a few hours. There are also problems, which are generally very difficult and for which the answer is unknown at the time of writing.

There are many important topics in random matrix theory that this book does not cover, and for which we refer the reader elsewhere:

(i) the orthogonal polynomial technique and Riemann–Hilbert theory, as considered by Deift in [56];
(ii) connections with analytic number theory as in [98, 50];
(iii) applications to von Neumann algebras, as developed by Voiculescu and others [163, 164, 165, 166, 83, 84, 85, 77];
(iv) applications to physics as in [89];
(v) joint distributions of pairs of random matrices as in [76];
(vi) random growth models, and similar applications.

Jessica Churchman first suggested this topic as the subject for a book. I am most grateful to Graham Jameson, François Bolley, Alex Belton, Stefan Olphert, Martin Cook and especially James Groves for reading sections and suggesting improvements. Finally, I express thanks to Roger Astley of Cambridge University Press for bringing the project to fruition.

1

Metric measure spaces

Abstract

The contents of this chapter are introductory and covered in many standard books on probability theory, but perhaps not all conveniently in one place. In Section 1.1 we give a summary of results concerning probability measures on compact metric spaces. Section 1.2 concerns the existence of invariant measure on a compact metric group, which we later use to construct random matrix ensembles. In Section 1.3, we resume the general theory with a discussion of weak convergence of probability measures on (noncompact) Polish spaces; the results here are technical and may be omitted on a first reading. Section 1.4 contains the Brunn–Minkowski inequality, which is our main technical tool for proving isoperimetric and concentration inequalities in subsequent chapters. The fundamental example of Gaussian measure and the Gaussian orthogonal ensemble appear in Section 1.5, then in Section 1.6 Gaussian measure is realised as the limit of surface area measure on the spheres of high dimension. In Section 1.7 we state results from the general theory of metric measure spaces. Some of the proofs are deferred until later chapters, where they emerge as important special cases of general results. A recurrent theme of the chapter is weak convergence, as defined in Sections 1.1 and 1.3, and which is used throughout the book. Section 1.8 shows how weak convergence gives convergence for characteristic functions, cumulative distribution functions and Cauchy transforms.

1.1 Weak convergence on compact metric spaces

Definition (*Polish spaces*). Let (Ω, d) be a metric space. Then (Ω, d) is said to be complete if every Cauchy sequence converges; that is, whenever a sequence (x_n) in Ω satisfies $d(x_n, x_m) \to 0$ as $n, m \to \infty$, there exists $x \in \Omega$ such that $d(x_n, x) \to 0$ as $n \to \infty$.

A metric space (Ω, d) is said to be separable if there exists a sequence $(x_n)_{n=1}^{\infty}$ in Ω such that for all $\varepsilon > 0$ and all $x \in \Omega$, there exists x_n such that $d(x, x_n) < \varepsilon$. Such a sequence (x_n) is said to be dense.

A complete and separable metric space (Ω, d) is called a Polish space. A map $\varphi : (\Omega_1, d_1) \to (\Omega_2, d_2)$ between metric spaces is an *isometry* if $d_2(\varphi(x), \varphi(y)) = d_1(x, y)$ for all $x, y \in \Omega$.

Let $C_b(\Omega; \mathbf{R})$ be the space of bounded and continuous functions $f : \Omega \to \mathbf{R}$ with the supremum norm $\|f\|_\infty = \sup\{|f(x)| : x \in \Omega\}$.

Definition (*Compact metric spaces*). A metric space is said to be (sequentially) compact if for any sequence (x_n) in Ω there exist $x \in \Omega$ and a subsequence (x_{n_k}) such that $d(x_{n_k}, x) \to 0$ as $n_k \to \infty$. The reader may be familiar with the equivalent formulation in terms of open covers. See [150].

Definition (*Total boundedness*). Let (Ω, d) be a metric space. An ε-net is a finite subset S of Ω such that for all $x \in \Omega$, there exists $s \in \Omega$ such that $d(x, s) < \varepsilon$. If (Ω, d) has an ε-net for each $\varepsilon > 0$, then (Ω, d) is totally bounded.

A metric space is compact if and only if it is complete and totally bounded. See [150].

Proposition 1.1.1 *Suppose that (K, d) is a compact metric space. Then $C(K; \mathbf{R})$ is a separable Banach space for the supremum norm.*

Proof. Let (x_n) be a dense sequence in K and let $f_n : K \to \mathbf{R}$ be the continuous function $f_n(x) = d(x, x_n)$. Then for any pair of distinct points $x, y \in K$ there exists n such that $f_n(x) \neq f_n(y)$. Now the algebra

$$A = \Big\{\beta(0)\mathbf{I} + \sum_{S: S \subset \mathbf{N}} \beta_S \prod_{j: j \in S} f_j(x) : \beta_S \in \mathbf{Q} \quad \text{for all} \quad S; \beta_S = 0$$

$$\text{for all but finitely many} \quad S; \quad S \quad \text{finite}\Big\} \tag{1.1.1}$$

that is generated by the f_n and the rationals is countable and dense in $C(K; \mathbf{R})$ by the Stone–Weierstrass theorem; hence $C(K; \mathbf{R})$ is separable. See [141]. □

Definition (*Dual space*). Let $(E, \|\cdot\|)$ be a real Banach space. A bounded linear functional is a map $\varphi : E \to \mathbf{R}$ such that:

(i) $\varphi(sx + ty) = s\varphi(x) + t\varphi(y)$ for all $x, y \in E$ and $s, t \in \mathbf{R}$;

(ii) $\|\varphi\| = \sup\{|\varphi(x)| : x \in E; \|x\| \leq 1\} < \infty$.

Let E^* be the space of all bounded linear functionals. Let $B = \{x \in E : \|x\| \leq 1\}$; then the product topology on $[-1, 1]^B$ is generated by the open sets

$$\{(x_b)_{b \in B} : |x_{b_j} - y_{b_j}| < \varepsilon_j; j = 1, \ldots, n\} \qquad (1.1.2)$$

given by $b_j \in B$, $y_{b_j} \in [-1, 1]$ and $\varepsilon > 0$ for $j = 1, \ldots, n$. Further, $B^* = \{\phi \in E^* : \|\phi\| \leq 1\}$ may be identified with a closed subspace of $[-1, 1]^B$ via the map $\phi \mapsto (\phi(x))_{x \in B}$. This is the weak* or $\sigma(E^*, E)$ topology on B^*. See [63, 141].

Theorem 1.1.2 (*Mazur*). *Let E be a separable Banach space. Then B^* is a compact metric space for the weak* topology. Further, E is linearly isometric to a closed linear subspace of $C(B^*; \mathbf{R})$.*

Proof. By Tychonov's theorem [141], $[-1, 1]^B$ is a compact topological space, and hence the closed subspace $\{(\phi(x))_{x \in B} : \phi \in B^*\}$ is also compact; this is known as Alaoglu's theorem. Now we show that B^* has a metric that gives an equivalent topology; that is, gives the same collection of open sets.

Let $(x_n)_{n=1}^{\infty}$ be a dense sequence in B and let

$$d(\psi, \varphi) = \sum_{n=1}^{\infty} 2^{-n} |\varphi(x_n) - \psi(x_n)| \qquad (\varphi, \psi \in B^*), \qquad (1.1.3)$$

so that d defines a metric on B^*. Now we check that d induces a compact Hausdorff topology on B^*, which must coincide with the weak* topology. Let (φ_j) be a sequence in B^*. We extract a subsequence $(\varphi_{j_1(k)})_{k=1}^{\infty}$ such that $\varphi_{j_1(k)}(x_1)$ converges as $j_1(k) \to \infty$; from this we extract a further subsequence $(\varphi_{j_2(k)})_{k=1}^{\infty}$ such that $\varphi_{j_2(k)}(x_2)$ converges as $j_2(k) \to \infty$; and so on. Generally we have $j_k : \mathbf{N} \to \mathbf{N}$ strictly increasing and $j_k(n) = j_{k-1}(m)$ for some $m \geq n$. Then we introduce the diagonal subsequence $(\varphi_{j_k(k)})$. By Alaoglu's theorem there exists $\phi \in B^*$ that is a weak* cluster point of the diagonal subsequence, and one checks that

$$d(\phi, \varphi_{j_k(k)}) = \sum_{n=1}^{\infty} 2^{-n} |\phi(x_n) - \varphi_{j_k(k)}(x_n)| \to 0 \qquad (1.1.4)$$

as $j_k(k) \to \infty$ since $\varphi_{j_k(k)}(x_n) \to \phi(x_n)$ as $j_k(k) \to \infty$ for each n.

Let $f \in E$. Then f gives a function $\hat{f} : B^* \to \mathbf{R}$ by $\varphi \mapsto \varphi(f)$ which is continuous by the definition of the weak* topology. Further, by the Hahn–Banach Theorem [141] we have

$$\|\hat{f}\|_\infty = \sup\{|\varphi(f)| : \varphi \in B^*\} = \|f\|, \qquad (1.1.5)$$

so $f \mapsto \hat{f}$ is a linear isometry $E \to C(B^*; \mathbf{R})$. The range of a linear isometry on a Banach space is complete and hence closed. $\qquad\square$

Definition (*Borel measures*). Let (Ω, d) be a Polish space. A σ-algebra $\mathcal{A}$ on Ω is a collection of subsets of Ω such that:

(σ1) $\Omega \in \mathcal{A}$;
(σ2) if $A \in \mathcal{A}$, then $\Omega \setminus A \in \mathcal{A}$;
(σ3) if $(A_n)_{n=1}^\infty$ satisfies $A_n \in \mathcal{A}$ for all n, then $A = \bigcup_{n=1}^\infty A_n$ has $A \in \mathcal{A}$.

The sets A in a σ algebra $\mathcal{A}$ are called events.

The open subsets of Ω generate the Borel σ-algebra $\mathcal{B}(\Omega)$ and $M_b(\Omega)$ is the space of bounded Borel measures $\mu : \mathcal{B}(\Omega) \to \mathbf{R}$ such that

(i) $\|\mu\|_{var} = \sup\{\sum_j |\mu(E_j)| : E_j \in \mathcal{B}(\Omega) \text{ mutually disjoint}, j = 1, \ldots, N\} < \infty$;
(ii) $\mu(\bigcup_{j=1}^\infty E_j) = \sum_{j=1}^\infty \mu(E_j)$ for all $(E_j)_{j=1}^\infty$ mutually disjoint $E_j \in \mathcal{B}(\Omega)$.

We write $M_b^+(\Omega)$ for the subspace of $\mu \in M_b(\Omega)$ such that $\mu(E) \geq 0$ for all $E \in \mathcal{B}(\Omega)$ and $Prob(\Omega)$ for the subspace $\{\mu \in M_b^+(\Omega) : \mu(\Omega) = 1\}$ of probability measures. Further, we write $M_1(\Omega) = \{\mu \in M_b(\Omega) : \|\mu\|_{var} \leq 1\}$. An event is a Borel-measurable subset of Ω. See [88]. For any Borel set A, $\mathbf{I}_A$ denotes the indicator function of A which is one on A and zero elsewhere, so $\mu(A) = \int_\Omega \mathbf{I}_A d\mu$.

A probability space $(\Omega, \mathbf{P})$ consists of a σ algebra $\mathcal{A}$ on Ω, and a probability measure $\mathbf{P} : \mathcal{A} \to \mathbf{R}$.

Theorem 1.1.3 (*Riesz representation theorem*). *Let (Ω, d) be a compact metric space and $\varphi : C_b(\Omega; \mathbf{R}) \to \mathbf{R}$ a bounded linear functional. Then there exists a unique $\mu \in M_b(\Omega)$ such that*

(iii) $\varphi(f) = \int f(x)\mu(dx)$ *for all* $f \in C_b(\Omega; \mathbf{R})$.

*Conversely, each $\mu \in M_b(\Omega)$ defines a bounded linear functional φ via
(iii) such that $\|\varphi\| = \|\mu\|_{var}$. Further, μ is a probability measure if and
only if*

(iv) $\|\varphi\| = 1 = \varphi(\mathbf{I})$.

Proof. See [88]. □

Definition (*Weak convergence*). Let $(\mu_j)_{j=1}^{\infty}$ be a sequence in $M_b(\Omega)$,
and let $\mu \in M_b(\Omega)$. If

$$\lim_{j \to \infty} \int f \, d\mu_j = \int f d\mu \qquad (f \in C_b(\Omega)), \qquad (1.1.6)$$

then (μ_j) converges weakly to μ. The term weak convergence is tradi-
tional in analysis; whereas the term weak* convergence would be more
suggestive, since we have convergence in the $\sigma(M_b(\Omega); C_b(\Omega))$ topology.

Proposition 1.1.4 *Let $E = C_b(\Omega)$ and let $J : Prob(\Omega) \to B^* \subset
[-1,1]^B$ be the map $J(\mu) = (\int f d\mu)_{f \in B}$. For a sequence (μ_j) in $Prob(\Omega)$
and $\mu \in Prob(\Omega)$,*

$$\mu_j \to \mu \quad weakly \Leftrightarrow J(\mu_j) \to J(\mu) \quad in \quad [-1,1]^B \qquad (j \to \infty). \quad (1.1.7)$$

Proof. This is immediate from the definitions. □

Proposition 1.1.5 *Let K be a compact metric space. Then $Prob$ (K)
with the weak topology is a compact metric space.*

Proof. This follows immediately from Theorems 1.1.2 and Theorem
1.1.3 since $Prob(K)$ is linearly isometric to a compact subset of
$C(K; \mathbf{R})^*$. □

Theorem 1.1.2 thus gives a metric for weak convergence on a compact
metric space so that *Prob* becomes a compact metric space. The definition of the metric in Theorem 1.1.2 is rather contrived, so in Section 3.3
we shall introduce a more natural and useful metric for the weak topology, called the Wasserstein metric.

Examples. (i) The metric space $(M_b(\Omega), \|\cdot\|_{var})$ is nonseparable when
Ω is uncountable. Indeed $\|\delta_y - \delta_x\|_{var} = 2$ for all distinct pairs $x, y \in \Omega$.

(ii) Whereas B^* is compact as a subspace of $[-1,1]^B$, $J(Prob(\Omega))$
is not necessarily compact when Ω is noncompact. For example, when
$\Omega = \mathbf{N}$ and δ_n is the Dirac unit mass at $n \in \mathbf{N}$, (δ_n) does not have any
subsequence that converges to any $\mu \in Prob(\mathbf{N})$. In Proposition 1.2.5

we shall introduce *tightness* criteria that ensure that measures do not leak away to infinity in this way.

In applications, one frequently wishes to move measures forward from one space to another by a continuous function. The following result defines the *push forward* or *induced* measure $\nu = \varphi\sharp\mu$ on a compact metric space. A more general version appears in Theorem 1.3.5.

Proposition 1.1.6 (*Induced measure*). *Let* $\varphi : (\Omega_0, d_0) \to (\Omega_1, d_1)$ *be a Borel map between metric spaces where* (Ω_1, d_1) *is compact. Then for each* $\mu \in M_b(\Omega_0)$ *there exists a unique* $\nu \in M_b(\Omega_1)$ *such that*

$$\int_{\Omega_1} f(y)\nu(dy) = \int_{\Omega_0} f(\varphi(x))\mu(dx) \qquad (f \in C_b(\Omega_1)). \quad (1.1.8)$$

Proof. For $f \in C_b(\Omega_1)$, the function $f \circ \varphi$ is also bounded and Borel, hence integrable with respect to μ. The right-hand side clearly defines a bounded linear functional on $C_b(\Omega_1)$, and hence by Theorem 1.1.3 there exists a unique measure ν that realizes this functional. □

The following result is very useful when dealing with convergence of events on probability space. See [73, 88].

Theorem 1.1.7 (*First Borel–Cantelli lemma*). *Let* $(A_n)_{n=1}^{\infty}$ *be events in a probability space* $(\Omega; \mathbf{P})$, *and let* C *be the event with elements given by:* $\omega \in C$ *if and only if* $\omega \in A_k$ *for infinitely many values of* k.
If $\sum_{n=1}^{\infty} \mathbf{P}(A_n) < \infty$, *then* $\mathbf{P}(C) = 0$.

Proof. We shall begin by checking that

$$C = \bigcap_{n=1}^{\infty} \bigcup_{k=n}^{\infty} A_k. \quad (1.1.9)$$

By axiom $(\sigma 3)$, $C_n = \bigcup_{k=n}^{\infty} A_k$ is an event for each integer $n \geq 1$; consequently, $\bigcap_{n=1}^{\infty} C_n$ is also an event. If x belongs to C, then for each n, there exists $k_n \geq n$ with $x \in A_{k_n}$, so $x \in C_n$. Consequently x belongs to $\bigcap_{n=1}^{\infty} C_n$. Conversely, if $x \in \bigcap_{n=1}^{\infty} C_n$, then for each n, x belongs to C_n; so there exists $k_n \geq n$ with $x \in A_{k_n}$. But then x belongs to infinitely many A_j, and hence x is an element of C.

We can estimate the probability of $C_n = \bigcup_{k=n}^{\infty} A_k$ by

$$\mathbf{P}(C_n) \leq \sum_{k=n}^{\infty} \mathbf{P}(A_k), \quad (1.1.10)$$

using the axioms of measure. By hypothesis, the right-hand side is the
tail sum of a convergent series, and hence $\sum_{k=n}^{\infty} \mathbf{P}(A_k) \to 0$ as $n \to \infty$.
Further, $C \subseteq C_n$, so we can form a sandwich

$$0 \leq \mathbf{P}(C) \leq \mathbf{P}(C_n) \leq \sum_{k=n}^{\infty} \mathbf{P}(A_k) \to 0 \qquad (n \to \infty). \qquad (1.1.11)$$

Hence $\mathbf{P}(C) = 0$. $\qquad\qquad\qquad\qquad\qquad\qquad\qquad\qquad\qquad\square$

Exercise 1.1.8 Let $\mu, \nu \in Prob(\Omega)$ be mutually absolutely continuous.

(i) Show that

$$\rho(\mu, \nu) = \int_{\Omega} \left(\frac{d\mu}{d\nu}\right)^{1/2} d\nu$$

satisfies $\rho(\mu, \nu) \leq 1$.

(ii) Now let $\delta(\mu, \nu) = -\log \rho(\mu, \nu)$. Show that:

(a) $\delta(\mu, \nu) \geq 0$;
(b) $\delta(\mu, \nu) = 0$ if and only if $\mu = \nu$ as measures;
(c) $\delta(\mu, \nu) = \delta(\nu, \mu)$.

(The triangle inequality does not hold for δ.)

1.2 Invariant measure on a compact metric group

• A compact metric group has a unique Haar probability measure.

Definition (*Compact metric group*). A topological group is a topological
space G that is a group with neutral element e such that multiplication
$G \times G \to G : (x, y) \mapsto xy$ and inversion $G \to G : x \mapsto x^{-1}$ are continuous.
Furthermore, if the topology on G is induced by a metric d, then (G, d)
is a metric group. Finally, if (G, d) is a metric group that is compact as
a metric space, then G is a *compact metric group*.

Metric groups can be characterized amongst topological groups by their
neighbourhoods of the identity as in [87, page 49], and we mainly use
metric groups for convenience. Our first application of Theorem 1.2.1 is
to show that a compact metric group has a unique probability measure
that is invariant under left and right translation. The proof given here
is due to von Neumann and Pontrjagin [130]. Compactness is essential
to several stages in the proof; the result is actually valid for compact
Hausdorff topological groups [87, 130].

Definition (*Haar measure*). Let G be a compact metric group. A Haar measure is a $\mu_G \in Prob(G)$ such that

$$\int_G f(gh)\mu_G(dh) = \int_G f(hg)\mu_G(dh)$$

$$= \int_G f(h)\mu_G(dh) \quad (g \in G, f \in C(G)), \quad (1.2.1)$$

and is also known as an invariant probability measure.

Examples 1.2.1 (i) Let G be a finite group, so $G = \{g_1, \ldots, g_n\}$ where the g_j are distinct. The Haar probability measure on G is $\mu_G = n^{-1} \sum_{j=1}^n \delta_{g_j}$. Then μ_G has the special property that for all $f : G \to \mathbf{R}$, the left and right averages with respect to μ_G, namely

$$\frac{1}{n} \sum_{j=1}^n f(g_j g) \quad \text{and} \quad \frac{1}{n} \sum_{j=1}^n f(gg_j) \quad (g \in G) \quad (1.2.2)$$

are constant.

(ii) The circle group $\mathbf{T} = \{e^{i\theta} : 0 \le \theta < 2\pi\}$ is an abelian compact metric group under the usual metric and multiplication on complex numbers. The normalized arclength measure $d\theta/2\pi$ is invariant under rotations of the circle.

(iii) The special orthogonal group $SO(3) = \{U \in M_3(\mathbf{R}) : U^t U = I, \det U = 1\}$ is a nonabelian compact metric group. The columns of U form a right-handed orthonormal triple Ue_1, Ue_2, Ue_3, where (e_1, e_2, e_3) is the standard orthonormal basis of $\mathbf{R}^3$. The eigenvalues of U are $1, e^{i\theta}, e^{-i\theta}$ for some $\theta \in [0, 2\pi)$, so we can identify U with a rotation of the unit sphere $S^2(1) = \{x \in \mathbf{R}^3 : \|x\| = 1\}$ in $\mathbf{R}^3$, where the eigenvector corresponding to eigenvalue 1 gives the axis of rotation.

Given $x \in S^2(1)$, there exists $U \in SO(3)$ such that $x = Ue_3$; indeed when x has colatitude θ and longitude ϕ, the coordinates of x with respect to (e_1, e_2, e_3) form the first column of

$$U = \begin{bmatrix} \sin\theta\cos\phi & -\sin\phi & -\cos\theta\cos\phi \\ \sin\theta\sin\phi & \cos\phi & -\cos\theta\sin\phi \\ \cos\theta & 0 & \sin\theta \end{bmatrix}. \quad (1.2.3)$$

When $x \neq \pm e_3$, there exists a unique great circle that passes through x and e_3 and the axis of rotation of U is perpendicular to this great circle. Further, $Ue_3 = Ve_3$ if and only if

$$V^{-1}U = \begin{bmatrix} \cos\phi & -\sin\phi & 0 \\ \sin\psi & \cos\psi & 0 \\ 0 & 0 & 1 \end{bmatrix}. \quad (1.2.4)$$

The map $\varphi : SO(3) \to S^2(1) : U \mapsto Ue_3$ induces the surface area $\hat{\sigma}_2$ on $S^2(1)$, normalized to be a probability measure, from the Haar measure on $SO(3)$, thus $SO(3)/SO(2) \cong S^2(1)$. In terms of the colatitude θ and longitude ϕ, we have

$$d\hat{\sigma}_2 = \frac{1}{4\pi} \sin\theta d\phi d\theta. \tag{1.2.5}$$

Theorem 1.2.2 (*Haar measure*). *Let G be a compact metric group. Then G has a unique Haar probability measure μ_G.*

Proof. Let $f \in C(G; \mathbf{R})$. For each finite list $A = \{a_1, \ldots, a_n\}$, possibly with repetitions, we form the left average

$$m_A(f)(x) = \frac{1}{n} \sum_{j=1}^{n} f(a_j x) \qquad (x \in G). \tag{1.2.6}$$

We can join lists A and $B = \{b_1, \ldots, b_k\}$ by concatenation to form

$$A \circ B = \{a_1, \ldots, a_n, b_1, \ldots, b_k\}; \tag{1.2.7}$$

then $m_A(m_B(f)) = m_{A \circ B}(f)$.

The oscillation of f is $\operatorname{osc}(f) = \sup f - \inf f$. Clearly, $\operatorname{osc}(m_A(f)) \leq \operatorname{osc}(f)$ since $\sup m_A(f) \leq \sup f$ and $\inf m_A(f) \geq \inf f$. Further, if g is continuous and not constant, then there exists A such that $\operatorname{osc}(m_A(g)) < \operatorname{osc}(g)$. To see this, we choose $K > 0$ such that $\inf g < K < \sup g$ and a nonempty open set U such that $g(x) < K$ for all $x \in U$. By compactness, there exists $A = \{a_1, \ldots, a_n\}$ such that $G = \bigcup_{j=1}^{n} a_j^{-1}U$; hence for each $x \in G$ there exists j such that $x \in a_j^{-1}U$. Now

$$\sup m_A(g) \leq \frac{1}{n}\Big(K + (n-1)\sup(g)\Big) < \sup g \tag{1.2.8}$$

and $\inf m_A(g) \geq \inf g$, so $\operatorname{osc}(m_A(g)) < \operatorname{osc}(g)$.

The collection of left averages $L(f) = \{m_A(f) : A \subseteq G, A \text{ finite}\}$ that is formed from each $f \in C(G; \mathbf{R})$ is uniformly bounded and uniformly equicontinuous. Given $\varepsilon > 0$, there exists by compactness an open neighbourhood V of e such that

$$|f(x) - f(y)| \leq \varepsilon \qquad (x^{-1}y \in V) \tag{1.2.9}$$

and hence

$$|m_A(f)(x) - m_A(f)(y)| \leq \varepsilon \qquad (x^{-1}y \in V) \tag{1.2.10}$$

for all A. By the Arzelà–Ascoli theorem [141], $L(f)$ is relatively compact for the supremum norm.

The next observation is that the closure of $L(f)$ must contain a constant function. We let $s = \inf_A \operatorname{osc} m_A(f)$, and by norm compactness can choose a sequence of A_n such that $\operatorname{osc} m_{A_n}(f) \to s$ as $n \to \infty$ and $m_{A_n}(f) \to g$ in supremum norm as $n \to \infty$. Clearly $s = \operatorname{osc}(g)$; suppose with a view to obtaining a contradiction that $s > 0$ so that g is continuous but not constant. Then there exists A such that $\operatorname{osc} m_A(g) < s$, and hence we could find A_n such that $m_A(m_{A_n}(f)) < s$; but this would contradict the definition of s; so g is constant. We define the left mean of f to be the constant value taken by g.

Likewise we can form the right average

$$m'_B(f) = \frac{1}{k} \sum_{j=1}^{k} f(xb_j) \qquad (x \in G); \tag{1.2.11}$$

clearly this has similar properties to the left average, so we can introduce a constant right mean g' analogously. Further, the operations of forming left and right averages commute in the sense that

$$m'_B(m_A(f)) = m_A(m'_B(f)). \tag{1.2.12}$$

Given $\varepsilon > 0$, there exist A and B such that

$$|m_A(f)(x) - g| < \varepsilon, \qquad |m'_B(f)(x) - g'| < \varepsilon \qquad (x \in G); \tag{1.2.13}$$

hence we have

$$|g - g'| \leq |m'_B(m_A(f))(x) - g| + |m_A(m'_B(f))(x) - g'| < 2\varepsilon. \tag{1.2.14}$$

Consequently $g = g'$, and we can define the left and right mean $\mu(f)$ of f to be their common value $\mu(f) = g = g'$.

By the construction, the translates $f_y(x) = f(yx)$ and $f^y(x) = f(xy)$ satisfy $\mu(f_y) = \mu(f^y) = \mu(f)$; consequently

$$\mu(m_A(f)) = \mu(m'_B(f)) = \mu(f) \tag{1.2.15}$$

for all A, B.

We now check the axioms of Theorem 1.1.1.

(i) Clearly $\mu(tf) = t\mu(f)$ holds for $t \in \mathbf{R}$. To see that μ is additive, we introduce $f, h \in C(G)$ and, for given $\varepsilon > 0$ introduce lists A, B such that

$$|m_A(f)(x) - \mu(f)| < \varepsilon, \qquad |m'_B(h)(x) - \mu(h)| < \varepsilon. \tag{1.2.16}$$

Now

$$|m_A m'_B (f+h)(x) - \mu(f) - \mu(h)| < 2\varepsilon, \qquad (1.2.17)$$

so $\mu(f+h) = \mu(f) + \mu(h)$.

(ii) If $-1 \le f(x) \le 1$, then $-1 \le m_A(f)(x) \le 1$, so $-1 \le \mu(f) \le 1$. Hence by the Riesz representation theorem there exists a unique $\mu_G \in Prob(G)$ such that

$$\mu(f) = \int_G f(x)\mu_G(dx). \qquad (1.2.18)$$

We have already seen that μ is invariant under left and right translation. The uniqueness of μ is immediate from the construction. $\qquad \square$

Definition (*Integer part*). For $x \in \mathbf{R}$, let $\lfloor x \rfloor = \max\{n \in \mathbf{Z} : n \le x\}$ be the integer part of x, so $x - \lfloor x \rfloor$ is the fractional part. Further, let A be a finite set, and $\sharp A$ the cardinality of A.

Exercise 1.2.3 (*Weyl's equidistribution theorem*). Let α be irrational and let $\mu_N \in Prob(\mathbf{T})$ be

$$\mu_N = \frac{1}{N} \sum_{j=0}^{N-1} \delta_{\exp(2\pi ij\alpha)}. \qquad (1.2.19)$$

Prove the following equivalent statements.

(i) μ_N converges weakly to normalized arclength $d\theta/2\pi$ as $N \to \infty$.
(ii) For all intervals $(a,b) \subseteq (0,1)$,

$$\frac{1}{N}\sharp\{k \in \mathbf{Z} : 0 \le k \le N-1; \alpha k - \lfloor \alpha k \rfloor \in (a,b)\} \to b - a \quad (N \to \infty). \qquad (1.2.20)$$

(iii) For all trigonometric polynomials $f(e^{2\pi i\theta}) = \sum_{j=-n}^{n} a_j e^{2\pi ij\theta}$,

$$\int_{\mathbf{T}} f d\mu_N \to \int_{\mathbf{T}} f(e^{2\pi i\theta})\frac{d\theta}{2\pi} \quad (N \to \infty). \qquad (1.2.21)$$

Conversely, if α is rational, then μ_N does not converge to $d\theta/2\pi$.

Definition (*Monogenic*). Let A be a compact metric group. Say that A is monogenic if there exists $g \in A$ such that A equals the closure of $\{g^n : n \in \mathbf{Z}\}$. The term topologically cyclic is also used for monogenic, and such a g is said to be a generator of A.

By continuity of multiplication, any monogenic group is abelian. Furthermore, a finite group is monogenic if and only if it is cyclic.

Proposition 1.2.4 (*Kronecker's theorem*).

(i) *The torus $\mathbf{T}^n$ is monogenic for each $n = 1, 2, \ldots$.*

(ii) *Let A be a compact abelian metric group with a closed subgroup A_0 such that A_0 is the connected component that contains the neutral element of A, the quotient group A/A_0 is cyclic and A_0 is continuously isomorphic as a group to $\mathbf{T}^n$ for some n. Then A is monogenic.*

Proof. (i) Let $\{1, \alpha_1, \ldots, \alpha_n\}$ be real numbers that are linearly independent over the rationals, and consider $g = (e^{2\pi i \alpha_1}, \ldots, e^{2\pi i \alpha_n})$. Then we show that for any $f \in C(\mathbf{T}^n; \mathbf{C})$,

$$\frac{1}{N} \sum_{j=0}^{N-1} f(g^j) \to \int_{\mathbf{T}^n} f(e^{2\pi i \theta}, \ldots, e^{2\pi i \theta_n}) \frac{d\theta_1}{2\pi} \cdots \frac{d\theta_n}{2\pi} \quad (N \to \infty).$$
$$(1.2.22)$$

When $f = \mathbf{I}$, this is clearly true. When $(m_1, \ldots, m_n) \in \mathbf{Z}^n \setminus \{0\}$, and

$$f(e^{2\pi i \theta_1}, \ldots, e^{2\pi i \theta_n}) = \exp(2\pi i \sum_{j=1}^{n} m_j \theta_j),$$
$$(1.2.23)$$

we can sum the geometric series

$$\frac{1}{N} \sum_{k=0}^{N-1} f(e^{2\pi i k \alpha_1}, \ldots, e^{2\pi i k \alpha_n}) = \frac{1}{N} \sum_{k=0}^{N-1} \exp\left(2\pi k i \sum_{j=1}^{n} m_j \alpha_j\right)$$

$$= \frac{1 - \exp(2\pi i N \sum_{j=1}^{n} m_j \alpha_j)}{N(1 - \exp(2\pi i \sum_{j=1}^{n} m_j \alpha_j))} \quad (1.2.24)$$

where the denominator is non zero by rational independence of the exponents. Hence

$$\frac{1}{N} \sum_{k=0}^{N-1} f(e^{2\pi i k \alpha_1}, \ldots, e^{2\pi i k \alpha_n}) \to 0 \quad (N \to \infty). \quad (1.2.25)$$

By Féjer's theorem, any $f \in C(\mathbf{T}^n; \mathbf{C})$ can be approximated in the uniform norm by a finite sum of $\exp(2\pi i \sum_{j=1}^{n} m_j \theta_j)$; hence the limit formula holds in general. As in Exercise 1.2.3, $\{g^j : j \in \mathbf{Z}\}$ is dense in $\mathbf{T}^n$.

(ii) Let $\sigma : A \to A/A_0$ be the quotient map, and observe that A/A_0 is compact and discrete, hence finite. By hypothesis, A/A_0 is isomorphic to the cyclic group of order k for some finite k, and we introduce $v \in A$ such that $\sigma(v)$ generates A/A_0, and by (i) we can introduce a generator $u \in A_0$ for A_0. Evidently the closure of the group generated by u and v

is A. Further, $v^k \in A_0$, so $v^k u^{-1} \in A_0$ and we can introduce $s \in A_0$ such that $s^k = v^k u^{-1}$. Now our generator for A is $g = vs^{-1}$; indeed, $u = g^k$ and $s \in A_0 = cl\langle g^k \rangle$, so $v = gs \in \langle g \rangle$.   □

Definition (*Actions*). Let G be a compact metric group and Ω a Polish space. An action of G on Ω is a continuous map $\alpha : G \times \Omega \to \Omega :$ $(g, \omega) \mapsto \alpha_g(\omega)$ such that $\alpha_e(\omega) = \omega$ for all $\omega \in \Omega$, and $\alpha_h \circ \alpha_g = \alpha_{hg}$ for all $g, h \in G$.

Definition (*Invariant measure*). Let μ be a measure on Ω. Say that μ is invariant for α if

$$\int_\Omega f(\alpha_h(\omega)) \, \mu(d\omega) = \int_\Omega f(\omega) \, \mu(d\omega) \qquad (h \in G)$$

for all continuous $f : \Omega \to \mathbf{R}$ of compact support.

Example. Let G be a compact metric group and let $G = \Omega$. Then there are three natural actions of G on G:

  (i) left translation, where $(g, h) \mapsto gh$;
 (ii) right translation, where $(g, h) \mapsto hg^{-1}$;
(iii) conjugation, where $(g, h) \mapsto ghg^{-1}$, for all $g, h \in G$.

Verify that Haar measure is invariant for all of these actions.

1.3 Measures on non-compact Polish spaces

In applications to random matrix theory we use Polish spaces (Ω, d) which are non-compact, and hence there are various natural topologies on $M_b(\Omega)$. In this section we aim to recover results which are analogous to those of Section 1.1, often by imposing stronger hypotheses. The contents of this section are rather technical, so the reader may wish to skip this on a first reading and proceed to 1.4.

For technical convenience, we sometimes wish to have a standard compact space into which we can map other spaces. Let

$$\Omega_\infty = [0,1]^{\mathbf{N}} = \{(x_j)_{j=1}^\infty : x_j \in [0,1]\} \qquad (1.3.1)$$

have the metric

$$d_\infty((x_j), (y_j)) = \sum_{j=1}^\infty 2^{-j} |x_j - y_j|. \qquad (1.3.2)$$

Let $B^\infty(\Omega)$ be the space of bounded and Borel measurable functions $f : \Omega \to \mathbf{R}$ with the supremum norm $\| \; \|_\infty$.

Proposition 1.3.1 (*Kuratowski*).

(i) $(\Omega_\infty, d_\infty)$ is compact.

(ii) For any Polish space (Ω, d), there exists a continuous and injective function $\varphi : \Omega \to \Omega_\infty$.

(iii) Let (Ω_0, d_0) be a compact metric space. Then there exists an isometric map $j : \Omega_0 \to B^\infty(\Omega_\infty)$.

Proof. (i) The metric space $(\Omega_\infty, d_\infty)$ is complete and totally bounded, hence compact.

(ii) Let $(\omega_j)_{j=1}^\infty$ be a dense sequence in (Ω, d), and let

$$\varphi(x) = \left(\frac{d(x, \omega_j)}{1 + d(x, \omega_j)} \right)_{j=1}^\infty. \qquad (1.3.3)$$

One can easily check that φ is injective and that $d(x_n, x) \to 0$ implies $d_\infty(\varphi(x_n), \varphi(x)) \to 0$ as $n \to \infty$.

(iii) Let $f_x : \Omega_0 \to \mathbf{R}$ be the function $f_x(y) = d_0(x, y)$. Then f_x is continuous, hence Borel measurable and by the triangle inequality, the map $x \mapsto f_x$ is an isometry $\Omega_0 \to B^\infty(\Omega_0)$. Now by (ii) there is an injective and continuous map $\varphi_0 : \Omega_0 \to \Omega_\infty$, which by compactness must give a homeomorphism $\Omega_0 \to \varphi_0(\Omega_0)$. Thus there is a linear isometric morphism $B^\infty(\Omega_0) \to B^\infty(\Omega_\infty)$ with range equal to

$$\{f \in B^\infty(\Omega_\infty) : f(x) = 0; x \in \Omega_\infty \setminus \varphi_0(\Omega_0)\}.$$

The composition of these maps gives the required isometry $\Omega_0 \to B^\infty(\Omega_\infty)$. $\qquad \square$

Definition (*Inner regularity*). Let $C_c(\Omega)$ be the space of continuous and compactly supported functions $f : \Omega \to \mathbf{R}$, and let $C_c(\Omega)^\sharp = \{g \in C_b(\Omega) : g = f + t\mathbf{I}; t \in \mathbf{R}, f \in C_c(\Omega)\}$. Suppose that (Ω, d) has infinite diameter and that closed and bounded subsets of Ω are compact. Then the norm closure of $C_c(\Omega)$ is $C_0(\Omega) = \{f \in C_b(\Omega) : f(x) \to 0$, as $d(X, X_0) \to \infty\}$ for some $X_0 \in \Omega$.

Let $\mu \in M_b^+(\Omega)$ where Ω is a non-compact Polish space. Then μ is inner regular if each $\varepsilon > 0$ there exists a compact subset K such that $\mu(\Omega \setminus K) < \varepsilon$. An inner regular Borel measure is called a Radon measure. See [88].

Lemma 1.3.2 (*Lusin*). Let (Ω, d) be a Polish space, let $\mu \in Prob(\Omega)$. Then μ is inner regular.

Proof. Given $\varepsilon > 0$, let $(a_n)_{n=1}^\infty$ be a dense sequence and let $B_{n,k} = \{x \in \Omega : d(x, a_n) \le 1/k\}$. Then $\bigcup_{n=1}^\infty B_{n,k} = \Omega$ so there exists $N(k) < \infty$

such that

$$\mu\left(\bigcup_{n=1}^{N(k)} B_{n,k}\right) > 1 - 2^{-k}\varepsilon. \qquad (1.3.4)$$

Now let $K = \bigcap_{k=1}^{\infty}\bigcup_{n=1}^{N(k)} B_{n,k}$, and observe that K is a closed and totally bounded subset of a Polish space, hence K is compact. Further,

$$\mu(\Omega \setminus K) \leq \sum_{k=1}^{\infty} \mu\left(\Omega \setminus \bigcup_{n=1}^{N(k)} B_{n,k}\right) < \sum_{k=1}^{\infty} 2^{-k}\varepsilon \leq \varepsilon. \qquad (1.3.5)$$

$\square$

Lemma 1.3.3 *The topology* $\sigma(Prob(\Omega), C_b(\Omega))$ *equals the topology* $\sigma(Prob(\Omega), C_c(\Omega)^{\sharp})$; *that is, for* $\mu \in M_b^+(\Omega)$ *a sequence* $(\mu_j)_{j=1}^{\infty}$ *satisfies*

$$(i) \qquad \int g\,d\mu_j \to \int g d\mu \qquad (g \in C_b(\Omega)) \qquad as \quad j \to \infty \qquad (1.3.6)$$

if and only if

$$(ii) \quad \mu_j(\Omega) \to \mu(\Omega)$$

$$and \int f d\mu_j \to \int f d\mu \quad as \quad j \to \infty \quad (f \in C_c(\Omega)). \qquad (1.3.7)$$

Proof. (i)$\Rightarrow$ (ii) This is clear.

(ii) $\Rightarrow$ (i) Let $g \in C_b(\Omega)$ and $\varepsilon > 0$. By Lusin's Lemma, there exists a compact set K such that $\int_{\Omega \setminus K} d\mu < \varepsilon$; hence by Urysohn's Lemma [141, p. 135] there exists $h \in C_c(\Omega)$ such that $0 \leq h \leq 1$ and $h(x) = 1$ on K. By hypothesis, there exists j_0 such that $\int(1-h)d\mu_j < \varepsilon$ for all $j \geq j_0$. Now

$$\int g d\mu_j - \int g d\mu = \left(\int ghd\mu_j - \int ghd\mu\right)$$
$$+ \int g(1-h)d\mu_j - \int g(1-h)d\mu \qquad (1.3.8)$$

where gh is of compact support, and

$$\left|\int g(1-h)d\mu_j\right| \leq \|g\| \int(1-h)d\mu_j < \varepsilon\|g\|; \qquad (1.3.9)$$

likewise

$$\left|\int g(1-h)d\mu\right| \leq \|g\| \int(1-h)d\mu < \varepsilon\|g\|, \qquad (1.3.10)$$

and hence

$$\lim_j \int g d\mu_j = \int g d\mu. \qquad (1.3.11)$$

$\square$

Lemma 1.3.4 *Let $(h_n)_{n=1}^{\infty}$ be a sequence in $C_c(\Omega)$ such that $0 \leq h_1 \leq h_2 \leq \cdots \leq 1$. Let $\varphi \in C_b(\Omega)^*$ and suppose that $0 \leq f \leq 1 \Rightarrow 0 \leq \varphi(f) \leq 1$ for $f \in C_b(\Omega)$ and $\varphi(h_n) \to 1$ as $n \to \infty$. Then there exists $\mu \in Prob(\Omega)$ such that $\varphi(f) = \int f d\mu$.*

Proof. Let $\varphi_n(f) = \varphi(h_n f)$; so that φ_n defines an element of $C_b(\Omega)^*$ such that $0 \leq \varphi_n(f) \leq 1$ when $0 \leq f \leq 1$. Since h_n has compact support, there exists a positive compactly supported measure μ_n such that $\|\mu_n\|_{var} = \varphi_n(\mathbf{I}) = \varphi(h_n)$ and $\varphi_n(f) = \int f d\mu_n$. Further, $\varphi_n \to \varphi$ as $n \to \infty$ in $[-1,1]^B$ since

$$|\varphi_n(f) - \varphi(f)| = |\varphi((1 - h_n)f)| \leq \|f\|_\infty \varphi(1 - h_n) \to 0. \qquad (1.3.12)$$

Now we define, for each Borel set E, $\mu(E) = \lim_{n \to \infty} \mu_n(E)$, where the limit exists by monotonicity. Given any sequence of disjoint subsets $(E_j)_{j=1}^{\infty}$ with $E = \bigcup_{j=1}^{\infty} E_j$, we have $\mu_n(E) = \sum_{j=1}^{\infty} \mu_n(E_j)$; by monotonicity, we deduce that $\mu(E) = \sum_{j=1}^{\infty} \mu(E_j)$, so μ is countably additive. By the Lemma, $\mu_n \to \mu$ weakly, and hence

$$\varphi(f) = \lim_n \varphi_n(f) = \lim_n \int f d\mu_n = \int f d\mu. \qquad (1.3.13)$$

We now have a satisfactory generalization of Proposition 1.1.6. $\qquad \square$

Theorem 1.3.5 (*Induced probability*). *Let (Ω_j, d_j) for $j = 1, 2$ be Polish spaces and let $\varphi : \Omega_1 \to \Omega_2$ be continuous. Then for each $\mu \in Prob(\Omega_1)$, there exists a unique $\nu \in Prob(\Omega_2)$ such that*

$$\int_{\Omega_2} f(x)\nu(dx) = \int_{\Omega_1} f(\varphi(x))\mu(dx) \qquad (f \in C_b(\Omega_2)). \qquad (1.3.14)$$

Proof. By Lusin's Lemma, for each $j \geq 2$ there exists $\mu_j \in M_b^+(\Omega_1)$ such that $\mu_j(E) = \mu(K_j \cap E)$ and $\|\mu_j\|_{var} \geq 1 - 1/j$, where K_j is some compact subset of Ω_1. Now $\varphi(K_j)$ is a compact subset of Ω_2 and $\varphi \sharp \mu_j$ is supported on $\varphi(K_j)$. Let $\phi_j \in C_b(\Omega_2)^*$ be

$$\phi_j(f) = \int_{K_j} f(\varphi(x))\mu(dx), \qquad (1.3.15)$$

and let ϕ be a weak* cluster point of $(\phi_j)_{j=1}^{\infty}$, as in Alaoglu's theorem. By Lemma 1.3.3, there exists $\nu \in Prob(\Omega_2)$ such that

$$\phi(f) = \int_{\Omega_2} f d\nu = \lim_j \int_{K_j} f(\varphi(x))\mu(dx) = \int_{\Omega_1} f(\varphi(x))\mu(dx). \qquad (1.3.16)$$

$\qquad \square$

Definition (*Induced probability*). In Theorem 1.3.5, ν is the probability measure that φ induces from μ, denoted $\nu = \varphi \sharp \mu$.

Whereas Proposition 1.1.5 does not have a simple generalization to non-compact metric spaces, the following is a usable analogue. The idea is to introduce a weight that diverges at infinity, and then restrict attention to families of probabilities that are 'tight' with respect to this weight.

Proposition 1.3.6 (*Tightness*). *Suppose that $w \in C(\Omega)$ has $w \geq 0$ and $K_M = \{x \in \Omega : w(x) \leq M\}$ compact for each $M < \infty$. Suppose moreover that $(\mu_j)_{j=1}^{\infty}$ is a sequence in $Prob(\Omega)$ such that $\int w(x)d\mu_j < L$ for $j = 1, 2, \ldots$ for some $L < \infty$. Then there exists $\mu \in Prob(\Omega)$ and a subsequence (μ_{n_k}) such that $\mu_{n_k} \to \mu$ weakly.*

Proof. We observe that $\mu_n(\Omega \setminus K_{ML}) \leq 1/M$ for all $n \geq 1$. For $M = 2$, we choose a subsequence $(\mu_{n_2(k)})_{k=1}^{\infty}$ and $\mu^{(2)}$ such that $\mu_{n_2} \to \mu^{(2)}$ weakly on K_{2L} where $1/2 \leq \mu^{(2)}(K_{2L}) \leq 1$ as $n_2(k) \to \infty$. Next we choose a further subsequence $(\mu_{n_3(k)})_{k=1}^{\infty}$ and a $\mu^{(3)}$ such that $\mu_{n_3} \to \mu^{(3)}$ weakly on K_{4L} where $3/4 \leq \mu^{(3)} \leq 1$ as $n_3(k) \to \infty$. By Lemma 1.3.3, the diagonal subsequence $(\mu_{n_k(k)})_{k=1}^{\infty}$ has $\mu_{jj} \to \mu$ for some probability measure μ. $\qquad\square$

Example 1.3.7 To see how measures can leak away, consider the co-ordinate projection $\varphi : \mathbf{R}^2 \to \mathbf{R} : \varphi(x,y) = x$ and the measures $\mu_n = \delta_{(0,n)}$ on $\mathbf{R}^2$. Then $\varphi \sharp \mu_n = \delta_0$, whereas $\mu_n \to 0$ in the weak topology $\sigma(Prob(\mathbf{R}^2), C_c(\mathbf{R}^2))$ as $n \to \infty$.

Examples. In subsequent sections we shall use the weights:

(i) $w(x) = d(x, x_0)^s$ where $s > 0$ and $x_0 \in \Omega$ is fixed;
(ii) $w(x) = 2\log_+ |x|$ on $\mathbf{R}$.

Definition (*Lebesgue spaces*). Let μ be a Radon measure on a Polish space (Ω, d). We identify Borel-measurable functions f and g if $f(x) = g(x)$ except on a set of μ measure zero. Then for $1 \leq p < \infty$, we define $L^p(\mu; \mathbf{C})$ to be the space of Borel-measurable $f : \Omega \to \mathbf{C}$ such that $\int_\Omega |f(x)|^p \mu(dx) < \infty$. Then $L^p(\mu; \mathbf{C})$ forms a Banach space for the norm $\|f\|_{L^p} = (\int_\Omega |f(x)|^p \mu(dx))^{1/p}$.

Proposition 1.3.7 (*Hölder's inequality*). *Let μ be a positive Borel measure on a Polish space Ω such that $\mu(K)$ is finite for each compact set*

K. Let $1 < p, q < \infty$ *satisfy* $1/p + 1/q = 1$. *Then for all* $f \in L^p(\mu; \mathbf{C})$
and $g \in L^q(\mu; \mathbf{C})$, *the product* fg *belongs to* $L^1(\mu; \mathbf{C})$ *and satisfies*

$$\left| \int_\Omega f(x)g(x)\mu(dx) \right| \leq \left(\int_\Omega |f(x)|^p \mu(dx) \right)^{1/p} \left(\int_\Omega |g(x)|^q \mu(dx) \right)^{1/q}.$$
(1.3.17)

Proof. See [71]. □

Exercise 1.3.8 (i) The set P of probability density functions f on $\mathbf{R}^n$ consists of integrable functions $f : \mathbf{R}^n \to \mathbf{R}$ such that $f \geq 0$ and $\int f(x)dx = 1$. There is a natural map $P \to Prob(\mathbf{R}^n) : f \mapsto f(x)dx$, such that the range is closed for $\|.\|_{var}$, but is not closed for the weak topology. Indeed, one can check that

$$\|f(x)dx - g(x)dx\|_{var} = \int |f(y) - g(y)|dy,$$

so the range is norm closed since L^1 is complete. For each $x_0 \in \mathbf{R}^n$ and $j = 1, 2, \ldots$, there exists $f_j \in P$ such that f_j is supported on $\{x : \|x - x_0\| \leq 1/j\}$; then $f_j(x)dx$ converges weakly to the Dirac point mass δ_{x_0} as $j \to \infty$. This construction gives rise to the term *Dirac delta function*, which is sometimes used for the measure δ_{x_0}.

(ii) Let $f_j \in P$ be supported on $\{x : \|x\| \leq 1\}$ and suppose that $\|f_j\|_{L^2} \leq M$ for some M and all j, and that $f_j(x)dx \to \mu$ weakly as $j \to \infty$ for some $\mu \in Prob(\mathbf{R}^n)$. Show that μ is absolutely continuous with respect to Lebesgue product measure, and hence that $\mu(dx) = f(x)dx$ for some $f \in P$.

Exercise 1.3.9 (i) Let $1 < p, q < \infty$ satisfy $1/p + 1/q = 1$. Use Hölder's inequality to show that, for $f \in L^p(\mathbf{R})$, $g \in L^1(\mathbf{R})$ and $h \in L^q(\mathbf{R})$, the function $f(x)g(x - y)h(y)$ is integrable with respect to $dxdy$ and

$$\left| \iint f(x)g(x - y)h(y)\, dxdy \right| \leq \|f\|_{L^p} \|g\|_{L^1} \|h\|_{L^q}.$$

(ii) Now let f, g and h be positive and integrable with respect to dx. By grouping the functions $f(x)^{1/3}$, $g(x - y)^{1/3}$ and $h(y)^{1/3}$ in suitable pairs, show that

$$\iint f(x)^{2/3} g(x-y)^{2/3} h(y)^{2/3}\, dxdy \leq \left(\int f(x)\, dx \int g(x)\, dx \int h(x)\, dx \right)^{2/3}.$$

1.4 The Brunn–Minkowski inequality

- The Brunn–Minkowski inequality has a functional form due to Prékopa and Leindler.
- In subsequent sections, these results are used to prove concentration inequalities.

Lebesgue product measure m on $\mathbf{R}^n$ is $dx = dx_1 \ldots dx_n$, so the restriction to $[0,1]^n$ is a probability measure. Let $\langle x, y \rangle = \sum_{j=1}^{n} x_j y_j$ be the usual bilinear pairing on $\mathbf{R}^n$, and let $\|x\| = \langle x, x \rangle^{1/2}$. For a function $f : \mathbf{R}^n \to \mathbf{R}$, write $[f > \lambda] = \{x \in \mathbf{R}^n : f(x) > \lambda\}$.

Definition (*Sum of sets*). For $A, B \subseteq \mathbf{R}^n$, let $A + B = \{a + b : a \in A, b \in B\}$ and $tA = \{ta : a \in A\}$.

Examples. (i) In many results it is convenient to use the quantity $m(A)^{1/n}$ since $m(tA)^{1/n} = t m(A)^{1/n}$ for $t > 0$ and A compact.
(ii) Let $A = B = [0,1]^n$. Then $A + B = [0,2]^n$.
(iii) For open A and $B = \{x : \|x\| < \varepsilon\}$, $\{x : d(x, A) < \varepsilon\} = A + B$.

Theorem 1.4.1 (*Brunn–Minkowski*). *For all compact subsets A and B of $\mathbf{R}^n$,*

$$m(A + B)^{1/n} \geq m(A)^{1/n} + m(B)^{1/n}. \tag{1.4.1}$$

The functional form of this inequality looks like a reverse of Hölder's inequality.

Theorem 1.4.2 (*Prékopa–Leindler*). *Let $f, g, h : \mathbf{R}^n \to \mathbf{R}_+$ be integrable functions and suppose that*

$$h((1-t)x + ty) \geq f(x)^{1-t} g(y)^t \qquad (x, y \in \mathbf{R}^n) \tag{1.4.2}$$

for some $t \in [0,1]$. Then

$$\int_{\mathbf{R}^n} h(x)\, dx \geq \left(\int_{\mathbf{R}^n} f(x)\, dx \right)^{1-t} \left(\int_{\mathbf{R}^n} g(x)\, dx \right)^t. \tag{1.4.3}$$

Proof. The following scheme of proof is due to Keith Ball and appears in Pisier's book [129]. We shall prove the Brunn–Minkowski inequality in the case $n = 1$, then prove by induction the Prékopa–Leindler inequality; finally we deduce the remaining cases of the Brunn–Minkowski inequality.

Suppose in Theorem 1.4.1 that $n = 1$, and that A and B are finite unions of disjoint open intervals. We prove the result by induction on the number of intervals, the result clearly being true for one or two intervals. Suppose then that $A = \bigcup_{j=1}^{N+1}(a_j, b_j)$ and $B = \bigcup_{k=1}^{L}(c_k, d_k)$ where $a < b_1 \leq a_2 < \cdots \leq b_{N+1}$. By simple geometry we have

$$m(A + B) \geq m\left(\bigcup_{j=1}^{N}(a_j, b_j) + B\right) + m(a_{N+1}, b_{N+1}),$$

and hence by the induction hypothesis

$$m(A + B) \geq m(B) + \sum_{j=1}^{N+1}(b_j - a_j) = m(B) + m(A), \qquad (1.4.4)$$

as required.

The map $(a, b) \mapsto a + b$ maps compact sets to compact sets, and open sets to open sets. Given compact sets A and B and $\varepsilon > 0$, since $A + B$ is compact, there exist finite unions $\tilde{A}$ and $\tilde{B}$ of open intervals such that $\tilde{A} \supseteq A$, $\tilde{B} \supseteq B$ and $m(\tilde{A} + \tilde{B}) \leq m(A + B) + \varepsilon$. Then by the preceding argument, we have

$$m(A) + m(B) \leq m(\tilde{A}) + m(\tilde{B}) \leq m(\tilde{A} + \tilde{B}) \leq m(A + B) + \varepsilon; \quad (1.4.5)$$

hence $m(A) + m(B) \leq m(A + B)$. Furthermore, for any Lebesgue measurable set A that has finite measure, there exists a compact subset K of A such that $m(K) \geq m(A) - \varepsilon$. Hence the result holds for Lebesgue measurable sets A and B that have finite measure.

We now consider Theorem 1.4.2 in the case $n = 1$. By the hypothesis on the functions, we have $[h > \lambda] \supseteq (1 - t)[f > \lambda] + t[g > \lambda]$, hence

$$m[h > \lambda] \geq (1 - t)m[f > \lambda] + tm[g > \lambda] \qquad (\lambda > 0) \qquad (1.4.6)$$

holds by the previous paragraph. Integrating this inequality, we obtain by the theorem of the arithmetic and geometric means

$$\int_0^\infty m[h > \lambda]\, d\lambda \geq (1 - t)\int_0^\infty m[f > \lambda]\, d\lambda + t\int_0^\infty m[g > \lambda]\, d\lambda$$
$$\geq \left(\int_0^\infty m[f > \lambda]\, d\lambda\right)^{1-t}\left(\int_0^\infty m[g > \lambda]\, d\lambda\right)^t.$$
$$(1.4.7)$$

This gives the case $n = 1$ of Theorem 1.4.2.

Now suppose Theorem 1.4.2 holds for some n, and consider $n + 1$. We write $x^{(n)} = (x_1, \ldots, x_n)$, so $x^{(n+1)} = (x^{(n)}, x_{n+1})$. By the induction hypothesis and the hypotheses on h, we have

$$\int_{\mathbf{R}^n} h(x^{(n)}, x_{n+1}) dx^{(n)} \geq \left(\int_{\mathbf{R}^n} f(y^{(n)}, y_{n+1}) dy^{(n)} \right)^{1-t}$$

$$\times \left(\int_{\mathbf{R}^n} g(z^{(n)}, z_{n+1}) dz^{(n)} \right)^t \quad (1.4.8)$$

whenever $x_{n+1} = (1-t)y_{n+1} + tz_{n+1}$. We now apply the inequality for $n = 1$ to the functions that are defined by the integrals in this inequality; hence the result.

Finally, we can deduce the general case of Theorem 1.4.1 from Theorem 1.4.2. Let A and B be any pair of compact sets that have positive measure, and introduce the scaled compact sets $A' = A/m(A)^{1/n}$ and $B' = B/m(B)^{1/n}$ so that A' and B' have unit volume. Then by Theorem 1.4.2 applied to the indicator functions $f - \mathbf{I}_{A'}$, $g = \mathbf{I}_{B'}$ and $h = \mathbf{I}_{tA'+(1-t)B'}$, we have

$$m(A')^{t/n} m(B')^{(1-t)/n} \leq m(tA' + (1-t)B')^{1/n}.$$

In particular, when we take $t = m(A)^{1/n}(m(A)^{1/n} + m(B)^{1/n})^{-1}$, we have

$$tA' + (1-t)B' = \frac{A+B}{m(A)^{1/n} + m(B)^{1/n}}, \quad (1.4.9)$$

so

$$m(A')^{t/n} m(B')^{(1-t)/n} \leq m\left(\frac{A+B}{m(A)^{1/n} + m(B)^{1/n}} \right)^{1/n} \quad (1.4.10)$$

and by homogeneity we recover the Brunn–Minkowski inequality

$$1 \leq \frac{m(A+B)^{1/n}}{m(A)^{1/n} + m(B)^{1/n}}. \quad (1.4.11)$$

$\square$

Corollary 1.4.3 (*Isoperimetric inequalities*). *Let* $B = \{x : \|x\| < 1\}$ *and for a bounded open set* A, *let* $A_\varepsilon = \{y : d(y, A) < \varepsilon\}$. *Then*

$$\liminf_{\varepsilon \to 0+} \frac{m_n(A_\varepsilon) - m_n(A)}{\varepsilon} \geq n m_n(A)^{(n-1)/n} m_n(B)^{1/n}. \quad (1.4.12)$$

Proof. We have $A_\varepsilon \subseteq A + \varepsilon B$, so the Brunn–Minkowski inequality gives

$$m_n(A_\varepsilon)^{1/n} \geq m_n(A)^{1/n} + \varepsilon m_n(B)^{1/n},$$

so with a little reduction one can obtain (1.4.12). $\square$

1.5 Gaussian measures

The standard $N(0,1)$ Gaussian probability density function on $\mathbf{R}$ is $\gamma(x) = e^{-x^2/2}/\sqrt{2\pi}$. Let dx be the usual Lebesgue product measure on $\mathbf{R}^n$, which we sometimes refer to as volume measure. Let $\langle , \rangle$ be the usual inner product on $\mathbf{R}^n$ associated with the Euclidean norm $\|x\|^2 = \langle x, x \rangle$. Let $b \in \mathbf{R}^n$, and let A be a real symmetric positive definite matrix, so A^{-1} is likewise. Let σ_1^2 be the largest eigenvalue of A. Then

$$\gamma_{b,A}(dx) = (2\pi)^{-n/2}\big(\det(A)\big)^{-1/2}\exp\Big(-\langle A^{-1}(x-b),(x-b)\rangle\Big)\,dx$$
(1.5.1)

is a Gaussian measure that is the joint distribution of $x = (x_1,\ldots,x_n)$ with mean vector b and covariance matrix A; we write $x \sim N(b,A)$. In particular, the standard Gaussian has $b = 0$ and $A = I$. Generally we can write

$$\gamma_{b,A}(dx) = (2\pi)^{-n/2}(\det(A))^{-1/2}$$
$$\times \exp\Big(-\frac{1}{2}\langle A^{-1}x,x\rangle + \langle A^{-1}b,x\rangle - \frac{1}{2}\langle A^{-1}b,b\rangle\Big)\,dx \quad (1.5.2)$$

where the dominant term in the exponent is $\langle A^{-1}x,x\rangle \geq \sigma_1^{-2}\|x\|^2$, hence the integrals converge. The coordinate projection $\pi_j : x \mapsto x_j$ induces a Gaussian measure on $\mathbf{R}$.

Lemma 1.5.1 (*Gaussian distribution*).

(*i*) *The Gaussian density $\gamma_{A,b}$ is correctly normalized so as to be a probability density function, and its moment generating function is*

$$\int_{\mathbf{R}^n} \exp(\langle x,y\rangle)\gamma_{A,b}(dx) = \exp\Big(\frac{1}{2}\langle Ay,y\rangle + \langle y,b\rangle\Big) \qquad (y \in \mathbf{C}^n). \quad (1.5.3)$$

(*ii*) *The standard normal distribution is invariant under orthogonal transformation $x \mapsto Ux$ for $x \in \mathbf{R}^n$ and $U \in O(n)$.*

Proof. (ii) Lebesgue measure is invariant under orthogonal transformation.

(i) Suppose temporarily that $b = 0$. The matrix A has eigenvalues $\sigma_1^2 \geq \sigma_2^2 \geq \cdots \geq \sigma_n^2 > 0$, and there exists a real orthogonal matrix U such that UAU^t is a diagonal matrix with these eigenvalues on the leading diagonal. The determinant of A satisfies $(\det A)^{1/2} = \sigma_1 \ldots \sigma_n$. We introduce the diagonal matrix $D^{1/2} = \mathrm{diag}(\sigma_1,\ldots,\sigma_n)$, and $z = D^{-1/2}Ux$ so that $dz = (\sigma_1 \ldots \sigma_n)^{-1}dx$ and $\langle A^{-1}x,x\rangle = \|z\|^2$. This reduces $\gamma_{0,A}$ to the standard Gaussian $\gamma_{0,I}$, and one can check that $\int \gamma_{0,I}(dz) = 1$.

Likewise, one can check the normalizing constant when $b = 0$, since Lebesgue measure in invariant under the transformation $x \mapsto b + x$.

Now we have the correct formula for the moment generating function when $y = 0$. More generally, when $y \in \mathbf{R}^n$, we simply replace b by $b + Ay$ in the exponent of $\gamma_{b,A}$ and adjust the normalizing constant accordingly. When y is complex, this formal transformation gives the correct result, and may be justified rigorously by Cauchy's theorem.   □

Definition (*The Gaussian orthogonal ensemble*). Let $X_{jk} : \Omega \to \mathbf{R}$ for $1 \leq j \leq k \leq n$ be mutually independent Gaussian random variables on $(\Omega, \mathbf{P})$, with $X_{jk} \sim N(0, 1/2n)$ for $j < k$ and $X_{jj} \sim N(0, 1/n)$; then let $X_{jk} = X_{kj}$ and form the real symmetric matrix $X = [X_{jk}]$. The Lebesgue measure on a real symmetric matrix X is the product of Lebesgue measure on the entries that are on or above the leading diagonal,

$$dX = dx_{11} \ldots dx_{1n} dx_{22} \ldots dx_{2n} \ldots dx_{nn}. \qquad (1.5.4)$$

Let $M_n^s(\mathbf{R})$ be the space of $n \times n$ real symmetric matrices, and let ν_n be the probability measure on $M_n^s(\mathbf{R})$ that is induced from $\mathbf{P}$ by the map $X : \Omega \to M_n^s(\mathbf{R})$; so that for all $F \in C_b(M_n^s(\mathbf{R}); \mathbf{R})$ we have

$$\int F(X)\nu_n(dX) = 2^{-n/2} \left(\frac{n}{\pi}\right)^{n(n+1)/4} \int_{M_n^s(\mathbf{R})} F(X)\exp\left(-\frac{n}{2}\sum_{j,k=1}^{n} x_{jk}^2\right) dX \qquad (1.5.5)$$

and in particular we have

$$\int X^2 \nu_n(dX) = \text{diag}\left[\frac{n+1}{2n}\right] \qquad (1.5.6)$$

and

$$\int \text{trace}(X^2)\nu_n(dX) = \frac{n+1}{2n}. \qquad (1.5.7)$$

Letting $\lambda_1, \ldots, \lambda_n$ be the eigenvalues of $X = [x_{jk}]$, we can rewrite the exponent as

$$\sum_{j,k=1}^{n} x_{jk}^2 = \text{trace}\, X^2 = \sum_{\ell=1}^{n} \lambda_\ell^2, \qquad (1.5.8)$$

where

$$\int \frac{1}{n}\sum_{\ell=1}^{n} \lambda_\ell^2 \nu_n(d\xi) = \frac{1}{n}\sum_{j,k=1}^{n} \int X_{j,k}^2 \nu_n(dX) = \frac{n+1}{2n}. \qquad (1.5.9)$$

The Gaussian orthogonal ensemble is the space $M_n^s(\mathbf{R})$ of real symmetric matrices, when endowed with this probability measure ν_n. See [116].

Remark. The term 'orthogonal' alludes to the fact that the measure is invariant under the conjugation action of the real orthogonal group $O(n)$ by $X \mapsto UXU^t$ for $X \in M_n^s(\mathbf{R})$ and $U \in O(n)$; the measure does not live on the space of orthogonal matrices. Whereas the matrix entries on or above the diagonal are mutually independent, the eigenvalues are correlated. Later we shall consider ensembles for which the entries of the random matrices are all correlated.

Definition (*Symmetric*). Let S_n be the symmetric group on $\{1, \ldots, n\}$. Then for $x^{(n)} = (x_1, \ldots, x_n) \in \mathbf{R}^n$, write $\sigma x^{(n)} = (x_{\sigma(1)}, \ldots, x_{\sigma(n)})$; now for $f : \mathbf{R}^n \to \mathbf{R}$, write $f_\sigma(x^{(n)}) = f(\sigma x^{(n)})$. Say that f is symmetric if $f_\sigma = f$ for all $\sigma \in S_n$.

Proposition 1.5.2 *Suppose that $F \in (M_n^s(\mathbf{R}); \mathbf{R})$ invariant under orthogonal conjugation, so that $F(UXU^t) = F(X)$ for all $U \in O(n)$ and all $X \in M_n^s(\mathbf{R})$. Then there exists a symmetric function $f \in C_b(\mathbf{R}^n; \mathbf{R})$ such that $f(\lambda) = F(X)$, where $\lambda = (\lambda_1, \ldots, \lambda_n)$ are the eigenvalues of X, listed according to multiplicity.*

Proof. For each X, there exists $U \in O(n)$ such that UXU^t is a diagonal matrix with diagonal entries in order $\lambda_1 \leq \cdots \leq \lambda_n$, namely the eigenvalues of X. Hence we can identify the invariant function F with a function on the real diagonal matrices. By elementary operator theory, the eigenvalues depend continuously on the matrix entries. The function f is symmetric since we can permute the diagonal entries by suitable orthogonal matrices. □

In particular, when X is chosen from the Gaussian orthogonal ensemble, we need to consider $f(\lambda)$ when λ lies close to the sphere $\{x \in \mathbf{R}^n : \|x\| = \sqrt{n/2}\}$.

1.6 Surface area measure on the spheres

The spheres in high dimensional Euclidean space are important:

- in the historical context;
- as geometrical objects on which groups act naturally;
- as models for high dimensional phenomena exhibited elsewhere.

The sphere $S^n(r) = \{x \in \mathbf{R}^{n+1} : \|x\| = r\}$ is the boundary of the open ball $B^{n+1}(r) = \{x \in \mathbf{R}^{n+1} : \|x\| < r\}$. For an open set E in $\mathbf{R}^{n+1}$, we define the Brunn–Minkowski outer upper content of the boundary of E to be

$$\sigma_n(\partial E) = \lim_{\varepsilon \to 0+} \sup \frac{1}{\varepsilon}\left(\text{vol}\left(E + B^{n+1}(\varepsilon)\right) - \text{vol}(E)\right). \quad (1.6.1)$$

In particular, by taking $E = B^{n+1}(r)$, we deduce that

$$\sigma_n(S^n(r)) = r^n \sigma_n(S^n(1)). \quad (1.6.2)$$

For an open subset F of $S^n(r)$, we introduce $F_\delta = \{rf : 1 < r < 1 + \delta; f \in F\}$ and define $\sigma_{n,r}(F) = \lim\sup_{\delta \to 0+} \text{vol}(F_\delta)/\delta$. This $\sigma_{n,r}$ gives surface area measure on $S^n(r)$.

Let $\Gamma(s) = \int_0^\infty t^{s-1} e^{-t}\, dt$ be Euler's Gamma function, as discussed in [154]. Letting the nth coordinate projection point towards the pole, we identify

$$C_\phi = \{x = (x^{(n-1)}\sin\theta, \cos\theta) : x^{(n-1)} \in S^{n-1}(1), 0 \le \theta \le \phi\} \quad (1.6.3)$$

with the polar cap of colatitude ϕ; so

$$\sigma_{n,1}(dx^{(n)}) = \sigma_{n-1,1}(dx^{(n-1)})\sin^{n-1}\theta\, d\theta \quad (1.6.4)$$

and so

$$\sigma_{n,1}(S^n(1)) = \frac{\Gamma(n/2)\Gamma(1/2)}{\Gamma((n+1)/2)}\sigma_{n-1,1}(S^{n-1}(1)). \quad (1.6.5)$$

We wish to consider the measure of polar caps in spheres as $n \to \infty$, and soon we find that most of the measure lies close to the equator; this is our first example of the concentration of measure phenomenon, as we discuss in Section 3.2. To make this precise, we consider the equatorial regions

$$E_\varepsilon = \{(x^{(n-1)}\sin\theta, \cos\theta) \in S^n(1) : x^{(n-1)} \in S^{n-1}(1),$$
$$\pi/2 - \varepsilon < \theta < \pi/2\}. \quad (1.6.6)$$

Proposition 1.6.1 *For each $\varepsilon > 0$,*

$$\frac{\sigma_{n,1}(E_{\varepsilon/\sqrt{n}})}{\sigma_{n,1}(S^n(1))} \to \frac{1}{\sqrt{2\pi}}\int_{-\varepsilon}^{\varepsilon} e^{-x^2/2}\, dx \quad (n \to \infty). \quad (1.6.7)$$

Proof. We need to estimate

$$\frac{\int_{\pi/2-\varepsilon/\sqrt{n}}^{\pi/2}\sin^{n-1}\theta\, d\theta}{\int_0^{\pi/2}\sin^{n-1}\theta\, d\theta} = \frac{\int_0^{\varepsilon/\sqrt{n}}\cos^{n-1}\theta\, d\theta}{\int_0^{\pi/2}\cos^{n-1}\theta\, d\theta}$$

$$= \frac{\int_0^{\varepsilon}(1 - \psi^2/2n + O(1/n^2))^{n-1}\, d\psi}{\sqrt{n}\int_0^{\pi/2}\cos^{n-1}\theta\, d\theta}. \quad (1.6.8)$$

By a familiar calculation, we have

$$\int_0^{\pi/2} \cos^{2n} \theta \, d\theta = \frac{(2n-1)(2n-3)\ldots 1}{(2n)(2n-2)\ldots 2} \frac{\pi}{2}, \quad (1.6.9)$$

where the right-hand side appears in Wallis's product

$$\frac{(2n-1)(2n-3)\ldots 3}{(2n-2)(2n-4)\ldots 2} \frac{\sqrt{\pi}}{2\sqrt{n}} \to 1 \quad (n \to \infty). \quad (1.6.10)$$

Thus we calculate the limit of the denominator of the previous quotient, while we apply the dominated convergence theorem to the numerator, thus obtaining

$$\int_0^\varepsilon (1 - \psi^2/2n + O(1/n^2))^{n-1} \, d\psi \to \int_0^\varepsilon e^{-x^2/2} dx \quad (n \to \infty).$$

Thus we obtain the said limit. □

Let

$$\hat{\sigma}_{n-1,\sqrt{n}} = \frac{\sigma_{n-1,\sqrt{n}}}{n^{(n-1)/2}\sigma_{n-1,1}(S^{n-1}(1))}, \quad (1.6.11)$$

so that $\hat{\sigma}_{n-1,\sqrt{n}}$ is the surface area measure on $S^{n-1}(\sqrt{n})$ normalized to be a probability measure, and let $\pi_{k,n} : \mathbf{R}^n \to \mathbf{R}^k \colon \pi_{k,n}(x_1,\ldots,x_n) = (x_1,\ldots,x_k)$ be the projection onto the first k coordinates.

Theorem 1.6.2 (*Poincaré–McKean [115]*). *As $n \to \infty$, the probability measure $\pi_{k,n\sharp}\hat{\sigma}_{n-1,\sqrt{n}}$ on the first k coordinates converges weakly to the standard Gaussian $\gamma_{0,I}$ on $\mathbf{R}^k$, so*

$$\int_{S^{n-1}(\sqrt{n})} g(x_1,\ldots,x_k)\hat{\sigma}_{n-1,\sqrt{n}}(dx) \to \frac{1}{(2\pi)^{k/2}} \int_{\mathbf{R}^k} g(x)e^{-\|x\|^2/2} \, dx$$

$$(1.6.12)$$

as $n \to \infty$ for all $g \in C_b(\mathbf{R}^k; \mathbf{R})$.

Proof. Let $\hat{\sigma}_{n,1}$ be the surface area measure on $S^{n-1}(1)$, normalized to be a probability. By geometrical considerations, we have

$$d\hat{\sigma}_{n-1,1} = |\sin\theta|^{n-2} d\hat{\sigma}_{n-2,1}$$

where θ is the colatitude.

Given $(\xi_1, \ldots, \xi_n) \in S^{n-1}(1)$, we introduce coordinates

$$\xi_1 = \cos\theta_1,$$
$$\xi_2 = \sin\theta_1 \cos\theta_2,$$
$$\xi_3 = \sin\theta_1 \sin\theta_2 \cos\theta_3,$$
$$\vdots \qquad \vdots$$
$$\xi_{n-1} = \sin\theta_1 \sin\theta_2 \ldots \sin\theta_{n-2} \cos\theta_{n-1},$$
$$\xi_n = \sin\theta_1 \ldots \sin\theta_{n-1}. \tag{1.6.13}$$

The Jacobian is lower triangular, so we can easily compute

$$\left| \frac{\partial(\xi_1, \ldots, \xi_{n-1})}{\partial(\theta_1, \ldots, \theta_{n-1})} \right| = |\sin^k\theta_1 \sin^{k-1}\theta_2 \ldots \sin\theta_k|. \tag{1.6.14}$$

By repeatedly applying (1.6.2), we deduce that

$$d\hat{\sigma}_{n-1,1} = |\sin\theta_1|^{n-2}|\sin\theta_2|^{n-3} \ldots |\sin\theta_k|^{n-k-1} d\hat{\sigma}_{n-k-1,1} \tag{1.6.15}$$

where

$$|\sin\theta_1 \ldots \sin\theta_k| = (1 - \xi_1^2 - \cdots - \xi_k^2)^{1/2}. \tag{1.6.16}$$

This one proves by induction, with the induction step

$$\sin^2\theta_1 \ldots \sin^2\theta_k = \sin^2\theta_1 \ldots \sin^2\theta_{k-1} - \xi_k^2$$
$$= 1 - \xi_1^2 - \cdots - \xi_k^2. \tag{1.6.17}$$

Now we have

$$|\sin\theta_1|^{n-2}|\sin\theta_2|^{n-3} \ldots |\sin\theta_k|^{n-k-1} d\theta_1 \ldots d\theta_k$$
$$= \left(|\sin\theta_1|^k |\sin\theta_2|^{k-1} \ldots |\sin\theta_k| \right) |\sin\theta_1 \ldots \sin\theta_k|^{n-k-2} d\theta_1 \ldots d\theta_k$$
$$= \left(1 - \xi_1^2 - \cdots - \xi_k^2 \right)^{(n-k-2)/2} d\xi_1 \ldots d\xi_k. \tag{1.6.18}$$

Now we scale up to $S^{n-1}(\sqrt{n})$, and find that for g depending only upon $\xi_1, \ldots, \xi_k$,

$$\int_{S^{n-1}(\sqrt{n})} g(\xi_1, \ldots, \xi_k) \hat{\sigma}_{n-1,\sqrt{n}}(d\xi)$$
$$= C_{k,n} \int_{B^k(\sqrt{n})} g(\xi_1, \ldots, \xi_k) \left(1 - \frac{1}{n} \sum_{j=1}^k \xi_j^2 \right)^{(n-k-2)/2} d\xi_1 \ldots d\xi_k \tag{1.6.19}$$

where $B^k(\sqrt{n}) = \{(\xi_j)_{j=1}^k : \sum_{j=1}^k \xi_j^2 \leq n\}$ and $C_{k,n}$ is some positive constant.

Hence

$$\int_{S^{n-1}(\sqrt{n})} g d\hat{\sigma}_{n-1,\sqrt{n}} \rightarrow \frac{1}{(2\pi)^{k/2}} \int_{\mathbf{R}^k} g(\xi_1,\ldots,\xi_k)$$

$$\times \exp\left(-\frac{1}{2}\sum_{j=1}^{k}\xi_j^2\right) d\xi_1 \ldots d\xi_k \quad (1.6.20)$$

as $n \rightarrow \infty$, where the limiting value of $C_{n,k}$ is given by the Gaussian constants in Section 1.5. $\square$

For each unit vector $e \in \mathbf{R}^k$, $x \mapsto \langle x,e \rangle$ gives a random variable x_e on $S^{n-1}(\sqrt{n})$ with probability distribution $\hat{\sigma}_{n-1,\sqrt{n}}$. The statistical interpretation is that for each collection of k orthogonal coordinate directions, the random variables $(x_{e_1},\ldots,x_{e_k})$ have an asymptotic $N(0,I_k)$ distribution; in particular they are asymptotically independent.

Remark. The previous results suggest an intuitively attractive picture of the probability measures $\hat{\sigma}_{n,\sqrt{n}}$ converging to some natural probability measure on an infinite-dimensional sphere that has Gaussian marginals; so one can think about $S^\infty(\sqrt{\infty})$. See [115]. In Chapter 11 we show how to construct Gaussian measure on Hilbert space H, but also learn there is no rotation-invariant Radon probability measure on the sphere of H. Nevertheless, the notion of spherical harmonics does make sense, and there is a self-adjoint operator that corresponds to the Laplace operator on the sphere. The Ornstein–Uhlenbeck process is a convenient way of uniting these ideas.

Exercise 1.6.5 Consider the unit sphere S^{n-1} in $\mathbf{R}^n$. Let $R_{jk} = x_j\frac{\partial}{\partial x_k} - x_k\frac{\partial}{\partial x_j}$ for $1 \leq j < k \leq n$, let $L = \sum_{jk} R_{jk}^2$ and $\Gamma(f,g) = 2^{-1}[L(fg) - fLg - gLf]$ where $f,g : S^{n-1} \rightarrow \mathbf{R}$ are twice continuously differentiable functions. Show that the R_{jk} generate rotations of the sphere, and that that $\Gamma(f,f) = \sum_{jk}(R_{jk}f)^2$.

1.7 Lipschitz functions and the Hausdorff metric

Detailed discussion of the concepts in this section is available in [40, 162].

Definition (*Lipschitz function*). Let (Ω_j, d_j) for $j = 1,2$ be Polish spaces. A function $f : \Omega_1 \rightarrow \Omega_2$ is L-Lipschitz if

$$d_2(f(x), f(y)) \leq Ld_1(x,y) \quad (1.7.1)$$

The infimum of such L is the Lipschitz seminorm $\|f\|_{Lip}$. Sometimes we use Lip to stand for the space of Lipschitz function between specified metric spaces.

Examples 1.7.1 (i) Let $f : \mathbf{R}^n \to \mathbf{R}$ be a differentiable function such that $\|\nabla f(x)\|_{\mathbf{R}^n} \leq L$ for all $x \in \mathbf{R}^n$. Then f and $|f|$ are L-Lipschitz. This follows from the mean value theorem.

(ii) The function $f(x) = x/(1 + |x|)$ is a Lipschitz bijection $\mathbf{R} \to (-1, 1)$, but the inverse function is certainly not Lipschitz.

Exercise 1.7.2 Let $f, g : (\Omega, d) \to \mathbf{R}$ be Lipschitz functions on a compact metric space. We define

$$\delta(f)(x) = \limsup_{y \to x} \frac{|f(x) - f(y)|}{d(x, y)} \qquad (x \in \Omega). \qquad (1.7.2)$$

Show that

$$\delta(tf + sg)(x) \leq |t|\delta(f)(x) + |s|\delta(g)(x) \qquad (s, t \in \mathbf{R}; x \in \Omega) \qquad (1.7.3)$$

$$\delta(fg)(x) \leq |f(x)|\delta(g)(x) + |g(x)|\delta(f)(x). \qquad (1.7.4)$$

By comparison with the previous examples, one regards $\delta(f)(x)$ as the analogue of $\|\nabla f(x)\|$, without attaching any particular interpretation to ∇f.

Definition (*Lipschitz equivalence*). Polish spaces (Ω_1, d_1) and (Ω_2, d_2) are said to be bi-Lipschitz equivalent if there exists a bijection $\varphi : \Omega_1 \to \Omega_2$ such that φ and its inverse φ^{-1} are Lipschitz maps. The Lipschitz distance between Ω_1 and Ω_2 is

$$d_L(\Omega_1, \Omega_2) = \inf_\varphi \big\{ \log(\|\varphi\|_{Lip} \|\varphi^{-1}\|_{Lip}) \big\}. \qquad (1.7.5)$$

One can prove that compact metric spaces (Ω_1, d_1) and (Ω_2, d_2) are isometric if and only if $d_L(\Omega_1, \Omega_2) = 0$; further, d_L defines a metric on each set of compact metric spaces that are bi-Lipschitz equivalent to a given (Ω, d).

Unfortunately, the metric d_L is too severe to give a useful notion of convergence for metric measure spaces with the weak topology. Gromov therefore introduced a more subtle concept which refines the classical notion of Hausdorff distance; see [40].

Definition (*Enlargement*). Given nonempty $A \subseteq \Omega$, let $d(b, A) = \inf\{d(b, a) : a \in A\}$ be the distance of b to A. Then the ε-enlargement of A is

$$A_\varepsilon = \{b \in \Omega : d(b, A) \leq \varepsilon\}. \qquad (1.7.6)$$

Definition (*Hausdorff distance*). The Hausdorff distance between subsets A and B of (Ω, d) is

$$d_H(A, B) = \inf\{\varepsilon > 0 : A \subseteq B_\varepsilon, \quad B \subseteq A_\varepsilon\}. \qquad (1.7.7)$$

Exercise 1.7.3 (*Blaschke*). Let (Ω, d) be a compact metric space and let $K(\Omega) = \{A \subseteq \Omega : A \quad \text{compact}\}$. Prove that $(K(\Omega), d_H)$ is a compact metric space by proving the following.

(i) Let N be a ε-net of Ω. Show that $2^N = \{S : S \subseteq N\}$ is a ε-net of $K(\Omega)$.

(ii) Given a Cauchy sequence (S_n) in $K(\Omega)$, let S be the set of $x \in \Omega$ such that, for each $\varepsilon > 0$, the set $\{y : d(x, y) < \varepsilon\}$ intersects S_n for infinitely many n. Show that $d_H(S_n, S) \to 0$ as $n \to \infty$.

Exercise 1.7.6 Let A_n $(n = 1, 2, \dots)$ and A be closed subsets of a compact metric space (Ω, d) such that $d_H(A_n, A) \to 0$ as $n \to \infty$. Suppose that $\mu_n \in Prob(A_n)$ and that $\mu_n \to \mu$ weakly as $n \to \infty$. Show that $\mu \in Prob(A)$.

1.8 Characteristic functions and Cauchy transforms

With a view to later generalization, we express here some basic facts about classical probability spaces in an unfamiliar style. Let $(\Omega, \mathbf{P})$ be a probability space and let X_j $(j = 1, \dots, n)$ be random variables on $(\Omega, \mathbf{P})$. Then the $(X_j)_{j=1}^n$ are mutually independent if and only if there exist $\mu_j \in Prob(\mathbf{R})$ such that

$$\mathbf{E} f(X_1, \dots, X_n) = \int \cdots \int_{\mathbf{R}^n} f(x_1, \dots, x_n) \mu_1(dx_1) \dots \mu_n(dx_n), \quad (1.8.1)$$

for any $f \in C_b(\mathbf{R}^n; \mathbf{R})$. We write $\mu_1 \otimes \cdots \otimes \mu_n$ for the product measure here.

Definition (*Convolution*). The convolution of $\mu_1, \mu_2 \in Prob(\mathbf{R})$ is the unique $\nu \in Prob(\mathbf{R})$ such that

$$\int f(x)\nu(dx) = \iint f(x + y)\mu_1(dx)\mu_2(dy) \qquad (1.8.2)$$

for any $f \in C_b(\mathbf{R})$. When X_1 and X_2 are independent random variables, $X_1 + X_2$ has distribution $\nu = \mu_1 * \mu_2$.

Definition (*Characteristic function*). Let X be a random variable on $(\Omega, \mathbf{P})$. Then the characteristic function of X is

$$\varphi_X(t) = \mathbf{E} \exp itX \qquad (t \in \mathbf{R}). \qquad (1.8.3)$$

We recall that if X_j on $(\Omega_j, \mathbf{P}_j)$ $(j = 1, 2)$ are random variables such that the respective characteristic functions satisfy $\varphi_{X_1}(t) = \varphi_{X_2}(t)$ for all $t \in \mathbf{R}$, then the respective cumulative distribution functions satisfy $\mathbf{P}_1[X_1 \leq x] = \mathbf{P}_2[X_2 \leq x]$ for all $x \in \mathbf{R}$.

Example 1.8.1 (*Gaussian characteristic function*). Let X be a Gaussian $N(b, \sigma^2)$ random variable. Then X has characteristic function

$$\varphi(t) = \exp(itb - \sigma^2 t^2 / 2) \qquad (t \in \mathbf{R}). \qquad (1.8.4)$$

This follows from Proposition 1.5.1 when one takes $y = it$ and $A = \sigma^2$. See [73, 88].

Definition (*Fourier transform*). For $f \in L^1(\mathbf{R})$, we introduce the Fourier transform by

$$\hat{f}(\xi) = \int_{-\infty}^{\infty} f(x) e^{-ix\xi} \, dx; \qquad (1.8.5)$$

we also use the Fourier–Plancherel transform

$$\mathcal{F}f(\xi) = \lim_{T \to \infty} \int_{-T}^{T} f(t) e^{-it\xi} \, dt / \sqrt{2\pi},$$

which gives a unitary operator on $L^2(\mathbf{R})$. Note that the sign in the exponent differs from that used in the definition of the characteristic function.

For completeness we include the following two results.

Theorem 1.8.2 (*Glivenko [73, 88]*). *Let X_n and X be random variables on $(\Omega, \mathbf{P})$, and let the corresponding characteristic functions be φ_n and φ. Then $\varphi_n(t) \to \varphi(t)$ as $n \to \infty$ for all $t \in \mathbf{R}$, if and only if the distribution of X_n converges weakly to the distribution of X as $n \to \infty$.*

Theorem 1.8.3 (*Kac*). *Random variables $X_1, \ldots, X_n$ on $(\Omega, \mathbf{P})$ are mutually independent if and only if*

$$\mathbf{E} \exp\left(\sum_{j=1}^{n} it_j X_j \right) = \prod_{j=1}^{n} \mathbf{E} \exp(it_j X_j) \qquad (t_j \in \mathbf{R}). \qquad (1.8.6)$$

Proof. See [88]. □

Definition (*Moments*). Let X be a random variable on $(\Omega, \mathbf{P})$. If $\mathbf{E}|X|^k$ is finite for some integer $k = 0, 1, 2, \ldots$, then the kth moment of X exists and we let $m_k = \mathbf{E}X^k$. The moment sequence associated with X is the sequence $(m_k)_{k=0}^{\infty}$ when all moments exist; otherwise we terminate the sequence.

Example (*Standard Gaussian*). By various arguments one can show that

$$\int_{-\infty}^{\infty} e^{-tx^2/2} \frac{dx}{\sqrt{2\pi}} = \frac{1}{\sqrt{t}}. \tag{1.8.7}$$

By differentiating this identity repeatedly with respect to t, and evaluating the derivatives at $t = 1$, one obtains

$$\int_{-\infty}^{\infty} x^{2n} e^{-x^2/2} \frac{dx}{\sqrt{2\pi}} = (2n-1)(2n-3)\ldots 3.1. \tag{1.8.8}$$

Example (*Stieltjes*). There exist random variables X and Y such that all moments $m_k(X)$ of X and $m_k(Y)$ of Y exist and satisfy $m_k(X) = m_k(Y)$ for $k = 0, 1, 2, \ldots$, but the distribution of X is not equal to the distribution of Y. In fact, the lognormal distribution is not uniquely determined by its moments.

Exercise. There exists a family of distinct probability distributions such that, for each integer $k \geq 0$, the kth moments of all the distributions are finite and are equal.

(i) Show that for integers $k = 0, 1, 2, \ldots$,

$$\int_0^{\infty} u^k u^{-\log u} \sin(2\pi \log u)\, du = 0. \tag{1.8.9}$$

This calculation can be carried out by successive changes of variable, the first being $u = e^v$.

(ii) Let $c = \int_0^{\infty} u^{-\log u}\, du$ Check that c is finite and that for each $0 \leq t \leq 1$ the function

$$p_t(u) = \begin{cases} c^{-1} u^{-\log u} \left(1 + t\sin(2\pi \log u)\right), & \text{if } u > 0; \\ 0, & \text{if } u \leq 0; \end{cases} \tag{1.8.10}$$

defines a probability density function.

(iii) Deduce that the distribution with density p_t has finite moments of all orders for $0 \leq t \leq 1$ and that

$$\int_{-\infty}^{\infty} u^k p_t(u)\, du = \int_{-\infty}^{\infty} u^k p_0(u)\, du \qquad (k = 0, 1, 2, \ldots). \tag{1.8.11}$$

We recover uniqueness in the moment problem by imposing extra integrability conditions on the random variables.

Proposition 1.8.4 (i) *Let X be a random variable and suppose that there exists $\delta > 0$ such that $\mathbf{E}\cosh(\delta X)$ is finite. Then the characteristic function $\varphi(t) = \mathbf{E}e^{itX}$ extends to define a analytic function on some* *open neighbourhood of zero. Further, the distribution of* X *is uniquely determined by the moments* $\mathbf{E}X^k = i^{-k}\varphi^{(k)}(0)k = 0, 1, \ldots$.

(ii) Equivalently, let ν *be a positive Radon measure on* $\mathbf{R}$ *such that* $\int(\cosh \delta x)\nu(dx) < \infty$ *for some* $\delta > 0$. *Then* $\phi(t) = \int_{-\infty}^{\infty} e^{-ixt}\nu(dx)$ *extends to define an analytic function on some neighbourhood of zero. Furthermore, the polynomials give a dense linear subspace of* $L^2(\mathbf{R}, \nu)$.

Proof. (i) Clearly we can introduce $\varphi(z) = \mathbf{E}(e^{izX})$ for $|z| < \delta$. By Fubini's theorem,

$$\int_T \varphi(z)\,dz = \mathbf{E}\int_T e^{izX}\,dz = 0 \qquad (1.8.12)$$

for any triangle T inside $\{z : |z| < \delta\}$, so $\varphi(z)$ is analytic by Morera's theorem [154]. Hence φ is uniquely determined by the coefficients in its power series about $z = 0$, which determine the moments of X. Indeed, we can equate the coefficients in the convergent power series

$$\varphi(z) = \sum_{n=0}^{\infty} \frac{i^n \varphi^{(n)}(0)z^n}{n!} \qquad (|z| < \delta) \qquad (1.8.13)$$

and obtain the stated result.

(ii) One can show likewise that if $f \in L^2(\mathbf{R}, \nu)$ satisfies $\int_{-\infty}^{\infty} f(x)x^k\nu(dx) = 0$ for all $k = 0, 1, 2, \ldots$, then $f = 0$ almost everywhere with respect to ν. Hence by the F. Riesz representation theorem, the polynomials give a dense linear subspace. □

Definition (*Cumulants*). Let X be a random variable and suppose that there exists $\delta > 0$ such that $\mathbf{E}\cosh(\delta X)$ is finite. Then the logarithmic moment generating function of X is $g(z) = \log \mathbf{E}\exp(zX)$, and the cumulants are the coefficients $(\kappa_n)_{n=1}^{\infty}$ in the power series

$$g(z) = \sum_{n=1}^{\infty} \frac{\kappa_n z^n}{n!} \qquad (|z| < \delta).$$

In particular, from the power series one easily checks that

$$\kappa_1 = \mathbf{E}(X), \qquad \kappa_2 = \mathrm{var}(X). \qquad (1.8.14)$$

Under the hypotheses of Proposition 1.8.4, the cumulants determine the distribution of X.

Proposition 1.8.5 (*Convergence of cumulants*). *Let* X *and* $(X_n)_{n=1}^{\infty}$ *be random variables on* $(\Omega, \mathbf{P})$ *and suppose that there exists* $\delta > 0$ *such that* $\mathbf{E}\cosh(\delta X) \leq M$ *and* $\mathbf{E}\cosh(\delta X_n) \leq M$ *for some* $M < \infty$ *and all* n. *Then the following are equivalent.*

(i) X_n *converges to X in distribution as $n \to \infty$.*

(ii) $\varphi_{X_n}(z) \to \varphi_X(z)$ *uniformly on $\{z : |z| \le \delta/2\}$ as $n \to \infty$.*

(iii) $\mathbf{E}X_n^k \to \mathbf{E}X^k$ *as $n \to \infty$ for all $k = 1, 2, \dots$;*

(iv) $g_{X_n}(z) \to g_X(z)$ *uniformly on $\{z : |z| \le \varepsilon\}$ as $n \to \infty$ for some*
   $\varepsilon > 0.$

(v) $\kappa_k(X_n) \to \kappa_k(X)$ *as $n \to \infty$.*

Proof. (ii) $\Rightarrow$ (i) This follows from Glivenko's theorem.

   (i) $\Rightarrow$ (ii) When (i) holds, $\varphi_{X_n}(t)$ converges uniformly to $\varphi_X(t)$ for $t \in [-a, a]$ as $n \to \infty$ for each $a > 0$; further, $|\varphi_{X_n}(z)| \le 2M$ for all n on $\{z : |z| \le \delta\}$. By Vitali's theorem on power series [154, p. 168], (ii) holds.

   (ii) $\Leftrightarrow$ (iii) Vitali's theorem and Cauchy's estimates show that these are equivalent.

   (ii) $\Rightarrow$ (iv) Since $\varphi_X(0) = 1$, there exists $0 < \varepsilon < \delta$ such that $|\varphi_{X_n}(z) - 1| < 1/2$ for all z such that $|z| \le \varepsilon$ and all sufficiently large n. Then one takes logarithms to deduce that $g_{X_n}(z) = \log \varphi(-iz)$ converges uniformly to $g_X(z)$ as $n \to \infty$.

   (iv) $\Rightarrow$ (ii) One takes exponentials.

   (iv) $\Leftrightarrow$ (v) Vitali's theorem and Cauchy's estimates give this equivalence. $\qquad\square$

In addition to the characteristic function, we shall use the Cauchy transform of measures. For convenience, we introduce the notion of a distribution sequence, and then consider the issue of whether this arises from the moments of any particular probability measure.

Definition (*Distribution sequence*). A distribution sequence is a complex sequence $(m_n)_{n=1}^{\infty}$ such that $\sum_{n=1}^{\infty} m_n / \zeta^{n+1}$ converge for some $\zeta \in \mathbf{C}$, hence

$$G(\zeta) = \frac{1}{\zeta} + \sum_{n=1}^{\infty} \frac{m_n}{\zeta^{n+1}} \qquad (1.8.15)$$

defines an analytic function on $\{\zeta \in \mathbf{C} : |\zeta| > K\}$ for some $K < \infty$.

Proposition 1.8.6 (*Uniqueness of distribution sequences*).

(i) *Any probability measure μ of compact support in $\mathbf{R}$ determines a distribution sequence by $m_n = \int_{[-a,a]} x^n \mu(dx)$.*

(ii) *If μ_1 and μ_2 are compactly supported probability measures that determine the same distribution sequence, then $\mu_1 = \mu_2$.*

Proof. When μ has support contained in $[-a, a]$,

$$G(z) = \int_{[-a,a]} \frac{\mu(dx)}{z - x} \qquad (1.8.16)$$

is analytic on $\mathbf{C} \setminus [-a, a]$ and hence has a Laurent series about infinity as in (1.8.16). The coefficients of the Laurent series, namely the (m_n), are the moments of μ.

Given $\varepsilon > 0$ and a continuous $f : [-a, a] \to \mathbf{R}$, there exists a rational function R with poles in $\mathbf{C} \setminus [-a, a]$ such that $|f(x) - R(x)| < \varepsilon$. Indeed, we can take $f(\infty) = 0$ and apply Runge or Mergelyan's theorem [176] to the set $[-a, a] \cup \{\infty\}$ in the Riemann sphere, where the set has empty interior. When R has partial fractions decomposition

$$R(x) = \sum_{j,k} \frac{a_{jk}}{(z_k - x)^j} \qquad (1.8.17)$$

we have by the Cauchy integral formula

$$\int_{[-a,a]} R(x)\mu_1(dx) = \sum_{j,k} (-1)^{j-1} \frac{a_{jk}}{(j-1)!} G_{\mu_1}^{(j-1)}(z_k)$$

$$= \int_{[-a,a]} R(x)\mu_2(dx) \qquad (1.8.18)$$

since $G_{\mu_1}(z) = G_{\mu_2}(z)$. Hence $|\int_{[-a,a]} f d\mu_1 - \int_{[-a,a]} f d\mu_2| < \varepsilon$, where $\varepsilon > 0$ is arbitrary, so $\int_{[-a,a]} f d\mu_1 = \int_{[-a,a]} f d\mu_2$.

Definition (*Cauchy transform* [1]). The Cauchy transform of $\mu \in Prob(\mathbf{R})$ is

$$G(z) = \int \frac{\mu(dx)}{z - x} \qquad (z \in \mathbf{C} \setminus \mathbf{R}). \qquad (1.8.19)$$

When μ has support inside $[-A, A]$, we regard the Cauchy transform as an analytic function on a region that includes $\mathbf{C} \setminus [-A, A]$. Note also that $G(\bar{z}) = \overline{G(z)}$, as in Schwarz's reflection principle, and that $G(z) \asymp 1/z$ as $|z| \to \infty$. Typically, G will be discontinuous across the support of μ.

Lemma 1.8.7 (*Convergence of distribution sequences*). *Let μ_n and μ be probability measures on $[-A, A]$. Then the following conditions are equivalent.*

(i) *$\mu_n \to \mu$ weakly as $n \to \infty$.*
(ii) *The respective Cauchy transforms converge, so $G_{\mu_n}(\zeta) \to G_\mu(\zeta)$ as $n \to \infty$ for all ζ on some arc in the upper half plane $\{\zeta : \Im \zeta > 0\}$.*

(iii) $G_{\mu_n}(\zeta) \to G_\mu(\zeta)$ *as* $n \to \infty$ *uniformly on compact subsets of* $\mathbf{C} \setminus$ $[-A, A]$.

(iv) The moments converge, so $\int x^k \mu_n(dx) \to \int x^k \mu(dx)$ *as* $n \to \infty$ *for each* $k = 1, 2, \ldots$.

Proof. $(i) \Rightarrow (ii)$ This is immediate from the definition of weak convergence applied to $f(x) = 1/(\zeta - x)$.

$(ii) \Rightarrow (iii)$ For any compact subset S in $\mathbf{C} \setminus [-A, A]$, there exists $\delta > 0$ such that $|\zeta - x| > \delta$ for all $x \in [-A, A]$ and $\zeta \in S$. Hence $|G_{\mu_n}(\zeta)| \le 1/\delta$ for all $\zeta \in S$, and $G_{\mu_n}(\zeta) \to G_\mu(\zeta)$ for all ζ on some arc. By Vitali's theorem [154], $G_{\mu_n} \to G_\mu$ uniformly on S as $n \to \infty$.

$(iii) \Rightarrow (i)$ By (iii) and Weierstrass's double series theorem [154], the jth derivatives converge, so $G^{(j)}_{\mu_n} \to G^{(j)}_\mu$ uniformly on compact subsets of $\mathbf{C} \setminus [-A, A]$ as $n \to \infty$. Given $\varepsilon > 0$ and a continuous $f : [-A, A] \to \mathbf{R}$, there exists a finite collection of coefficients a_{jk} and points $z_j \in \mathbf{C} \setminus [-A, A]$ such that

$$\left| f(x) - \sum_{j,k} \frac{a_{jk}}{(z_k - x)^j} \right| < \varepsilon, \tag{1.8.20}$$

hence

$$\left| \int f d\mu_n - \sum_{j,k} \frac{(-1)^{j-1} a_{jk}}{(j-1)!} G^{(j)}_{\mu_n}(z_k) \right| < \varepsilon,$$

and likewise with μ_n instead of μ. Hence $\int f d\mu_n \to \int f d\mu$ as $n \to \infty$.

$(iv) \Rightarrow (i)$ This follows from Weierstrass's approximation theorem [154].

$(i) \Rightarrow (iv)$. This is immediate from the definition of weak convergence. $\quad\square$

Definition *(Hölder continuity [1])*. For $0 < \alpha \le 1$, a function $f : [a, b] \to \mathbf{C}$ is α-Hölder continuous at x if there exists $C < \infty$ such that

$$|f(x) - f(y)| \le C|x - y|^\alpha \qquad (y \in [a, b]).$$

Definition *(Principal value integral)*. Let $f : [-K, L] \setminus \{a\} \to \mathbf{C}$ be continuous, and suppose that the limit

$$\text{p.v.} \int_{-K}^{L} f(x)dx = \lim_{\varepsilon \to 0+} \left(\int_{-K}^{a-\varepsilon} f(x)dx + \int_{a+\varepsilon}^{L} f(x)dx \right) \tag{1.8.21}$$

exists. Then we call this the principal value integral of f over $[-K, L]$. We do not assert that f is (absolutely) integrable in Lebesgue's sense.

Definition (*Hilbert transform*). The Hilbert transform is the linear operator $\mathcal{H} : L^2(\mathbf{R}) \to L^2(\mathbf{R})$ that is given by

$$\mathcal{H}f(x) = \text{p.v.} \int \frac{f(y)}{x - y} \frac{dy}{\pi}, \qquad (1.8.22)$$

or equivalently by the Fourier multiplier $\hat{f}(\xi) \mapsto -i\xi \hat{f}(\xi)/|\xi|$.

Theorem 1.8.8 (*M. Riesz*). *For $1 < p < \infty$ the Hilbert transform defines a continuous linear operator $L^p(\mathbf{R}) \to L^p(\mathbf{R})$, so there exists A_p such that*

$$\|\mathcal{H}f\|_{L^p} \le A_p \|f\|_{L^p} \qquad (f \in L^p(\mathbf{R})). \qquad (1.8.23)$$

There exists $A > 0$ such that the optimal constant satisfies

$$A_2 = 1; \qquad A_q \sim A/(q-1) \quad \text{as} \quad q \to 1+; \qquad A_p \sim Ap \quad \text{as} \quad p \to \infty. \qquad (1.8.24)$$

Proof. See [71]. □

Proposition 1.8.9 (*Plemelj, [1, 119]*). *Suppose that $\mu \in \text{Prob}(\mathbf{R})$ has compact support Ω, and that $\mu = \mu_p + \mu_{ac}$ where μ_p is atomic and μ_{ac} is absolutely continuous with respect to Lebesgue measure. Then $G = G_p + G_{ac}$ where G_p and G_{ac} are the Cauchy transforms of μ_p and μ_{ac}. Further, G is analytic on $(\mathbf{C} \cup \{\infty\}) \setminus \Omega$,*

$$\mu_p(\{x\}) = \lim_{h \to 0+} \frac{ih}{2}\big(G(x + ih) - G(x - ih)\big) \qquad (1.8.25)$$

and

$$\frac{d\mu_{ac}}{dx} = \lim_{h \to 0+} \frac{G_{ac}(x + ih) - G_{ac}(x - ih)}{-2\pi i} \qquad (1.8.26)$$

at the points where $d\mu_{ac}/dx$ is Hölder continuous.

Proof. We have

$$\frac{ih}{2}\big(G(x + ih) - G(x - ih)\big) = \int \frac{h^2}{(x - t)^2 + h^2}\mu(dt),$$

so the dominated convergence theorem gives the first formula. The properties of the Cauchy integral are local, so to obtain the second formula, we can suppose that $f = d\mu_{ac}/dx$ is compactly supported and Hölder continuous. Then

$$G_{ac}(z) = \int \frac{f(x)}{z - x}dx \qquad (1.8.27)$$

has limits $G_{ac}^+(x) = \lim_{h\to 0+} G_{ac}(x + ih)$ and $G_{ac}^-(x) = \lim_{h\to 0+} G_{ac}(x - ih)$ which satisfy

$$G_{ac}^\pm(x) = \mp \pi i f(x) - \text{p.v.} \int \frac{f(t)}{t - x} dt; \qquad (1.8.28)$$

so $G_{ac}^\pm(x)$ satisfy the pair of equations

$$G_{ac}^+(x) - G_{ac}^-(x) = -2\pi i f(x), \quad G_{ac}^+(x) + G_{ac}^-(x) = 2\pi \mathcal{H}f(x), \quad (1.8.29)$$

by Plemelj's formula. Thus the pair of functions G_{ac}^+ and G_{ac}^- determine and are determined by the pair f and $\mathcal{H}f$. □

The Cauchy transform G is uniquely determined by the pair of functions $G^\pm$ on account of the following result.

Proposition 1.8.10 *Suppose that $g : \mathbf{C} \to \mathbf{C}$ is a continuous function that is analytic on $\mathbf{C} \setminus [-A, A]$ for some $A < \infty$, that $g(\bar{z}) = \overline{g(z)}$ for all $z \in \mathbf{C}$ and that $g(z) \to 0$ as $|z| \to \infty$. Then $g(z) = 0$ for all z.*

Proof. By Schwarz's reflection principle [1, 154], g extends to an analytic function on $\mathbf{C}$. Further, g is bounded on $\mathbf{C}$ and vanishes at infinity, so by Liouville's theorem is identically zero.

Exercise 1.8.11 Let p be a probability density function on $\mathbf{R}$, and let $p_-(x) = p(-x)$.

(i) Show that $k = p * p_-$ has Fourier transform $\hat{k} = |\hat{p}|^2$; so that $0 \leq \hat{k}(\xi) \leq 1$ for all $\xi \in \mathbf{R}$.

(ii) Let $K : L^2(\mathbf{R}) \to L^2(\mathbf{R})$ be the convolution operator $Kf = k * f$ for $f \in L^2(\mathbf{R})$. Using Plancherel's formula, show that $0 \leq K \leq I$ as an operator.

2

Lie groups and matrix ensembles

Abstract

The purpose of this chapter is to give a brief introduction to the theory of
Lie groups and matrix algebras in a style that is suited to random matrix
theory. Ensembles are probability measures on spaces of random matrices that are invariant under the action of certain compact groups, and
the basic examples are known as the orthogonal, unitary and symplectic
ensembles according to the group action. One of the main objectives is
the construction of Dyson's circular ensembles in Sections 2.7–2.9, and
the generalized ensembles from the affine action of classical compact Lie
groups on suitable matrix spaces in Section 2.5. As our main interest is
in random matrix theory, our discussion of the classification is patchy
and focuses on the examples that are of greatest significance in RMT.
We present some computations on connections and curvature, as these
are important in the analysis in Chapter 3. The functional calculus of
matrices is also significant, and Section 2.2 gives a brief treatment of
this topic. The chapter begins with a list of the main examples and
some useful results on eigenvalues and determinants.

2.1 The classical groups, their eigenvalues and norms

Throughout this chapter,

- $\mathbf{R}$ = real numbers;
- $\mathbf{C}$ = complex numbers;
- $\mathbf{H}$ = quaternions;
- $\mathbf{T}$ = unit circle.

By a well-known theorem of Frobenius, $\mathbf{R}, \mathbf{C}$ and $\mathbf{H}$ are the only finite-
dimensional division algebras over $\mathbf{R}$, and the dimensions are $\beta = 1, 2$

42

and 4 respectively; see [90]. From these division algebras, we introduce the classical Lie groups on which we later construct the matrix ensembles.

Definition (*Involutions*). The natural involution on $\mathbf{C}$ is conjugation $z \mapsto \bar{z}$. Now let i, j, k be the natural units in the quaternions, and for real q_0, q_1, q_2, q_3, let

$$(q_0 + q_1 i + q_2 j + q_3 k)^R = (q_0 - q_1 i - q_2 j - q_3 k). \qquad (2.1.1)$$

The transpose of a square matrix $X = [x_{jk}]$ is $X^t = [x_{kj}]$; the Hermitian conjugate of $X = [x_{jk}] \in M_n(\mathbf{C})$ is $X^* = [\bar{x}_{kj}]$. We extend the operation of R to $M_n(\mathbf{H})$ by $[W_{jk}] \mapsto [W_{kj}^R]$.

Using these involutions, we briefly define the classical groups, and then we discuss the reasoning behind the definitions.

Definition (*Classical groups*). The main examples are:

- real symmetric matrices $M_n^s(\mathbf{R}) = \{X \in M_n(\mathbf{R}) : X = X^t\}$;
- hermitian matrices $M_n^h(\mathbf{C}) = \{X \in M_n(\mathbf{C}) : X = X^*\}$; and
- self-dual hermitian matrices $M_n^h(\mathbf{H}) = \{X \in M_{2n}(\mathbf{C}) : X = X^R = X^*\}$.

The main compact Lie groups, known as invariance groups, are:

- the real orthogonal group $O(n) = \{U \in M_n(\mathbf{R}) : U^t U = I\}$, with
- special orthogonal subgroup $SO(n) = \{U \in O(n) : \det U = 1\}$;
- the unitary group $U(n) = \{U \in M_n(\mathbf{C}) : U^* U = I\}$, with
- special unitary subgroup $SU(n) = \{U \in U(n) : \det U = 1\}$ and
- the symplectic group $Sp(n) = \{X \in M_{2n}(\mathbf{C}) : X^R X = X^* X = I\}$.

The standard basis for $M_2(\mathbf{C})$ is

$$X = q_0 \begin{bmatrix} 1 & 0 \\ 0 & 1 \end{bmatrix} + q_1 \begin{bmatrix} i & 0 \\ 0 & -i \end{bmatrix} + q_2 \begin{bmatrix} 0 & 1 \\ -1 & 0 \end{bmatrix} + q_3 \begin{bmatrix} 0 & i \\ i & 0 \end{bmatrix}, \qquad (2.1.2)$$

where $q_0, q_1, q_2, q_3 \in \mathbf{C}$, and the standard hermitian conjugation on $M_2(\mathbf{C})$ is

$$X \mapsto X^* = \bar{q}_0 \begin{bmatrix} 1 & 0 \\ 0 & 1 \end{bmatrix} - \bar{q}_1 \begin{bmatrix} i & 0 \\ 0 & -i \end{bmatrix} - \bar{q}_2 \begin{bmatrix} 0 & 1 \\ -1 & 0 \end{bmatrix} - \bar{q}_3 \begin{bmatrix} 0 & i \\ i & 0 \end{bmatrix}. \qquad (2.1.3)$$

There is another involution

$$X \mapsto X^R = \bar{q}_0 \begin{bmatrix} 1 & 0 \\ 0 & 1 \end{bmatrix} - q_1 \begin{bmatrix} i & 0 \\ 0 & -i \end{bmatrix} - q_2 \begin{bmatrix} 0 & 1 \\ -1 & 0 \end{bmatrix} - q_3 \begin{bmatrix} 0 & i \\ i & 0 \end{bmatrix}.$$

(2.1.4)

Now $\mathbf{H}$ may be identified with the subspace of such X where q_0, q_1, q_2 and q_3 are all real, or equivalently as the subspace of $X \in M_2(\mathbf{C})$ such that $X^* = X^R$.

We extend the involution R to $M_{2n}(\mathbf{C})$ by transposing matrices and conjugating matrix elements $[W^R]_{jk} = [W]_{kj}^R$; we can identify $M_n(\mathbf{H})$ with those $X \in M_{2n}(\mathbf{C})$ such that $X^R = X^*$.

A $X \in M_n(\mathbf{C})$ is symmetric when its entries satisfy $[X]_{jk} = [X]_{kj}$, so $X^t = X$; further, X is Hermitian when $X^* = X$; so if $X \in M_n(\mathbf{C})$ is both symmetric and Hermitian, then X is real symmetric. Further, $X \in M_{2n}(\mathbf{C})$ matrix is self-dual when $X^R = X$; so if $X \in M_{2n}(\mathbf{C})$ is self-dual and Hermitian, then X may be identified with a $n \times n$ symmetric matrix with quaternion entries.

Proposition 2.1.1 *When viewed as real vector spaces, the dimensions of the ensembles are*

$$N = n + \frac{\beta}{2}n(n-1)$$

(2.1.5)

where: $\beta = 1$ *for* $M_n^s(\mathbf{R})$; $\beta = 2$ *for* $M_n^h(\mathbf{C})$; $\beta = 4$ *for* $M_n^h(\mathbf{H})$.

Proof. The matrices are symmetric with n real entries on the leading diagonal, and $n(n-1)/2$ entries from $\mathbf{R}$, $\mathbf{C}$ or $\mathbf{H}$ above the leading diagonal. Hence one can find a basis over $\mathbf{R}$ with the dimension as listed.   □

Definition (*Invariant metrics*). Let $G = O(n), SO(n), U(n), SU(n)$ or $Sp(n)$. Then a metric d on G is *bi-invariant* if

$$d(UV, UW) = d(VU, WU) = d(V, W) \qquad (U, V, W \in G). \qquad (2.1.6)$$

Exercise. Show that for any bi-invariant metric, we have

$$d(V, W) = d(D, I),$$

where D is the diagonal matrix that has entries given by the eigenvalues of VW^{-1}.

Let $\langle (z_j), (w_j) \rangle = \sum_{j=1}^{n} z_j \bar{w}_j$ be the usual inner product on $\mathbf{C}^n$.

Definition (*Positive semidefinite*). A matrix $[a_{jk}]$ in $M_n^h(\mathbf{C})$ is positive semidefinite if

$$\sum_{j,k=1}^{n} a_{jk} z_j \bar{z}_k \geq 0 \qquad ((z_j) \in \mathbf{C}^n). \tag{2.1.7}$$

A minor of a matrix is the determinant of any square submatrix. A principal minor is the determinant of any square submatrix that has the same index set for the rows as it has for columns; see [4].

Proposition 2.1.2 (*Characterisations of positive semidefiniteness*). *The following are equivalent for $A = [a_{jk}] \in M_n^h(\mathbf{C})$.*

 (i) *A is positive semidefinite.*
 (ii) *There exists $U \in U(n)$ and a diagonal matrix $D = \mathrm{diag}(\lambda_j)$ such that $\lambda_j \geq 0$ and $A = UDU^*$.*
 (iii) *There exist $\xi_j \in \mathbf{C}^n$ such that $a_{jk} = \langle \xi_j, \xi_k \rangle$ for $j, k = 1, \ldots, n$.*
 (iv) *The principal minors $\det[a_{jk}]_{j,k=1,\ldots,\ell}$ are non-negative for $\ell = 1, \ldots, n$.*
 (v) *The principal minors $\det[a_{s,t}]_{s,t \in S}$ indexed by $S \subseteq \{1, \ldots, n\}$ are all nonnegative.*

Proof. (i) $\Rightarrow$ (ii) This is the spectral theorem applied to $A = A^*$.

(ii) $\Rightarrow$ (iii) Let (e_j) be the usual orthonormal basis for $\mathbf{C}^n$, and let $\xi_j = \sqrt{D}U^* e_j$.

(iii) $\Rightarrow$ (i) By an easy calculation,

$$\sum_{j,k} a_{jk} z_j \bar{z}_k = \langle \sum_j z_j \xi_j, \sum_k z_k \xi_k \rangle \geq 0. \tag{2.1.8}$$

(i) $\Rightarrow$ (iv) Evidently $\det A \geq 0$ whenever A is positive semidefinite on account of (ii). Further, PAP is positive semidefinite for all projections P, in particular for the projection P_ℓ onto $\mathrm{span}\{e_1, \ldots, e_\ell\}$, so $\det P_\ell A P_\ell \geq 0$.

(iv) $\Rightarrow$ (i) One uses induction on dimension.

(v) $\Rightarrow$ (iv) This is clear.

(iv)&(ii) $\Rightarrow$ (v) One uses a permutation matrix to take S into the top left corner of the matrix, then one invokes (iv). $\qquad\square$

Corollary 2.1.3 *Let $[a_{jk}]$ and $[b_{jk}]$ be positive semidefinite $n \times n$ matrices. Then the Schur product $[a_{jk}b_{jk}]$ is likewise positive semidefinite.*

Proof. By (iii) we can introduce vectors ξ_j and η_k such that $a_{jk} = \langle \xi_j, \xi_k \rangle$ and $b_{jk} = \langle \eta_j, \eta_k \rangle$; then we obtain $a_{jk}b_{jk} = \langle \xi_j \otimes \eta_j, \xi_k \otimes \eta_k \rangle$. $\qquad\square$

We can impose a common metric structure upon all of these examples, since they are presented as closed subspaces of $M_n^h(\mathbf{C})$. Let X be any normal matrix in $M_n(\mathbf{C})$, so that $XX^* = X^*X$. Then there exist a unitary matrix U and a diagonal matrix D such that $UXU^* = D$; further, the diagonal entries of D are the eigenvalues of X, listed according to algebraic multiplicity.

For any $Y \in M_{m,n}(\mathbf{C})$, we observe that Y^*Y is Hermitian, and that any eigenvector ξ with corresponding eigenvalue s satisfies $s\|\xi\|^2 = \langle s\xi, \xi \rangle = \langle Y^*Y\xi, \xi \rangle = \|Y\xi\|^2$; so $s \geq 0$. Hence we can write $VY^*YV^* = \Delta^2$, where $\Delta = \text{diag}(s_1, \ldots, s_n)$ and V is so chosen that $s_1 \geq \cdots \geq s_n \geq 0$. The s_j are the singular numbers of Y.

Definition (*Matricial norms*). A matricial norm is a norm $\|\cdot\|$ on $M_n(\mathbf{C})$ such that $\|ABC\| \leq \|A\|_{op}\|B\|\|C\|_{op}$ for all $A, B, C \in M_n(\mathbf{C})$ where $\|\cdot\|_{op}$ denotes the usual operator norm on matrices. See [143].

Proposition 2.1.4 *(i) When U, V are unitary matrices, the matricial norm satisfies $\|UBV\| = \|B\|$; indeed, $\|B\|$ depends only upon the singular numbers of B.*

(ii) The singular numbers satisfy

$$s_j(ABC) \leq \|A\|_{op}s_j(B)\|C\|_{op} \qquad (j = 1, 2, \ldots). \tag{2.1.9}$$

(iii) For $m \times n$ matrices A and B the singular numbers satisfy

$$s_{j+k-1}(A + B) \leq s_j(A) + s_k(B) \quad (j, k = 1, 2, \ldots). \tag{2.1.10}$$

Proof.

(i) Given B, we can choose unitary matrices U and V so that UBV is a diagonal matrix with diagonal entries $S_1 \geq \cdots \geq S_n \geq 0$.

(ii) & (iii) See [143, Theorem 1.6]. □

Definition (*Schatten norms*). The Schatten norms are

$$\|Y\|_{c^p(n)} = \left(\frac{1}{n}\sum_{j=1}^{n} s_j^p\right)^{1/p} \tag{2.1.11}$$

for $1 \leq p < \infty$ and

$$\|Y\|_{M_{m,n}} = s_1 = \|Y\|_{op} = \sup\{\Re\langle Y\xi, \eta \rangle : \|\xi\| = \|\eta\| = 1\} \tag{2.1.12}$$

is the usual operator norm. The factor of $1/n$ is introduced so that the identity matrix satisfies $\|I\|_{c^p(n)} = 1$. When $m = n$, the normalized trace of $Y = [y_{jk}]$ is

$$\text{trace}_n Y = \frac{1}{n}\sum_{j=1}^{n} y_{jj}. \tag{2.1.13}$$

Definition (*Standard metrics on compact groups*). The standard metric on $O(n), SO(n), U(n)$ and $SU(n)$ is the $c^2(n)$ norm via $d(U,V) = \|U - V\|_{c^2(n)}$; the standard metric on $Sp(n)$ is given by the $c^2(2n)$ norm.

Note that the diameter of the said groups is less than or equal to $\sqrt{2}$, a bound which is independent of n.

If $X = X^*$, then the eigenvalues are real; whereas if $X^*X = I$ then the eigenvalues lie on the unit circle. For any function $v : \mathbf{C} \to \mathbf{C}$, we can form $v(D) = \mathrm{diag}(v(\lambda_1), \ldots, v(\lambda_n))$ and hence $v(X) = U^* v(D) U$. By checking through the definitions, one can verify that

$$\|X\|_{c^p(n)} = \left(\frac{1}{n} \sum_{j=1}^n |\lambda_j|^p\right)^{1/p} = \left(\mathrm{trace}_n |X|^p\right)^{1/p} = \|(\lambda_j)\|_{\ell^p(n)}. \quad (2.1.14)$$

In particular, the $c^2(n)$ is the scaled Hilbert–Schmidt norm and satisfies

$$\|Y\|_{c^2(n)}^2 = \mathrm{trace}_n Y^* Y = \frac{1}{n} \sum_{j,k=1}^n |y_{jk}|^2. \quad (2.1.15)$$

For consistency with these definitions, we introduce

$$\Delta^n = \{\lambda = (\lambda_1, \ldots, \lambda_n) \in \mathbf{R}^n : \lambda_1 \leq \cdots \leq \lambda_n\} \quad (2.1.16)$$

with scaled norms $\|\lambda\|_{\ell^p(n)} = (\frac{1}{n} \sum_{j=1}^n |\lambda_j|^p)^{1/p}$. Note that we usually work with eigenvalues instead of singular numbers, and we take the eigenvalues in increasing order.

Definition (*Empirical eigenvalue distribution*). Let $X \in M_n^s(\mathbf{R})$ have eigenvalues $\lambda_1, \ldots, \lambda_n$. Then the empirical eigenvalue distribution of X is the probability measure $\mu_X = n^{-1} \sum_{j=1}^n \delta_{\lambda_j}$.

Let $\Delta^n = \{(x_j) \in \mathbf{R}^n : x_1 \leq x_2 \leq \cdots \leq x_n\}$; observe that the simplex $\Delta^n \cap [0,1]^n$ has volume $1/n!$. Note that there is a natural correspondence between empirical eigenvalue distributions and points in Δ^n.

Example. Let X and Y be real symmetric matrices with eigenvalues $\lambda_1 \leq \cdots \leq \lambda_n$ and $\rho_1 \leq \cdots \leq \rho_n$, and empirical eigenvalue distributions μ_n^X and μ_n^Y respectively. Then the following measure the distance between the spectra:

(i) The Hausdorff distance $d_H(\mathrm{spec}(X), \mathrm{spec}(Y))$;
(ii) $(n^{-1} \sum_{j=1}^n |\lambda_j - \rho_j|^p)^{1/p}$;
(iii) the Wasserstein distance $W_s(\mu_n^X, \mu_n^Y)$ (as defined in Section 3.3).

We generally prefer (ii) and (iii) to (i), since it takes account of the spectral multiplicity.

Lemma 2.1.5 (*Lidskii*). *The map* $\Lambda : X \mapsto \lambda$ *that associates to each hermitian matrix its ordered list* $\lambda = (\lambda_1 \leq \lambda_2 \leq \cdots \leq \lambda_n)$ *of eigenvalues, is Lipschitz from* $(M_n^h(\mathbf{C}), c^p(n)) \rightarrow (\Delta^n, \ell^p(n))$ *with constant one.*

Proof. We require to show that, for any pair of hermitian matrices X and Y, the eigenvalues satisfy

$$\Big(\frac{1}{n}\sum_{j=1}^{n}|\lambda_j(X) - \lambda_j(Y)|^p\Big)^{1/p} \leq \Big(\frac{1}{n}\sum_{j=1}^{n}|\lambda_j(X - Y)|^p\Big)^{1/p}. \qquad (2.1.17)$$

Lidskii [143 Theorem 120(b)] showed that there exists a real matrix $A = [a_{jk}]$ that is doubly substochastic [71], so that

$$\sum_{j=1}^{n}|a_{jk}| \leq 1, \qquad \sum_{m=1}^{n}|a_{km}| \leq 1 \qquad (k = 1, \ldots, n), \qquad (2.1.18)$$

and such that

$$\lambda_j(X) - \lambda_j(Y) = \sum_{j=1}^{n} a_{jk}\lambda_k(X - Y). \qquad (2.1.19)$$

The doubly substochastic matrix A can be written as $A = \sum_j \alpha_j P_j$, where α_j are real numbers such that $\sum_j |\alpha_j| \leq 1$ and P_j are permutation matrices, so

$$\|\lambda(X) - \lambda(Y)\|_{\ell^p(n)} \leq \|\lambda(X - Y)\|_{\ell^p(n)} = \|X - Y\|_{c^p(n)} \qquad (2.1.20)$$

follows by convexity of the norm on $\ell^p(n)$. $\qquad\qquad\qquad\qquad\square$

More generally, several of our results apply to other norms on the matrices.

Exercise. (i) Show that $\mathrm{trace}_n(XX^*) \geq |\mathrm{trace}_n X|^2$ holds for all $X \in M_n(\mathbf{C})$.

(ii) Deduce the consequence when

$$X = \Big[\frac{\partial^2 f}{\partial x_j \partial x_k}\Big]_{j,k=1}^{n} \qquad (2.1.21)$$

is a Hessian matrix of partial derivatives.

Some of these ideas are useful for infinite matrices.

Definition (*Operator convergence*). Let H be a Hilbert space and $B(H)$ the bounded linear operators on H with the operator norm. Let (T_n) be a sequence of operators in $B(H)$, and $T \in B(H)$. Say that:

(1) $T_n \to T$ in norm, if $\|T_n - T\|_{B(H)} \to 0$ as $n \to \infty$;
(2) $T_n \to T$ strongly, if $\|T_n\xi - T\xi\|_H \to 0$ as $n \to \infty$ for each $\xi \in H$;
(3) $T_n \to T$ weakly, if $|\langle T_n\xi, \eta \rangle - \langle T\xi, \eta \rangle| \to 0$ as $n \to \infty$ for each $\xi, \eta \in H$.

Definition (*Hilbert–Schmidt norm*). Let H and K be Hilbert spaces, with complete orthonormal bases (e_j) and (f_j) respectively. A linear operator $T : H \to K$ is said to be Hilbert–Schmidt if

$$\|T\|_{HS}^2 = \sum_{j=1}^{\infty} \|Te_j\|^2 < \infty; \qquad (2.1.22)$$

this expression is independent of the choice of the basis. The matrix that represents T with respect to these bases is

$$[T]_{jk} = \langle Te_j, f_k \rangle_K \qquad (2.1.23)$$

with the property that

$$\sum_{j,k} |[T]_{jk}|^2 = \|T\|_{HS}^2. \qquad (2.1.24)$$

Any Hilbert–Schmidt operator is necessarily bounded, indeed compact.

Definition (*von Neumann–Schatten norm* [171, 143]). Let T be a compact operator on H, and let $s_1 \geq s_2 \geq \cdots$ be the eigenvalues of $(T^*T)^{1/2}$, listed according to multiplicity. For $1 \leq p < \infty$ the von Neumann–Schatten norm of T is

$$\|T\|_{c^p} = \Big(\sum_{j=1}^{\infty} s_j^p \Big)^{1/p}$$

and c^p is the Banach space of all operators T such that this norm is finite. In particular, c^2 coincides with the space of Hilbert–Schmidt operators as in; whereas c^1 is the space of trace-class operators, namely the set of products $\{ST : S, T \in c^2\}$.

2.2 Determinants and functional calculus

Definition (*Vandermonde determinant*). Let x_j be commuting indeterminates, and let the Vandermonde determinant be

$$V_n(x_1, \ldots, x_n) = \det[x_j^{k-1}]_{j,k=1,\ldots,n}. \qquad (2.2.1)$$

Lemma 2.2.1 (*Vandermonde's determinant*).

$$(i) \quad \prod_{j,k:1\le j<k\le n} (x_k - x_j) = V_n(x_1,\ldots,x_n). \qquad (2.2.2)$$

(*Cauchy's determinants*) [4, 138].

$$(ii) \quad \det\left[\frac{1}{x_j + y_k + 1}\right]_{j,k=1,\ldots,n} = \frac{V_n(x_1,\ldots,x_n)V_n(y_1,\ldots,y_n)}{\prod_{j,k=1}^{n}(1 + x_j + y_k)}. \qquad (2.2.3)$$

$$(iii) \quad \det\left[\frac{1}{1 - x_j y_k}\right]_{j,k=1,\ldots,n} = \frac{V_n(x_1,\ldots,x_n)V_n(y_1,\ldots,y_n)}{\prod_{j,k=1}^{n}(1 - x_j y_k)}. \qquad (2.2.4)$$

Proof. (i) This is a standard induction argument in which one uses row and column operations on the determinant.

Both (ii) and (iii) may be proved by induction on n, and the induction step consists of clearing the entries in the first column that are below the top left entry. One applies the column operations $C_k \mapsto C_k - C_1$, for $k = 2,\ldots,n$; then one takes out common factors; then one applies the row operations $R_j \mapsto R_j - R_1$ for $j = 2,\ldots,n$. □

Definition (*Pfaffian*). Let A be a skew-symmetric $2N \times 2N$ matrix, and define the Pfaffian of A to be

$$\text{Pf}(A) = \frac{1}{2^N N!} \sum_{\sigma \in S_{2N}} \text{sgn}(\sigma) \prod_{j=1}^{N} a_{\sigma(2j-1),\sigma(2j)}. \qquad (2.2.5)$$

Let $n = 2N$ be even. The standard non-singular skew matrix is $C = [c_{jk}]$, where

$$c_{12} = c_{34} = \cdots = c_{n-1,n} = 1,$$
$$c_{21} = c_{43} = \cdots = c_{n,n-1} = -1;$$

then $\text{Pf}(C) = 1$ and $\det(C) = 1$. See (2.3.17) for an example. Any nonsingular skew matrix A can be written as $A = BCB^t$. The identity

$$\text{Pf}(QAQ^t) = \text{Pf}(A)$$

reduces to the formula

$$\text{Pf}(A)^2 = \det(A).$$

Lemma 2.2.2 *(i)* (*Andréief's identity*). *Let f_j and g_j be continuous complex functions on some bounded interval J. Then*

$$\frac{1}{n!} \int \cdots \int_{J^n} \det[f_j(x_k)]_{1 \leq j,k \leq n} \det[g_\ell(x_k)]_{1 \leq \ell,k \leq n} dx_1 \ldots dx_n$$
$$= \det\left[\int_J f_j(x)g_k(x)\,dx\right]_{1 \leq j,k \leq n}. \tag{2.2.6}$$

(ii) (*de Bruijn's identity*). *For N even,*

$$\int_{\Delta^N \cap J^N} \det[f_j(x_k)]_{j,k=1}^N dx_1 \ldots dx_N$$
$$= \text{Pf}\left(\iint_{J \times J} \text{sgn}(y - x)f_j(x)f_k(y)\,dy dx\right)_{j,k=1}^N. \tag{2.2.7}$$

Proof. (i) See [50]. Multiplying out the determinants and gathering terms, we obtain

$$\int_{J^n} \det[f_j(x_k)] \det[g_\ell(x_k)] dx_1 \ldots dx_n$$
$$= \sum_{\sigma,\tau \in S_n} \text{sgn}(\sigma)\text{sgn}(\tau) \prod_{j=1}^n \int_J f_{\sigma(j)}(x_j)g_{\tau(j)}(x_j)dx_j$$

so we introduce $\alpha = \sigma\tau^{-1}$ and express this as

$$\int_{J^n} \det[f_j(x_k)] \det[g_\ell(x_k)] dx_1 \ldots dx_n$$
$$= \sum_{\tau \in S_n} \sum_{\alpha \in S_n} \text{sgn}(\alpha) \prod_{j=1}^n \int_J f_{\alpha\tau(j)}(x)g_{\tau(j)}(x)\,dx$$
$$= \sum_{\tau \in S_n} \det\left[\int_J f_{\tau(k)}(x)g_{\tau(j)}(x)\,dx\right]$$
$$= n! \det\left[\int_J f_j(x)g_k(x)\,dx\right]. \tag{2.2.8}$$

(ii) Omitted. □

Exercise. (i) For some real function v, let $f_j(x) = e^{-v(x)+ijx}$ and $g_j(x) = e^{-ijx}$, then simplify the resulting version of Andréief's identity.

(ii) Apply de Bruijn's identity to $f_j(x) = e^{-v(x)}x^{j-1}$ and simplify the left-hand side.

Definition (*Unitary representation*). Let G be any group, and $U(H)$ the group of unitary operators on separable Hilbert space H. A unitary representation of G on H is a map $\pi : G \to U(H)$ such that $\pi(gh) = \pi(g)\pi(h)$ for all $g, h \in G$. Suppose further that (G, d) is a metric group. Then π is continuous if $\|\pi(g_n)\xi - \xi\|_H \to 0$ if $d(g_n, e) \to 0$ as $n \to \infty$. A vector $\xi \in H$ is cyclic if the closed linear span of $\{\pi(g)\xi : g \in G\}$ is norm dense in H.

Proposition 2.2.3 *Let (G, d) be a metric group with a continuous unitary and injective representation $\pi : G \to U(H)$ on a separable Hilbert space H. Then there exists a metric ρ on G such that*

(i) $\rho(gh, gk) = \rho(h, k)$, for all $g, h, k \in G$;
(ii) $\rho(g_n, e) \to 0$ if and only if $\|\pi(g_n)\xi - \xi\|_H \to 0$ as $n \to \infty$ for all $(g_n)_{n=1}^{\infty}$ in G.

Proof. Let (e_n) be a complete orthonormal basis for H, and let

$$\rho(g, h) = \sum_{n=1}^{\infty} 2^{-n} \|\pi(g)e_n - \pi(h)e_n\|_H. \qquad (2.2.9)$$

(i) By unitary invariance $\|\pi(gh)e_n - \pi(gk)e_n\|_H = \|\pi(h)e_n - \pi(k)e_n\|_H$.
(ii) Evidently $\rho(g_n, e) \to 0$ if and only if $\|\pi(g_n)e_n - e_n\|_H \to 0$ as $n \to \infty$, and we can approximate a typical x in norm by a linear combination of e_j. $\qquad \square$

Example. Let $Y \in M_n(\mathbf{C})$. Then there is a natural algebra homomorphism $\mathbf{C}[x] \to M_n(\mathbf{C})$ given by $f(x) \mapsto f(Y)$. In this section we extend this in various natural ways.

Definition (*Modules*). Let A be an algebra over $\mathbf{C}$. A left A module M is a vector space and a left multiplication $A \times M \to M$ such that:

$a(\lambda m + \mu n) = \lambda am + \mu an$;
$(\lambda a + \mu b)m = \lambda am + \mu bm$;
$(ab)m = a(bn)$; for all $\lambda, \mu \in \mathbf{C}$; $a, b \in A$ and $m, n \in M$.

When A is unital, we further require that M is unitary, so that $1m = m$ for all $m \in M$. Each $a \in A$ gives rise to a linear map $M \to M : m \mapsto am$.

Right A-modules are defined analogously; bimodules have left and right multiplications that commute, so that $(am)b = a(mb)$.

Definition (*Derivations*). Let A be an algebra over $\mathbf{C}$ and let M be an A-bimodule. A derivation is a linear map $\delta : A \to M$ that satisfies Leibniz's rule

$$\delta(fg) = (\delta f)g + f(\delta g).$$

Example. Let G be a finite group, and let $A = \{\sum_g \alpha_g g : \alpha_g \in \mathbf{C}\}$ be the group algebra. Then for any unitary representation $\pi : G \to U(H)$, the Hilbert space H becomes a left A module for the multiplication $A \times H \to H : (\sum_g \alpha_g g, \xi) \mapsto \sum_g \alpha_g \pi(g)\xi$.

The spectrum of a Hermitian matrix is real, so the Weierstrass approximation theorem provides a natural means for extending the functional calculus from polynomials in a single variable to spaces of continuous and differentiable functions on a compact subset of the real line. To convert these results into conventional calculus notation, we introduce a natural system of coordinates.

Definition (*Hessian*). The space $(M_n^s(\mathbf{R}), c^2(n))$ has orthonormal basis

$$E_{jj} = \sqrt{n}e_{jj}, E_{km} = \sqrt{\frac{n}{2}}(e_{km} + e_{mk}) \qquad (2.2.10)$$

for $j = 1, \ldots, n$ and $1 \le k < m \le n$, where e_{km} are the usual matrix units in $M_n^s(\mathbf{R})$. We define the gradient of $F : M_n^s(\mathbf{R}) \to \mathbf{R}$ by

$$\langle \nabla F(X), Y \rangle_{c^2(n)} = \left(\frac{d}{dt}\right)_{t=0} F(X + tY) \qquad (2.2.11)$$

and express this in terms of the coordinates $X = \sum_{1 \le j \le k \le n} x_{jk} E_{jk}$ as

$$\nabla F(X) = \sum_{1 \le j \le k \le n} \frac{\partial F}{\partial x_{jk}} E_{jk}. \qquad (2.2.12)$$

The Hessian matrix is

$$\text{Hess}\, F = \sum_{1 \le j \le k \le n} \sum_{1 \le l \le m \le n} \frac{\partial^2 F}{\partial x_{jk} \partial x_{lm}} E_{jk} \otimes E_{lm} \qquad (2.2.13)$$

and the Laplacian is

$$\Delta F = \text{trace}(\text{Hess}\, F) = \sum_{1 \le j \le k \le n} \frac{\partial^2 F}{\partial x_{jk}^2}. \qquad (2.2.14)$$

In some significant cases, it is possible to express these derivatives in terms of the eigenvalues of the matrices.

Proposition 2.2.4 *Let $v : \mathbf{R} \to \mathbf{R}$ be continuously differentiable. Then $V(X) = \mathrm{trace}_n v(X)$ is a differentiable function on $M_n^h(\mathbf{C})$ such that*

$$\left(\frac{d}{dt}\right)_{t=0} V(X + tY) = \mathrm{trace}_n\left(v'(X)Y\right). \qquad (2.2.15)$$

Proof. Suppose that $\|X\|_{op}, \|Y\|_{op} \leq L$ and let (p_j) be a sequence of polynomials such that $p_j \to v$ and $p'_j \to v'$ uniformly on $[-2L, 2L]$ as $j \to \infty$. Now introduce $f_j(t) = \mathrm{trace}_n p_j(X + tY)$ and observe that $f'_j(t) = \mathrm{trace}_n(p'_j(X+tY)Y)$; this follows from the expansion of a typical monomial as

$$(X + tY)^m - X^m = tX^{m-1}Y + tX^{m-2}YX + \cdots + tYX^{m-1} + O(t^2).$$

$$(2.2.16)$$

Now $f'_j(t) \to \mathrm{trace}_n(v'(X + tY)Y)$ uniformly on $[-2L, 2L]$ as $j \to \infty$ and $f_j(0) \to V(X)$; hence $f(t) = \mathrm{trace}_n v(X + tY))$ is differentiable with $f'(t) = \mathrm{trace}_n(v'(X + tY)Y)$. $\qquad\square$

Theorem 2.2.5 (*Rayleigh–Schrödinger*). *Let $v : \mathbf{R} \to \mathbf{R}$ be twice continuously differentiable. Let $X \in M_n^s(\mathbf{R})$ have distinct eigenvalues $\lambda_1, \ldots, \lambda_n$ with corresponding unit eigenvectors $\xi_1, \ldots, \xi_n$. Then $V(X) = \mathrm{trace}_n v(X)$ is a twice differentiable function on $M_n^s(\mathbf{R})$ such that*

$$\langle \mathrm{Hess}\, V, Y \otimes Y \rangle = \frac{1}{n} \sum_{j=1}^{n} v''(\lambda_j) \langle Y\xi_j, \xi_j \rangle^2$$

$$+ \frac{2}{n} \sum_{1 \leq j < k \leq n} \frac{v'(\lambda_j) - v'(\lambda_k)}{\lambda_j - \lambda_k} \langle Y\xi_j, \xi_k \rangle^2. \qquad (2.2.17)$$

Proof. As in the proof of the preceding proposition, we suppose that $v(x) = x^m$ and observe that

$$\langle \mathrm{Hess}\, V, Y \otimes Y \rangle = \left(\frac{d^2}{dt^2}\right)_{t=0} V(X + tY)$$

$$= \sum_{\ell=0}^{m-1} m\,\mathrm{trace}_n(X^\ell Y X^{m-\ell-2} Y), \qquad (2.2.18)$$

where by the spectral theorem

$$\mathrm{trace}_n(X^\ell Y X^{m-\ell-2} Y) = \frac{1}{n} \sum_{j=1}^{n} \langle X^\ell Y X^{m-\ell-2} Y\xi_j, \xi_j \rangle$$

$$= \frac{1}{n} \sum_{j,k=1}^{n} \lambda_k^\ell \lambda_j^{m-\ell-2} \langle Y\xi_j, \xi_k \rangle^2. \qquad (2.2.19)$$

Now we have

$$\sum_{\ell=0}^{m-1} m\lambda_j^\ell \lambda_k^{m-\ell-2} = m\frac{\lambda_j^{m-1} - \lambda_k^{m-1}}{\lambda_j - \lambda_k} = \frac{v'(\lambda_j) - v'(\lambda_k)}{\lambda_j - \lambda_k}, \qquad (2.2.20)$$

for $\lambda_j \neq \lambda_k$; whereas when $\lambda_j = \lambda_k$ we have

$$\sum_{\ell=0}^{m-1} m\lambda_j^\ell \lambda_k^{m-\ell-2} = m(m-1)\lambda_j^{m-2} = v''(\lambda_j). \qquad (2.2.21)$$

$\square$

Definition (*Functional calculus*). Let F be analytic on $\{z : |z| < R\}$ and suppose that $Y \in M_n(\mathbf{C})$ has $\|Y\|_{op} < R$. Let

$$F(Y) = \sum_{j=0}^\infty \frac{F^{(j)}(0)}{j!} Y^j. \qquad (2.2.22)$$

Examples 2.2.6 (i) The exponential $\exp Y$ is defined for any $Y \in M_n(\mathbf{C})$. When $Y = Y^*$, $\exp(iY)$ is unitary. We shall use this in the context of Lie groups later.

(ii) When $Y \in M_n(\mathbf{C})$ has $\|Y\| < 1$, we can define

$$\log(I + Y) = Y - Y^2/2 + Y^3/3 - \cdots. \qquad (2.2.23)$$

(iii) Suppose that $X \in M_n(\mathbf{C})$ has $\lambda I + X$ invertible for all $\lambda \geq 0$. Then

$$\log X = \int_0^\infty \left((\lambda+1)^{-1}I - (\lambda I + X)^{-1}\right) d\lambda$$

defines a logarithm of X such that $\exp \log X = X$.

(iv) For either of the above definitions, an invertible Y satisfies

$$\log \det Y = \operatorname{trace} \log Y.$$

Proposition 2.2.7 *Let $f \in C^2(\mathbf{R})$ have $f, f'' \in L^2(\mathbf{R})$. Then there exists a constant L_f, depending only on f, such that $X \mapsto f(X)$ is a L_f-Lipschitz function from $(M_n^h(\mathbf{C}), \|\cdot\|) \to (M_n^h(\mathbf{C}), \|\cdot\|)$ for any matricial norm.*

Proof. This proof is suggested by Duhamel's formula. First we prove that $X \mapsto e^{iX}$ is 1-Lipschitz $(M_n^h, \|\cdot\|) \to (M_n^h, \|\cdot\|)$. We connect X and Y

by a straight line segment in $M_n^h(\mathbf{C})$ and write

$$
e^{iX} - e^{iY} = \int_0^1 \frac{d}{dt} e^{i(tX+(1-t)Y)}\, dt
$$
$$
= \int_0^1 \int_0^1 e^{iu(tX+(1-t)Y)} i(X-Y) e^{i(1-u)(tX+(1-t)Y)}\, dt\, du,
$$

so by convexity we have

$$
\|e^{iX} - e^{iY}\| \le \|X - Y\|. \qquad (2.2.24)
$$

Now we use the Fourier inversion formula to write

$$
f(X) - f(Y) = \frac{1}{2\pi} \int_{-\infty}^{\infty} \hat{f}(\xi)\left(e^{i\xi X} - e^{i\xi Y}\right) d\xi, \qquad (2.2.25)
$$

so by the triangle inequality and the preceding observations we obtain

$$
\|f(X) - f(Y)\|
$$
$$
\le \frac{1}{2\pi} \int_{-\infty}^{\infty} |\hat{f}(\xi)| \|e^{i\xi X} - e^{i\xi Y}\| d\xi
$$
$$
\le \frac{1}{2\pi} \int_{-\infty}^{\infty} |\xi| |\hat{f}(\xi)| d\xi \, \|X - Y\|
$$
$$
\le \left(\frac{1}{2\pi} \int_{-\infty}^{\infty} \frac{d\xi}{1+\xi^2}\right)^{1/2} \left(\int_{-\infty}^{\infty} (1+\xi^2)^2 |\hat{f}(\xi)|^2 d\xi\right)^{1/2} \|X - Y\|
$$
$$
\le \frac{1}{2} \left(\int_{-\infty}^{\infty} (|f(x)|^2 + 2|f'(x)|^2 + |f''(x)|^2) d\xi\right)^{1/2} \|X - Y\|.
$$

$$(2.2.26)$$

□

2.3 Linear Lie groups

Definition (*Lie group*). A Lie group is a finite-dimensional analytic manifold that is a group such that multiplication $(g, h) \mapsto gh$ is analytic $G \times G \to G$ and the inversion $g \mapsto g^{-1}$ is analytic $G \to G$. We often write e_G for the neutral element of the group G. A linear Lie group G is a closed subset of $M_N(\mathbf{R})$ for some N that is a group for multiplication. See [134].

Whereas one can develop the theory axiomatically without reference to $M_N(\mathbf{R})$, we find it convenient to have an ambient space in which to carry out calculations. In particular, the group G has a natural manifold structure and a global system of coordinates such that the multiplication and inversion are differentiable. The connection between the topological structure of G, the analytic and differentiable structures and the algebraic operations has been investigated in great detail. The general theory shows that any compact metric group is the limit, in a certain sense, of a sequence of Lie groups. We refer the reader to [87, 130] for an account of the structure theorem.

Examples. The main examples of interest in random matrix theory are $O(n), U(n)$ and $Sp(n)$ with the standard multiplication, as mentioned above; these are all compact. The group $GL(n) = \{X \in M_n(\mathbf{C}) : X \text{ invertible}\}$ is a noncompact example. The additive group $M_n(\mathbf{R})$ becomes a linear Lie group when we embed it in $GL(N)$ with the usual multiplication for a suitably large N.

Definition (*Lie algebra*). The Lie bracket on $M_N(\mathbf{R})$ is $[X, Y] = XY - YX$. A linear Lie algebra is a real linear subspace g of $M_N(\mathbf{R})$ such that $[X, Y] \in$ g for all $X, Y \in$ g. The product on the Lie algebra is nonassociative and nonunital, and $[X, X] = 0$.

The exponential map provides the link between the Lie algebra and the Lie group.

Definition (*Generator*). The set $\{\exp(tX) : t \in \mathbf{R}\}$ forms a connected one parameter abelian group under matrix product, and X is its generator, since it satisfies

$$\left(\frac{d}{dt}\right)_{t=0} \exp(tX) = X. \tag{2.3.1}$$

Example. In $G = SO(2)$ we have a one-parameter group

$$\begin{bmatrix} \cos\theta & \sin\theta \\ -\sin\theta & \cos\theta \end{bmatrix} = \exp \begin{bmatrix} 0 & \theta \\ -\theta & 0 \end{bmatrix} \qquad (\theta \in \mathbf{R}), \tag{2.3.2}$$

where the right-hand side may be computed as a power series of matrices, as in (2.2.22).

The following result identifies the Lie algebra g of G with the set of connected, one-parameter abelian subgroups of G. The Campbell–Baker–Hausdorff formula states that $\exp(tX)\exp(tY) = \exp(tZ)$, where Z is given by an explicit universal series that involves only iterated Lie brackets in X and Y. By considering power series, one can easily obtain the first few terms in the series for Z; the first couple of terms are used in the following proof.

Proposition 2.3.1 *For a connected linear Lie group G, the set*

$$g = \{X \in M_N(\mathbf{R}) : \exp(tX) \in G \quad \text{for all} \quad t \in \mathbf{R}\} \qquad (2.3.3)$$

is a linear Lie algebra. Each $X \in g$ gives a tangent vector to G at e_G.

Proof. Let $X, Y \in G$. Clearly $sX \in g$ for all $s \in \mathbf{R}$. By power series expansions we have

$$\exp(tX)\exp(tY) = \exp\left(t[X,Y] + \frac{t^2}{2}[X,Y] + O(t^3)\right), \qquad (2.3.4)$$

so

$$\exp\left(t(X+Y)\right) = \lim_{n\to\infty}\left(\exp(tX/n)\exp(tY/n)\right)^n \qquad (2.3.5)$$

belongs to G for all $t \subset \mathbf{R}$; hence $X + Y \in g$.

Further, the series expansions give

$$\exp(tX)\exp(tY)\exp(-tX)\exp(-tY) = \exp\left(t^2[X,Y] + O(t^3)\right), \quad (2.3.6)$$

so

$$\exp(t[X,Y]) = \lim_{n\to\infty}\left(\exp\left(\sqrt{\frac{t}{n}}X\right)\exp\left(\sqrt{\frac{t}{n}}Y\right)\right.$$
$$\left. \times \exp\left(-\sqrt{\frac{t}{n}}X\right)\exp\left(-\sqrt{\frac{t}{n}}Y\right)\right)^n \qquad (2.3.7)$$

belongs to G for $t > 0$ and by inversion also for $t < 0$; hence $[X, Y] \in g$.

We have $X = \frac{d}{dt}\exp(tX)$ at $t = 0$, so X is tangential to G at e_G. This matches with the usual notion of tangents from differential geometry since we can take a differentiable function $f : G \to \mathbf{R}$ and write $Xf(e_G) = \frac{d}{dt}f(\exp(tX))$ at $t = 0$. By the preceding calculations, we have $(X + Y)f(e_G) = Xf(e_G) + Yf(e_G)$. $\qquad \square$

Exercise. Show that the subgroups $\exp(tX)$ and $\exp(tY)$ commute if and only if $[X, Y] = 0$.

Definition (*Lie algebra of subgroup*). Let H be a closed subgroup of a linear Lie group G that is also a linear Lie group. Then the Lie algebra of H is the set

$$h = \{Y \in g : \exp(tY) \in H \quad \text{for all} \quad t \in \mathbf{R}\}. \qquad (2.3.8)$$

By the proof of Proposition 2.3.1, h is a subalgebra of g.

Example. Any closed subalgebra h of g gives rise to an immersed subgroup $H = \{\exp(Y) : Y \in h\}$ of G, but such an H need not be closed as a subset of G. For instance, let $G = \mathbf{T}^2$, which may be identified with the torus $(\mathbf{R}/2\pi\mathbf{Z})^2$. The Lie algebra is $g = \mathbf{R}^2$ with the trivial Lie product $[X, Y] = 0$ for all $X, Y \in g$. For α irrational, the element $X = 2\pi(1, \alpha)$ generates the line of irrational slope in the torus, which gives a connected subgroup which is not closed.

Mainly we are interested in compact linear Lie groups, so we wish to formulate a criterion on the Lie algebra for compactness of the group.

Definition (*Adjoint action*). There is a natural action of G on g by $Ad : G \rightarrow Aut(g) : g \mapsto Ad(g)$ where $Ad(g)X = gXg^{-1}$; evidently

$$Ad(g)X = \left(\frac{d}{dt}\right)_{t=0} g \exp(tX)g^{-1}. \qquad (2.3.9)$$

The adjoint satisfies

$$Ad(gh) = Ad(g) \circ Ad(h); \quad Ad(I) = Id; \quad Ad(g^{-1}) = (Ad(g))^{-1}.$$
$$(2.3.10)$$

The adjoint action $ad : g \rightarrow End(g)$ is $X \mapsto ad(X)$ where $ad(X)(Y) = [X, Y]$.

Remark. One can regard $Ad(g) : g \rightarrow g$ as the derivative of the conjugation $c(g) : G \rightarrow G : c(g)h = ghg^{-1}$. The conjugation satisfies

$$c(gh) = c(g) \circ c(h); \quad c(I) = Id; \quad c(g^{-1}) = c(g)^{-1}. \qquad (2.3.11)$$

The kernel of Ad is a closed subgroup of G, and hence a Lie group.

Since $End(g)$ is an algebra of linear transformations on the vector space g, the exponential function operates on $End(g)$ and one can define $\exp(ad(X))$ by power series. The operations are natural in the sense of the following result, which shows that the derivation $ad(X)$ generates the endomorphism $Ad(\exp(tX))$.

Proposition 2.3.2 (*i*) *The adjoint $ad(X)$ gives a derivation $g \rightarrow g$.*
(*ii*) *The exponential and adjoint act naturally, so that*

$$Ad(\exp(X)) = \exp(ad(X)). \qquad (2.3.12)$$

Proof. (i) We can regard g as a bimodule over g. The property

$$ad(X)[Y, Z] = [ad(X)(Y), Z] + [Y, ad(X)Z] \qquad (2.3.13)$$

is equivalent to Jacobi's identity

$$[X, [Z, Y]] + [[X, Y], Z] + [Y, [X, Z]] = 0, \qquad (2.3.14)$$

which follows from a simple calculation.

(ii) A more explicit version of this formula is $Ad(\exp(tX))Y = \exp(ad(tX))(Y)$, and where $\exp(tX)$ belongs to the group G and $\exp(ad(tX))$ is calculated in $End(\mathfrak{g})$. Both sides of this formula give Y_t, where Y_t is the solution of

$$\frac{d}{dt} Y_t = [X, Y_t]; \quad Y_0 = Y \qquad (2.3.15)$$

in $\mathfrak{g}$. By uniqueness of solutions of ordinary differential equations, the expressions are therefore equal. $\qquad \square$

Proposition 2.3.3 *Let G be a connected linear Lie group. The Lie algebra of the centre $Z(G)$ of G equals the centre z of the Lie algebra $\mathfrak{g}$, namely $z = \{X \in \mathfrak{g} : ad(X) = 0\}$.*

Proof. Now X belongs to the Lie algebra of $Z(G)$ if and only if $\exp(tX) \in Z(G)$ for all t; that is $\exp(tX) \exp(Y) \exp(-tX) = \exp(Y)$ for all $Y \in \mathfrak{g}$. Differentiating this identity, we obtain $[X, Y] = 0$ for all $Y \in \mathfrak{g}$, so that X belongs to the centre z of $\mathfrak{g}$, or equivalently $ad(X) = 0$. $\qquad \square$

Definition (*Killing's form*). The Killing form is the bilinear map $B : \mathfrak{g} \times \mathfrak{g} \to \mathbf{R}$

$$B(X, Y) = \text{trace}(ad(X)ad(Y)), \qquad (2.3.16)$$

where the trace may be calculated with respect to any basis of $\mathfrak{g}$. Suppose that B is nondegenerate so that $B(X, Y) = 0$ for all $Y \in \mathfrak{g}$ implies $X = 0$. Then the Lie algebra $\mathfrak{g}$ is said to be semisimple and by extension, G is also said to be semisimple.

Definition (*Ideals*). An ideal $\mathfrak{h}$ in a Lie algebra $\mathfrak{g}$ is a subalgebra such that $[X, Y] \in \mathfrak{h}$ for all $X \in \mathfrak{h}$ and $Y \in \mathfrak{g}$. By convention, a simple Lie algebra $\mathfrak{g}$ has dimension greater than one and the only ideals are $\{0\}$ and $\mathfrak{g}$ itself.

Proposition 2.3.4 (*Cartan's criterion*). *Any semisimple linear Lie algebra is the direct sum of simple ideals.*

The following list gives the main examples of semisimple linear Lie algebras, as described in terms of compact groups. There are finitely many other exceptional Lie algebras, which we do not describe here.

- $A_{n-1} : su(n) = \{X \in M_n(\mathbf{C}) : X^* = -X, \mathrm{trace}(X) = 0\}$, $n = 2, 3, \ldots$;
- $B_n : so(2n+1) = \{X \in M_{2n+1}(\mathbf{R}) : X^t = -X\}$, $n = 2, 3, \ldots$;
- $C_n : sp(n) = \{X \in M_{2n}(\mathbf{C}) : X^* = -X, JX^tJ = X\}$, $n = 3, 4, \ldots$;
- $D_n : so(2n) = \{X \in M_{2n}(\mathbf{R}) : X^t = -X\}$, $n = 4, 5, \ldots \ldots$;

where

$$J = \bigoplus_{j=1}^{n} \begin{bmatrix} 0 & 1 \\ -1 & 0 \end{bmatrix}. \qquad (2.3.17)$$

The ranges of n start from different places so as to avoid redundancy in the listings. There are alternative descriptions in terms of the Lie algebras of noncompact groups, and the lists may be classified by Dynkin diagrams as in [144].

Example 2.3.5 For $su(n)$ we take inner product $\langle A, B \rangle = -2^{-1}$ $\mathrm{trace}(AB)$ to get an invariant inner product on the Lie algebra. Then $su(n)$ has orthonormal basis

$$\{e_{jk} - e_{kj}, ie_{jk} + ie_{kj}, iF_\ell : 1 \le j < k \le n, 1 \le \ell \le n - 1\} \qquad (2.3.18)$$

where e_{jk} has 1 in place (j, k) and zeros elsewhere, and the F_ℓ are chosen by the Gram–Schmidt process to give an orthonormal basis for the real diagonal matrices with zero trace. The dimension equals $n^2 - 1$.

Proposition 2.3.6 *The exponential map gives a surjection in each of the cases:*

(A_{n-1}) $su(n) \to SU(n)$;
(B_n) $so(2n+1) \to SO(2n+1)$;
(C_n) $sp(n) \to Sp(n)$;
(D_n) $so(2n) \to SO(2n)$.

Proof. We verify this by diagonalizing the matrices involved. We take $u \in SU(n)$, and introduce $v \in SU(n)$ such that $u = v\mathrm{diag}(e^{i\theta_1}, \ldots, e^{i\theta_n})v^*$, where we can suppose that $\theta_1 + \cdots + \theta_n = 0$. Then $Y = v\mathrm{diag}(\theta_1, \ldots, \theta_n)v^*$ belongs to $su(n)$, and satisfies $u = \exp(Y)$. The other cases are similar. $\qquad \square$

Theorem 2.3.7 *Let G be a connected, semisimple and linear Lie group with finite centre. Then G is compact if and only if the Killing form is negative definite.*

Remark. This result is true without the hypothesis that the centre is finite, but the final step in the proof involves more differential geometry than we wish to discuss here.

Proof. Suppose that G is compact, and let μ_G be its Haar probability measure. Suppose that $\dim g = n$, and let $(\,,\,)$ be any inner product on g. Then

$$\langle X, Y \rangle = \int_G (Ad(g)X, Ad(g)Y)\mu_G(dg) \qquad (2.3.19)$$

given an inner product on g such that

$$\langle Ad(g)X, Ad(g)Y \rangle = \langle X, Y \rangle \qquad (g \in G; X, Y \in g). \quad (2.3.20)$$

Thus $Ad(g)$ can be represented by an element of $O(n)$. Now the Lie algebra of $O(n)$ is the space of real skew-symmetric matrices, so by Proposition 2.3.2, each $ad(X)$ can be represented by some $A = [a_{jk}]$ where $a_{jk} = -a_{kj}$; hence

$$
\begin{aligned}
B(X,X) &= \operatorname{trace} ad(X)^2 \\
&= \operatorname{trace} A^2 \\
&= \sum_{j,k=1}^{n} a_{jk} a_{kj} = -\sum_{j,k=1}^{n} a_{jk}^2 \\
&\leq 0. \qquad\qquad\qquad\qquad\qquad (2.3.21)
\end{aligned}
$$

By hypothesis, the Killing form is nondegenerate, and now known to be negative semidefinite, hence B is negative definite.

The converse uses a similar idea. Since g is semisimple, the kernel of $ad : g \to End(g)$ is $\{0\}$, so the kernel of $Ad : G \to Aut(G)$ is a zero dimensional Lie group. Hence $KerAd$, namely the centre of G, is a discrete subgroup Z of G. We shall prove that G/Z is compact.

Suppose that B is negative definite, and introduce the (positive definite) inner product $\langle X, Y \rangle = -B(X, Y)$ on g. We shall prove that

$$\langle Ad(g)X, Ad(g)Y \rangle = \langle X, Y \rangle \qquad (X, Y \in g; g \in G)$$

so that $Ad(g)$ is an orthogonal transformation on $(g, \langle\,,\,\rangle)$. Let $\sigma = Ad(g)$, and note that $ad(\sigma X) = \sigma ad(X)\sigma^{-1}$ and likewise $ad(\sigma Y) = \sigma ad(Y)\sigma^{-1}$;

hence

$$\begin{aligned}
B(\sigma X, \sigma Y) &= \operatorname{trace}\big(ad(\sigma X)ad(\sigma Y)\big) \\
&= \operatorname{trace}\big(\sigma ad(X)ad(Y)\sigma^{-1}\big) \\
&= \operatorname{trace}\big(ad(X)ad(Y)\big) \\
&= B(X, Y).
\end{aligned} \tag{2.3.22}$$

Hence $Ad : G \to O(n)$ has image $Ad(G)$ which is a closed subgroup of the compact Lie group $O(n)$, and hence is compact; so G/Z is homeomorphic to a compact subgroup of $O(n)$. Since Z is finite by hypothesis, we deduce that G itself is compact. $\square$

2.4 Connections and curvature

See [69, 47, 128] for more discussion of the concepts mentioned in this section.

Suppose that G is a compact and connected semisimple linear Lie group with finite centre, and let $\langle \cdot \rangle$ be an invariant inner product on g as in (2.3.20). For each $g \in G$ there is a left translation $L_g : G \to G : h \mapsto gh$, which is differentiable. The derivative of L_g induces an isomorphism between the tangent space to G at e and the tangent space to G at g, namely $dL_g : T_e G \to T_g G \ dL_g(Y)(h) = dL_g(X(g^{-1}h))$. Now $T_e G$ may be identified with the Lie algebra g of G by Proposition 2.3.1 since exp is surjective. For $V, W \in T_g G$, we introduce $V_e, W_e \in$ g such that $V = dL_g V_e$ and $W = dL_g W_e$; then we use the invariant inner product to define an inner product on $T_g G$ by $\langle V, W \rangle = \langle V_e, W_e \rangle$.

Definition (*Curvatures*). An *affine connection* is a rule that associates to each $X \in$ g a linear map $\nabla_X :$ g $\to$ g such that

$$\nabla_{fX+hY} = f\nabla_X + h\nabla_Y, \quad \nabla_X(fY) = f\nabla_X Y + (Xf)Y, \tag{2.4.1}$$

for $X, Y \in$ g and $f, h \in C^\infty(G)$. Further, for each Riemannian metric, there is a unique affine connection such that

$$\begin{aligned}
X\langle Y, Z \rangle &= \langle \nabla_X Y, Z \rangle + \langle Y, \nabla_X Z \rangle, \\
\nabla_X Y - \nabla_Y X - [X, Y] &= 0.
\end{aligned} \tag{2.4.2}$$

For this connection, we introduce the *curvature tensor* $R(X, Y) :$ g $\to$ g by

$$R(X, Y)Z = \nabla_X \nabla_Y Z - \nabla_Y \nabla_X Z - \nabla_{[X,Y]} Z, \tag{2.4.3}$$

and the *Ricci curvature tensor* Ric by

$$\text{Ric}(X, Y) = \sum_{j=1}^{n} \langle R(E_j, X)Y, E_j \rangle \qquad (2.4.4)$$

where $(E_j)_{j=1}^{n}$ is an orthonormal basis of g for the invariant inner product given by the negative of Killing's form; the *sectional curvature* of the plane spanned by $X, Y \in$ g is

$$\frac{\langle R(X, Y)X, Y \rangle}{\langle X, X \rangle \langle Y, Y \rangle - \langle X, Y \rangle^2}. \qquad (2.4.5)$$

Proposition 2.4.1 (*Curvature of compact Lie groups*).

  (*i*) *The Riemannian affine connection is given by* $\nabla_X Y = [X, Y]/2$.
  (*ii*) *The curvature tensor satisfies*

$$\langle R(X, Y)Z, W \rangle = 4^{-1} \langle [X, W], [Y, Z] \rangle - 4^{-1} \langle [X, Z], [Y, W] \rangle. \qquad (2.4.6)$$

(*iii*) *The sectional curvature of the plane spanned by* $X, Y \in$ g *is*

$$\frac{1}{4} \frac{\|[X, Y]\|^2}{\langle X, X \rangle \langle Y, Y \rangle - \langle X, Y \rangle^2}, \qquad (2.4.7)$$

  and in particular is non-negative.
(*iv*) *The Ricci curvature satisfies*

$$\text{Ric}(X, X) = \frac{1}{4} \sum_j \langle [X, E_j], [X, E_j] \rangle \qquad (2.4.8)$$

  where $(E_j)_{j=1}^{n}$ *is any orthonormal basis of* g.

Proof. (i) Since the inner product is invariant, we have

$$\langle Ad(\exp(tZ))X, Ad(\exp(tZ))Y \rangle = \langle X, Y \rangle, \qquad (2.4.9)$$

so that, when we take the derivative at $t = 0$ we obtain

$$\langle [Z, X], Y \rangle + \langle X, [Z, Y] \rangle = 0 \qquad (X, Y, Z \in \text{g}). \qquad (2.4.10)$$

Again by left invariance, we obtain relations such as

$$0 = X \langle Y, Z \rangle = \langle \nabla_X Y, Z \rangle + \langle Y, \nabla_X Z \rangle,$$

and similarly with X, Y and Z permuted. When we combine these relations with the Riemannian metric condition, we obtain

$$\langle \nabla_X Y, Z \rangle = \frac{1}{2} \big(\langle [X, Y], Z \rangle - \langle Y, [X, Z] \rangle - \langle X, [Y, Z] \rangle \big) \qquad (2.4.11)$$

which on account of the preceding identity reduces to

$$\langle \nabla_X Y, Z \rangle = \frac{1}{2}\langle [X, Y], Z \rangle; \qquad (2.4.12)$$

the identity $\nabla_X Y = [X, Y]/2$ follows.

(ii) By the definition of the curvature tensor and invariance we have

$$\begin{aligned}
\langle R(X,Y)Z, W \rangle &= \langle \nabla_X \nabla_Y Z, W \rangle - \langle \nabla_Y \nabla_X Z, W \rangle - \langle \nabla_{[X,Y]} Z, W \rangle \\
&= -\langle \nabla_Y Z, \nabla_X W \rangle + \langle \nabla_X Z, \nabla_Y W \rangle - \langle \nabla_{[X,Y]} Z, W \rangle \\
&= -\frac{1}{4}\langle [Y,Z], [X,W] \rangle + \frac{1}{4}\langle [X,Z], [Y,W] \rangle \\
&\quad - \frac{1}{2}\langle [X,Y]Z, W \rangle, \qquad (2.4.13)
\end{aligned}$$

where the last step follows from (i). Now Jacobi's identity in the form

$$[[X,Y],Z] = [Y,[Z,X]] + [X,[Y,Z]] \qquad (2.4.14)$$

and invariance leads to a simplified from of the last term, namely

$$\langle [X,Y]Z, W \rangle = -\langle [Z,X], [Y,W] \rangle - \langle [Y,Z], [X,W] \rangle. \qquad (2.4.15)$$

This choice of affine connection is consistent with the Riemannian metric on the group.

(iii) The identity

$$\langle R(X,Y)X, Y \rangle = \frac{1}{4}\langle [X,Y], [X,Y] \rangle \qquad (2.4.16)$$

follows from (ii) by substitution, and then (iii) follows.

(iv) This is immediate from (ii). □

Exercise 2.4.2 (Ricci curvature of $SO(n)$). (See [48]). On $so(n)$, take the inner product $\langle X, Y \rangle = -2^{-1}\text{trace}(XY)$ and introduce the orthonormal basis $\{\hat{E}_{j,k} = e_{jk} - e_{kj} : 1 \le j < k \le n\}$, where e_{jk} has one in place (j, k) and zeros elsewhere.

(i) Show that when $X = [x_{jk}] \in so(n)$,

$$\langle [X, \hat{E}_{j,k}], [X, \hat{E}_{j,k}] \rangle = \sum_{\ell=1}^{n}(x_{j\ell}^2 + x_{k\ell}^2) - x_{jk}^2 - x_{kj}^2 \qquad (1 \le j < k \le n).$$

(ii) By using the formulæ of Proposition 2.4.1, deduce that

$$\text{Ric}(X, X) = \frac{1}{4}\sum_{1 \le j < k \le n} \langle [X, \hat{E}_{j,k}], [X, \hat{E}_{j,k}] \rangle = \frac{n-2}{2}\langle X, X \rangle.$$

$$(2.4.17)$$

(iii) Use a similar technique to show that for $SU(n)$ there exists $c > 0$ such that Ric $\geq cnI$.

The Ricci curvature plays an important role in the next chapter, where it is the main hypothesis of the isoperimetric inequalities that are used in subsequent convergence theory.

2.5 Generalized ensembles

In this section we are concerned with probability measures on the real symmetric, Hermitian or quaternion matrices. Let G be a compact linear Lie group with Haar probability measure μ_G. We suppose that M is a submanifold of $M_N(\mathbf{R})$ and thus inherits a differentiable structure from $M_N(\mathbf{R})$; likewise G inherits a differentiable structure.

Let α be a continuous action of G on M. Since the action is continuous, each orbit $X^G = \{\alpha_U(X) : U \in G\}$ is a compact subset of M, and we form the quotient space

$$\Delta = M/G = \{X^G : X \in M\} \tag{2.5.1}$$

with the quotient topology. The natural quotient map $\Lambda : M \to \Delta$ is open.

Example 2.5.1 (i) Let $G = O(n)$, $\beta = 1$, and $M = M_n^s(\mathbf{R})$, and let G act on M by conjugation; so $\alpha_U(X) = UXU^*$. Each $X \in M$ is conjugate to a unique real diagonal matrix that has eigenvalues in increasing order; so $\Delta = \{\lambda : \lambda_1 \leq \cdots \leq \lambda_n\}$ and $\Lambda : M \to \Delta$ is the eigenvalue map $X \mapsto \lambda$.

(ii) Let $G = U(n)$, $\beta = 2$, and $M = M_n^h(\mathbf{C})$, and let G act on M by conjugation; so $\alpha_U(X) = UXU^*$. Then Λ is the eigenvalue map as in (i).

(iii) Let $G = Sp(n)$, $\beta = 4$ and $M = M_n^h(\mathbf{H})$, and let G act on M by conjugation; so $\alpha_U(X) = UXU^*$. Then Λ is the eigenvalue map as in (i).

In each of these cases, the conjugation action satisfies

$$\alpha_U(tX + sY) = t\alpha_U(X) + s\alpha_U(Y), \tag{2.5.2}$$

and hence α is a linear representation of G as a group of linear automorphisms of the real vector space M; that is, there is a group homomorphism $\alpha : G \to GL(M)$.

(iv) In Sections 2.8 and 2.9, we consider actions on a group M by a subgroup G that are given by left translation $\alpha_U : V \mapsto UV$ where $U \in G$ and $V \in M$. Then Δ is a space of cosets.

Definition (*Invariant measure*). Let α be an action of G on M and let ν be a probability measure on M such that $\alpha_U \sharp \nu = \nu$ for all $U \in G$. Then ν is invariant for the action.

Lemma 2.5.2 (*i*) *The map* $\Lambda : M \to \Delta$ *induces*

$$\delta(d\lambda) = \frac{1}{n!} \prod_{1 \le j < k \le n} \left(\lambda_k - \lambda_j \right)^{\beta} d\lambda_1 \dots d\lambda_n \qquad (2.5.3)$$

from the Lebesgue product measure on M.

(*ii*) *The map* $C_c(M; \mathbf{R}) \to C_c(\Delta; \mathbf{R})$ *specified by* $f \mapsto \bar{f}$ *where*

$$\bar{f}(\lambda(X)) = \int_G f(\alpha_U(X)) \mu_G(dU) \qquad (2.5.4)$$

satisfies

$$\int_\Delta \bar{f}(\lambda) \delta(d\lambda) = \int_M f(X) dX. \qquad (2.5.5)$$

Proof. (i) Weyl calculated this formula. In Theorem 2.6.5 we give a detailed proof of a similar result.

(ii) Let

$$C_b^G(M) = \{ f \in C_b(M; \mathbf{R}) : f(X) = f(\alpha_U(X)) : U \in G; X \in M \} \qquad (2.5.6)$$

be the space of invariant functions, and observe that the linear map $P : C_b(M) \to C_b(M)$:

$$(Pf)(X) = \int_G f(\alpha_U(X)) \mu_G(dU) \qquad (2.5.7)$$

satisfies $\|P\| = 1$, $P^2 = P$ and has range $C_b^G(M)$. The map $C_b(\Delta) \to C_b^G(M) : g \mapsto g \circ \Lambda$ is a linear isometric isomorphism which identifies functions on Δ with G-invariant functions on M; further, $g \circ \Lambda$ has compact support if and only if g has compact support. The adjoint of $C_c(\Delta) \to C_c(M)$ is the map $M_b(M) \to M_b(\Delta) : \mu \mapsto \Lambda \sharp \mu$.

Since dX is invariant under α_U, we have

$$\int_M \bar{f}(\lambda(X)) dX = \int_M \int_G f(\alpha_U(X)) \mu_G(dU) dX \qquad (2.5.8)$$

so

$$\int_\Delta \bar{f}(\lambda) \Lambda \sharp dX = \int_M f(X) dX. \qquad (2.5.9)$$

$\square$

Definition (*Generalized ensembles*). The following result proves the existence of the generalized ensembles, which consist of a probability measure $\nu_n^{(\beta)}$ on the space $n \times n$ real symmetric matrices for $\beta = 1$, complex Hermitian matrices for $\beta = 2$ or self-dual hermitian matrices for $\beta = 4$ that is determined by a suitable potential function v. The invariance group is: $O(n)$ for $\beta = 1$, $U(n)$ for $\beta = 2$ and $Sp(n)$ for $\beta = 4$; hence the ensembles are called the generalized orthogonal, unitary and symplectic ensembles. The action of the invariance group and the specific probability measure ν_n^β determine a probability measure σ_n^β on the space of ordered eigenvalues.

Theorem 2.5.3 (*Generalized ensembles*). *Suppose that* $v : \mathbf{R} \to \mathbf{R}$ *is a continuous function such that* $v(x) \geq \kappa x^2/2$ *all sufficiently large* $|x|$ *for some* $\kappa > 0$. *Then with* $V(X) = \mathrm{trace}_n \, v(X)$, *there exists* $Z_n < \infty$ *such that*

$$\nu_n(dX) = Z_n^{-1} \exp(-n^2 V(X)) \, dX \qquad (2.5.10)$$

defines a probability measure on M *where* $M = M_n^s(\mathbf{R})$; *likewise with* $M_n^h(\mathbf{C})$ *and* $M_n^s(\mathbf{H})$ *for suitable* Z_n.

The probability measure induced by $\Lambda : M \to \Delta^n$ *is*

$$\sigma_n^{(\beta)}(d\lambda) = (Z_n^{(\beta)})^{-1} \exp\left(-n \sum_{j=1}^{n} v(\lambda_j)\right) \prod_{1 \leq j < k \leq n} (\lambda_k - \lambda_j)^\beta \, d\lambda_1 \ldots d\lambda_n$$

with $\beta = 1$ *for* $M_n^s(\mathbf{R})$; $\beta = 2$ *for* $M_n^h(\mathbf{C})$; $\beta = 4$ *for* $M_n^h(\mathbf{H})$. *The normalizing constant* $Z_n^{(\beta)}$ *is stated in Proposition 2.5.4.*

Proof. By the Lemma, there is a bijective correspondence between probability measures on Δ^n with respect to $\delta(d\lambda)$ and probability measures on M with respect to dX that are invariant with respect to G. The density $\exp(-n^2 V(X))$ is invariant under the action of G.

In particular, when $v(x) = \kappa x^2/2$, we have the Gaussian measure on M, which defines a probability measure when we choose a suitable constant. Generally, when $v(x) \geq \kappa x^2/2$, we can choose Z so that ν_n defines a probability measure. □

ensemble	orthogonal	hermitian	symplectic
M	$M_n^s(\mathbf{R})$	$M_n^h(\mathbf{C})$	$M_n^h(\mathbf{H})$
G	$O(n)$	$U(n)$	$Sp(n)$
action	$\beta_U : X \mapsto UXU^*$	$\beta_U : X \mapsto UXU^*$	$\beta_U : X \mapsto UXU^*$
M/G	Δ^n	Δ^n	Δ^n
Haar measure on M	dX	dX	dX
Gibbs measure	$\nu_n^{(1)}(dX)$	$\nu_n^{(2)}(dX)$	$\nu_n^{(4)}(dX)$
σ	$\sigma_n^{(1)}$	$\sigma_n^{(2)}$	$\sigma_n^{(4)}$

$$(2.5.11)$$

We now express the normalizing constants from Proposition 2.5.3 in terms of one or two dimensional integrals.

Proposition 2.5.4 *Let $Z_n^{(\beta)}$ be the normalizing constant for $\sigma_n^{(\beta)}$.*

(i) For n even, the generalized orthogonal ensemble has

$$Z_n^{(1)} = \mathrm{Pf}\left[\int_{-\infty}^{\infty}\int_{-\infty}^{\infty}(\mathrm{sgn}(y-x))x^{j-1}y^{k-1}e^{-nv(x)-nv(y)}\,dxdy\right]_{j,k=1,\ldots,n}.$$

$$(2.5.12)$$

(ii) For all n, the generalized unitary ensemble has

$$Z_n^{(2)} = \det\left[\int_{-\infty}^{\infty}\lambda^{j+k-2}e^{-nv(\lambda)}\,d\lambda\right]_{j,k=1,\ldots,n}. \qquad (2.5.13)$$

Proof. (i) By Lemma 2.2.1(i), the expression for $\sigma_n^{(1)}$ integrates to one when

$$Z_n^{(1)} = \int_{\Delta^n}\det\left[e^{-nv(\lambda_j)}\lambda_j^{k-1}\right]_{j,k=1,\ldots,n}d\lambda_1\ldots d\lambda_n, \qquad (2.5.14)$$

and this reduces to the stated formula by Lemma 2.2.2(ii).

(ii) The $(\lambda_1,\ldots,\lambda_n)$ in Proposition 2.5.3 range over Δ^n, but we can permute the λ_j so that integrals are taken over $\mathbf{R}^n$; of course, we must introduce a factor of $n!$ when we do this. We observe that by Lemma 2.2.1(i),

$$n!Z_n^{(2)} = \int_{\mathbf{R}^n}\det[\lambda_j^{k-1}]_{j,k=1,\ldots,n}\det[\lambda_j^{k-1}e^{-nv(\lambda_j)}]_{j,k=1,\ldots,n}d\lambda_1\ldots d\lambda_n,$$

$$(2.5.15)$$

so the result follows from Lemma 2.2.2(i). $\qquad\square$

Remarks. (i) In the generalized orthogonal ensemble, we have a specific form for the potential, namely $V(X) = \frac{1}{n}\sum_{j=1}^{n} v(\lambda_j)$. One can construct invariant ensembles where the potential has the more general form

$$W(X) = \sum_{j=1}^{n} w_1(\lambda_j) + \sum_{j\neq k:j,k=1}^{n} w_2(\lambda_j, \lambda_k)$$

$$+ \sum_{j,k,\ell=1;j,\ell\neq k;j,k\neq\ell}^{n} w_3(\lambda_j, \lambda_k, \lambda_\ell) + \cdots.$$

(ii) The entries of random matrices under the $\nu_n^{(\beta)}$ distribution are typically correlated; the Gaussian ensembles are an exception to this. In all cases, the eigenvalues are correlated, and their interaction is described by powers of the Vandermonde determinant

$$\prod_{1\leq j<k\leq n} (\lambda_k - \lambda_j)^\beta = \exp\Big(\beta \sum_{j,k:1\leq j<k\leq n} \log(\lambda_k - \lambda_j)\Big). \qquad (2.5.16)$$

Clearly, the product and hence the density of σ_n are small when eigenvalues are close together, so eigenvalues tend to repel one another. In Chapter 4 we shall consider this type of logarithmic interaction term in detail for any $\beta > 0$, using ideas borrowed from electrostatics.

(iii) For the generalized unitary ensembles, there are special results from the theory of orthogonal polynomials which can be used to analyse $\sigma_n^{(2)}$; see [56, 116] and Chapter 8 below.

Exercise 2.5.5 (*Characteristic functions*). Let $e_{jk} \in M_n(\mathbf{R})$ have entry one in (j,k) and zeros elsewhere, let $T \in M_n(\mathbf{R})$ and let

$$\varphi(T) = \int_{M_n^s(\mathbf{R})} \exp\big(i\text{trace}(TX)\big)\, \nu_n^{(1)}(dX) \qquad (2.5.17)$$

be the characteristic function of the matrix-valued random variable X on $(M_n^s(\mathbf{R}), \nu_n^{(1)})$. Let X_{jk} be the $(j,k)^{th}$ entry of X, which is a random variable with characteristic function

$$\varphi(t e_{jk}) = \int_{M_n^s(\mathbf{R})} \exp\big(it\,\text{trace}(e_{jk}X)\big)\, \nu_n^{(1)}(dX) \qquad (t \in \mathbf{R}).$$

Show that

$$\varphi\Big(\sum_{1\leq j\leq k\leq n} t_{jk} e_{jk}\Big) = \prod_{1\leq j\leq k\leq n} \varphi(t_{jk} e_{jk})$$

if and only if the entries on and above the leading diagonal of X are mutually independent.

The generalized orthogonal ensemble is invariant under the action of $O(n)$ on $M_n^s(\mathbf{R})$, and it is natural to consider the distribution of blocks and entries within the matrix. For $0 < m < n$, there is a natural action of the subgroup $O(m) \times O(n-m)$ on $M_n^s(\mathbf{R})$ by left and right translations

$$\begin{bmatrix} X & Z \\ Z^t & Y \end{bmatrix} \mapsto \begin{bmatrix} UXU^t & UZV^t \\ VZ^tU^t & VYV^t \end{bmatrix} \qquad (2.5.18)$$

where $U \in O(m)$, $V \in O(n-m)$, $X \in M_m^s(\mathbf{R})$, $Y \in M_{n-m}^s(\mathbf{R})$ and $Z \in M_{m,n-m}(\mathbf{R})$.

Proposition 2.5.6 *Suppose that the block matrix in (2.5.18) is random, subject to the generalized orthogonal ensemble.*

(i) *The marginal distribution of X is invariant under conjugation by $O(m)$, and hence depends only upon the eigenvalues of X.*

(ii) *The marginal distribution of Y is invariant under conjugation by $O(n-m)$, and hence depends only upon the eigenvalues of Y.*

(iii) *The marginal distribution of Z is invariant under the left and right translation action of $O(m) \times O(n-m)$, and hence depends only upon the singular numbers of Z.*

Proof. (i) and (ii). These statements follow from Lemma 2.4.2(ii).

(iii) Suppose that $m \geq n-m$. For each $Z \in M_{m,n-m}(\mathbf{R})$, there exists $U \in O(n-m)$ such that $U^t(Z^tZ)^{1/2}U = S$, where S is a diagonal matrix that has the nonnegative eigenvalues of $(Z^tZ)^{1/2}$ on the diagonal. Now $(Z^tZ)^{1/2} : \mathbf{R}^{n-m} \to \mathbf{R}^{n-m} \subseteq \mathbf{R}^m$ and $Z : \mathbf{R}^{n-m} \to \mathbf{R}^m$ satisfy $\|Z\xi\| = \|(Z^tZ)^{1/2}\xi\|$ for all $\xi \in \mathbf{R}^{n-m}$, so there exists $V \in O(m)$ such that $V = I$ on the orthogonal complement of the range of $(Z^tZ)^{1/2}$ in $\mathbf{R}^{n-m}$, and $Z\xi = V(Z^tZ)^{1/2}\xi$ for all $\xi \in \mathbf{R}^{n-m}$, hence we have an identity between $m \times m$ matrices as follows:

$$[Z \quad 0] = V \begin{bmatrix} U & 0 \\ 0 & I \end{bmatrix} \begin{bmatrix} S & 0 \\ 0 & 0 \end{bmatrix} \begin{bmatrix} U^t & 0 \\ 0 & I \end{bmatrix}. \qquad (2.5.19)$$

Hence the orbits of Z under the action of $O(m) \times O(m-n)$ are parametrized by the matrices S, which are determined by the singular numbers of Z.

When $m \leq n-m$, we can apply a similar argument to Z^t; hence the result. $\qquad \square$

2.6 The Weyl integration formula

- Complex semisimple Lie algebras can be classified in terms of root systems.
- The Weyl integration formula gives the measure on the maximal torus that is induced from Haar measure by the eigenvalue map.

The purpose of this section is to prove the Weyl integration formula, which we use later to obtain the eigenvalue distributions of unitary matrices in terms of the Haar probability measure. Readers who are knowledgeable about Lie groups, or interested only in the specific examples, may wish to proceed to Section 2.7. The approach that is taken in the results in this section resembles that of Steer's lectures and uses proofs due to Bott [36].

Definition (*Maximal torus*). A torus T of rank k in a linear Lie group G is a connected abelian Lie group that is continuously isomorphic to $\mathbf{T}^k$. A maximal torus T is a torus that is not contained in any strictly larger torus in G. The normalizer of T is $N(T) = \{u \in G : uTu^{-1} = T\}$, which contains T as a normal subgroup, and the Weyl group is $W = N(T)/T$.

Lemma 2.6.1 *Suppose that G is a compact linear Lie group such that* $\exp : \mathrm{g} \to G$ *is surjective. Then*

(i) each $g \in G$ is an element of some maximal torus;
(ii) the Weyl group is finite.

Proof. In the specific examples of the classical groups, one can easily verify (i) by following the proof of Proposition 2.3.5. Not every maximal abelian subgroup is a maximal torus.

(i) First we observe that each $x \in G$ lies in some torus. Indeed, since $\exp$ is surjective, $x = \exp X$ for some $X \in \mathrm{g}$ and the closure of $\{\exp(tX) : t \in \mathbf{R}\}$ is a torus that contains x.

A connected abelian subgroup T of G is a maximal torus if and only if $T = Z(T)$, where

$$Z(T) = \{u : utu^{-1} = t \quad \text{for all} \quad t \in T\} \qquad (2.6.1)$$

is a closed submanifold.

First suppose that T is a maximal torus. Clearly T is a subgroup of $Z(T)$, so to prove that reverse containment, we suppose that $g \in G$ satisfies $gtg^{-1} = t$ for all $t \in T$. Then $A = \{g^n t : n \in \mathbf{Z}, t \in T\}$ is an abelian group, so the connected component A_0 that contains the identity of G is a closed and connected abelian group which contains T, and such that A/A_0 is finite by compactness. Unfortunately, it is not clear that g belongs to A_0, so we introduce the abelian group $A_1 = \{g^n a : n \in \mathbf{Z}, a \in A_0\}$ which has A_1/A_0 cyclic and A_0 a torus; hence A_1 is monogenic by Proposition 1.2.2, so A_1 is the closure of $\{x^n : n \in \mathbf{Z}\}$ for some $x \in G$. Now x belongs to some torus T_1, and so T_1 contains A_1 and hence the maximal torus T; so $T = T_1$ and $g \in T$.

Conversely, suppose that $T = Z(T)$. Then T is a closed and connected abelian subgroup of a Lie group, hence is a torus. If T is not a maximal torus, then there exists a torus T' that contains T; but any $x \in T'$ satisfies $x \in Z(T)$, so $x \in T$. Hence T is a maximal torus.

(ii) The normalizer subgroup $N(T)$ acts on T by conjugation, and since T is isomorphic to $\mathbf{R}^k/\mathbf{Z}^k$ for some k, we thus obtain a homomorphism $N(T) \to GL(k, \mathbf{Z})$ with kernel T by the proof of (i). The Weyl group $W = N(T)/T$ is therefore compact and discrete, hence finite.

$\square$

Theorem 2.6.2 *Suppose that G is a compact, semisimple, and linear Lie group such that* $\exp : \mathrm{g} \to G$ *is surjective and let T_1 and T_2 be maximal tori in G. Then T_1 and T_2 are conjugate, so there exists $g \in G$ such that $T_2 = gT_1g^{-1}$.*

Proof. (In the specific examples of the classical groups, one can easily verify this by following the proof of Proposition 2.3.5.) Since tori are monogenic, there exists $x, y \in G$ such that T_1 is the closure of $\{x^n : n \in \mathbf{Z}\}$ and T_2 is the closure of $\{y^n : n \in \mathbf{Z}\}$; further, $x = \exp(X)$ and $y = \exp(Y)$ for some $X, Y \in \mathrm{g}$. By Theorem 2.3.7, g has an inner product associated with the Killing form.

We introduce the adjoint orbit of X by $\xi_X = \{Ad(g)X; g \in G\}$, which is compact and contains X. If $Y \in \xi_X$, then $Y = Ad(g)X$ for some $g \in G$ and since the exponential map has the natural property that

$$y = \exp(Y) = c(g)\exp(X) = c(g)(x), \qquad (2.6.2)$$

one deduces that $T_2 = gT_1g^{-1}$.

Suppose now that Y lies outside of ξ_X, and choose g that attains the infimum

$$Ad(g)(X) = \inf\{\langle W, Y \rangle : W \in \xi_X\}.$$

In particular, we can consider $W = Ad(\exp(tZ)g)X$ for any $Z \in$ g. Then the derivative of $\langle Ad(\exp(tZ)g)X, Y \rangle$ vanishes at $t = 0$, so

$$0 = \langle [Z, Ad(g)X], Y \rangle = \langle Z, [Ad(g)X, Y] \rangle, \qquad (2.6.3)$$

and hence $[Ad(g)X, Y] = 0$ since Z was arbitrary. Now

$$A = \{\exp(tAd(g)(X) + sY) : t, s \in \mathbf{R}\} \qquad (2.6.4)$$

is a connected abelian group such that the closure $\bar{A}$ of A contains the maximal torus T_2, so $\bar{A} = T_2$. In particular, $\exp(tAd(g)X)$ belongs to T_2 for all $t \in \mathbf{R}$; so $c(g)x \in T_2$ and hence gT_1g^{-1} is contained in T_2; hence $gT_1g^{-1} = T_2$. $\qquad \square$

Let $g_{\mathbf{C}}$ be a finite-dimensional complex Lie algebra, so for each $Z \in g_{\mathbf{C}}$, $ad(Z)$ is a linear transformation of $g_{\mathbf{C}}$. To motivate further work, we make an informal definition. A root is an eigenvalue of $ad(Z)$, or equivalently a solution of the characteristic equation

$$\det(\lambda I - ad(Z)) = 0.$$

The root space is the eigenspace that corresponds to the root.

Proposition 2.6.3 *Let G be a compact and semisimple Lie groups with Lie algebra* g. *Then the roots of $ad(Z)$ are purely imaginary for all $Z \in$ g.*

Proof. By Proposition 2.3.2, $\exp(ad(Z)) = Ad(\exp Z)$. By Theorem 2.3.7, we can regard $Ad(\exp Z)$ as an orthogonal transformation of g with the inner product given by the negative Killing form; hence $Ad(\exp Z)$ has all its eigenvalues in $\mathbf{T}$, so $ad(Z)$ has all its eigenvalues in $i\mathbf{R}$. $\qquad \square$

We now construct representations from roots. A typical element of $\mathbf{T}^k$ is $u = \exp(2\pi i \sum_{j=1}^{k} \theta_j)$. The Lie algebra

$$\tau = \{\Theta = (\theta_1, \dots, \theta_k) : \theta_j \in \mathbf{R}\}$$

has dual

$$\tau^* = \{r^* = (r_1, \dots, r_k) : r_j \in \mathbf{Z}\}; \qquad (2.6.5)$$

so that $r^*(\Theta) = \sum_j \theta_j r_j$ belongs to $\mathbf{Z}$ whenever $\Theta \in \mathbf{Z}^k$. There is a natural map $\exp : \tau \to T : \Theta \mapsto \exp(2\pi i \sum_{j=1}^{k} \theta_j)$; for each $r^* \in \tau^*$ there is an irreducible representation of $\mathbf{T}$ on $\mathbf{R}^2$ by $\mathbf{T} \to SO(2)$:

$$\exp\left(2\pi i \sum_{j=1}^{k} \theta_j\right) \mapsto \begin{bmatrix} \cos 2\pi r^*(\Theta) & \sin 2\pi r^*(\Theta) \\ -\sin 2\pi r^*(\Theta) & \cos 2\pi r^*(\Theta) \end{bmatrix}. \qquad (2.6.6)$$

Lemma 2.6.4 *Let G be a compact, connected and semisimple Lie group. Let $g_{\mathbb{C}}$ be the complexified version of the Lie algebra of G, and let τ be the Lie algebra of some maximal torus T in G. Then there exist nonzero linear functionals $\alpha : \tau \to \mathbf{R}$ such that $-i\alpha(Z)$ is an eigenvalue of $ad(Z) : g_{\mathbb{C}} \to g_{\mathbb{C}}$ for each $Z \in \tau$. Each such α gives rise to an irreducible representation of T on $\mathbf{R}^2$.*

Proof. By Theorem 2.3.7, the Killing form is negative definite on g, so we extend the inner product $\langle X, Y \rangle = -B(X, Y)$ to a sesquilinear form on $g_{\mathbb{C}}$. Then by Theorem 2.3.7, $ad(Z)$ is a skew-symmetric linear operator on $g_{\mathbb{C}}$ for each $Z \in$ g, so $\{iad(Z); Z \in \tau\}$ is a commuting family of selfadjoint operators on $g_{\mathbb{C}}$. Eigenvectors corresponding to distinct eigenvalues of a selfadjoint operator are orthogonal. Hence there is a simultaneous diagonalization, and for each $Z \in \tau$ there exists a non-zero $Y \in g_{\mathbb{C}}$ such that $iad(Z)(Y) = -\alpha(Z)Y$, where $\alpha(Z)$ is real; evidently the map $Z \mapsto \alpha(Z)$ is a linear functional $\tau \to \mathbf{R}$.

Given such an α, let the eigenvector be $Y = U + iV$ where $U, V \in$ g. Then the eigenvalue equation leads via Proposition 2.3.2 to

$$\exp(2\pi ad(Z))(U + iV) = e^{2\pi i \alpha(Z)}(U + iV) \qquad (2.6.7)$$

and in terms of the real and imaginary subspaces of $g + ig$ we have a rotation matrix

$$\exp(2\pi ad(Z)) \begin{bmatrix} U \\ V \end{bmatrix} = \begin{bmatrix} \cos 2\pi\alpha(Z) & \sin 2\pi\alpha(Z) \\ -\sin 2\pi\alpha(Z) & \cos 2\pi\alpha(Z) \end{bmatrix} \begin{bmatrix} U \\ V \end{bmatrix}, \qquad (2.6.8)$$

so $\exp(2\pi Z) \mapsto \exp(2\pi ad(Z))$ is a representation $T \to SO(2)$. $\qquad \square$

Definition (*Roots*). A root α is a nonzero linear functional $\alpha : \tau \to \mathbf{R}$ such that $i\alpha(Z)$ is an eigenvalue of $ad(Z)$ for each $Z \in \tau$.

Let G be a compact, connected and semisimple Lie group. Then by an extension of Proposition 2.3.4, the Lie algebra of G has the form

$$g = \tau \oplus g_{r_1^*} \oplus \cdots \oplus g_{r_m^*} ;$$

where $g_{r_j^*}$ is the eigenspace that corresponds to r_j^* and here the $r_j^* \in \tau^*$ are said to form a fundamental system of positive real roots.

Each positive real root gives rise to a distinct irreducible representations of T as above. When viewed as a real vector space, g has dimension $k + 2m$.

Definition (*Weyl denominator*). For $u = \exp 2\pi i \sum_{j=1}^{k} \theta_j$, let the Weyl denominator be

$$\Delta(u) = \prod_{j=1}^{m} \left(e^{2\pi i r_j^*(\Theta)} - 1 \right). \tag{2.6.9}$$

Theorem 2.6.5 (*Weyl integration formula*). *Let G act upon itself by conjugation. Then the space of orbits may be identified with a maximal torus $\mathcal{T}$ in G. Let $|W|$ be the order of the Weyl group. Suppose that F is a continuous class function, so that F is constant on orbits. Then the natural probability measure on the orbits associated with μ_G satisfies*

$$\int_G F(g)\mu_G(dg) = \frac{1}{|W|} \int_{\mathcal{T}} F(u)|\Delta(u)|^2 \mu_{\mathcal{T}}(du). \tag{2.6.10}$$

Proof. We introduce the cosets $g\mathcal{T} = \{yt : t \in \mathcal{T}\}$ and the coset space $G/\mathcal{T} = \{g\mathcal{T} : g \in G\}$; the quotient map $\pi : G \to G/\mathcal{T}$ induces $\pi \sharp \mu_G$ from the Haar measure on G; whereas $\mathcal{T}$ has its Haar measure $\mu_{\mathcal{T}}(dt)$. The map $\Phi : G \times \mathcal{T} \to G : (g, u) \mapsto gug^{-1}$ is surjective since every element of G is conjugate to some point in the maximal torus by Theorem 2.6.2; further, Φ induces a surjective map $\varphi : G/\mathcal{T} \times \mathcal{T} \to G$ since $gug^{-1} = huh^{-1}$ whenever $h^{-1}g \in \mathcal{T}$. Hence φ is well defined, and the number of times that φ covers each $g \in G$ equals $|W|$ since

$$gtg^{-1} = huh^{-1} \Leftrightarrow (h^{-1}g)t(h^{-1}g)^{-1} = u \Leftrightarrow h^{-1}g \in N(\mathcal{T}). \tag{2.6.11}$$

One can show that φ is differentiable, except on a set E which has zero measure with respect to $\pi \sharp \mu_G \otimes \mu_{\mathcal{T}}$; further, $\varphi(E)$ has measure zero with respect to μ_G. By the change of variables formula, we have

$$|W| \int_G f(g)\mu_G(dg) = \int_{\mathcal{T}} \int_{G/\mathcal{T}} f \circ \varphi(g\mathcal{T}, u) J_\varphi(g\mathcal{T}, u) \pi \sharp \mu_G(dg\mathcal{T}) \mu_{\mathcal{T}}(du)$$
$$\tag{2.6.12}$$

for all $f \in C(G)$ where J_φ is the Jacobian; in particular

$$\int_G F(g)\mu_G(dg) = \frac{1}{|W|} \int_{\mathcal{T}} F(u) \left(\int_G J_\varphi(g\mathcal{T}, u) \mu_G(dg) \right) \mu_{\mathcal{T}}(du) \tag{2.6.13}$$

where $F \in C(G)$ depends only upon the conjugacy class.

Now we calculate the Jacobian. The Lie algebra decomposes as $\mathfrak{g} = \tau \oplus \tau^{\perp}$, where τ is the Lie algebra of $\mathcal{T}$ and $\tau^{\perp}$ is the orthogonal complement with respect to the invariant inner product that is given by the Killing form as in Theorem 2.3.7. We have shown that the intersection of each adjoint orbit ξ_X with τ is an orbit of the Weyl group.

To calculate the derivative at (g, u), we shift to a neighbourhood of the identity and consider

$$\Psi(X, Z) = gu^{-1}g^{-1}(g\exp(X)\exp(Z)u\exp(-X)g^{-1}) \quad (X \in \tau^{\perp}, Z \in \tau)$$
(2.6.14)

which satisfies $\Psi(0,0) = I$ and

$$\left(\frac{d}{dt}\right)_{t=0}\Psi(tX, tZ) = gu^{-1}Zug^{-1} + gu^{-1}Xug^{-1} - gXg^{-1}$$
$$= Ad(g)(Z + Ad(u^{-1})X - X). \quad (2.6.15)$$

Now $\det Ad(g) = 1$, so we need to find $\det(Ad(u^{-1}) - I)$, where $u \in T$ and $Ad(u^{-1}) - I$ operates on $\tau^{\perp}$. Suppose that $u = \exp(-2\pi Z)$, so

$$Ad(u^{-1}) = \exp(ad(2\pi Z))$$
$$= \bigoplus_{j=1}^{m} \begin{bmatrix} \cos 2\pi r_j^*(Z) & \sin 2\pi r_j^*(Z) \\ -\sin 2\pi r_j^*(Z) & \cos 2\pi r_j^*(Z) \end{bmatrix} \quad (2.6.16)$$

and hence the determinant is

$$\left|\det(Ad(u^{-1}) - I)\right| = \prod_{j=1}^{m} \det \begin{bmatrix} \cos 2\pi r_j^*(Z) - 1 & \sin 2\pi r_j^*(Z) \\ -\sin 2\pi r_j^*(Z) & \cos 2\pi r_j^*(Z) - 1 \end{bmatrix}$$
$$= \prod_{j=1}^{m}\left(e^{2\pi r_j^*(Z)} - 1\right)\left(e^{-2\pi r_j^*(Z)} - 1\right)$$
$$= \Delta(u)\bar{\Delta}(u). \quad (2.6.17)$$

$\square$

One can classify the complex semisimple Lie algebras by their roots. The axiomatic approach involves specifying a root system as in [110].

Definition (*Root system*). Let V be an ℓ-dimensional real vector space with inner product, and R a finite collection of non-zero vectors in V, such that R spans V. Let $\check{\alpha} = 2\alpha/\langle\alpha, \alpha\rangle$ be the co-root that corresponds to $\alpha \in R$, and let $s_\alpha : V \to V$ be the linear map $s_\alpha(\beta) = \beta - \langle\check{\alpha}, \beta\rangle\alpha$; so s_β gives the reflection in the hyperplane $V_\alpha = \{v \in V : \langle v, \alpha\rangle = 0\}$. We say that R is a root system when:

(1) $s_\alpha(R) = R$ for all $\alpha \in R$;
(2) $\langle\check{\alpha}, \beta\rangle \in \mathbf{Z}$ for all $\alpha, \beta \in R$;

and the root system is reduced when:

(3) If $\alpha, \beta \in R$ are proportional, then $\alpha = \pm\beta$.

The Weyl group W that corresponds to R is the group generated by the reflections $\{s_\alpha : \alpha \in R\}$, which by (1) is a group of permutations of R. One can show that this is consistent with the definition given above. The hyperplanes V_α divide up V into cones, which are known as Weyl chambers, and the Weyl group permutes the Weyl chambers. One can select such a chamber C which is bounded by hyperplanes V_{α_j} for $j = 1, \ldots, \ell$, where the $\alpha_j \in R$ are so chosen that $\alpha_j(x) > 0$ for all $x \in C$; then C is called the fundamental chamber and the α_j $(j = 1, \ldots, \ell)$ give a basis of R. The root system is determined by the Cartan matrix, namely $[\langle \check{\alpha}_j, \alpha_k \rangle]_{j,k=1,\ldots,\ell}$; on account of (2) the entries are integers, and one can show that the off-diagonal entries are less than or equal to zero. Let $\phi_{\alpha,\beta}$ be the angle between α and β. Then $4\cos^2 \phi_{\alpha,\beta} = \langle \check{\beta}, \alpha \rangle \langle \check{\alpha}, \beta \rangle$.

- The Cartan matrix determines the complex semisimple Lie algebra up to isomorphism.

Remark. (*Killing–Cartan classification*). Killing classified the semisimple Lie algebras that give compact Lie groups of rank n, and Cartan filled the lacunae in his proof. There are four infinite series A_n, B_n, C_n and D_n that give Lie groups of rank n, and five exceptional algebras G_2, F_4, E_6, E_7 and E_8 which we shall not discuss further. Dynkin showed how to classify the algebras in terms of fundamental root systems, for which the Cartan matrices have a diagrammatic representation; see [110].

Cartan type	Lie algebra	$\dim_{\mathbf{R}}$	$\lvert W \rvert$	
A_n	$su(n+1)$	$n(n+2)$	$(n+1)!$	
B_n	$so(2n+1)$	$n(2n+1)$	$2^n n!$	(2.6.18)
C_n	$sp(n)$	$n(2n+1)$	$2^n n!$	
D_n	$so(2n)$	$n(2n-1)$	$2^{n-1} n!$	

Exercise. Let G and H be compact and semisimple Lie groups, and let $\varphi : G \to H$ be a bijective and differentiable map such that $\varphi(e_G) = e_H$.

(i) Show that the derivative $D\varphi(e_G)$ gives a nonsingular linear map between the Lie algebras $\mathrm{g} \to \mathrm{h}$.

(ii) Let dx be some version of Lebesgue product measure on g and dy be some version of Lebesgue product measure on h. Prove that the Jacobian of $D\varphi(e_G)$ is $\lvert \det D\varphi(e_G) \rvert$.

2.7 Dyson's circular ensembles

- Circular ensembles live on tori.
- The circular ensembles arise from actions on the unitary group by various compact groups.

- The probability measure for the circular unitary ensemble is given by Weyl's integration formula.
- CUE describes integer spin and asymmetric time reversal.

Dyson [64, 116] introduced three ensembles based upon classical compact Lie groups, namely the orthogonal, circular and symplectic circular ensembles. Whereas their properties are similar, their definitions are subtly different. So we begin by stating how they can be defined in terms of the general construction via group actions, and then consider each one in turn.

The circular ensemble consists of the space M with the probability measure ν. The joint eigenvalue distribution is given by a probability measure of the form

$$\sigma_n^{(\beta)}(d\Theta) = c_n^{(\beta)} \prod_{j,k:1\leq j<k\leq n} |e^{i\theta_j} - e^{i\theta_k}|^\beta d\theta_1 d\theta_2 \cdots d\theta_n. \qquad (2.7.1)$$

ensemble	circular orthogonal	circular unitary	circular symplectic
M	$U(n)$	$U(n)$	$U(2n)$
G	$O(n)$	$U(n)$	$Sp(n)$
action:	$\beta_h : g \mapsto hg$	$\alpha_h : g \mapsto hgh^{-1}$	$\beta_h : g \mapsto hg$
M/G	$O(n)\backslash U(n)$ symmetric $U(n)$	$\mathbf{T}^n$ maximal torus	$Sp(n)\backslash U(2n)$ self-dual $U(2n)$
ν	Haar measure on $U(n)$	Haar measure on $U(n)$	Haar measure on $U(2n)$
invariance under:	$O(n)$ action on symmetric matrices in $U(n)$	conjugation	$Sp(n)$ action on self-dual $U(2n)$ matrices
σ	$\sigma_n^{(1)}$	$\sigma_n^{(2)}$	$\sigma_n^{(4)}$

$$(2.7.2)$$

As an application of Theorem 2.6.2, we can now introduce the unitary circular ensemble. For each $U \in U(n)$ there exists $V \in U(n)$ such that $V^*UV = \mathrm{diag}(e^{i\theta_j})$, where $e^{i\theta_j}$ are eigenvalues of U with $0 \leq \theta_1 \leq \cdots \leq \theta_n \leq 2\pi$.

Definition (*Circular unitary ensemble*). The unitary circular ensemble is the probability measure on $\mathbf{T}^k$ that is induced from the Haar probability measure on $U(n)$ by the eigenvalue map Λ.

The circular ensemble is used to describe some physical systems that have integer spin and are not symmetric under time reversal.

Proposition 2.7.1 (*Existence of CUE*). *The group $SU(n)$ has rank $k = n - 1$, the Weyl group has order $n!$ and with $\theta_n = -\sum_{j=1}^{k} \theta_j$, the natural measure on the maximal torus associated with Haar measure on the group is*

$$\sigma_{SU(n)}(d\Theta) = \frac{1}{n!} \prod_{1 \leq j < \ell \leq n} |e^{i\theta_j} - e^{i\theta_\ell}|^2 d\theta_1 \ldots d\theta_k. \qquad (2.7.3)$$

Proof. Each $U \in SU(n)$ has eigenvalues on $\mathbf{T}$, and $\det U = 1$, hence a maximal torus is

$$\mathcal{T} = \Big\{ \text{diag}(e^{2\pi i \theta_1}, \ldots, e^{2\pi i \theta_n}) : \sum_{j=1}^{n} \theta_j \equiv 0 \mod 2\pi \Big\}. \qquad (2.7.4)$$

Evidently $\mathcal{T}$ has rank $n-1$, and $N(\mathcal{T})$ is generated by $\mathcal{T}$ and the permutation matrices which swap around the diagonal entries, hence $|W| = n!$.
    The Lie algebra is

$$su(n) = \{ A \in M_n(\mathbf{C}) : A^* = -A; \ \text{trace}(A) = 0 \}, \qquad (2.7.5)$$

and hence $su(n)$ has dimension $n^2 - 1$ as a real vector space. The exponential map $A \mapsto \exp A$ is surjective and open, so $su(n)$ has dimension $n^2 - 1$ as a real manifold.
    Let (e_j) be the standard unit vector basis of $\mathbf{R}^n$; then e_j gives rise to the linear functional $(\theta_1, \ldots, \theta_n) \mapsto \theta_j$. Hence each pair $\{j, \ell\}$ with $1 \leq j < \ell \leq n$ gives a root $r_{j,\ell}^* = e_j - e_\ell$ or

$$\text{diag}(e^{2\pi i \theta_j}) \mapsto e^{2\pi i (\theta_j - \theta_\ell)}, \qquad (2.7.6)$$

and there are $\binom{n}{2}$ choices of such pairs. Hence we obtain $\Delta(u)$ in the Weyl denominator. By applying the formula $n = 2m + k$, we recover the reassuring identity

$$n^2 - 1 = 2\binom{n}{2} + n - 1.$$

$\square$

$$M = G \qquad G \text{ compact semisimple}$$

$$G \text{ acts on } M \qquad \beta_h : g \mapsto hgh^{-1}$$

$$M/G \qquad \mathbf{T}^k = \text{maximal torus}$$

$$\text{phase space} \qquad \text{periods} \qquad\qquad (2.7.7)$$

$$\nu \text{ on } M \qquad \text{Haar measure of } G$$

$$\sigma \text{ on } M/G \qquad \frac{\bar{\Delta}\Delta(\Theta)}{|W|}d\Theta$$

$$\text{invariance} \qquad \text{conjugation}$$

The Weyl denominator is

$$\Delta(X) = \prod_{\alpha \in P_+} \left(e^{i\pi\alpha(X)} - e^{-i\pi\alpha(X)} \right), \qquad (2.7.8)$$

P_+ is the set of positive roots and W is the Weyl group.

Example 2.7.2 Let

$$Q_N(e^{i\theta}) = \sum_{j=-N}^{N} e^{i\theta} = \frac{\sin(2N+1)\theta/2}{\sin\theta/2},$$

and let U be the $(2N+1) \times (2N+1)$ matrix $U = [e^{i\theta_j k}]_{j=1,\dots,2N+1; k=-N,\dots,N}$. Then by a simple calculation we have

$$UU^* = [Q_N(e^{i(\theta_j-\theta_k)})]$$

and hence

$$\det[Q_N(e^{i(\theta_j-\theta_k)})] = |\det U|^2 = \prod_{1 \le j < k \le 2N+1} |e^{i\theta_j} - e^{i\theta_k}|^2.$$

The operator

$$f \mapsto \int_{\mathbf{T}} Q_N(e^{i(\theta-\phi)})f(e^{i\phi})\frac{d\phi}{2\pi}$$

gives the orthogonal projection onto span$\{e^{ij\theta} : j = -N,\dots,N\}$, and hence has rank $2N+1$.

2.8 Circular orthogonal ensemble

- COE describes integer spin and symmetric time reversal.

The circular orthogonal ensemble was introduced to describe some physical systems that have integer spin and are symmetrical under time reversal. The COE is not quite the joint eigenvalue distribution of eigenvalues from the groups $O(n)$. In this section only we write $S(n) = \{u \in U(n) : u^t = u\}$ and let $U(n)$ act on $S(n)$ by $\beta_u(s) = usu^t$. Dyson showed that there exists a unique $\nu_{S(n)} \in Prob(S(n))$ that is invariant under β_u. See [18].

Definition (*Circular orthogonal ensemble*). The orthogonal circular ensemble is the probability measure $\sigma_n^{(1)} = \Lambda \sharp \nu_{S(n)}$ that is induced from $\nu_{S(n)}$ by the eigenvalue map $\Lambda : S(n) \to \mathbf{T}^n$.

We observe that $S(n)$ is not a group, so we construct $\sigma_n^{(1)}$ by introducing a suitable coset space. For each $s \in S(n)$, there exists $w \in U(n)$ such that $w^t w = s$; furthermore, $w^t w = v^t v$ with $w, v \in U(n)$ if and only if there exists $r \in O(n)$ such that $w = rv$.

Let $H = O(n)$ and let $G = U(n)$, so that H is a closed subgroup of G, and let H act on G by left multiplication $\beta_h(g) = hg$ to give orbits the right cosets $Hg = \{hg : h \in H\}$ and the space of right cosets $H\backslash G$. There is a natural bijective correspondence between $S(n)$ and $H\backslash G$ given by $u^t u \leftrightarrow Hu$; hence any continuous function $f : S(n) \to \mathbf{C}$ may be identified with a continuous function $F : G \to \mathbf{C}$ that is invariant under the action of H, and conversely.

Proposition 2.8.1 (*Existence of COE*). *Let $\nu_{S(n)}$ be the probability measure on $S(n)$ that is induced from $\mu_{U(n)}$ by the quotient map $U(n) \to O(n)\backslash U(n)$ followed by the natural isomorphism $O(n)\backslash U(n) \cong S(n)$. Then $\nu_{S(n)}$ is invariant under the action β of $U(n)$ on $S(n)$.*

Proof. Since $s = w^t w$ corresponds to Hw, $u^t su$ corresponds to Hwu; hence

$$\int_{S(n)} f(u^t su)\nu(du) = \int_{H\backslash G} F(Hwu)\nu(dHw)$$

$$= \int_G F(wu)\mu_G(dw)$$

$$= \int_G F(w)\mu_G(dw)$$

$$= \int_{S(n)} f(s)\nu(ds). \qquad (2.8.1)$$

$\square$

2.9 Circular symplectic ensemble

• CSE describes half integer spin.

By following the construction of the preceding section, we can introduce the circular symplectic ensemble. The CSE was introduced to describe physical systems that have half integer spin. A complex matrix v is symplectic if $v^R v = I$. The symplectic group $Sp(n)$ is the subgroup of $U(2n)$ that consists of symplectic matrices. Let $P(n) = \{U \in U(2n) : U^R = U\}$, and let $U(n)$ act on $P(n)$ by $\beta_u(s) = usu^R$. As we shall show shortly, there exists a unique $\nu_{P(n)} \in \text{Prob}(P(n))$ that is invariant under β.

Definition (*Circular symplectic ensemble*). The symplectic circular ensemble $\sigma^{(4)}$ is the probability measure that is induced from $\nu_{P(n)}$ by $\Lambda : P(n) \to \mathbf{T}^n$.

For each $s \in P(n)$, there exists $w \in U(2n)$ such that $w^R w = s$; further, $w^R w = v^R v$ with $w, w \in u(2n)$ if and only if there exists $r \in Sp(n)$ such that $w = bv$.

Let $G = u(2n)$ and $H = Sp(n)$, and let H act on G by left multiplication $\beta_h(g) = hg$ to give right cosets Hg and space of right cosets $H\backslash G = \{Hg : g \in G\}$. There is a bijective correspondence between $P(n)$ and $H\backslash G$ given by $u^R u \leftrightarrow Hu$; hence any $f \in C(P(n))$ may be identified with a $F \in C(H\backslash G)$ that is invariant under β, and conversely.

Proposition 2.9.2 (*Existence of CSE*). *Let $\nu_{P(n)}$ be the probability measure on $P(n)$ that is induced from $\mu_{U(2n)}$ by the quotient map $U(2n) \to Sp(n)\backslash U(2n)$ followed by the natural isomorphism $Sp(n)\backslash U(2n) \cong P(n)$. Then $\nu_{P(n)}$ is invariant under the action β of $Sp(n)$ on $P(n)$.*

Proof. This follows by a similar argument to Proposition 2.8.1. □

Exercise 2.9.3 (i) Show that $G = \prod_{n=1}^{\infty} U(n)$ is a compact metric group.
(ii) Describe the Haar probability measure on G.

3

Entropy and concentration of measure

Abstract

In this chapter we introduce the notion of entropy, and then relative entropy. We then introduce the fundamental concept of concentration of measure in the context of Lipschitz functions on metric probability spaces. We consider the transportation problem for probability measures on metric spaces. The dual form of a concentration inequality is a transportation inequality. In order to prove such transportation inequalities, we use the Brunn–Minkowski inequality on Euclidean space. We formulate concentration inequalities for matrix ensembles with uniformly convex potentials, including Gaussian ensembles. We state some concentration inequalities for measures on compact groups which follow from the Gromov–Lévy isoperimetric inequality in differential geometry. Surface area measure on the spheres in $\mathbf{R}^n$ provides a contrasting example of the concentration of measure phenomenon in high dimension. An important aspect of this approach is that we can move measures and inequalities from one space to another; for instance, we can deduce concentration for eigenvalue distributions from concentration of the corresponding measures on the random matrices.

3.1 Relative entropy

Definition (*Entropy*). Let ρ be a probability density function on $\mathbf{R}^n$. Then the entropy of ρ is

$$S(\rho) = -\int \rho(x) \log \rho(x)\, dx \qquad (3.1.1)$$

when this integral is absolutely convergent. In this case, we say that ρ has finite entropy. Shannon entropy involves the $(-)$ sign, whereas Boltzmann entropy involves the $(+)$ sign.

84

For probability densities ρ_1 on $\mathbf{R}^m$ and ρ_2 on $\mathbf{R}^n$, the tensor product $\rho = \rho_1 \otimes \rho_2$ is $\rho(x,y) = \rho_1(x)\rho_2(y)$. An immediate consequence of the definition is that ρ satisfies

$$S(\rho) = S(\rho_1) + S(\rho_2); \qquad (3.1.2)$$

an identity which corresponds to the thermodynamic property that entropy is an extensive variable [3].

Simplifying notation, we write γ_1 for the probability density function of the $N(0,1)$ random variable.

Proposition 3.1.1 *Amongst all probability density functions on $\mathbf{R}$ with finite second moments that have variance one, the standard Gaussian density γ_1 is a maximizer of the entropy. Furthermore, $S(\gamma_1) = \log \sqrt{2\pi e}$.*

Proof. We shall introduce the Gaussian as the candidate maximizer, and then prove rigorously that γ_1 attains the maximum. The entropy is invariant under translation, so we seek to maximise $S(\rho)$ subject to the constraints

$$\rho \geq 0; \qquad (3.1.3)$$

$$\int \rho(x)\,dx = \int x^2 \rho(x)\,dx = 1 \qquad (3.1.4)$$

$$\int x\rho(x)\,dx = 0. \qquad (3.1.5)$$

By calculus we have $\xi \log \xi \geq -1/e$ for $\xi \geq 0$, so with $\xi = \rho e^{-v}$ any such ρ satisfies

$$-\int \rho \log \rho\, dx \leq -\int v\rho\, dx + \int e^{v-1}dx; \qquad (3.1.6)$$

in particular with $v = 1 - x^2/2 - \log\sqrt{2\pi}$ we deduce

$$S(\rho) \leq \frac{1}{2}\int x^2 \rho(x)\,dx + \log\sqrt{2\pi} = \log\sqrt{2\pi e}. \qquad (3.1.7)$$

When $\rho = \gamma_1$, we have $\xi = 1/e$ and there is equality throughout. $\quad\square$

In the next result, we consider another notion of the average of $\log \rho(x)$.

Proposition 3.1.2 *Let ρ be a probability density function on $\mathbf{R}^n$ that has finite entropy, and for $\varepsilon > 0$, let*

$$E_{N,\varepsilon} = \left\{ (x_j)_{j=1}^N \in (\mathbf{R}^n)^N : \left| \frac{1}{N}\sum_{j=1}^N \log \rho(x_j) + S(\rho) \right| < \varepsilon \right\}. \qquad (3.1.8)$$

Then

$$\frac{1}{N} \log \mathrm{vol}\,(E_{N,\varepsilon}) - \varepsilon \leq S(\rho) \leq \frac{1}{N} \log \mathrm{vol}\,(E_{N,\varepsilon}) + \varepsilon \qquad (3.1.9)$$

holds for all sufficiently large N.

Proof. See [152]. We regard the x_j as mutually independent random variables with identical distribution specified by the probability density function ρ. By hypothesis, $\log \rho$ is absolutely integrable with respect to ρ; hence the $\log \rho(x_j)$ have finite expectation. Let $\rho^{\otimes N}$ be the probability measure $\rho(x_1)\ldots\rho(x_N)dx_1\ldots dx_N$. By the strong law of large numbers [88],

$$\frac{1}{N}\sum_{j=1}^{N} \log \rho(x_j) \rightarrow \int \log \rho(x)\,\rho(x)\,dx \qquad (N \rightarrow \infty) \quad (3.1.10)$$

in probability; hence

$$\rho^{\otimes N}\,(E_{N,\varepsilon}) \rightarrow 1 \qquad (N \rightarrow \infty). \qquad (3.1.11)$$

Now for $(x_j)_{j=1}^{N} \in E_{N,\varepsilon}$, we have

$$e^{N\varepsilon}\rho(x_1)\ldots\rho(x_N) \geq e^{-NS(\rho)} \geq e^{-N\varepsilon}\rho(x_1)\ldots\rho(x_N) \quad (3.1.12)$$

hence by integration with respect to Lebesgue product measure we have

$$e^{N\varepsilon}\rho^{\otimes N}\,(E_{N,\varepsilon}) \geq e^{-NS(\rho)}\mathrm{vol}(E_{N,\varepsilon}) \geq e^{-N\varepsilon}\rho^{\otimes N}\,(E_{N,\varepsilon}), \qquad (3.1.13)$$

so

$$\exp(S(\rho)+\varepsilon)\big(\rho^{\otimes N}\,(E_{N,\varepsilon})\big)^{1/N} \geq \mathrm{vol}(E_{N,\varepsilon})^{1/N}$$
$$\geq \exp(S(\rho)-\varepsilon)\big(\rho^{\otimes N}\,(E_{N,\varepsilon})\big)^{1/N}, \qquad (3.1.14)$$

hence result. $\qquad\square$

Definition (*Convexity*). Let Ω be nonempty subset of $\mathbf{R}^n$. Say that Ω is convex if $tx + (1-t)y$ belongs to Ω for all $x, y \in \Omega$ and $0 < t < 1$. We define a continuous function $V : \Omega \rightarrow \mathbf{R}$ to be convex if

$$tV(x) + (1-t)V(y) - V(tx + (1-t)y) \geq 0 \qquad (3.1.15)$$

holds for all distinct $x, y \in \Omega$ and $0 < t < 1$, and strictly convex if the inequality is always strict.

Exercise. (i) Show that the functions $\varphi(x) = x\log x$ for $x > 0$ and e^x on $\mathbf{R}$ are convex.

(ii) Show that a strictly convex function on $\mathbf{R}$ cannot have a strict local maximum.

(iii) Show that a positive strictly convex function on **R** has at most one strict local minimum.

Lemma 3.1.3 (*Jensen's inequality*). *Suppose that* $\mu \in Prob\,(\Omega)$, *where* (Ω, d) *is a Polish space, and that* $\varphi : (a, b) \to \mathbf{R}$ *is convex. Then for any* $f : \Omega \to (a, b)$ *such that* f *and* $\varphi \circ f$ *are integrable with respect to* μ,

$$\varphi\left(\int_{\Omega} f(x)\mu(dx)\right) \leq \int_{\Omega} \varphi(f(x))\mu(dx). \qquad (3.1.16)$$

Proof. See [71]. □

Definition (*Relative entropy*). Let μ and ν be probability measures on a Polish space (Ω, d), with μ absolutely continuous with respect to ν and let $d\mu/d\nu$ be the Radon–Nikodym derivative. The relative entropy of μ with respect to ν is

$$\mathrm{Ent}(\mu \mid \nu) = \int \log \frac{d\mu}{d\nu} d\mu; \qquad (3.1.17)$$

when this integral converges absolutely. We take $\mathrm{Ent}(\mu \mid \nu) = \infty$ when the integral diverges or when μ is not absolutely continuous with respect to ν.

Proposition 3.1.4 *Relative entropy has the following properties:*

(i) $0 \leq \mathrm{Ent}(\mu \mid \nu)$, *with equality only if* $\mu = \nu$;
(ii) for $\mu_1, \mu_2, \nu_1, \nu_2 \in Prob\,(\Omega)$ *the product measures satisfy*

$$\mathrm{Ent}(\mu_1 \otimes \mu_2 \mid \nu_1 \otimes \nu_2) = \mathrm{Ent}(\mu_1 \mid \nu_1) + \mathrm{Ent}(\mu_2 \mid \nu_2); \quad (3.1.18)$$

(iii) the variational formula

$$\mathrm{Ent}(\mu \mid \nu) = \sup\left\{\int_{\Omega} g\,d\mu : \int_{\Omega} e^g\,d\nu \leq 1; g \in C(\Omega; \mathbf{R})\right\}. \quad (3.1.19)$$

Proof. (i) The function exp is convex, and hence by Jensen's inequality,

$$\exp \int \log \frac{d\nu}{d\mu}\,d\mu \leq \int \frac{d\nu}{d\mu} d\mu = 1, \qquad (3.1.20)$$

hence $\mathrm{Ent}(\mu \mid \nu) \geq 0$. Further, we have equality only if $d\nu/d\mu = 1$ holds on a set of measure one with respect to μ; that is, $\mu = \nu$.

(ii) One integrates the identity

$$\log \frac{d(\mu_1 \otimes \mu_2)}{d(\nu_1 \otimes \nu_2)} = \log \frac{d\mu_1}{d\nu_1} + \log \frac{d\mu_2}{d\nu_2} \qquad (3.1.21)$$

with respect to $\nu_1 \otimes \nu_2$.

(iii) By calculus we have $\xi \log \xi \geq -1/e$ for $\xi \geq 0$, with equality only at $\xi = 1/e$; hence,

$$uv \leq u \log u + e^{v-1} \quad (u > 0, v \in \mathbf{R}). \tag{3.1.22}$$

Let $v = g + 1$ and $u = d\mu/d\nu$ and obtain

$$\int g \, d\mu \leq \int \log \frac{d\mu}{d\nu} d\nu + \int e^g \, d\nu - 1, \tag{3.1.23}$$

with equality only when $g = \log d\mu/d\nu$. $\qquad\square$

Exercise (Lower semicontinuity of relative entropy). Let $\sigma \in Prob\,(\Omega)$ and let f and f_n $(n = 1, 2, \dots)$ be probability density functions such that $f_n \sigma \to f\sigma$ weakly as $n \to \infty$. Show that

$$\mathrm{Ent}(f\sigma \mid \sigma) \leq \lim_{n \to \infty} \inf \mathrm{Ent}(f_n \sigma \mid \sigma). \tag{3.1.24}$$

Exercise (Joint convexity of relative entropy).

(i) Let $U(x, y) = x \log(x/y)$ for $x, y > 0$. Show that $\mathrm{Hess}\, U \geq 0$.

(ii) Let f_0, f_1, g_0 and g_1 be probability density functions with respect to σ where $\sigma \in Prob\,(\Omega)$, and let $f_t = (1 - t)f_0 + tf_1$ and $g_t = (1 - t)g_0 + tg_1$ for $0 < t < 1$. Show that

$$U(f_t, g_t) \leq (1 - t)U(f_0, g_0) + tU(f_1, g_1) \quad (0 < t < 1) \tag{3.1.25}$$

and hence that

$$\mathrm{Ent}(f_t \mid g_t) \leq (1 - t)\mathrm{Ent}(f_0 \mid g_0) + t\mathrm{Ent}(f_1 \mid g_1) \quad (0 < t < 1). \tag{3.1.26}$$

Definition (*Gibbs measure*). Let $V : \mathbf{R} \to \mathbf{R}$ be a measurable function such that, for some $\beta > 0$ there exists a constant $Z(\beta)$ such that

$$\mu_V(dx) = Z(\beta)^{-1} \exp(-\beta V(x)) \, dx \tag{3.1.27}$$

defines a probability measure on $\mathbf{R}$. Then μ_V is called a Gibbs measure with potential V and partition function Z at inverse temperature β.

Proposition 3.1.5 (*Partition functions*). *Let V_1 and V_2 be potentials that give rise to Gibbs measures μ_1 and μ_2 respectively. Then the corresponding partition functions satisfy*

$$\beta \int (V_2(x) - V_1(x))\mu_1(dx) \leq \log \frac{Z_1(\beta)}{Z_2(\beta)} \leq \beta \int (V_2(x) - V_1(x))\mu_2(dx). \tag{3.1.28}$$

Proof. By Proposition 3.1.4(i), we have

$$0 \leq \mathrm{Ent}(\mu_1 \mid \mu_2) = \int \left(\beta V_2(x) - \beta V_1(x) + \log Z_2(\beta) - \log Z_1(\beta)\right) \mu_2(dx)$$
$$(3.1.29)$$

which implies the right-hand inequality, and one can obtain the left-hand inequality by interchanging μ_1 and μ_2. $\square$

The property Proposition 3.1.4(ii) is special to product measures, and admits of some generalization.

Definition (*Marginal densities*). Let Ω^n be a product of Polish spaces, let $\sigma \in \mathrm{Prob}\,(\Omega^n)$ and let f be a probability density function with respect to σ. We define f_j to be the jth marginal probability density functions of f; so that

$$\int_{\Omega^n} f_j(x_j) h_j(x_j) \sigma(dx) = \int_{\Omega^n} h_j(x_j) f(x) \,\sigma(dx) \qquad (3.1.30)$$

for any bounded and continuous $h_j : \Omega \to \mathbf{R}$ that depends only upon the jth coordinate x_j in $(x_k)_{k=1}^n \in \Omega^n$. The jth marginal measure σ_j for $j = 1, \ldots, n$ of σ is induced by the map $\Omega^n \to \Omega : (x_k)_{k=1}^n \mapsto x_j$, and likewise the jth marginal measure of $f d\sigma$ is $f_j d\sigma_j$.

Theorem 3.1.6 (*Subadditivity of relative entropy*). *Let* $\sigma \in \mathrm{Prob}\,(\Omega^n)$ *and let* f *be a probability density function such that* $fd\omega$ *has finite relative entropy with respect to* $d\sigma$. *Suppose that* $1 \leq p < \infty$ *satisfies*

$$\int_{\Omega^n} \prod_{j=1}^n f_j(x_j)^{1/p} \sigma(dx) \leq \left(\prod_{j=1}^n \int_{\Omega^n} f_j(x_j)\,\sigma(dx)\right)^{1/p} = 1. \qquad (3.1.31)$$

(i) Then

$$p\,\mathrm{Ent}(f\sigma \mid \sigma) \geq \sum_{j=1}^n \mathrm{Ent}(f_j\sigma \mid \sigma). \qquad (3.1.32)$$

In particular, the inequality (3.1.31) holds:

(ii) with $p = 1$ *when* $\sigma = \otimes_{j=1}^n \sigma_j$ *is a product of* $\sigma_j \in \mathrm{Prob}\,(\Omega)$;
(iii) with $p = n$ *for all* $\sigma \in \mathrm{Prob}\,(\Omega^n)$;
(iv) with $p = n/(n-1)$ *and* $n \geq 2$ *when* σ *is a standard Gaussian measure on the hyperplane*

$$P = \left\{(x_j)_{j=1}^n : \sum_{j=1}^n x_j = 0\right\} \subset \mathbf{R}^n; \qquad (3.1.33)$$

(v) with $p = 2$, when σ is normalized surface area measure on the sphere S^{n-1} in $\mathbf{R}^n$.

Proof. (i) We let

$$C = \int_{\Omega^n} \prod_{j=1}^{n} f_j(x_j)^{1/p} \omega(dx) \qquad (3.1.34)$$

so that $g(x) = C^{-1} \prod_{j=1}^{n} f_j(x_j)^{1/p}$ is a probability measure, and so by Proposition 3.1.3(i) we have

$$0 \leq \mathrm{Ent}(f\sigma \mid g\sigma)$$
$$= \int_{\Omega^n} f(x) \log f(x) \sigma(dx) - \int f(x) \log g(x) \sigma(dx)$$
$$= \int_{\Omega^n} f(x) \log f(x) \sigma(dx) + (\log C) \int_{\Omega^n} f(x) \sigma(dx)$$
$$- \sum_{j=1}^{n} \int_{\Omega^n} f(x) \log f_j(x_j)^{1/p} \sigma(dx) \qquad (3.1.35)$$

and hence by the definition of the marginal measures we have

$$\frac{1}{p} \sum_{j=1}^{n} \int_{\Omega^n} f_j(x_j) \log f_j(x) \sigma(dx) \leq \int_{\Omega^n} f(x) \log f(x) \sigma(dx) + \log C.$$
$$(3.1.36)$$

Now $C \leq 1$ by the main hypothesis, so

$$\frac{1}{p} \sum_{j=1}^{n} \mathrm{Ent}(f_j\sigma \mid \sigma) \leq \mathrm{Ent}(f\sigma \mid \sigma). \qquad (3.1.37)$$

(ii) The jth marginal measure σ_j for $j = 1, \ldots, n$ of ω is induced by the map $\Omega^n \to \Omega : (x_k)_{k=1}^{n} \mapsto x_j$, and likewise the jth marginal measure of $f d\sigma$ is $f_j d\sigma_j$. Hence $f_1(x_1) \ldots f_n(x_n)$ is a probability density function with respect to $\sigma(dx) = \otimes_{j=1}^{n} \sigma_j(dx_j)$. Hence the inequality holds with $p = 1$. This $p = 1$ is the optimal constant.

(iii) By Hölder's inequality we have

$$\int_{\Omega^n} \prod_{j=1}^{n} f_j(x_j)^{1/n} \sigma(dx) \leq \left(\prod_{j=1}^{n} \int_{\Omega^n} f_j(x_j) \sigma(dx) \right)^n = 1 \qquad (3.1.38)$$

since the f_j are probability density functions. This result is in some sense trivial, and indicates that some should strive for $p < n$, and preferably p independent of n.

(iv) A special case of the Brascamp–Lieb inequality [45, (1.15)] gives

$$\int_P \prod_{j=1}^n f_j(x_j)^{(n-1)/n} \sigma(dx) \leq \prod_{j=1}^n \left(\int_P f_j(x_j) \sigma(dx) \right)^{(n-1)/n}. \quad (3.1.39)$$

Villani [162] gives a detailed discussion of how to prove such inequalities by the techniques that we shall discuss later in this chapter, so we omit the proof. The coordinates $(x_j)_{j=1}^n \in P$ are dependent, but only weakly so; hence the constant $n/(n-1) \to 1$ as $n \to \infty$.

(v) By a theorem of Carlen, Lieb and Loss [42],

$$\int_{S^{n-1}} \prod_{j=1}^n f_j(x_j)^{1/2} \sigma(dx) \leq \prod_{j=1}^n \left(\int_{S^{n-1}} f_j(x_j) \sigma(dx) \right)^{1/2}. \quad (3.1.40)$$

In this case the coordinates of $x = (x_1, \ldots, x_n)$ are clearly not mutually independent since $\sum_{j=1}^n x_j^2 = 1$, and σ is not a product measure in $\mathbf{R}^n$. In fact 2 is the smallest constant that works in all dimensions n, so the mutual dependence of coordinates does not appear to diminish with increasing dimension. $\qquad \square$

Exercises 3.1.7 (a) Let ω be Haar probability measure on $SO(n)$, and let $U = [\xi_1, \ldots, \xi_n] \in SO(n)$ have j^{th} column $\xi_j \in S^{n-1}$. The distribution of ξ_j is the jth marginal measure of ω, namely normalized area measure on S^{n-1}.

(b) The following probability measures arise in joint eigenvalue distributions from random matrices. Suppose that $v : \mathbf{R} \to \mathbf{R}$ is continuous and grows like x^2 as $x \to \pm\infty$. Then there exists $\zeta > 0$ such that

$$\omega_j(d\lambda_j) = \zeta^{-1} \exp\big(-nv(\lambda_j)\big)d\lambda_j \quad (3.1.41)$$

is a probability measure on $\mathbf{R}$; further for $\beta > 0$ there exists $Z_n > 0$ such that

$$\sigma_n^{(\beta)} = \frac{1}{Z_n} \prod_{1 \leq j,k \leq n : j \neq k} |\lambda_j - \lambda_k|^{\beta/2} \exp\Big(-\sum_{j=1}^n nv(\lambda_j)\Big)d\lambda_1 \ldots d\lambda_n$$
$$(3.1.42)$$

defines a probability measure on $\mathbf{R}^n$.

(i) Show that the first marginal probability density of $\sigma_n^{(\beta)}$ with respect to the product measure $\omega(d\lambda) = \otimes_{j=1}^n \omega_j(d\lambda_j)$ is

$$f_1(\lambda_1) = \frac{\zeta}{Z_n} \int_{\mathbf{R}^{n-1}} \prod_{1 \le j,k \le n: j \ne k} |\lambda_j - \lambda_k|^{\beta/2} \exp\left(-\sum_{j=2}^n nv(\lambda_j)\right) d\lambda_2 \ldots d\lambda_n,$$

$$(3.1.43)$$

and deduce that

$$\mathrm{Ent}(\sigma_n^{(\beta)} \mid \omega) \ge n\mathrm{Ent}(f_1\omega_1 \mid \omega_1). \qquad (3.1.44)$$

(ii) Let p be a continuous probability density function on $\mathbf{R}$ that has compact support, and let $z > 0$ be such that

$$q_j(d\lambda_j) = z^{-1} \exp\left(-nv(\lambda_j) + n\beta \int \log|\lambda_j - \mu|p(\mu)d\mu\right) d\lambda_j \quad (3.1.45)$$

defines a probability measure on $\mathbf{R}$, and let $Q(d\lambda) = \otimes_{j=1}^n q_j(d\lambda_j)$. Repeat (i) with Q in place of ω.

Problem. What is the appropriate version of Theorem 3.1.6 for the joint eigenvalue distribution of generalized orthogonal ensembles? Is there a version for Haar measure on $SO(n)$?

Proposition 3.1.8 (*Csiszár's inequality*). *Suppose that f and g are probability density functions on $\mathbf{R}^n$ such that f is of finite relative entropy with respect to g. Then*

$$\int_{\mathbf{R}^n} |f(x) - g(x)|dx \le 2\Big(\mathrm{Ent}(f \mid g)\Big)^{1/2}. \qquad (3.1.46)$$

Proof. Let $\varphi(u) = u \log u$ for $u > 0$; then by the mean value theorem there exists a v between 1 and u such that

$$\varphi(u) = \varphi(1) + (u-1)\varphi'(1) + \frac{1}{2}(u-1)^2\varphi''(v), \qquad (3.1.47)$$

so

$$u \log u = u - 1 + \frac{(u-1)^2}{2v}. \qquad (3.1.48)$$

Hence with $u = f(x)/g(x)$, we have

$$\frac{f(x)}{g(x)} \log \frac{f(x)}{g(x)} \ge \frac{f(x)}{g(x)} - 1 + \frac{1}{2}\Big(\frac{f(x)}{g(x)} - 1\Big)^2 \min\Big\{1, \frac{g(x)}{f(x)}\Big\}, \qquad (3.1.49)$$

so after multiplying by $g(x)$ and integrating, we find that

$$\int f(x) \log \frac{f(x)}{g(x)} dx \ge \frac{1}{2} \int (f(x) - g(x))^2 \min\Big\{\frac{1}{g(x)}, \frac{1}{f(x)}\Big\} dx. \qquad (3.1.50)$$

We observe that

$$\int \left(\min\left\{\frac{1}{f(x)}, \frac{1}{g(x)}\right\}\right)^{-1} dx = \int \max\{f(x), g(x)\}\, dx \le 2, \quad (3.1.51)$$

and so by the Cauchy–Schwarz inequality

$$\left(\int |f(x) - g(x)|\, dx\right)^2$$

$$\le \int \left(\min\left\{\frac{1}{f(x)}, \frac{1}{g(x)}\right\}\right)^{-1} dx \int (f(x) - g(x))^2 \min\left\{\frac{1}{g(x)}, \frac{1}{f(x)}\right\} dx$$

$$\le 4 \int f(x) \log \frac{f(x)}{g(x)}\, dx. \qquad (3.1.52)$$

□

Remarks. (i) The result actually holds with an optimal constant $\sqrt{2}$ instead of 2, but the proof we give is more natural.

(ii) Suppose that (f_k) is a sequence of probability densities such that $\text{Ent}(f_k \mid g) \to 0$ as $k \to \infty$. Then we say that f_k converges to g in relative entropy. By Csiszár's inequality, the limit g is uniquely determined by the sequence (f_k). However, relative entropy does not define a metric since it is asymmetrical and does not satisfy the triangle inequality; nevertheless, relative entropy dominates the L^1 metric on the probability density functions.

(iii) We have occasion to use three measures of distance on probability densities:

- L^1 norm $\int |f - g| dx$;
- relative entropy $\text{Ent}(f \mid g)$;
- transportation cost $W_1(f(x)dx, g(x)dx)$.

In Section 3.3 we compare the relative entropy with transportation cost.

3.2 Concentration of measure

- The moment generating function is used to define concentration inequalities.

Definition (*Logarithmic moment generating function*). The logarithmic moment generating function of $\mu \in Prob\,(\mathbf{R})$ is $\Lambda(t) = \log \int_{\mathbf{R}} e^{tx}\mu(dx)$, where this is finite.

Definition (*Legendre transform*). Let $\Lambda : (a,b) \to \mathbf{R}$ be a convex function. Define the Legendre transform $\Lambda_* : \mathbf{R} \to \mathbf{R} \cup \{\infty\}$ by

$$\Lambda_*(s) = \sup_{t \in (a,b)} \{st - \Lambda(t)\}.$$

Choose a slope s and find the point on the graph of Λ such that the tangent T_s has slope s; then let $-\Lambda_*(s)$ be the intercept of T_s with the axis $t = 0$. By convexity, $\Lambda(t) \geq ts - \Lambda_*(s)$ for all $t \in (a,b)$ and $s \in \mathbf{R}$.

Exercise. Let $\mu \in Prob(\mathbf{R})$ have infinite support and mean zero, and suppose that $\Lambda(t) = \log \int_{\mathbf{R}} e^{tx}\mu(dx)$ is finite for all $t \in (-\delta, \delta)$.

(i) Show that $\Lambda : (-\delta, \delta) \to [0, \infty)$ is strictly convex.
(ii) Let $\delta' = \lim_{t \to \delta-} \Lambda'(t)$, which may be infinite, and define the Legendre transform of Λ as above. Show that the supremum is uniquely attained at the point t such that $s = \Lambda'(t)$, and that $\Lambda_* : [0, \delta') \to [0, \infty)$ is convex.
(iii) Compute the following table of Legendre transforms.

$\Lambda(t)$		$\Lambda_*(s)$	
$t^2/2$	$t \in (-\infty, \infty)$	$s^2/2$	$s \in (-\infty, \infty)$
$t \log t$	$t \in [0, \infty)$	e^{s-1}	$s \in (-\infty, \infty)$
$\sqrt{1+t^2} - 1$	$-\infty < t < \infty$	$1 - \sqrt{1-s^2}$	$s \in (-1, 1)$
$t - \log(1+t)$	$t \in (-1, \infty)$	$-s - \log(1-s)$	$s \in (-\infty, 1)$
$-\log(1-t^2)$	$t \in (-1,1)$	$\sqrt{1+s^2} - 1 + \log \frac{2}{1+\sqrt{s^2+1}}$	$s \in (-\infty, \infty)$

The domains of the functions have been adjusted to make them more natural; note that $t \log t < 0$ for some $t > 0$.

Definition (*Concentration inequality*). Suppose that (Ω, d) is a Polish space with $\sigma \in Prob(\Omega)$ such that there exists $\alpha > 0$ such that

$$\int e^{tf(x)}\sigma(dx) \leq e^{t^2/(2\alpha)} \qquad (t \in \mathbf{R}) \qquad (3.2.1)$$

holds for all 1-Lipschitz functions $f : \Omega \to \mathbf{R}$ with $\int f(x)\sigma(dx) = 0$. Then σ satisfies the *concentration inequality* $C(\alpha)$.

Examples 3.2.1 Given a L-Lipschitz function g with mean a, we can apply this inequality to $f = (g - a)/L$. The form of this inequality is suggested by (1.5.3) for Gaussian measure, where f is a linear function. Lévy showed that the probability surface area measure on the sphere $(S^n(r), \|\cdot\|_{R^{n+1}})$ satisfies the concentration inequality $C(\alpha)$ with $\alpha = (n-1)/r^2$. Gromov and Milman considered a more general phenomenon, which they described in terms of Lévy families and isoperimetric inequalities; see [74, 117]. For suitable Riemannian manifolds, α coincides with Ricci curvature.

Exercise 3.2.2 Suppose that $\mu \in Prob\,(\Omega)$ satisfies $C(\alpha)$, and that $F : \Omega \to \mathbf{R}$ is L-Lipschitz, where $L^2 < \alpha$ and that $\int F(x)\mu(dx) = 0$. Let γ_1 be the probability density function of the Gaussian $N(0,1)$ distribution. By considering

$$\int_\Omega \int_{\mathbf{R}} e^{\xi F(x)} \gamma_1(d\xi)\mu(dx), \qquad (3.2.2)$$

show as in [61] that

$$\int_\Omega e^{F(x)^2/2} \mu(dx) < \infty. \qquad (3.2.3)$$

The concentration inequality expresses the fact that most of the σ measure is concentrated near to the mean value of f namely zero. The following result gives an essentially equivalent formulation.

Lemma 3.2.3 *Suppose that σ satisfies $C(\alpha)$, and let $f : \Omega \to \mathbf{R}$ be a 1-Lipschitz function such that $\int f(x)\sigma(dx) = 0$. Then*

$$\sigma\{x \in \Omega : |f(x)| < \varepsilon\} \geq 1 - 2\exp\left(-\frac{\alpha\varepsilon^2}{2}\right) \qquad (\varepsilon > 0). \quad (3.2.4)$$

Proof. By Chebyshev's inequality we have

$$e^{t\varepsilon}\sigma\{x \in \Omega : f(x) > \varepsilon\} \leq \int_\Omega e^{tf(x)}\sigma(dx) \leq e^{t^2/(2\alpha)} \qquad (\varepsilon, t > 0) \tag{3.2.5}$$

and we choose $t = \varepsilon\alpha$ to optimize this inequality. Hence

$$\sigma\{x \in \Omega : f(x) > \varepsilon\} \leq e^{-\alpha\varepsilon^2/2} \qquad (3.2.6)$$

and likewise

$$\sigma\{x \in \Omega : f(x) < -\varepsilon\} \leq e^{-\alpha\varepsilon^2/2}; \qquad (3.2.7)$$

these inequalities combine to give (3.2.4). $\qquad\square$

The following result is in the same spirit as the isoperimetric inequality for $\mathbf{R}^n$ as shown in 1.4.3, although the functions involved are Gaussian rather than powers that depend upon dimension. See [172].

Proposition 3.2.4 (*Isoperimetric form of concentration*). *Suppose that* σ *satisfies* $C(1/\alpha)$. *Then any closed subset* A *of* Ω *such that* $\sigma(A) \geq 1/2$ *and any enlargement* A_ε *with* $\varepsilon > 0$ *satisfy*

$$\sigma(A_\varepsilon) \geq 1 - \exp\bigl(-\alpha/(2\varepsilon^2)\bigr). \qquad (3.2.8)$$

Proof. Let B^c be the complement of a set B. The distance function $d(x, A) = \inf\{d(x, a) : a \in A\}$ is 1-Lipschitz, so $f : \Omega \to [0, 1]$ defined by

$$f(x) = \min\Bigl\{ \frac{d(x, A)}{\varepsilon}, 1 + \frac{d(x, (A_\varepsilon)^c)}{\varepsilon} \Bigr\} \qquad (3.2.9)$$

is $2/\varepsilon$-Lipschitz. Further, $f(x) = 0$ on A and $f(x) = 1$ on $(A_\varepsilon)^c$, so $\int f d\sigma \leq 1/2$, hence

$$(A_\varepsilon)^c \subseteq \Bigl[f - \int f d\sigma \geq 1/2 \Bigr]. \qquad (3.2.10)$$

By the proof of Lemma 3.2.3, the measures of these sets satisfy

$$\sigma\bigl((A_\varepsilon)^c\bigr) \leq \exp\bigl(-\alpha/(2\varepsilon^2)\bigr), \qquad (3.2.11)$$

hence the result. □

In applications to random matrix theory it is very important to know how the constant α depends upon the size of the matrices. To describe this we make a definition that was proposed by Gromov and Milman in [74].

Definition (*Concentration of measure phenomenon*). Let (Ω_n, d_n) be Polish spaces and σ_n probability measures on Ω_n that satisfy $C(\alpha_n)$ for $n = 1, 2, \ldots$. If $\alpha_n \to \infty$ as $n \to \infty$, then we say that $(\sigma_n)_{n=1}^\infty$ exhibits the concentration of measure phenomenon.

Suppose that $F_n : \Omega_n \to \mathbf{R}$ are L-Lipschitz and that $\int F_n d\sigma_n = 0$. We introduce the logarithmic moment generating functions

$$\Lambda_n(t) = \log \int_{\Omega_n} e^{t F_n(x)} \sigma_n(dx) \qquad (t \in \mathbf{R}) \qquad (3.2.12)$$

and their Legendre transforms

$$\Lambda_n^*(s) = \sup_{t \in \mathbf{R}} \bigl(st - \Lambda_n(t) \bigr) \qquad (s \in \mathbf{R}, n = 1, 2, \ldots). \qquad (3.2.13)$$

Proposition 3.2.5 *The* σ_n *satisfy concentration of measure with* $\alpha_n \to \infty$ *if and only if*

$$\Lambda_n^*(s) \geq \frac{\alpha_n s^2}{2L^2} \qquad (s \in \mathbf{R}). \qquad (3.2.14)$$

Proof. The inequality (3.2.14) is equivalent to

$$st - \Lambda_n(t) \geq \frac{\alpha_n s^2}{2L^2} \qquad (s, t \in \mathbf{R}) \tag{3.2.15}$$

which is equivalent to

$$\frac{L^2 t^2}{2\alpha_n} \geq \Lambda_n(t) \qquad (t \in \mathbf{R}). \tag{3.2.16}$$

$\square$

Remark 3.2.6 (*Concentration and deviations*). In the theory of large deviations [59], one considers quantities such as

$$\limsup_{n \to \infty} \alpha_n^{-1} \Lambda_n^*(s), \quad \liminf_{n \to \infty} \alpha_n^{-1} \Lambda_n^*(s) \tag{3.2.17}$$

and obtains asymptotic estimates on probabilities; whereas in the theory of concentration of measure, one wishes to have uniform inequalities for all s and n.

Example 3.2.7 As in Proposition 1.6.1, the normalized surface area measures $\sigma_{n,1}$ on the spheres $S^n(1)$ satisfy concentration of measure with $\alpha_n \geq cn$. We shall sort out the details in Theorem 3.8.2.

Whereas the definition does not require any specific connection between the Ω_n, in applications the Ω_n are typically spaces with a common geometrical structure and with dimension increasing to infinity as $n \to \infty$. The merit of this definition is that it applies to a wide class of objects such as graphs, manifolds or spaces of measures. The significance in random matrix theory is that the main matrix ensembles involve matrices of increasing size, and the measures sometimes exhibit concentration of measure.

There are several routes towards concentration inequalities with many subtle connections between them. The main approaches involve:

• isoperimetric inequalities in the style of Gromov, Lévy and Milman, as in [117] and Theorem 3.8.3 below;
• logarithmic Sobolev inequalities as introduced by Gross [75], as in Theorem 6.3.2;
• transportation inequalities, as considered by Marton and Talagrand [162], as in Theorem 3.4.4 and Corollary 3.5.4.

In Section 3.5 we shall systematically prove concentration inequalities by taking the Prékopa–Leindler inequality as fundamental and introduce transportation inequalities, which are dual to concentration inequalities.

This line of development is effective for the generalized orthogonal, unitary and symplectic ensembles; whereas for the circular ensembles we shall use results from Riemannian geometry. In Section 6.8 we shall ultimately reconcile these two approaches.

Remark 3.2.8 (*Scalings*). (i) There is a cheap way of replacing the metric on (Ω, d), namely by introducing $\delta(x, y) = \tau d(x, y)$ where $\tau > 0$ is fixed. This does not materially affect the concentration inequality, since the Lipschitz constant of a given f is scaled likewise. To avoid trivial examples, we often scale the metrics on compact spaces (Ω_n, d_n) so that the diameter is independent of n. In particular, the diameter of $(U(n), c^2(n))$ is 2 for $n = 1, 2, \ldots$.

(ii) For Riemannian manifolds, we adopt a slightly different scaling, whereby the Ricci curvature is computed in natural coordinates and the metric $d(x, y)$ is the infimum of lengths of all piecewise smooth curves between x and y. For each $y \in \Omega$, the function $f(x) = d(x, y)$ is 1-Lipschitz. Myers showed that if Ω_n is complete n-dimensional Riemannian manifold with $Ric \geq \kappa I$ for some $\kappa > 0$, then Ω_n is compact and diameter $(\Omega_n) \leq \pi \sqrt{n-1}/\sqrt{\kappa}$. So we often want the curvature to grow with increasing n. In Theorem 3.8.3 we state a suitable concentration inequality.

Another merit of the formulation in terms of Lipschitz functions is that we can move easily from one space to another. In particular, we can replace (Ω, d) by any Polish space (Φ, δ) that is bi-Lipschitz equivalent to (Ω, d).

Lemma 3.2.9 *Let (Ω_j, d_j) for $j = 1, 2$ be Polish spaces and let $\Lambda : \Omega_1 \to \Omega_2$ be a L-Lipschitz function. If $\sigma_1 \in Prob(\Omega_1)$ satisfies $C(\alpha_1)$, then $\Lambda \sharp \sigma_1$ satisfies $C(\alpha_2)$ where $\alpha_2 \geq \alpha_1 / L^2$.*

Proof. We observe that any 1-Lipschitz function $g : \Omega_2 \to \mathbf{R}$ gives rise to a 1-Lipschitz function $f : \Omega_1 \to \mathbf{R}$, namely $f(x) = g(\Lambda(x))/L$. Then we use the definition of induced measure to obtain the required inequality for (3.2.1) applied to f.   □

Exercise 3.2.10 Suppose that (Ω_j, d_j) are Polish spaces and that $\sigma_j \in Prob(\Omega_j)$ satisfy the concentration of measure phenomenon with constants $\alpha_j \to \infty$ as $j \to \infty$. Now let (Φ_j, δ_j) be Polish spaces and $\varphi_j : \Omega_j \to \Phi_j$ be maps such that

$$d_j(x_j, y_j) \leq \delta_j(\varphi_j(x_j), \varphi_j(y_j)) \leq L d_j(x_j, y_j) \qquad (x_j, y_j \in \Omega_j)$$

for some $L < \infty$ and all $j = 1, 2, \ldots$. Use Lemma 3.2.9 to show that $(\varphi_j \natural \sigma_j)$ satisfy the concentration of measure phenomenon with constants α_j / L^2.

3.3 Transportation

We now introduce metrics on the probability measures on a Polish space, motivated by the following question.

Problem (*Monge*). Consider a pile of sand with unit volume and shape represented by a probability measure μ, and a hole in the ground of unit volume and shape represented by a probability measure ν. Suppose that the cost of moving a grain of sand from x to y is proportional to $|x - y|^s$. The problem is to devise a strategy for filling the hole with the sand, at the lowest possible cost.

In one dimension, a possible strategy is to take a monotonically increasing function φ that induces ν from μ and take a grain of sand from x to $\varphi(x)$. Depending upon the value of s, this may or may not be a good strategy. See [162].

The reader will be aware that the costs of transportation involve both the cost of loading the vehicle and the cost of moving the vehicle. We can account for this by defining a cost function c on a Polish space (Ω, d) to be any function such that

(i) $c : \Omega \times \Omega \to \mathbf{R}_+ \cup \{\infty\}$ is lower semicontinuous.

Now suppose that $c_1 : \Omega \times \Omega \to \mathbf{R} \cup \{\infty\}$ is any cost function such that

(ii) $c_1(x, x) = 0$ for all $x \in \Omega$;
(iii) and that $L : \Omega \to \mathbf{R} \cup \{\infty\}$ is lower semicontinuous, and represents the loading and unloading cost at x.

Then $c(x, y) = c_1(x, y) + L(x) + L(y)$ is a cost function, which takes account of both the loading and the unloading cost. Suppose that $\pi \in Prob(\Omega \times \Omega)$ has marginals $\pi_1 = \mu$ and $\pi_2 = \nu$ and represents a strategy for moving distribution μ to ν; then the cost is

$$\iint_{\Omega \times \Omega} c(x, y) \pi(dxdy) = \iint_{\Omega \times \Omega} c_1(x, y) \pi(dxdy)$$
$$+ \int_\Omega L(x) \mu(dx) + \int_\Omega L(y) \nu(dy).$$

Evidently the loading and unloading costs depend only upon L, μ and ν, and not upon the strategy which we adopt for transportation; so we simply accept these costs, and concentrate on minimizing $\iint_{\Omega \times \Omega} c_1(x,y)\pi$ $(dxdy)$. A possible choice of cost function is $c_1(x,d) = d(x,y)^s$ for $s > 0$.

Definition (*Transportation cost*). Let (Ω, d) be a Polish space, let $Prob_s(\Omega)$ be the set of all $\mu \in Prob(\Omega)$ such that $\int_\Omega d(x_0, x)^s \mu(dx)$ is finite for some, or equivalently all, $x_0 \in \Omega$. Given $\mu, \nu \in Prob_s(\Omega)$, the cost of transporting μ to ν with respect to the cost function $d(x,y)^s$ is

$$W_s(\mu, \nu)^s = \inf_\pi \left\{ \iint_{\Omega \times \Omega} d(x,y)^s \pi(dxdy) : \pi_1 = \mu, \pi_2 = \nu \right\} \quad (3.3.1)$$

where $\pi \in Prob_s(\Omega \times \Omega)$ has marginals μ and ν. When $1 \leq s < \infty$, W_s gives the Wasserstein metric on $Prob_s(\Omega)$. The names of Tanaka, Kantorovich and Monge are also associated with these metrics.

Example. Let $\Omega = \{1, \ldots, n+1\}$, with the metric $|m-k|$, and introduce $\mu = n^{-1} \sum_{j=1}^n \delta_j$ and $\nu = n^{-1} \sum_{j=2}^{n+1} \delta_j$.

(i) If $[a_{jk}]$ is an $n \times n$ doubly stochastic matrix such that $a_{jk} \geq 0$, $\sum_{\ell=1}^n a_{j\ell} = 1$ and $\sum_{\ell=1}^n a_{\ell k} = 1$ for $j, k = 1, \ldots, n$, then $[a_{jk}]$ takes μ to μ. In particular, we can consider permutation matrices.

(ii) Let G be the group of permutations on Ω, and consider the subgroups

$$\{g \in G : g\sharp\mu = \mu\} = \{g \in G : g(n+1) = n+1\},$$
$$\{h \in G : h\sharp\nu = \nu\} = \{h \in G : h(1) = 1\}.$$

One way of transporting μ to ν is by the cyclic permutation $\sigma = (1, 2, \ldots, n+1)$; then others are $h \circ \sigma \circ g$ where $h(1) = 1$ and $g(n+1) = n+1$, and a notable example is the transposition $\tau = (1, n+1)$. Indeed, for the cost function $|m - k|$, the strategies σ and τ both have cost 1; whereas for $|m - k|^2$, σ costs 1, while τ costs n.

Proposition 3.3.1 (*Upper bound on Wasserstein metric*). *Let* $\mu \in Prob_s(\mathbf{R})$ *have cumulative distribution function* F, *and* $\nu \in Prob_s(\Omega)$ *have cumulative distribution function* G, *where* F *and* G *are continuous and strictly increasing. Then for* $1 \leq s < \infty$, *the inverse functions satisfy*

$$W_s(\mu, \nu)^s \leq \int_0^1 |F^{-1}(x) - G^{-1}(x)|^s \, dx. \quad (3.3.2)$$

(*In fact equality holds here.*)

Proof. Let φ be a strictly increasing and continuous function such that $F(x) = G(\varphi(x))$, namely $\varphi = G^{-1} \circ F$; then $\varphi : \mathbf{R} \to \mathbf{R}$ is bijective with inverse $F^{-1} \circ G$. Further, φ induces ν from μ since

$$\mu(a,b) = F(b) - F(a) = G(\varphi(b)) - G(\varphi(a)) = \nu(\varphi(a), \varphi(b)). \quad (3.3.3)$$

The map $x \mapsto (x, \varphi(x))$ induces a measure π on $\mathbf{R}^2$ which has marginals $\pi_1 = \mu$ and $\pi_2 = \nu$, hence

$$W_s(\mu, \nu)^s \leq \iint |x - y|^s \pi(dxdy)$$

$$= \int |\varphi(x) - x|^s \mu(dx). \quad (3.3.4)$$

Now F^{-1} induces μ from the uniform distribution on $(0,1)$, so

$$W_s(\mu, \nu)^s \leq \int_0^1 |\varphi(F^{-1}(t)) - F^{-1}(t)|^s \, dt$$

$$= \int_0^1 |G^{-1}(t) - F^{-1}(t)|^s \, dt. \quad (3.3.5)$$

The transportation strategy takes points to points monotonically. One can show that this gives the optimal transportation strategy for W_s when $s \geq 1$; whereas for $s < 1$, this strategy is not optimal. Later we shall use a deep result of Brenier and McCann which shows that for probability density functions on $\mathbf{R}^n$ monotone transport gives the optimal strategy for W_s when $1 \leq s < \infty$. $\quad \square$

Theorem 3.3.2 (*Kantorovich*). *Let* (Ω, d) *be a Polish space. Then for all* $\mu, \nu \in Prob_s(\Omega)$

$$W_s(\mu, \nu)^s = \sup \Big\{ \int_\Omega f(x)\mu(dx) - \int_\Omega g(y)\nu(dy) : f, g \in C_b(\Omega),$$

$$f(x) - g(y) \leq d(x,y)^s \Big\} \quad (3.3.6)$$

Proof. Villani [162] provides a detailed discussion of this result. $\quad \square$

There are some special cases worthy of note.

Proposition 3.3.3 (*Formulae for the Wasserstein metric.*)

(i) Let (Ω, d) *be a Polish space. Then for* $\mu, \nu \in Prob_1(\Omega)$,

$$W_1(\mu, \nu) = \sup \Big\{ \int_\Omega f(x)(\mu(dx) - \nu(dx)) : f \in C_b(\Omega); \|f\|_{Lip} \leq 1 \Big\}.$$

$$(3.3.7)$$

(ii) Let $\mu \in Prob_1(\mathbf{R})$ have cumulative distribution function F, and let $\nu \in Prob_1(\mathbf{R})$ have cumulative distribution function G. Then

$$W_1(\mu, \nu) = \int_{-\infty}^{\infty} |F(x) - G(x)| \, dx. \qquad (3.3.8)$$

(iii) Let $d(x,y)^2 = (x-y)^2/2$ for $x, y \in \mathbf{R}$; let $F(x) = x^2/2 - f(x)$ and $G(y) = y^2/2 + g(y)$. Then $f(x) - g(y) \le d(x,y)^2$ if and only if $F(x) + G(y) \ge xy$, as in the definition of the Legendre transform.

Proof. (i) See [162].

(ii) We have

$$\int_{-\infty}^{\infty} |F(x) - G(x)| dx$$

$$= \sup_{\varphi} \left\{ \int_{-\infty}^{\infty} \varphi'(x)(F(x) - G(x)) \, dx : \varphi \in C_b(\mathbf{R}); \|\varphi\|_{Lip} \le 1 \right\}$$

$$(3.3.9)$$

while

$$W_1(\mu, d\nu) = \int \varphi(x)(\mu(dx) - \nu(dx)). \qquad (3.3.10)$$

(iii) This is clear.   □

Proposition 3.3.4 (*Characterization of weak convergence [88]*). *Let $\rho_n, \rho \in Prob([a, b])$, and let $F_n(t) = \rho_n([a, t])$ and $F(t) = \rho([a, t])$ be their cumulative distribution functions. Then the following are equivalent.*

(i) $\rho_n \to \rho$ weakly as $n \to \infty$.
(ii) $F_n(t) \to F(t)$ as $n \to \infty$ at all points of continuity of F.
(iii) $W_1(\rho_n, \rho) \to 0$ as $n \to \infty$.

Proof. (i) ⇒ (ii) Let $\delta > 0$. We choose continuous functions $\varphi, \psi : \mathbf{R} \to [0, 1]$ such that

$$\mathbf{I}_{(-\infty, t-\delta]}(x) \le \varphi(x) \le \mathbf{I}_{(-\infty, t]}(x) \le \psi(x) \le \mathbf{I}_{(-\infty, t+\delta]}(x), \qquad (3.3.11)$$

and by integrating against ρ_n, we deduce that

$$F_n(t - \delta) \le \int_{-\infty}^{\infty} \varphi(x)\rho_n(dx) \le F_n(t) \le \int_{-\infty}^{\infty} \psi(x)\rho_n(dx) \le F_n(t + \delta).$$

$$(3.3.12)$$

Similarly, we deduce that

$$F(t - \delta) \leq \int_{-\infty}^{\infty} \varphi(x)\rho_n(dx) \leq F(t) \leq \int_{-\infty}^{\infty} \psi(x)\rho(dx) \leq F(t + \delta).$$
(3.3.13)

Letting $n \to \infty$, we have $\int \varphi d\rho_n \to \int \varphi d\rho$ and $\int \psi d\rho_n \to \int \psi d\rho$, hence

$$F(t - \delta) \leq \lim_{n \to} \inf F_n(t) \leq \lim_{n \to \infty} \sup F_n(t) \leq F(t + \delta);$$
(3.3.14)

so at points of continuity we have

$$F(t) = \lim_{n \to \infty} \inf F_n(t) = \lim_{n \to \infty} \sup F_n(t).$$
(3.3.15)

$(ii) \Rightarrow (iii)$ Since the cumulative distributions functions are increasing, there exists a countable set E such that F_n and F are continuous on $[a, b] \setminus E$, hence $F_n(x) \to F(x)$ as $n \to \infty$ on $[a, b] \setminus E$. Then by the bounded convergence theorem,

$$W_1(\rho_n, \rho) = \int_a^b |F_n(x) - F(x)| \, dx \to 0$$
(3.3.16)

as $n \to \infty$.

$(iii) \Rightarrow (i)$ Suppose that $\rho_n \to \rho$ in W_1 metric. Let $f \in C[a, b]$, and given $\varepsilon > 0$ let p be a polynomial such that $\|f - p\|_\infty < \varepsilon$ on $[a, b]$. Now let $L = \|p\|_{Lip}$ and observe that

$$\left| \int_a^b f d\rho_n - \int_a^b f d\rho \right| \leq \left| \int_a^b p(d\rho_n - d\rho) \right| + \int_a^b |f - p|(d\rho_n + d\rho)$$

$$\leq LW_1(\rho_n, \rho) + 2\varepsilon$$
(3.3.17)

by Proposition 3.3.3(i). Hence $\rho_n \to \rho$ weakly as $n \to \infty$. $\qquad \square$

3.4 Transportation inequalities

- Transportation inequalities are dual to concentration inequalities.
- Transportation inequalities bound transportation cost between probability measures by their relative entropy.

Definition (*Transportation inequality*). We say that $\nu \in Prob\,(\Omega)$ satisfies a transportation inequality for cost $d(x, y)^s$ if there exists $\alpha = \alpha(\nu, s)$ such that

$$(T_s(\alpha)) \qquad W_s(\mu, \nu) \leq \left(\frac{2}{\alpha} \mathrm{Ent}(\mu \mid \nu) \right)^{1/2}$$
(3.4.1)

for all $\mu \in Prob\,(\Omega)$.

Whereas the left-hand side is symmetrical in μ and ν, the validity of this inequality represents a special property of ν as we shall see below.

Examples 3.4.1 Talagrand showed that the standard Gaussian γ_1 satisfies $T_2(1)$, and used the following result to generate further examples; see [162].

Proposition 3.4.2 *Suppose that ν satisfies $T_2(\alpha)$. Then the product measure $\nu^{\otimes n}$ satisfies $T_s(\alpha)$ for all $s \in [1,2]$ and $n \geq 1$.*

This follows from repeated applications of the next Lemma.

Lemma 3.4.3 *(i) Suppose that ν satisfies $T_s(\alpha)$. Then ν also satisfies $T_r(\alpha)$ for $1 \leq r \leq s$.*

(ii) Suppose that ν_j on (Ω_j, d_j) satisfies $T_s(\alpha_j)$ for $j = 1, 2$. Then $\nu_1 \otimes \nu_2$ satisfies $T_s(\alpha)$ on (Ω, d) where $\Omega = \Omega_1 \times \Omega_2$, $\alpha = 2^{s/2-1} \min\{\alpha_1, \alpha_2\}$ and

$$d((x_1, x_2), (y_1, y_2))^s = d_1(x_1, y_1)^s + d_2(x_2, y_2)^s. \tag{3.4.2}$$

Proof. (i) By Hölder's inequality applied to (3.3.1), we have

$$W_r(\mu, \nu) \leq W_s(\mu, \nu). \tag{3.4.3}$$

(ii) If we are so fortunate as to have $\mu = \mu_1 \otimes \mu_2$, then

$$W_s(\mu, \nu)^s = W_s(\mu_1, \nu_1)^s + W_s(\mu_2, \nu_2)^s \tag{3.4.4}$$

and we can use Lemma 5.3.1 to conclude the proof. In general we introduce the marginal $\pi_1 \sharp \mu$ on Ω_1 and disintegrate μ as an integral

$$\iint f(x, y)\, \mu(dxdy) = \iint f(x, y)\mu_2(dy \mid x)\mu_1(dx) \tag{3.4.5}$$

where $\mu_2(dy \mid x) \in Prob(\Omega_2)$ is the conditional measure. The theory of measures on Polish spaces ensures the existence of this decomposition. Then we take $f = \log d\mu/d(\nu_1 \otimes \nu_2)$ and write

$$\log f(x, y) = \log \frac{d\mu_1(x)}{d\nu_1(x)} + \log \frac{d\mu_2(y \mid x)}{d\nu_2(y)} \tag{3.4.6}$$

hence

$$\mathrm{Ent}(\mu \mid \nu_1 \otimes \nu_2) = \mathrm{Ent}(\mu_1 \mid \nu_1) + \int \mathrm{Ent}(\mu_2(\cdot \mid x) \mid \nu_2)\mu_1(dx). \tag{3.4.7}$$

There is a matching inequality for the transportation costs

$$W_s(\mu, \nu_1 \otimes \nu_2)^s \leq W_s(\mu_1, \nu_1)^s + \int W_s(\mu_2(\cdot \mid x), \nu_2)^s \mu_1(dx) \tag{3.4.8}$$

and hence by the transportation inequality

$$W_s(\mu, \nu_1 \otimes \nu_2)^s \leq \left(\frac{2}{\alpha_1}\right)^{s/2} \text{Ent}(\mu_1 \mid \nu_1)^{s/2}$$

$$+ \left(\frac{2}{\alpha_2}\right)^{s/2} \left(\int \text{Ent}(\mu_2(\cdot \mid x) \mid \nu_2)^{s/2} \mu_1(dx)\right)$$

$$\leq \left(\frac{2}{\alpha_1}\right)^{s/2} \text{Ent}(\mu_1 \mid \nu_1)^{s/2}$$

$$+ \left(\frac{2}{\alpha_2}\right)^{s/2} \left(\int \text{Ent}(\mu_2(\cdot \mid x) \mid \nu_2) \mu_1(dx)\right)^{s/2} \quad (3.4.9)$$

by Hölder's inequality. By applying the elementary inequality $a^{s/2} + b^{s/2} \leq 2^{(2-s)/2}(a+b)^{s/2}$, one can obtain the stated result. $\square$

Theorem 3.4.4 (*Bobkov–Götze [25]*). *Let (Ω, d) be a Polish space and let $\nu \in \text{Prob}(\Omega)$. Then ν satisfies $T_1(\alpha)$ if and only if ν satisfies the concentration inequality $C(\alpha)$.*

Proof. Suppose that ν satisfies $C(\alpha)$ and that $f : \Omega \to \mathbf{R}$ is 1-Lipschitz with mean $\int f(x)\nu(dx) = 0$. Then

$$\int e^{tf(x) - t^2/(2\alpha)} \, \nu(dx) \leq 1 \qquad (t \in \mathbf{R}) \qquad (3.4.10)$$

and hence by Proposition 3.1.4 we have

$$\text{Ent}(\mu \mid \nu) \geq \int (tf(x) - t^2/(2\alpha))\mu(dx) \qquad (t \in \mathbf{R}) \qquad (3.4.11)$$

so

$$\int f d\mu \leq \frac{t}{2\alpha} + \frac{1}{t}\text{Ent}(\mu \mid \nu) \qquad (t > 0). \qquad (3.4.12)$$

Optimizing this inequality, we obtain

$$\int f \, d\mu \leq \left(\frac{2}{\alpha}\text{Ent}(\mu \mid \nu)\right)^{1/2} \qquad (3.4.13)$$

and hence by Kantorovich–Rubinstein theorem Proposition 3.3.3(i)

$$W_1(\mu, \nu) \leq \left(\frac{2}{\alpha}\text{Ent}(\mu \mid \nu)\right)^{1/2}; \qquad (3.4.14)$$

so ν satisfies $T_1(\alpha)$. This argument also works in reverse, and we obtain the stated result. $\square$

The following refinement of Theorem 3.4.4 shows that all probability measures that have Gaussian decay at infinity satisfy a T_1 inequality; in particular, all probability measures on compact metric spaces satisfy a T_1 inequality.

Theorem 3.4.5 (*Djellout, Guillin, Wu, [61]*). *Let* (Ω, d) *be a Polish space and let* $\nu \in Prob\,(\Omega)$. *Then* ν *satisfies* $T_1(\alpha)$ *for some* $\alpha > 0$ *if and only if there exists* $\varepsilon > 0$ *such that*

$$\iint_{\Omega^2} \exp\big(\varepsilon d(x,y)^2\big)\nu(dx)\nu(dy) < \infty. \qquad (3.4.15)$$

Proof. See [61].   □

In Section 3.5 we provide examples of measures on $\mathbf{R}^n$ that satisfy the concentration of measure inequality.

3.5 Transportation inequalities for uniformly convex potentials

- The Prékopa–Leindler inequality implies transportation and concentration inequalities for Gibbs measures with uniformly convex potentials.
- The standard Gaussian measure on $\mathbf{R}^n$ satisfies a concentration inequality with constants independent of dimension.

In this section we produce examples of Gibbs probability measures that satisfy transportation and concentration inequalities. The method originates in [27].

Definition (*Uniform convexity*). Let Ω be a convex subset of $\mathbf{R}^n$ and let $V : \Omega \to \mathbf{R}$ be continuous. Then V is uniformly convex if there exists $\alpha > 0$ such that

$$tV(x) + (1-t)V(y) - V(tx + (1-t)y) \geq \frac{\alpha}{2} t(1-t) \|x - y\|^2 \qquad (3.5.1)$$

holds for all $x, y \in \Omega$ and $0 < t < 1$.

Proposition 3.5.1 *Suppose that* $V : \Omega \to \mathbf{R}$ *is twice continuously differentiable with Hessian matrix*

$$\mathrm{Hess}V = \Big[\frac{\partial^2 V}{\partial x_j \partial x_k}\Big]. \qquad (3.5.2)$$

Then V *is uniformly convex with constant* $\alpha > 0$ *if and only if*

$$\mathrm{Hess}\,V \geq \alpha I \qquad (3.5.3)$$

as operators on ℓ^2.

Proof. By Taylor's theorem, we have

$$V(x+z) = V(x) + \langle \nabla V(x), z \rangle + \frac{1}{2}\langle \text{Hess}V z, z \rangle, \qquad (3.5.4)$$

where the Hessian is computed at some point between x and $x+z$. Hence $\varphi(t) = V(tx + (1-t)y)$ is convex with $\varphi''(t) \geq \alpha$ if and only if

$$\langle \text{Hess}V\,(x-y), (x-y) \rangle \geq \alpha \|x-y\|^2. \qquad (3.5.5)$$

$\square$

Theorem 3.5.2 *Let* $V : \mathbf{R}^n \to \mathbf{R}$ *be uniformly convex with constant* $\alpha > 0$. *Then there exists* $Z < \infty$ *such that*

$$\nu(dx) = Z^{-1}\exp(-V(x))dx \qquad (3.5.6)$$

defines a probability measure. Further, ν *satisfies* $T_2(\alpha)$ *for the* ℓ^2 *metric, so*

$$W_2(\rho, \nu)^2 \leq \frac{2}{\alpha}\text{Ent}(\rho \mid \nu). \qquad (3.5.7)$$

Proof. By (3.5.5), $\liminf V(x)/\alpha\|x\|^2 > 0$ as $\|x\| \to \infty$, so V can be normalized. We write $s + t = 1$, where $0 < s, t < 1$ and introduce the expression

$$L_s(x,y) = \frac{1}{st}\big(tV(x) + sV(y) - V(tx + sy)\big). \qquad (3.5.8)$$

By uniform convexity, there exists ε_s such that $\varepsilon_s \to 0$ as $s \to 0+$ and

$$L_s(x,y) \geq \frac{\alpha}{2t}(1 + \varepsilon_s)\|x-y\|^2.$$

Now suppose that $F, G : \mathbf{R}^n \to \mathbf{R}$ are continuous and bounded functions such that

$$F(y) - G(x) \leq \frac{\alpha}{2}\|x-y\|^2. \qquad (3.5.9)$$

Then by the Prékopa–Leindler inequality

$$1 = \int e^{-V(x)}dx/Z \geq \left(\int e^{-sG-V}dx/Z\right)^t \left(\int e^{tF-V}dx/Z\right)^s \qquad (3.5.10)$$

Taking power $1/s$ and letting $s \to 0+$, we deduce that

$$1 \geq \left(\exp\left(-\int G\,d\nu\right)\right)\int e^F d\nu, \qquad (3.5.11)$$

so $h(x) = F(x) - \int G d\nu$ has $\int e^h d\nu \leq 1$. Hence by Proposition 3.1.4(iii) we have

$$\int F(x)\rho(dx) - \int G(y)\nu(dy) \leq \text{Ent}(\rho \mid \nu).$$

Finally, by the Kantorovich duality theorem we have

$$\frac{\alpha}{2}W_2(\rho,\nu)^2 \leq \mathrm{Ent}(\rho \mid \nu). \qquad (3.5.12)$$

$\square$

Corollary 3.5.3 *Let ν be as in Theorem 3.5.2. Then ν satisfies the concentration inequality $C(\alpha)$.*

Proof. By Proposition 3.4.2 and Theorem 3.5.2, ν satisfies $T_1(\alpha)$. Hence ν satisfies the concentration inequality $C(\alpha)$ by the Bobkov–Götze theorem. $\square$

Corollary 3.5.4 *(Talagrand). Let γ_n be the standard Gaussian on $\mathbf{R}^n$. Then γ_n satisfies the transportation inequality $T_2(1)$ and the concentration incquality $C(1)$.*

Proof. Here $V(x) = \|x\|^2/2$ has Hessian equal to the identity, so Theorem 3.5.2 and Corollary 3.5.3 apply with $\alpha = 1$. $\square$

Our first application to random matrix theory covers a variety of models, including the Gaussian orthogonal ensemble.

Proposition 3.5.5 *(Concentration for GOE). Suppose that $X \in M_n^s(\mathbf{R})$ has entries X_{jk}, and that X is random subject to probability measure μ such that*

(i) X_{jk} are mutually independent for $1 \leq j \leq k \leq n$;
(ii) the distribution μ_{jk} of X_{jk} satisfies $T_2(\alpha)$ for some $\alpha > 0$ and all $1 \leq j \leq k \leq n$.

Let $F : (M_n^s(\mathbf{R}), c^2(n)) \to \mathbf{R}$ be an L-Lipschitz function such that $\int F(X)\mu(dX) = 0$. Then

$$\int \exp\bigl(tF(X)\bigr)\mu(dX) \leq \exp(t^2 L^2/(n\alpha)) \qquad (t \in \mathbf{R}).$$

Proof. The product measure $\omega = \otimes_{1 \leq j \leq k \leq n}\mu_{jk}$ on $\mathbf{R}^{n(n-1)/2}$ satisfies $T_2(\alpha)$ and hence $T_1(\alpha)$ by Proposition 3.4.2. Let $X = \sum_{j=1}^{n} X_{jj}e_{jj} + \sum_{1 \leq j < k \leq n} X_{jk}\,(e_{jk} + e_{kj})$. Then the map $(\mathbf{R}^{n(n-1)/2}, \ell^2) \to \mathbf{R}$ given by $(X_{jk}) \mapsto F(X)$ is $L\sqrt{(2/n)}$-Lipschitz; so by Theorem 3.4.4, the concentration inequality holds. $\square$

In particular, Proposition 3.5.5 applies with $\alpha = 2$ when the X_{jj} have a $N(0,1)$ distribution and the X_{jk} have a $N(0,1/2)$ distribution for $j < k$. In Section 2.2, there are examples of Lipschitz functions on $(M_n^s(\mathbf{R}), c^2(n))$.

Exercise 3.5.6 Suppose that X is as in Proposition 3.5.5 and let $F(X) = \operatorname{trace}_n f(X)$ where $f : \mathbf{R} \to \mathbf{R}$ is 1-Lipschitz. Obtain a concentration inequality for F that improves with increasing dimension.

3.6 Concentration of measure in matrix ensembles

We now have the analytical tools that we require to prove the main concentration theorems concerning concentration of measure for matrix ensembles. The cases of the orthogonal, hermitian and symplectic ensembles are all treated likewise. For the orthogonal ensembles, there are four main levels:

- generalized orthogonal ensemble $\nu_n(dX)$ on $M_n^s(\mathbf{R})$;
- joint eigenvalue distribution $\sigma_n(d\lambda)$ on Δ^n;
- empirical eigenvalue distribution $\mu_n = \frac{1}{n}\sum_{j=1}^n \delta_{\lambda_j}$ on $\mathbf{R}$;
- integrated density of states ρ_n on $\mathbf{R}$ (which we introduce in the next chapter).

We begin at the level of eigenvalue distributions. The space $(\Delta^n, \ell^2(n))$ is noncompact, and the scaling is chosen so that $\|(1,\ldots,1)\|_{\ell^2(n)} = 1$. We scale up the potential v to nv so as to balance the n summands from the λ_j with the number of pairs $\lambda_j < \lambda_k$ of distinct eigenvalues which contribute to the Vandermonde determinant.

Theorem 3.6.1 *Suppose that $v : \mathbf{R} \to \mathbf{R}$ is twice continuously differentiable with $v''(x) \geq \alpha$ for all x and some $\alpha > 0$; let $\beta > 0$. Then there exists $Z(n, \beta) < \infty$ such that*

$$\sigma_n^{(\beta)}(d\lambda) = Z(n,\beta)^{-1} \exp\left(-n\sum_{j=1}^n v(\lambda_j)\right) \prod_{1 \leq j < k \leq n}(\lambda_k - \lambda_j)^\beta d\lambda_1 \ldots d\lambda_n$$

defines a probability measure on $(\Delta^n, \ell^2(n))$ and $\sigma_n^{(\beta)}$ satisfies $T_2(\alpha n^2)$ and $C(\alpha n^2)$.

Proof. We can apply Theorem 3.5.2 to the potential

$$V(\lambda) = n\sum_{j=1}^n v(\lambda_j) + \beta \sum_{1 \leq j < k \leq n} \log\frac{1}{\lambda_k - \lambda_j} \tag{3.6.1}$$

on $(\Delta^n, \ell^2(n))$, which is uniformly convex with constant αn^2. To check that the potential is convex, we introduce $\lambda = (\lambda_1, \ldots, \lambda_n)$ and

$\mu = (\mu_1, \ldots, \mu_n)$ in Δ^n, and join them by the straight line segment $\sigma(t) = (1-t)\mu + t\lambda$. Then for indices $j < k$, the components satisfy

$$\sigma_k(t) - \sigma_j(t) = (1-t)(\mu_k - \mu_j) + t(\lambda_k - \lambda_j) > 0 \quad (0 < t < 1) \quad (3.6.2)$$

so $t \mapsto \log 1/(\sigma_k(t) - \sigma_j(t))$ is convex. □

Hence the sequence of ensembles $(\sigma_n^{(\beta)})$ on $(\Delta^n, \ell^2(n))$ exhibits the concentration of measure phenomenon since $\alpha n^2 \to \infty$ as $n \to \infty$. We recall the special cases:

- $\beta = 1$: the eigenvalues from the generalized orthogonal ensemble;
- $\beta = 2$: the eigenvalues from the generalized unitary ensemble;
- $\beta = 4$: the eigenvalues from the generalized symplectic ensemble.

We now obtain a concentration inequality for the empirical eigenvalue distribution.

Corollary 3.6.2 *Let $f : \mathbf{R} \to \mathbf{R}$ be an L-Lipschitz function, and let $F(\lambda) = \frac{1}{n}\sum_{j=1}^{n} f(\lambda_j)$. Then*

$$\sigma_n^{(\beta)}\left\{\lambda \in \Delta^n : \left|F(\lambda) - \int F(\xi)\sigma_n^{(\beta)}(d\xi)\right| > \varepsilon\right\} \le 2\exp\left(-\frac{\alpha n^2 \varepsilon^2}{2L^2}\right) \quad (\varepsilon > 0).$$
$$(3.6.3)$$

Proof. The function $F : (\Delta^n, \ell^2(n)) \to \mathbf{R}$ is L-Lipschitz since

$$|F(\lambda) - F(\xi)| \le \frac{1}{n}\sum_{j=1}^{n}|f(\lambda_j) - f(\xi_j)|$$

$$\le \frac{L}{n}\sum_{j=1}^{n}|\lambda_j - \xi_j|$$

$$\le L\left(\frac{1}{n}\sum_{j=1}^{n}|\lambda_j - \xi_j|^2\right)^{1/2}. \quad (3.6.4)$$

Now we can use Lemma 3.2.3 and Theorem 3.4.4 to obtain the stated result. □

Moving up to the level of random matrices themselves, we have a counterpart of the previous result. Recall that the metrics are scaled so that $\|I\|_{c^2(n)} = 1$.

Theorem 3.6.3 *Let $v : \mathbf{R} \to \mathbf{R}$ be twice continuously differentiable, suppose that $v''(x) \ge \alpha$ for all $x \in \mathbf{R}$ where $\alpha > 0$ and let*

$V(X) = \operatorname{trace} v(X)/n$. Then there exists $Z_n < \infty$ such that

$$\nu_n(dX) = Z_n^{-1} \exp\left(-n^2 V(X)\right) dX \qquad (3.6.5)$$

defines a probability measure on $(M_n^s(\mathbf{R}), c^2(n))$ and ν_n satisfies $T_2(\alpha n^2)$ and $C(\alpha n^2)$.

Proof. We shall use Proposition 3.4.2 to verify that $n^2 V$ is uniformly convex with constant αn^2, and then Theorem 3.6.2 to establish the concentration inequality.

Let (ξ_j) be an orthonormal basis of eigenvectors of X that correspond to eigenvalues $\lambda_1, \ldots, \lambda_n$; the eigenvalues are distinct, except on a set of Lebesgue measure zero. Then

$$\frac{d}{dt} V(X + tY) = \frac{1}{n} \sum_{j=1}^{n} v'(\lambda_j)\langle Y\xi_j, \xi_j \rangle \qquad (3.6.6)$$

by Proposition 2.2.4. Now the Rayleigh–Schrödinger formula Theorem 2.2.5 gives the variation of λ_j and ξ_j with respect to X, and in particular leads to the formula

$$\left(\frac{d^2}{dt^2}\right)_{t=0} V(X + tY) = \frac{1}{n} \sum_{j=1}^{n} v''(\lambda_j)\langle Y\xi_j, \xi_j \rangle^2$$

$$+ \frac{2}{n} \sum_{1 \le j < k \le n} \frac{v'(\lambda_j) - v'(\lambda_k)}{\lambda_j - \lambda_k} \langle Y\xi_j, \xi_k \rangle^2. \qquad (3.6.7)$$

By the mean value theorem $(v'(\lambda_j) - v'(\lambda_k))/(\lambda_j - \lambda_k) \ge \alpha$ and hence

$$\left(\frac{d^2}{dt^2}\right)_{t=0} V(X + tY) \ge \frac{1}{n} \sum_{j,k=1}^{n} \langle Y\xi_j, \xi_k \rangle^2 = \alpha \|Y\|_{c^2(n)}^2. \qquad (3.6.8)$$

The result now follows from Theorem 3.5.2. $\qquad\square$

Remark. Theorem 3.6.3 is formally stronger than Theorem 3.6.1, since in the cases $\beta = 1, 2, 4$ we can use the eigenvalue map Λ to obtain concentration inequalities for $\sigma_n^{(\beta)}$ from the corresponding results for ν_n on $M_n^s(\mathbf{R})$, $M_n^h(\mathbf{C})$ and $M_n^h(\mathbf{H})$. For other $\beta > 0$, the 'eigenvalue distribution' $\sigma_n^{(\beta)}$ does not emerge naturally from a matrix ensemble, so this approach is not available.

A fundamental property of Gaussian measure γ is that $\int \exp(\varepsilon\|x\|^2) \gamma(dx)$ is finite for some $\varepsilon > 0$. The following result generalizes this fact to ensembles with uniformly convex potentials.

Theorem 3.6.4 *Let v be a continuously differentiable and even function such that $v''(x) \geq \alpha$ for all real x and some $\alpha > 0$. Then for $0 < \varepsilon < \alpha/8$ there exists $M < \infty$, independent of n, such that*

$$\int_{M_n^s(\mathbf{R})} \exp\bigl(\varepsilon \|X\|_{op}^2\bigr) \nu_n(dX) \leq M. \tag{3.6.9}$$

Proof. We have a simple bound $\|X\|_{op} \leq \sqrt{n} \|X\|_{c^2(n)}$ which we need to refine. The unit sphere $S^n(1) = \{\xi \in \mathbf{R}^n : \|\xi\| = 1\}$ is compact and hence has a finite 2δ net for each $\delta > 0$. We can select a maximal sequence $\{\xi_j : j = 1, \ldots, N\}$ in $S^n(1)$ such that $\|\xi_j - \xi_k\| \geq 2\delta$ for all distinct pairs $j, k \in \{1, \ldots, N\}$. Now the sets $B(\xi_j, \delta) = \{\eta \in \mathbf{R}^n : \|\eta - \xi_j\| < \delta\}$ are mutually disjoint, and all contained in $B(0, 1 + \delta)$; hence

$$N \leq \frac{\mathrm{vol}(B(0, 1 + \delta))}{\mathrm{vol}(B(0, \delta))} = \left(\frac{1 + \delta}{\delta}\right)^n. \tag{3.6.10}$$

Now we choose ξ to be a unit real eigenvector of X, corresponding either to λ_1 or λ_n, so that $\|X\|_{op} = |\langle X\xi, \xi \rangle|$; then we choose j such that $\|\xi_j - \xi\| \leq 2\delta$. The last step is possible, since we could otherwise enlarge the maximal collection $\{\xi_k\}$. By simple estimates we have

$$\|X\|_{op} \leq |\langle X\xi_j, \xi_j \rangle| + |\langle X\xi_j, \xi_j \rangle - \langle X\xi, \xi \rangle|$$
$$\leq 2\|\xi_j - \xi\| \|X\|_{op} + |\langle X\xi_j, \xi_j \rangle|, \tag{3.6.11}$$

so

$$\|X\|_{op} \leq \frac{|\langle X\xi_j, \xi_j \rangle|}{1 - 4\delta}.$$

The function $f_j : (M_n^s(\mathbf{R}), c^2(n)) \to \mathbf{R}$ given by $f_j(X) = \langle X\xi_j, \xi_j \rangle$ satisfies

$$\int f_j(X) \nu_n(dX) = 0 \tag{3.6.12}$$

by symmetry, and has $\|f_j\|_{Lip} = \sqrt{n}$; hence by Theorem 3.6.3 we have

$$\nu_n[|f_j(X)| > s] \leq 2 \exp\bigl(-\alpha n s^2/2\bigr) \qquad (s > 0).$$

Taking $\delta = 1/8$ in (3.6.10), we deduce that $N \leq 9^n$ and so

$$\nu_n[\|X\|_{op} > 2s] \leq \sum_{j=1}^{N} \nu_n[|f_j(X)| > s]$$
$$\leq 2 \cdot 9^n \exp\bigl(-\alpha n s^2/2\bigr). \tag{3.6.13}$$

Now we choose K so large that $\alpha K^2 > 8 \log 9$, and split the integral

$$\int_{M_n^s(\mathbf{R})} \exp\!\big(\varepsilon \|X\|_{op}^2\big) \nu_n(dX) = \int_0^\infty 2\varepsilon s e^{\varepsilon s^2} \nu_n[\|X\|_{op} > s]\, ds$$

$$\leq \int_0^K 2\varepsilon s e^{\varepsilon s^2}\, ds + 4.9^n \varepsilon \int_K^\infty s e^{(\varepsilon - n\alpha/8)s^2}\, ds$$

$$= e^{\varepsilon K^2} - 1 + \frac{16\varepsilon 9^n}{n\alpha - 8\varepsilon} e^{-(n\alpha - 8\varepsilon)K^2/8}.$$

$$(3.6.14)$$

We observe that the right-hand side is bounded as $n \to \infty$, hence we can select M independently of n. $\qquad\square$

For a large random matrix, there is the remote possibility that some eigenvalue is very large. The following result shows that even this occurs with small probability. We introduce the space $\prod_{n=1}^\infty M_n^s(\mathbf{R})$ of sequences $(X_n)_{n=1}^\infty$ of real symmetric matrices of increasing size, endowed with the probability measure $\otimes_{n=1}^\infty \nu_n$. The corresponding sequence of empirical eigenvalue distributions in $(\mu_n)_{n=1}^\infty$.

Corollary 3.6.5 *Let v be as in Theorem 3.6.4. Then for all $\delta > 0$ and $A > 0$, the event*

$$[\mu_n \quad \text{has support in} \quad [-n^\delta A, n^\delta A] \quad \text{for all but finitely many } n]$$
$$(3.6.15)$$

has probability one with respect to $\otimes_{n=1}^\infty \nu_n$.

Proof. Clearly the support of μ_n is contained in $[-\|X_n\|_{op}, \|X_n\|_{op}]$, so we use Theorem 3.6.4 to bound the probability that μ_n has support outside $[-n^\delta A, n^\delta A]$, thus obtaining

$$\exp\!\big(\varepsilon n^{2\delta} A^2\big) \nu_n[\|X_n\|_{op} > n^\delta A] \leq \int \exp\!\big(\varepsilon \|X_n\|_{op}^2\big) \nu_n(dX_n) \leq M.$$
$$(3.6.16)$$

Hence

$$\sum_{n=1}^\infty \nu_n[\|X_n\|_{op} > n^\delta A] \leq \sum_{n=1}^\infty M \exp\!\big(-\varepsilon n^{2\delta} A^2\big) < \infty; \quad (3.6.17)$$

so by the first Borel–Cantelli Lemma (Theorem 1.1.7), the event

$$[\|X_n\|_{op} > n^\delta A \quad \text{for infinitely many distinct } n]$$
$$(3.6.18)$$

has probability zero. Taking complements, we obtain the stated result. $\qquad\square$

3.7 Concentration for rectangular Gaussian matrices

- Wishart considered rectangular matrices with IID Gaussian entries.
- The joint distribution of the singular numbers is known.
- Concentration inequalities hold for the joint distribution.
- The distribution of the smallest singular number is hard to deal with.

In previous sections we have considered normal matrices, which are unitarily similar to diagonal matrices. In this section we consider probability measures on the space $M_{m \times n}(\mathbf{R})$ of real $m \times n$ matrices, for which the singular numbers are important. In particular, we obtain some results for rectangular Gaussian matrices which are consequences of the preceding theorems on symmetric Gaussian matrices. Rectangular Gaussian matrices were introduced by Wishart and others in the context of multivariate statistics, and have been studied by many authors.

We begin with an elementary case from [104]. Suppose that Y is an $m \times m$ real matrix with entries that are mutually independent $N(0,1)$ random variables Y_{jk}; then certainly we have $\|Y\| \geq \max_j |Y_{jj}|$. The asymptotic distribution of the right-hand side is given as follows.

Definition (*Gumbel distribution*). The Gumbel distribution is the probability distribution on $\mathbf{R}$ that has cumulative distribution

$$G(x) = \exp(-e^{-x}) \qquad (x \in \mathbf{R}).$$

Proposition 3.7.1 *Let Z_j $(j = 1, \ldots, m)$ be mutually independent $N(0,1)$ random variables, let $M_m = \max\{Z_1, \ldots, Z_m\}$ be their maximum and introduce the scaling constants $a_m = (2 \log m)^{1/2}$ and*

$$b_m = (2 \log m)^{1/2} - 2^{-1}(2 \log m)^{-1/2}(\log \log m + \log 4\pi). \qquad (3.7.1)$$

Then

$$\mathbf{P}[a_m(M_m - b_m) \leq x] \to G(x) \qquad (m \to \infty).$$

Proof. We observe that the tail of the Gaussian distribution satisfies

$$\Psi(s) = \int_s^\infty \frac{e^{-s^2/2} ds}{\sqrt{2\pi}}$$
$$= \frac{e^{-s^2/2}}{s\sqrt{2\pi}}\left(1 + O(s^{-2})\right) \qquad (s \to \infty) \qquad (3.7.2)$$

due to elementary and familiar estimates. Likewise, one can easily show that

$$\mathbf{P}\left[M_m \leq b_m + \frac{x}{a_m}\right] = \mathbf{P}\left[Z_j \leq b_m + \frac{x}{a_m}; j = 1, \ldots, m\right]$$

$$= \left(1 - \Psi\left(b_m + \frac{x}{a_m}\right)\right)^m \qquad (3.7.3)$$

where

$$b_m + \frac{x}{a_m} = (2\log m)^{1/2} + \frac{2x - \log\log m - \log 4\pi}{2(2\log m)^{1/2}}$$

has $b_m + x/a_m \to \infty$ as $m \to \infty$ and

$$\left(b_m + \frac{x}{a_m}\right)^2 = 2\log m + 2x - \log\log m - \log 4\pi$$

$$+ \frac{(2x - \log\log m - \log 4\pi)^2}{8\log m}, \qquad (3.7.4)$$

hence

$$\frac{\exp\left(-\frac{1}{2}\left(b_m + \frac{x}{a_m}\right)^2\right)}{\left(b_m + \frac{x}{a_m}\right)\sqrt{2\pi}} = \frac{e^{-x+o(1)}}{m} \qquad (m \to \infty),$$

so

$$\left(1 - \Psi\left(b_m + \frac{x}{a_m}\right)\right)^m \to \exp(-e^{-x}) \qquad (m \to \infty). \qquad (3.7.5)$$

$\square$

Let Y be a $m \times n$ real matrix with mutually independent standard Gaussian $N(0,1)$ random variables as entries, and let $S = Y^t Y$ have nonnegative eigenvalues $s_1 \leq s_2 \leq \cdots \leq s_n$. If $m < n$, then rank $S \leq m$, so at least $n-m$ of the s_j are zero. We suppose henceforth that $m = \lfloor \lambda n \rfloor$ where $\lambda > 1$, so $m \geq n$. Let

$$\Delta_+^n = \{s = (s_1, \ldots, s_n) \in \mathbf{R}^n : 0 \leq s_1 \leq \cdots \leq s_n\},$$

be endowed with the $\ell^2(n)$ norm.

Proposition 3.7.2 *(i)* $\mathbf{E}\sum_{j=1}^n s_j = mn$.

(ii) There exists $\varepsilon_0 > 0$ such that for $0 < \varepsilon < \varepsilon_0$ there exists M_ε, independent of n and λ, such that

$$\mathbf{E}\exp\left(\frac{\varepsilon s_n}{(\lambda+1)n}\right) \leq M_\varepsilon.$$

(iii) For $\delta > 0$, the largest eigenvalue of S satisfies

$$\mathbf{P}[s_n > n^{1+\delta}] \leq M_\varepsilon \exp\left(\frac{-\varepsilon n^\delta}{1+\lambda}\right). \qquad (3.7.6)$$

Proof. (i) We have

$$\sum_{j=1}^{n} s_j = \operatorname{trace} S = \operatorname{trace} Y^t Y = \sum_{j=1}^{m}\sum_{k=1}^{n}[Y_{jk}]^2 \qquad (3.7.7)$$

where the entries $[Y_{jk}]$ have a $N(0,1)$ distribution; hence the result.

(ii) Let $X \in M^s_{m+n}(\mathbf{R})$ be a matrix from the Gaussian orthogonal ensemble as in Section 1.5. Then we obtain $Y/\sqrt{2(m+n)}$ as the top right corner of X, so

$$\frac{s_n}{2(m+n)} = \frac{1}{2(m+n)}\|Y^t Y\|_{M_n(\mathbf{R})}$$

$$= \frac{1}{2(m+n)}\left\| \begin{bmatrix} 0 & Y \\ Y^t & 0 \end{bmatrix} \right\|^2_{M_{m+n}(\mathbf{R})}$$

$$\leq \|X\|^2_{M^s_{m+n}(\mathbf{R})}. \qquad (3.7.8)$$

The result now follows directly from Theorem 3.6.4.

(iii) This follows from (ii) by Chebyshev's inequality. $\qquad\square$

Olkin [123] provides a survey of the distributions of random matrices, using algebraic arguments similar to the following proof.

Lemma 3.7.3 (*Hua*). *Let $f(X^t X)$ be a probability density function with respect to Lebesgue product measure dX on the entries of $X \in M_{m\times n}(\mathbf{R})$ for $m \geq n$; let $S = X^t X$. Then there exists $C(m,n) > 0$ such that S has distribution*

$$C(m,n)f(S)\det(S)^{(m-n-1)/2}\mathbf{I}_{[S\geq 0]}\,dS \qquad (3.7.9)$$

where dS is the product of Lebesgue measure on entries on or above the leading diagonal of $S \in M^s_n(\mathbf{R})$.

Proof. The joint distribution of the elements of S on and above the leading diagonal is $f(S)h(S)\,dS$, where h is a function to be determined.

For each row ξ_j of X and nonsingular $A \in M_{n\times n}(\mathbf{R})$, the linear transformation $\xi_j = \eta_j A$ of the row η_j has Jacobian $|\det A|$; hence the transformation $X = YA$ has Jacobian $|\det A|^m$. With $V = Y^t Y$, we find that the distribution of the entries of V is $f(A^t V A)h(V)|\det A|^m\,dV$.

Now the scaling transformation $S = a^2 V$ where $a > 0$ gives $dS = a^{n(n+1)}dV$ since there are $n(n+1)/2$ entries of S on and above the leading diagonal. More generally, the linear transformation $S = A^t V A$, with $A \in M_{n\times n}(\mathbf{R})$ nonsingular, gives the joint distribution

$$f(S)h(S)dS = f(A^t V A)h(A^t V A)|\det A|^{n+1}dV. \qquad (3.7.10)$$

Equating these expressions, we find that

$$h(V)|\det A|^m = h(A^t V A)|\det A|^{n+1}; \qquad (3.7.11)$$

so that with $V = I$ and $B = A^t A$ we have $h(B) = C(m,n)$ $(\det B)^{(m-n-1)/2}$ where $C(m,n) = h(I)$. $\qquad \square$

Definition (*Chi-squared distribution*). A random variable ξ has a $\chi^2(k)$ distribution when it has probability density function

$$p_k(x) = \frac{x^{(k/2)-1}e^{-x/2}}{2^{k/2}\Gamma(k/2)}, \qquad x > 0;$$

$$0, \qquad x \le 0. \qquad (3.7.12)$$

A standard result states that if ζ_j $(j = 1, \ldots, k)$ are mutually independent $N(0,1)$ random variables, then $\sum_{j=1}^{k} \zeta_j^2$ has a $\chi^2(k)$ distribution.

Exercise. (i) Show that $p_k(x)$ is largest at $x = k - 2$ when $k > 2$.
(ii) Use integration by parts to show that

$$\int_\alpha^\infty p_k(x)\,dx \le \frac{\alpha^{k/2}e^{-\alpha/2}}{(\alpha - k + 2)2^{(k/2)-1}\Gamma(k/2)} \qquad (\alpha > k - 2 > 0).$$
$$(3.7.13)$$

(iii) Suppose that $\xi = X^s$ where $X \sim \chi^2(k)$. Show that

$$\mathbf{E}\,\xi = 2^s \frac{\Gamma((k+s)/2)}{\Gamma(k/2)} \asymp k^s$$

as $k \to \infty$.

The following gives a noncommutative χ^2 distribution.

Proposition 3.7.4 (*Wishart*). *Let S be as in Proposition 3.7.2. Then the joint eigenvalue distribution of S is*

$$\sigma_n(ds) = Z(n,\lambda)^{-1} \exp\left(-\frac{1}{2}\sum_{j=1}^n s_j\right)\left(\prod_{j=1}^n s_j\right)^{(m-n-1)/2}$$

$$\times \prod_{1\le j<k\le n} (s_k - s_j)\mathbf{I}_{\Delta_+^n}(s)ds_1 \ldots ds_n \qquad (3.7.14)$$

where the normalizing constant is

$$Z(n,\lambda) = \left(\frac{2^m}{\pi}\right)^{n/2} \prod_{j=1}^n \Gamma\left(\frac{m+1-j}{2}\right)\Gamma\left(\frac{n+1-j}{2}\right). \qquad (3.7.15)$$

Proof. In Lemma 3.7.3 we let $S = X^t X$ and

$$f(X^t X) = (2\pi)^{-mn/2} \exp\left(-2^{-1}\text{trace}(X^t X)\right) \qquad (3.7.16)$$

so that the entries of $X \in M_{m \times n}(\mathbf{R})$ are mutually independent Gaussian $N(0,1)$ random variables. The ordered eigenvalues of S satisfy $\mathrm{trace}(X^t X) = \sum_{j=1}^n s_j$ and $\det S = \prod_{j=1}^n s_j$; furthermore, the Jacobian of the map $S \mapsto (s_j)$ contributes a factor $\prod_{1 \leq j < k \leq n}(s_k - s_j)$. These give the variable terms in the probability density function. We omit the lengthy calculations that are needed to compute the numerical constants.   □

We change variables to $s_j = n x_j$ and introduce the potential function

$$V(x) = \frac{n}{2} \sum_{j=1}^n \left(x_j - (\lambda - 1)\log x_j \right) + \frac{\beta}{2} \sum_{\substack{j \neq k; j,k=1}}^n \log \frac{1}{|x_j - x_k|}. \qquad (3.7.17)$$

In view of Proposition 3.7.2(ii) the scaled eigenvalues x_j are concentrated on a bounded interval of the real line, so there is no great loss in truncating the distribution to a bounded interval $[0, K]$ where we can use previous theory to obtain a more precise concentration inequality.

Theorem 3.7.5 (*Eigenvalue concentration for the Wishart distribution [170]*). *For $0 < K, \beta < \infty$ and $n = 1, 2, \dots$ there exists $Z > 0$ such that*

$$\nu(dx) = Z^{-1} \exp\!\left(-V(x)\right) \mathbf{I}_{\Delta_+^n \cap [0,K]^n}(x)\,dx_1 \dots dx_n \qquad (3.7.18)$$

defines a probability measure. Furthermore, ν satisfies the transportation inequality on $(\Delta_+^n \cap [0, K]^n, \ell^2(n))$

$$W_2(\rho, \nu) \leq \frac{4K^2}{(\lambda - 1)n^2} \mathrm{Ent}(\rho, \nu). \qquad (3.7.19)$$

Proof. The function $v(x) = (x - (\lambda - 1)\log x)/2$ satisfies

$$v''(x) = \frac{\lambda - 1}{2x^2} \geq \frac{\lambda - 1}{2K^2} \qquad (x \in (0, K]). \qquad (3.7.20)$$

Hence V is uniformly convex on $(\Delta_+^n \cap [0, K]^n, \ell^2(n))$ with constant $n^2(\lambda - 1)/(2K^2)$, so we can apply Theorem 3.5.2.   □

Proposition 3.7.6 (*Polar decomposition*). *Let Y be a $n \times n$ Gaussian random matrix with mutually independent $N(0,1)$ entries. Then with probability one, Y has a polar decomposition $Y = U S^{1/2}$ where S is a $n \times n$ real symmetric matrix with eigenvalues jointly distributed according to σ_n, and U is a real $n \times n$ orthogonal matrix that is distributed according to Haar measure on $O(n)$.*

Proof. We choose $S = Y^t Y$, which is invertible with probability one, and the eigenvalue distribution of S is specified in Proposition 3.7.4. Now we observe that $U = Y(Y^tY)^{-1/2}$ satisfies $UU^t = I$; so there exists a Borel set $K \subseteq M_n(\mathbf{R})$ of full Gaussian measure such that $\varphi : K \rightarrow O(n)$: $\varphi(Y) = Y(Y^tY)^{-1/2}$ is well defined. For $V \in O(n)$, the matrices VY and Y have equal distributions, and hence the maps $Y \mapsto \varphi(Y)$ and $Y \mapsto \varphi(VY)$ induce the same probability measure on $O(n)$ from the Gaussian measure on K; but $\varphi(VY) = V\varphi(Y)$, so the induced probability measure is invariant under translation and hence must be Haar measure on $O(n)$. □

Remarks. In Proposition 3.7.6, we do not assert that U and S are independent. In Exercise 3.9.7, we give an alternative approach to polar decomposition.

The following result was used by Silverstein [140] to analyze the smallest singular values of large Wishart matrices. General concentration theorems are not ideally suited for dealing with smallest eigenvalues and singular numbers, so his proof involves a reduction process that makes use of the special invariance properties of Gaussians.

Lemma 3.7.7 (*Silverstein*). *Let Y be a $n \times n$ Gaussian random matrix with mutually independent $N(0,1)$ entries. Then there exist random matrices $U, V \in O(n)$ and nonnegative random variables ξ_j and η_k such that:*

(i) the ξ_j and η_k are mutually independent;
(ii) $\xi_j^2 \sim \chi^2(j)$ for $j = 1, \ldots, n$, and $\eta_j^2 \sim \chi^2(j)$ for $j = 1, \ldots, n-1$;
(iii) UYV is lower triangular, with the ξ_j on the leading diagonal and η_j on the diagonal below, with all other entries zero; so

$$UYV = \begin{bmatrix} \xi_n & 0 & 0 & 0 & \cdots & 0 \\ \eta_{n-1} & \xi_{n-1} & 0 & 0 & \cdots & 0 \\ 0 & \eta_{n-2} & \xi_{n-2} & 0 & \cdots & 0 \\ \vdots & & & & & \vdots \\ 0 & 0 & \cdots & \eta_2 & \xi_2 & 0 \\ 0 & 0 & \cdots & 0 & \eta_1 & \xi_1 \end{bmatrix}. \tag{3.7.21}$$

Proof. We select U and V by the following iterative process. We let R_1 be the first row of Y, and let $\xi_n = \|R_1\|_{\ell^2}$; so $\xi_n^2 \sim \chi^2(n)$. Now we let V_1 be an orthogonal matrix with first column $R_1^t/\|R_1\|_{\ell^2}$ and the other columns chosen by the Gram–Schmidt process; so that V_1 depends statistically on the row R_1, but is independent of the other rows in Y.

Then

$$YV_1 = \begin{bmatrix} \xi_n & 0 \\ C_1 & Z_1 \end{bmatrix} \tag{3.7.22}$$

where C_1 is a $(n-1) \times 1$ random matrix with mutually independent $N(0,1)$ entries, Z_1 is a $(n-1) \times (n-1)$ random matrix with mutually independent $N(0,1)$ entries, and C_1, Z_1 and ξ_n are mutually independent. Next we let $\eta_{n-1} = \|C_1\|_{\ell^2}$ so that $\eta_{n-1}^2 \sim \chi^2(n-1)$ and form

$$U_1 = \begin{bmatrix} 1 & 0 \\ 0 & \hat{U}_1 \end{bmatrix}$$

where $\hat{U}_1 \in O(n-1)$ is chosen with first row $C_1^t / \|C_1\|_{\ell^2}$ and the remaining rows given by the Gram Schmidt process. Then we have

$$U_1 Y V_1 = \begin{bmatrix} \xi_n & 0 \\ \eta_{n-1} & \\ 0 & Y_1 \end{bmatrix} \tag{3.7.23}$$

where Y_1 is a $(n-1) \times (n-1)$ random matrix with $N(0,1)$ entries, such that the entries of Y_1, ξ_n and η_{n-1} are mutually independent.

We repeat the construction for Y_1 by choosing $V_2 \in O(n-1)$ and $\hat{U}_2 \in O(n-2)$ such that

$$\begin{bmatrix} 1 & 0 \\ 0 & \hat{U}_2 \end{bmatrix} Y_1 V_2 = \begin{bmatrix} \xi_{n-1} & 0 \\ \eta_{n-2} & \\ 0 & Y_2 \end{bmatrix}$$

where Y_2 is a $(n-2) \times (n-2)$ random matrix with $N(0,1)$ entries, such that the entries of Y_2, ξ_{n-1} and η_{n-2} are mutually independent.

By repeating this process, we obtain smaller and smaller matrices Y_j and unitaries of the form

$$U_j = \begin{bmatrix} I & 0 \\ 0 & \hat{U}_j \end{bmatrix} \tag{3.7.24}$$

such that $U_{n-1} \ldots U_1 Y V_1 \ldots V_n$ has the required form. □

Exercise. Let Y be a $n \times n$ Gaussian random matrix with mutually independent $N(0,1)$ entries.

(i) Use Proposition 3.7.6 to show that for $s > -1/2$,

$$\mathbf{E}|\det Y|^{2s} = 2^{ns} \prod_{j=1}^{n} \frac{\Gamma((j+s)/2)}{\Gamma(j/2)}.$$

(ii) By differentiating this formula at $s = 0$, deduce that

$$\mathbf{E} \log |\det Y| = \frac{n}{2} \log 2 + \frac{1}{2} \sum_{j=1}^{n} \frac{\Gamma'(j/2)}{\Gamma(j/2)}.$$

Remark. Szarek [151] observed that the ξ_j and η_k in (3.7.21) are close to their mean values, so the singular numbers of Y are with high probability close to the singular numbers of the constant lower triangular matrix

$$A = \begin{bmatrix} n^{1/2} & 0 & 0 & 0 & 0 \\ (n-1)^{1/2} & (n-1)^{1/2} & 0 & 0 & 0 \\ 0 & (n-2)^{1/2} & (n-2)^{1/2} & 0 & 0 \\ \vdots & \vdots & & & \vdots \\ 0 & 0 & 2^{1/2} & 2^{1/2} & 0 \\ 0 & 0 & 0 & 1 & 1 \end{bmatrix}$$

so

$$A = [\operatorname{diag}(n - j + 1)^{1/2}](I + R) \qquad (3.7.25)$$

where R is the operator

$$R = \begin{bmatrix} 0 & 0 & 0 & \cdots & 0 & 0 \\ 1 & 0 & 0 & \cdots & 0 & 0 \\ 0 & 1 & 0 & \cdots & 0 & 0 \\ \vdots & & & & & \vdots \\ 0 & 0 & 0 & \cdots & 1 & 0 \end{bmatrix}. \qquad (3.7.26)$$

Proposition 3.7.8 *The singular numbers of A satisfy*

$$n^{-1}(k+1)^{1/2} \leq s_{n-k}(A) \leq (k+2)\pi/\sqrt{n} \qquad (k = 0, \ldots, n-1).$$

Proof. We express the previous equality as $A = D(I + R)$, and observe that $\|(I + R)\| \leq 2$ and

$$\|(I + R)^{-1}\| = \|I - R + R^2 - \cdots + (-1)^{n-1}R^{n-1}\| \leq n. \qquad (3.7.27)$$

Hence by Proposition 2.1.3, we have $n^{-1}s_{n-k}(D) \leq s_{n-k}(A) \leq 2s_{n-k}(D)$ where $s_{n-k}(D) = (k+1)^{1/2}$. For small k, we can improve the right-hand side of this inequality.

We let B be the matrix that arises by replacing 0 by $n^{1/2}$ in the top right corner of A, and let T be the permutation matrix that arises by replacing 0 by 1 in the top right corner of R; then $B = D(I + T)$.

Since $B - A$ has rank one, the theory of singular numbers gives $s_n(A) \leq s_{n-1}(B)$. Further, T is a unitary matrix with eigenvalues given by the roots of unity of order n; so $I - T$ is unitarily equivalent to the operator $[\mathrm{diag}(1 + e^{2\pi i k/n})]$. Now

$$|1 + e^{2\pi i k/n}| = 2|\cos(\pi k/n)| \qquad (k = 0, \ldots, n-1)$$

gives the sequence of singular values of $I + T$ in some order, so by considering cases of odd and even n, one can determine the singular numbers and show that $s_{n-k}(I + T) \leq \pi(k+1)/n$ for $k = 0, \ldots, n-1$. Hence

$$s_{n-k}(A) \leq s_{n-1-k}(B) \leq \|D\| s_{n-1-k}(I + T) \leq (k+2)\pi/\sqrt{n}.$$

$\square$

From these and similar estimates, Szarek obtains detailed results on the distribution of the upper bounds on singular values of Y. Lower bounds on the singular values are more difficult to achieve, and are important in applications to random numerical algorithms. At the time of writing, the following result of Rudelson seems to be the best known; the proof involves subtle arguments from the geometry of Banach spaces relating to almost Euclidean sections of convex bodies. Note that the lower bound on s_n of n^{-1} in Proposition 3.7.8 deteriorates to $n^{-3/2}$ once we return from the constant matrix A to the random matrix Y. For geometrical applications of these results, the reader can consult [53].

Proposition 3.7.9 (*Rudelson [136]*). *Let Y be a $n \times n$ Gaussian random matrix with mutually independent $N(0,1)$ entries. Then there exist absolute positive constants c_j such that for all $k = 1, \ldots, n$ and $c_1 n^{-3/2} \leq t \leq c_2 k/n$, the singular numbers satisfy*

$$s_{n-k}(Y) \geq \frac{c_3 t k}{\sqrt{n}} \tag{3.7.28}$$

with probability greater than $1 - c_4 \exp(-c_5 n) - (c_6 tn/k)^k$.
Furthermore, for $c_1 n^{-1/2} \leq \varepsilon \leq c_2$,

$$s_n(Y) \geq c_7 \varepsilon n^{-3/2}$$

with probability greater than $1 - c_4 \exp(-c_5 n) - \varepsilon$.

Proof. See [136]. $\square$

3.8 Concentration on the sphere

- The Gaussian concentration inequality gives the concentration of measure phenomenon for the spheres $S^n(1)$.
- Riemannian manifolds with uniformly positive Ricci curvature satisfy a concentration inequality.

In Section 1.4 we obtained the Prékopa–Leindler inequality for Euclidean space. Cordero–Erausquin proved an analogue of this result for the rotation invariant measure on the spheres in Euclidean space by considering the notion of optimal transport on the sphere. Given a suitable Prékopa–Leindler inequality, one can follow the same path as in Section 3.5 and deduce concentration inequalities. However, in this section we take a short-cut: the concentration inequality for Gaussian measure implies the concentration inequality for the Euclidean spheres.

The orthogonal group $O(n)$ acts naturally on the sphere $S^{n-1}(1)$ by $x \mapsto Ux$; hence there is a single orbit $S^{n-1}(1) = \{Ue : U \in O(n)\}$ for any $e \in S^{n-1}(1)$ and the stability group $\{U \in O(n) : Ue = e\}$ is $O(n-1)$. Thus the map $U \mapsto Ue$ induces surface measure $\sigma_{n-1,1}$ from Haar probability measure $\mu_{O(n)}$ on $O(n)$. Let $(e_k)_{k=1}^n$ be the standard orthonormal basis for the ℓ^2 norm on $\mathbf{R}^n$.

Proposition 3.8.1 *Under the natural actions of the classical groups on the spheres, given by $U \mapsto Ue$, there are the following homeomorphisms:*

$$SO(n)/SO(n-1) \cong S^{n-1}, \quad SU(n)/SU(n-1) \cong S^{2n-1},$$
$$Sp(n)/Sp(n-1) \cong S^{4n-1}. \tag{3.8.1}$$

Proof. See [36]. □

We continue with some comments on the metric structure. On $S^{n-1}(1)$, there are several natural metrics that are invariant under this action; in particular:

- $\|x - y\|_{\ell^2}$, the chordal distance between $x, y \in S^{n-1}(1)$;
- $\rho(x, y)$, the geodesic distance, namely the distance along the shorter arc of the great circle that passes through x and y.

By simple geometry we have

$$\|x - y\|_{\ell^2} \leq \rho(x, y) \leq \frac{\pi}{2}\|x - y\|_{\ell^2}, \tag{3.8.2}$$

with $\|Ux - Uy\|_{\ell^2} = \|x - y\|_{\ell^2}$, $\rho(Ux, Uy) = \rho(x, y)$ and $\rho(x, y)/\|x - y\|_{\ell^2} \to 1$ as $x \to y$.

Definition (*Metrics*). The standard metric on $S^{n-1}(1)$ is the chordal metric for the ℓ^2 norm on $\mathbf{R}^n$.

Theorem 3.8.2 (*Lévy*). *Let $f : S^{n-1}(1) \to \mathbf{R}$ be a Lipschitz function such that $\|f\|_{Lip} \leq L$. Then there exists a universal constant $c > 0$ such that*

$$\sigma_{n-1,1}\Big\{x \in S^{n-1}(1) : \Big|f(x) - \int f d\sigma_{n-1,1}\Big| > t\Big\} \leq 4e^{-ct^2 n/L^2} \qquad (t \in \mathbf{R}).$$
$$(3.8.3)$$

Proof. Following [117], we deduce this result from the concentration of measure for the standard Gaussian γ_n on $\mathbf{R}^n$. Let $\xi = (\xi_1, \ldots, \xi_n)$ be mutually independent $N(0,1)$ random variables; then ξ has the same distribution as $U\xi$ for any $U \in O(n)$, by the basic invariance properties of Gaussian measure as in Lemma 1.5.1. Hence we pick $e_0 \in S^{n-1}(1)$ and integrate with respect to U and obtain for any continuous $h : S^{n-1}(1) \to \mathbf{R}$:

$$\int_{\mathbf{R}^n} h(\xi/\|\xi\|)\gamma_n(d\xi) = \int_{O(n)} \int_{\mathbf{R}^n} h(U\xi/\|\xi\|)\gamma_n(d\xi)\mu_{O(n)}(dU)$$

$$= \int_{O(n)} h(Ue_0)\mu_{O(n)}(dU); \qquad (3.8.4)$$

hence we obtain the identity

$$\int_{\mathbf{R}^n} h(\xi/\|\xi\|)\gamma_n(d\xi) = \int_{S^{n-1}(1)} h(e)\sigma_{n-1,1}(de). \qquad (3.8.5)$$

Suppose without loss that $\int f d\sigma_{n-1,1} = 0$; then the f vanishes at some point on $S^{n-1}(1)$ and hence f is bounded with upper bound $L \operatorname{diam}(S^{n-1}(1))$, so $|f| \leq 2L$. Next we introduce the function

$$g(x) = \|x\| f\Big(\frac{x}{\|x\|}\Big), \qquad (3.8.6)$$

which is Lipschitz with constant $4L$ since

$$g(x) - g(y) = \|x\|\Big(f\Big(\frac{x}{\|x\|}\Big) - f\Big(\frac{y}{\|y\|}\Big)\Big) + (\|x\| - \|y\|)f\Big(\frac{y}{\|y\|}\Big) \quad (3.8.7)$$

and hence

$$|g(x) - g(y)| \leq L\|x\| \Big\|\frac{x}{\|x\|} - \frac{y}{\|y\|}\Big\| + 2L\|x - y\|$$

$$\leq L\|x\| \Big\|\frac{x}{\|x\|} - \frac{y}{\|x\|} + y\Big(\frac{1}{\|x\|} - \frac{1}{\|y\|}\Big)\Big\| + 2L\|x - y\|$$

$$\leq 4L\|x - y\|. \qquad (3.8.8)$$

Further, one can check by calculation in polar coordinates that

$$\int \|x\| \gamma_n(dx) = \frac{\int_0^\infty r^n e^{-r^2/2} dr}{\int_0^\infty r^{n-1} e^{-r^2/2} dr}$$

$$= \frac{\sqrt{2} \Gamma((n+1)/2)}{\Gamma(n/2)}$$

$$\asymp \sqrt{n} \quad (n \to \infty). \tag{3.8.9}$$

Now we have, for $t > 0$ and $0 < \delta < 1$, the inclusion of events

$$\left[|f(x/\|x\|)| > t\right] \subseteq \left[|g(x)| > t(1-\delta)\sqrt{n}\right] \cup \left[\|x\| < (1-\delta)\sqrt{n}\right], \tag{3.8.10}$$

and hence by Corollary 3.5.4,

$$\sigma_{n-1,1}[|f| > t] \le \gamma_n[|g| > t(1-\delta)\sqrt{n}] + \gamma_n[\|x\| < (1-\delta)\sqrt{n}]$$
$$\le 2\exp(-nt^2(1-\delta)^2/(32L^2/\alpha)) + \exp(-\alpha n \delta^2/2). \tag{3.8.11}$$

By choosing $\delta = t/(4L + t)$ we obtain the inequality

$$\sigma_{n-1,1}[|f| > t] \le 3\exp(-nt^2\alpha/(2(t+4L)^2)) \quad (t > 0);$$

where only the values $0 < t < 2L$ are of interest due the upper bound on $|f|$. $\qquad\square$

The sphere $S^{n-1}(R)$ has constant sectional curvature R^{-2}. The preceding result extends to other manifolds with positive curvature.

In Section 2.4, we defined the Ricci curvature tensor for Lie groups. There is a corresponding definition for a Riemannian manifold M, where the Lie algebra g is replaced by the tangent space to M at each point of M. We shall state the isoperimetric inequality in general, and apply it only in the case of Lie groups. The normalized Riemannian measure on a compact Lie group coincides with Haar measure, since Haar measure is the unique probability measure that is invariant under left translation. In the next section we compare the Riemannian metric with the standard metric on the classical compact Lie groups.

Theorem 3.8.3 (*Gromov–Lévy*). *Let M be a smooth compact Riemannian manifold, and let σ be the Riemannian volume form, normalized so as to be a probability measure. Suppose further that there exists $\kappa > 0$ such that the Ricci curvature satisfies $Ric \ge \kappa I$ in the sense that the Ricci tensor $R_{ijk\ell}$ and the Riemannian metric g_{jk} satisfy $\sum_\ell R_{\ell j\ell k} \ge \kappa g_{jk}$ at all points on M. Then any 1-Lipschitz function $f : M \to \mathbf{R}$ such* *that $\int_M f(x)\sigma(dx) = 0$ satisfies*

$$\int_M \exp(tf(x))\sigma(dx) \leq \exp(t^2/\kappa) \qquad (t \in \mathbf{R}). \qquad (3.8.12)$$

Proof. This statement appears in D. Cordero–Erausquin *et al.* [51]. See also [69]. □

Corollary 3.8.4 (*Concentration for Riemannian manifolds*). *Let M_n be smooth compact Riemannian manifolds with normalized Riemannian measures σ_n, and suppose that $Ric(M_n) \geq \kappa_n I$, where $\kappa_n \to \infty$ as $n \to \infty$. Then (σ_n) satisfies the concentration of measure phenomenon.*

The sharp form of the Sobolev inequality for the sphere was obtained by Beckner by a delicate symmetrization argument. Note that as $n \to \infty$, the constant in the inequality improves, whereas the permitted range of values of p contracts.

Theorem 3.8.5 *Suppose that $F \in L^2(S^n)$ has gradient $\nabla F \in L^2(S^n)$. Then*

$$\|F\|^2_{L^p(S^n)} \leq \|F\|^2_{L^2(S^n)} + \frac{p-2}{n}\|\nabla F\|^2_{L^2(S^n)} \qquad (3.8.13)$$

where $2 \leq p < \infty$ for $n = 1$ and $n = 2$, and $2 \leq p \leq 2n/(n-2)$ for $n \geq 3$.

Proof. See [12]. □

Exercise 3.8.6 By differentiating this inequality at $p = 2+$, deduce that

$$\int_{S^n} F^2 \log\left(F^2 / \int F^2 d\sigma\right) d\sigma \leq \frac{2}{n} \int_{S^n} \|\nabla F\|^2 d\sigma.$$

These inequalities lead to concentration inequalities for Lipschitz functions on the spheres, and to related isoperimetric inequalities.

3.9 Concentration for compact Lie groups

- The Riemannian metric is equivalent to the standard metric on compact Lie groups.
- The Gromov–Lévy concentration inequality implies concentration of measure for the families of classical compact Lie groups.
- Their eigenvalue distributions also exhibit concentration of measure.

We introduce the normalized Hilbert–Schmidt metric of $c^2(n)$ on $M_n(\mathbf{C})$ by

$$\|X - Y\|_{c^2(n)} = \left(\frac{1}{n}\text{trace}(X - Y)(X - Y)^*\right)^{1/2}, \qquad (3.9.1)$$

and the matching metric of $\ell^2(n)$ of $\mathbf{C}^n$ by $\|\lambda - \mu\|_{\ell^2(n)} = (\frac{1}{n}\sum_{j=1}^{n}|\lambda_j - \mu_j|^2)^{1/2}$.

Definition (*Standard metrics on the classical groups*). The standard metric on $O(n)$, $U(n)$, $SU(n)$ and $SO(n)$ is the $c^2(n)$ norm; the standard metric on $Sp(n)$ is the $c^2(2n)$ norm.

There are several invariant metrics on $O(n)$, and the following give equivalent formulae for the normalized Hilbert–Schmidt norm:

$$\|U - V\|_{c^2(n)}^2 = \text{trace}_n \int_{O(n)} W^*(U - V)^*(U - V)W \mu_{O(n)}(dW)$$

$$= \int_{S^{n-1}(1)} \|(U - V)e\|_{\ell^2}^2 \sigma_{n-1,1}(de)$$

$$= \frac{1}{n}\sum_{k=1}^{n} \|Ue_k - Ve_k\|_{\ell^2}^2.$$

Lemma 3.9.1 *Let $G = SO(n), U(n), SU(n)$ or $Sp(n)$, and let g be the corresponding Lie algebra. Then for all $U, V \in G$, there exists $A \in$ g such that $UV^{-1} = \exp A$ and*

$$\frac{2}{\pi}\|A\|_{c^2(n)} \leq \|U - V\|_{c^2(n)} \leq \|A\|_{c^2(n)}. \qquad (3.9.2)$$

Hence the Riemannian metric $d(U, V)$ is Lipschitz equivalent to the standard metric, so

$$\frac{4}{\pi\sqrt{n}}d(U, V) \leq \|U - V\|_{c^2(n)} \leq \frac{2}{\sqrt{n}}d(U, V). \qquad (3.9.3)$$

Proof. Suppose that $G = U(n)$. We introduce $W \in G$ and a diagonal matrix D such that $UV^{-1} = WDW^{-1}$ and $D = \text{diag}(e^{i\theta_j})$ where $\theta_j \in (-\pi, \pi]$. Then $UV^{-1} = \exp A$ where $A = WBW^{-1}$ with $B = i\text{diag}(\theta_j)$ and

$$\left(\frac{2}{\pi}\right)^2 \frac{1}{n}\sum_{j=1}^{n}|\theta_j|^2 \leq \frac{1}{n}\sum_{j=1}^{n}|e^{i\theta_j} - 1|^2 \leq \frac{1}{n}\sum_{j=1}^{n}|\theta_j|^2, \qquad (3.9.4)$$

so that

$$\frac{2}{\pi}\|A\|_{c^2(n)} \le \|D - I\|_{c^2(n)} \le \|A\|_{c^2(n)}, \qquad (3.9.5)$$

and the desired inequality follows.

In the case of $G = SO(n)$, we replace D by some element of the maximal torus, and B by an element of the corresponding Lie algebra as in Proposition 2.3.6.

With the standard Riemannian structure on G from Proposition 2.4.1, $d(U, V)$ is the distance along the one-parameter subgroup $\exp tA$ from I to UV^{-1}. Further, the Riemannian metric is given by the invariant inner product

$$\langle A, B \rangle = -\frac{1}{2}\text{trace}(AB) = \frac{1}{2}\text{trace}(AB^*) = \frac{n}{2}\langle A, B \rangle_{c^2(n)} \qquad (A, B \in \mathfrak{g}).$$
$$(3.9.6)$$

$\square$

Theorem 3.9.2 *Let μ_n be Haar probability measure on $U(n)$ and let $F : (U(n), c^2(n)) \to \mathbf{R}$, be an L-Lipschitz function such that $\int_{U(n)} F(U)\mu_n(dU) = 0$. Then there exists an absolute constant $c > 0$ such that*

$$\int_{U(n)} \exp\{tF(U)\}\mu_n(dU) \le \exp\{ct^2L^2/n^2\} \qquad (t \in \mathbf{R}). \qquad (3.9.7)$$

Similar statements hold for $SO(n), SU(n)$ and $Sp(n)$.

Proof. By Lemma 3.9.1, we have

$$|F(u) - F(v)| \le \frac{2L}{\sqrt{n}}d(U, V) \qquad (U, V \in U(n)), \qquad (3.9.8)$$

so F is $(2L/\sqrt{n})$-Lipschitz for the Riemannian metric. The Ricci curvature of $U(n)$ satisfies $\text{Ric} \ge cn$ by Exercise 2.4.2, hence the result by Theorems 3.8.3. The Ricci curvature of $SO(n)$ is $(n-2)I/4$ by Exercise 2.4.2. $\square$

For compact Lie groups G, there are probability measures at four main levels:

- Haar measure on $\mu_G(dg)$ on G;
- the joint eigenvalue distribution $\sigma_G(d\lambda)$ on the maximal torus T^k;
- empirical eigenvalue distribution $\frac{1}{k}\sum_{j=1}^{k}\delta_{\exp(i\theta_j)}$ on $\mathbf{T}$;
- equilibrium measure $d\theta/2\pi$ on $\mathbf{T}$.

Given a unitary matrix U, there is a unique way of listing the eigenvalues according to multiplicity $e^{i\theta_1}, \ldots, e^{i\theta_n}$, so that $0 \le \theta_1 \le \cdots \le \theta_n < 2\pi$; hence we can introduce the eigenvalue map $\Lambda : U(n) \to \mathbf{T}^n : U \mapsto (e^{i\theta_j})_{j=1}^n$. Let μ_n be a probability measure on $U(n)$, and let σ_n be the joint eigenvalue distribution; that is, the measure that is induced on $\mathbf{T}^n$ from μ_n by Λ so

$$\int_{\mathbf{T}^n} f(\Lambda(U))\sigma_n(dU) = \int_{U(n)} f(U)\mu_n(dU). \qquad (3.9.9)$$

When convenient, we can relax the ordering and permute the eigenvalues on $\mathbf{T}$.

Corollary 3.9.3 [18] *Any 1-Lipschitz function* $F : (\mathbf{T}^n, \ell^2(n)) \to \mathbf{R}$ *such that* $\int_{\mathbf{T}^n} F(\Theta)\sigma_n(d\Theta) = 0$ *satisfies*

$$\int_{\mathbf{T}^n} \exp\{tF(e^{i\theta_1}, \ldots, e^{i\theta_n})\}\sigma_n(d\Theta) \le \exp\{ct^2/n^2\} \qquad (t \in \mathbf{R}).$$
$$(3.9.10)$$

Proof. On account of Theorem 3.9.2 and Proposition 2.7.1, this follows directly from the following variant on Lidskii's Lemma. $\qquad\square$

Lemma 3.9.4 *(Lidskii's Lemma for unitaries).* *The map* $\Lambda : (U(n), c^2(n)) \to (\mathbf{T}^n, \ell^2(n))$ *is* K-*Lipschitz for some finite* K *that is independent of* n.

Proof of Lemma 3.9.4. Let u and v be unitary matrices, and let A and B be self-adjoint matrices such that $u = e^{iA}$ and $v = e^{iB}$. We suppose that the eigenvalues of A and B are

$$0 \le \theta_1 \le \theta_2 \le \cdots \le \theta_n \le 2\pi \quad \text{and} \quad 0 \le \phi_1 \le \phi_2 \le \cdots \le \phi_n \le 2\pi$$
$$(3.9.11)$$

respectively. By Lidskii's Lemma [144], we have

$$\left(\frac{1}{n}\sum_{j=1}^n |e^{i\theta_j} - e^{i\phi_j}|^2\right)^{1/2} \le \left(\frac{1}{n}\sum_{j=1}^n |\theta_j - \phi_j|^2\right)^{1/2} \le \|A - B\|_{c^2(n)}.$$
$$(3.9.12)$$

To convert this into a condition involving $u - v$, we employ a functional calculus argument. We form a partition of unity on $\mathbf{T}$ by taking the overlapping quadrants $Q_j = \{e^{i\theta} : \theta \in [j\pi/4, (j+2)\pi/4]\}$ for $j = 1, \ldots, 8$ and introduce twice continuously differentiable functions $h_j : \mathbf{T} \to [0,1]$ such that $h_j(e^{i\theta}) = 0$ for $e^{i\theta}$ outside I_j and $\sum_{j=1}^8 h_j(e^{i\theta}) = 1$. Next we let $g_j(e^{i\theta}) = \theta h_j(e^{i\theta})$ and extend it to give a 2π periodic function with an absolutely convergent Fourier expansion $g_j(\theta) = \sum_{k \in \mathbf{Z}} \hat{g}_j(k)e^{ik\theta}$. Then we define

$$g_j(A) = \sum_{k=-\infty}^{\infty} \hat{g}_j(k)e^{ikA} = \sum_{k=-\infty}^{\infty} \hat{g}_j(k)u^k; \qquad (3.9.13)$$

likewise we define $g_j(B)$. Then we estimate

$$\left\| g_j(A) - g_j(B) \right\|_{c^2(n)} = \left\| \sum_{k=-\infty}^{\infty} \hat{g}_j(k)(u^k - v^k) \right\|_{c^2(n)}$$

$$\leq \left(\sum_{k=-\infty}^{\infty} k^4 |\hat{g}_j(k)|^2 \right)^{1/2}$$

$$\times \left(\sum_{k \neq 0; k=-\infty}^{\infty} k^{-4} \| u^k - v^k \|_{c^2(n)}^2 \right)^{1/2}$$

$$(3.9.14)$$

by the Cauchy–Schwarz inequality. Using the fact that $c^2(n)$ is a matricial norm, one can easily check that $\| u^k - v^k \|_{c^2(n)} \leq |k| \| u - v \|_{c^2(n)}$ for all $k \in \mathbf{Z}$, hence

$$\| g_j(A) - g_j(B) \|_{c^2(n)} \leq \frac{\pi}{\sqrt{3}} \left(\int_0^{2\pi} |g_j''(\theta)|^2 \frac{d\theta}{2\pi} \right)^{1/2} \| u - v \|_{c^2(n)} \quad (3.9.15)$$

since $\pi^2/6 = \sum_{k=1}^{\infty} k^{-2}$.

Since $\theta = \sum_{j=1}^{8} g_j(\theta)$, we have $A = \sum_{j=1}^{8} g_j(A)$ and we can use the preceding estimates to obtain a constant K such that

$$\| A - B \|_{c^2(n)} \leq \sum_{j=1}^{8} \| g_j(A) - g_j(B) \| \leq K \| u - v \|_{c^2(n)}.$$

$\square$

Exercise 3.9.5 (*Alternative proof of concentration of measure on the spheres*). Let $S^{n-1}(1)$ be the unit sphere in $\mathbf{R}^n$ with the ℓ^2 metric, and $SO(n)$ the special orthogonal group with the Riemannian metric d, as Lemma 3.9.1.

(i) Let $e \in \mathbf{R}^n$ be a unit vector, and let $\lambda : SO(n) \to S^{n-1}(1)$ be the map $\lambda(U) = Ue$. Show that λ is 1-Lipschitz, and that λ induces the normalized surface areas measure σ_n on $S^{n-1}(1)$ from Haar probability measure on $SO(n)$.

(ii) Deduce from Theorem 3.8.3 that if $f : (S^{n-1}(1), \ell^2) \to \mathbf{R}$ is 1-Lipschitz, and that $\int f(x)\sigma_n(dx) = 0$, then

$$\int_{S^{n-1}(1)} \exp(tf(x))\sigma_n(dx) \leq \exp(ct^2/n) \qquad (t \in \mathbf{R}),$$

where c is a constant that is independent of t and n.

Problem 3.9.6 Can one deduce Theorem 3.9.2 from Corollary 3.5.4, and thereby avoid the more difficult Theorem 3.8.3?

Exercise 3.9.7 In the polar factorization of matrices, the orthogonal factor can be expressed as the closest point in the orthogonal matrices; see [162, p. 123]. Let X be an invertible $n \times n$ real matrix. Then there exists a unique $R \in M_n^s(\mathbf{R})$ and $U \in O(n)$ such that $R \geq 0$ and $X = RU$. Further, U is uniquely characterized by the condition

$$\|X - U\|_{c^2(n)} \leq \|X - V\|_{c^2(n)} \qquad (V \in O(n)). \qquad (3.9.16)$$

(i) Show that the condition (3.9.16) is equivalent to trace$(XU^t) \geq$ trace(XV^t) for all $V \in O(n)$. By considering $V = \exp(S)U$ with S skew-symmetric, deduce that XU^t is symmetric, and hence that the eigenvalues of XU^t are nonnegative.
(ii) Let $U = P(X)$; show that $P(XW) = P(X)W$ for all $W \in O(n)$.
(iii) Let γ_{jk} $(1 \leq j,k \leq n)$ be mutually independent $N(0,1)$ random variables, and let $X = [\gamma_{jk}/\sqrt{n}]$. Show that $P(X)$ is distributed according to Haar measure on $O(n)$.

Exercise 3.9.8 Let μ_n be Haar measure on $\mathbf{T}^n$. Use Lemma 3.4.3 and the concentration inequality for $\mathbf{T} = SO(2)$ to show that any 1-Lipschitz function $F : (\mathbf{T}^n, \ell^2(n)) \to \mathbf{R}$ such that $\int_{\mathbf{T}^n} F(\Theta)\mu_n(d\Theta) = 0$ satisfies

$$\int_{\mathbf{T}^n} \exp\{tF(e^{i\theta_1}, \ldots, e^{i\theta_n})\}\mu_n(d\Theta) \leq \exp\{ct^2/n\} \qquad (t \in \mathbf{R}).$$

Remark. In Exercise 3.9.8, the concentration inequality does not improve with increasing n since the factor of $1/n$ in the exponent arises solely from the scaling of the metric. Hence the measure σ_n in Corollary 3.9.3 has better concentration constant than does Haar measure μ_n. At the start of Chapter 4, we shall see why this is the case. In Section 10.6, we shall show that the dependence of the concentration constant upon n in Theorem 3.9.2 is optimal.

4
Free entropy and equilibrium

Abstract

In this chapter we are mainly concerned with one-dimensional electrostatic problems; that is, with measures on the circle or the real line that represent charge distributions subject to logarithmic interaction and an external potential field. First we consider configurations of electrical charges on the circle and their equilibrium configuration. Then we review some classical results of function theory and introduce the notion of free entropy for suitable probability densities on the circle; these ideas extend naturally to spheres in Euclidean space. The next step is to introduce free entropy for probability distributions on the real line, and show that an equilibrium distribution exists for a very general class of potentials. For uniformly convex potentials, we present an effective method for computing the equilibrium distribution, and illustrate this by introducing the semicircle law. Then we present explicit formulæ for the equilibrium measures for quartic potentials with positive and negative leading term. Finally we introduce McCann's notion of displacement convexity for energy functionals, and show that uniform convexity of the potential implies a transportation inequality.

4.1 Logarithmic energy and equilibrium measure

Suppose that N unit positive charges of strength $\beta > 0$ are placed upon a circular conductor of unit radius, and that the angles of the charges are $0 \leq \theta_1 < \theta_2 < \cdots < \theta_N < 2\pi$. Then the electrostatic energy is

$$E = \beta \sum_{1 \leq j < k \leq N} \log \frac{1}{|e^{i\theta_j} - e^{i\theta_k}|}. \tag{4.1.1}$$

As $N \to \infty$, one would expect the charges to seek a configuration that minimizes the energy, and to move to a configuration in which they are uniformly distributed around the unit circle. We can calculate the energy exactly when the charges are equally spaced around the unit circle.

Proposition 4.1.1 *Suppose that* $\theta_k = 2\pi(k-1)/N$ *for* $k = 1, \ldots, N$. *Then*

$$E = -\frac{\beta}{2} N \log N. \qquad (4.1.2)$$

Proof. The points $e^{i\theta_j}$ are equally spaced around the unit circle, and $|e^{i\theta_j} - e^{i\theta_k}|$ depends only upon $|j - k|$, so we have

$$-\beta \sum_{1 \le j < k \le N} \log |e^{i\theta_j} - e^{i\theta_k}| = -\frac{\beta N}{2} \sum_{k=2}^{N} \log |e^{i\theta_k} - 1|$$

$$= -\frac{\beta N}{2} \sum_{k=1}^{N-1} \log \left| 2 \sin \frac{\pi k}{N} \right|. \qquad (4.1.3)$$

The Gauss–Legendre multiplication formula [169]

$$\Gamma(z)\Gamma(z + 1/N)\Gamma(z + 2/N) \ldots \Gamma(z + (N-1)/N)$$
$$= (2\pi)^{(N-1)/2} N^{1/2 - Nz} \Gamma(Nz) \qquad (4.1.4)$$

and the familiar identity $\Gamma(z)\Gamma(1 - z) = \pi \operatorname{cosec} \pi z$ combine to give

$$\prod_{k=1}^{N-1} \left(2 \sin \frac{k\pi}{N} \right) = N; \qquad (4.1.5)$$

hence

$$-\beta \sum_{1 \le j < k \le N} \log |e^{i\theta_j} - e^{i\theta_k}| = \frac{-\beta N \log N}{2}. \qquad (4.1.6)$$

$\square$

We note that, as $N \to \infty$,

$$\frac{1}{N} \sum_{k=1}^{N-1} \log \left| \sin \frac{\pi k}{N} \right| \to \int_0^1 \log |\sin \pi \theta| d\theta = -\log 2, \qquad (4.1.7)$$

where the value of the integral is due to Euler [35, p. 245].

When we choose a random element of $SU(N + 1)$, the eigenvalues $(e^{i\theta_j})_{j=1}^{N+1}$ have a distribution specified by Proposition 2.7.1, so that the

θ_j are random, subject to the joint distribution

$$\sigma_{SU(N)}^{(\beta)}(d\Theta) = Z^{-1} \prod_{1 \leq j < k \leq N+1} |e^{i\theta_j} - e^{i\theta_k}|^{\beta} d\theta_1 d\theta_2 \dots d\theta_N \qquad (4.1.8)$$

with $\beta = 2$ and $\sum_{j=1}^{N+1} \theta_j = 0$. Then we would expect that the θ_j would tend to form a uniform distribution round the circle as $N \to \infty$ since the uniform distribution appears to minimize the energy. In Chapter 5, we investigate this further.

The logarithmic term in the electrostatic energy is sometimes called a Coulomb interaction, and some authors refer to the study of E as the statistical mechanics of a Coulomb gas [94].

Definition (*Logarithmic energy*). Let ν be a probability measure on $\mathbf{R}^n$ that has no atoms and is such that

$$\iint_{\mathbf{R}^n \times \mathbf{R}^n} \log \frac{1}{\|x - y\|} \nu(dx)\nu(dy) \qquad (4.1.9)$$

converges absolutely. Then we say that ν has finite logarithmic energy.

The logarithmic energy satisfies, for suitable probability density functions, the formula

$$\iint_{\mathbf{R}^n \times \mathbf{R}^n} \log \frac{1}{\|x - y\|} f(x)f(y)\, dxdy$$
$$= \left(\frac{d}{dp}\right)_{p=0} \iint_{\mathbf{R}^n \times \mathbf{R}^n} \frac{f(x)f(y)}{\|x - y\|^p} dxdy, \qquad (4.1.10)$$

which contrasts with the formula for entropy

$$\int_{\mathbf{R}^n} f(x) \log f(x)\, dx = \left(\frac{d}{dp}\right)_{p=1} \int_{\mathbf{R}^n} f(x)^p\, dx. \qquad (4.1.11)$$

4.2 Energy spaces on the disc

The best place from which to view the circle is the unit disc $\mathbf{D} = \{z \in \mathbf{C} : |z| < 1\}$. In this section, we review some basic results concerning function theory on $\mathbf{D}$ which are involved in random matrix theory, specifically the definition of free entropy. Some of these are particularly relevant to circular ensembles; others provide models of abstract constructions that we require later.

Let $L^2(\mathbf{T})$ be the space of measurable functions $f : \mathbf{T} \to \mathbf{C}$ such that $\int_0^{2\pi} |f(e^{i\theta})|^2 d\theta/2\pi$ is finite, with the inner product

$$\langle f, g \rangle_{L^2} = \int_0^{2\pi} f(e^{i\theta})\bar{g}(e^{i\theta}) \frac{d\theta}{2\pi}; \qquad (4.2.1)$$

often we use the complete orthonormal basis $(e^{in\theta})_{n=-\infty}^{\infty}$ of trigonometric characters.

Definition (*Poisson kernel*). The Poisson kernel is

$$P_r(\theta - \phi) = \frac{1 - r^2}{1 - 2r\cos(\theta - \phi) + r^2} = \sum_{n=-\infty}^{\infty} r^{|n|}e^{in(\theta-\phi)}, \qquad (4.2.2)$$

and the Poisson extension of $u \in L^2(\mathbf{T})$ is

$$u(z) = P_z u = \int_0^{2\pi} P_r(\theta - \phi)u(e^{i\theta})\frac{d\phi}{2\pi} \qquad (z = re^{i\theta} \in \mathbf{D}). \qquad (4.2.3)$$

Theorem 4.2.1 (*Fatou, [102]*). *(i) For each $u \in L^2(\mathbf{T})$, the Poisson extension of u gives a harmonic function on $\mathbf{D}$ such that*

$$\sup_{0<r<1} \int_0^{2\pi} |u(re^{i\theta})|^2 \frac{d\theta}{2\pi} = \int_0^{2\pi} |u(e^{i\theta})|^2 \frac{d\theta}{2\pi}, \qquad (4.2.4)$$

and $u(re^{i\theta}) \to u(e^{i\theta})$ almost surely and in L^2 as $r \to 1-$.
(ii) Conversely, let U be a harmonic function on $\mathbf{D}$ such that

$$\sup_{0<r<1} \int_0^{2\pi} |U(re^{i\theta})|^2 d\theta/(2\pi) \qquad (4.2.5)$$

is finite. Then there exists $u \in L^2(\mathbf{T})$ such that $U(z) = P_z u$ for all $z \in \mathbf{D}$.

Proof. We omit the proof, which may be found in [102], but note the formula

$$u(re^{i\theta}) = \sum_{n=-\infty}^{\infty} r^{|n|}a_n e^{in\theta} \qquad (4.2.6)$$

where $(a_n)_{n=-\infty}^{\infty}$ are the Fourier coefficients of u. $\qquad \square$

Thus we can identify $L^2(\mathbf{T})$ with the boundary values of harmonic functions that satisfy (ii). For many purposes, $\mathbf{D}$ is a more convenient domain.

Definition (*Reproducing kernels*). A Hilbert space of functions on a set Ω is a Hilbert space H such that each $f \in H$ defines a function $f : \Omega \to \mathbf{C}$, and such that $f \mapsto f(w)$ is a continuous linear functional on H for each $w \in \Omega$. By the Riesz–Fréchet theorem, there exists a unique $k_w \in H$ such that $f(w) = \langle f, k_w \rangle_H$; furthermore $K(z,w) = \langle k_w, k_z \rangle_H$ is called the reproducing kernel for H.

Evidently K satisfies $K(z,w) = \overline{K(w,z)}$ and $K(z,w) = k_w(z)$, so may be viewed either as a function $\Omega \times \Omega \to \mathbf{C}$ or as a family of elements of H parametrized by Ω. When Ω is a metric space, we normally require $w \mapsto k_w$ to be injective and continuous for the weak topology on H.

Example 4.2.2 (*Harmonic functions*). Consider $L^2(\mathbf{T})$ as a function space on $\mathbf{D}$ via Theorem 4.2.1, and observe that $u(z) = \langle u, k_z \rangle_{L^2}$ where $k_z(\psi) = P_r(\psi - \theta)$ when $z = re^{i\theta}$. Evidently u is continuous on $\mathbf{D}$ as a harmonic function.

Definition (*Hardy space*). Let H^2 be the Hardy space of analytic functions $f : \mathbf{D} \to \mathbf{C}$ such that $\sup_{0<r<1} \int |f(re^{i\theta})|^2 d\theta/(2\pi)$ is finite; evidently each f has a Taylor series $f(z) = \sum_{n=0}^{\infty} b_n z^n$ such that

$$\sum_{n=0}^{\infty} |b_n|^2 = \sup_{0<r<1} \int |f(re^{i\theta})|^2 \frac{d\theta}{2\pi}. \qquad (4.2.7)$$

Example 4.2.3 (*The Cauchy kernel for Hardy space*). By Theorem 4.2.1, we can identify H^2 as the closed linear subspace of $L^2(\mathbf{T})$ spanned by $(e^{in\theta})_{n=0}^{\infty}$. Furthermore, by Cauchy's integral formula, H^2 is a reproducing kernel Hilbert space on $\mathbf{D}$ with kernels $h_w(z) = 1/(1 - \bar{w}z)$ since

$$f(w) = \frac{1}{2\pi i} \int_{\mathbf{T}} \frac{f(z)dz}{z - w} = \frac{1}{2\pi} \int_0^{2\pi} \frac{f(e^{i\theta})d\theta}{1 - se^{i(\psi-\theta)}} \qquad (w = se^{i\psi}). \quad (4.2.8)$$

Definition (*Hilbert transform on the circle* [102]). Let $\text{sgn}(n) = n/|n|$ for $n \neq 0$ and $\text{sgn}(0) = 0$. Then the conjugate function operator $\mathcal{H} : L^2 \to L^2$ is defined by

$$\mathcal{H} : \sum_{n=-\infty}^{\infty} a_n e^{in\theta} \to \sum_{n=-\infty}^{\infty} -i\,\text{sgn}(n) a_n e^{in\theta}. \qquad (4.2.9)$$

Clearly $\mathcal{H}$ is a linear isometry such that $\mathcal{H}^2 = P_0 - I$; furthermore, $R_+ = (I + P_0 + i\mathcal{H})/2$ gives the orthogonal projection $L^2 \to H^2$.

Proposition 4.2.4 (*Conjugate function*). (i) *Let* $u \in L^2(\mathbf{T})$, *and let*

$$v(re^{i\theta}) = \int_{-\pi}^{\pi} \frac{2r\sin(\theta - t)}{1 - 2r\cos(\theta - t) + r^2} u(e^{it}) \frac{dt}{2\pi}. \qquad (4.2.10)$$

Then $f = u + iv$ *belongs to* H^2, *and the boundary values of* v *are*

$$v(e^{i\theta}) = \text{p.v.} \int \cot\frac{\theta - t}{2} u(e^{it}) \frac{dt}{2\pi}. \qquad (4.2.11)$$

(ii) Now suppose that $u \in L^1(\mathbf{T})$. Then the principal value integral for v exists almost everywhere, and there exists an analytic function f on $\mathbf{D}$ such that $f(re^{i\theta}) \to u(e^{i\theta}) + iv(e^{i\theta})$ almost everywhere as $r \to 1-$.

Proof. (i) By summing geometric series, one can prove that

$$\sum_{n=-\infty}^{\infty} -i\operatorname{sgn}(n) r^{|n|} e^{in\theta} = \frac{i2r\sin\theta}{1 - 2r\cos\theta + r^2}, \qquad (4.2.12)$$

and that as $r \to 1-$, this gives the formal identity

$$\sum_{n=-\infty}^{\infty} -i\operatorname{sgn}(n) e^{in\theta} = i\cot\frac{\theta}{2}. \qquad (4.2.13)$$

Now $v = \mathcal{H}u$ is the harmonic conjugate of u in the sense that $u + i\mathcal{H}u$ extends by the Poisson kernel to define the analytic function $f(z) = a_0 + \sum_{n=1}^{\infty} 2a_n z^n$.

(ii) See [102, p. 78]. □

Definition (*Dirichlet space*). Let $\nabla u \in \mathbf{C}^2$ be the gradient of a continuously differentiable function of two variables $u : \mathbf{D} \to \mathbf{C}$. Let D_0 be the space of harmonic functions $u : \mathbf{D} \to \mathbf{C}$ such that $u(0) = 0$ and the area integral is finite, so

$$\frac{1}{2\pi} \iint_{\mathbf{D}} \|\nabla u(z)\|^2 dxdy < \infty. \qquad (4.2.14)$$

Dirichlet spaces play an important role in the theory of orthogonal polynomials on the unit circle, and hence in random matrix theory for compact Lie groups. See [145].

Proposition 4.2.5 *The harmonic function u has finite area integral if and only if*

$$\sum_{n=-\infty}^{\infty} |n| |\hat{u}(n)|^2 < \infty; \qquad (4.2.15)$$

in this case

$$\frac{1}{2\pi} \iint_{\mathbf{D}} \|\nabla u\|^2 rdrd\theta = \iint_{\mathbf{T}^2} \left| \frac{u(e^{i\theta}) - u(e^{i\phi})}{e^{i\theta} - e^{i\phi}} \right|^2 \frac{d\theta}{2\pi} \frac{d\phi}{2\pi}. \qquad (4.2.16)$$

Proof. We can express the gradient in polar coordinates $\nabla = (\partial/\partial r, -i\partial/r\partial\theta)$, and then calculate

$$\nabla u(re^{i\theta}) = \sum_{n=-\infty}^{\infty} r^{|n|-1} \hat{u}(n) e^{in\theta} (|n|, n) \qquad (4.2.17)$$

so

$$\int_0^{2\pi} \|\nabla u(re^{i\theta})\|^2 \frac{d\theta}{2\pi} = \sum_{n=-\infty}^{\infty} 2n^2 |\hat{u}(n)|^2 r^{2|n|-2} \qquad (4.2.18)$$

and

$$\frac{1}{2\pi} \iint_{\mathbf{D}} \|\nabla u(re^{i\theta})\|^2 r \, dr \, d\theta = \sum_{n=-\infty}^{\infty} |n| |\hat{u}(n)|^2. \qquad (4.2.19)$$

Using $\sum'$ to stand for a sum excluding the index $n = 0$, we observe that

$$\frac{u(e^{i\theta}) - u(e^{i\phi})}{e^{i\theta} - e^{i\phi}} = {\sum_{}}' \hat{u}(n) \frac{e^{in\theta} - e^{in\phi}}{e^{i\theta} - e^{i\phi}}$$

$$= \sum_{n=1}^{\infty} \hat{u}(n) \sum_{k=0}^{n-1} e^{ik\theta + i(n-k-1)\phi}$$

$$+ \sum_{n=-\infty}^{-1} \hat{u}(n) \sum_{k=0}^{\infty} e^{-ik\theta - i(|n|-1-k)\phi}, \qquad (4.2.20)$$

wherein all of these characters $e^{i\ell\theta + im\phi}$ are orthogonal with respect to $d\theta d\phi/(2\pi)^2$ hence

$$\iint_{\mathbf{T}^2} \left| \frac{u(e^{i\theta}) - u(e^{i\phi})}{e^{i\theta} - e^{i\phi}} \right|^2 \frac{d\theta}{2\pi} \frac{d\phi}{2\pi} = \sum_{n=-\infty}^{\infty} |n| |\hat{u}(n)|^2. \qquad (4.2.21)$$

□

Example 4.2.6 The inner product on D_0 is

$$\langle u, v \rangle_D = \frac{1}{\pi} \iint_{\mathbf{D}} \nabla u(z) \overline{\nabla v(z)} \, dx \, dy = \sum_{n=-\infty}^{\infty} |n| a_n \bar{b}_n \qquad (4.2.22)$$

when $v(re^{i\theta}) = \sum_{n=-\infty}^{\infty} r^{|n|} b_n e^{in\theta}$ and $u(re^{i\theta}) = \sum_{n=-\infty}^{\infty} r^{|n|} a_n e^{in\theta}$. Evidently the subspace of D_0 that consists of analytic functions is a reproducing kernel Hilbert space on $\mathbf{D}$ with kernels $g_w(z) = \log 1/(1 - \bar{w}z)$.

The identity $h_w(z) = e^{g_w(z)}$ that relates the reproducing kernels of H^2 and D_0 is no coincidence, for in the next result we show that the exponential function operates on analytic elements to D_0 to give elements of H^2. We prove that $\|e^f\|_{H^2}^2 \leq \exp \|u\|_{D_0}^2$.

Theorem 4.2.7 (*Lebedev–Milin [124]*). *Suppose that* $u : \mathbf{D} \to \mathbf{R}$ *belongs to* D_0. *Then*

$$\int_0^{2\pi} e^{2u(\theta)} \frac{d\theta}{2\pi} \leq \exp \left(\frac{1}{\pi} \iint_{\mathbf{D}} \|\nabla u(z)\|^2 \, dx \, dy \right). \qquad (4.2.23)$$

Proof. We prove the inequality

$$\int_0^{2\pi} \exp\Big(2 \sum_{n=-N}^{N} a_n e^{in\theta}\Big) \frac{d\theta}{2\pi} \le \exp\Big(\sum_{n=-N}^{N} |n||a_n|^2\Big), \qquad (4.2.24)$$

where $a_{-n} = \bar{a}_n$. Let $\alpha_k = \sqrt{2}|a_k|$; then let $\varphi(z) = \sum_{k=1}^{\infty} \alpha_k z^k$ and $\psi(z) = \sum_{k=0}^{\infty} \beta_k z^k$, where the β_k are chosen so that $\psi(z) = e^{\varphi(z)}$. We need to show that $\|\psi\|_{H^2}^2 \le \exp \|\varphi\|_{D_0}^2$.

Then $\psi'(z) = e^{\varphi(z)}\varphi'(z) = \psi(z)\varphi'(z)$, so by equating coefficients in the power series, we have

$$\beta_n = \frac{1}{n} \sum_{k=0}^{n-1} (n-k)\alpha_{n-k}\beta_k. \qquad (4.2.25)$$

Hence by the Cauchy–Schwarz inequality, we have

$$|\beta_n|^2 \le \frac{1}{n} \sum_{k=0}^{n-1} (n-k)|\alpha_{n-k}|^2 |\beta_k|^2. \qquad (4.2.26)$$

We write $A_k = k|\alpha_k|^2$ and observe that $|\beta_k|^2 \le B_k$, where (B_k) is the solution of the recurrence relation

$$B_n = \frac{1}{n} \sum_{k=0}^{n-1} (n-k)A_{n-k}B_k \qquad (4.2.27)$$
$$B_0 = 1 = \beta_0.$$

Now this has a similar form to (4.2.25), so we deduce that

$$\sum_{k=0}^{\infty} B_k z^k = \exp\Big(\sum_{k=1}^{\infty} A_k z^k\Big); \qquad (4.2.28)$$

setting $z = 1$, we obtain

$$\sum_{k=0}^{\infty} |\beta_k|^2 \le \sum_{k=0}^{\infty} B_k = \exp\Big(\sum_{k=1}^{\infty} A_k\Big) = \exp\Big(\sum_{k=1}^{\infty} k|\alpha_k|^2\Big), \qquad (4.2.29)$$

as required. $\square$

Clearly there is a natural contractive linear inclusion map $D_0 \to L^2$ since $\|f\|_{D_0} \ge \|f\|_{L^2}$; hence each $h \in L^2$ defines a bounded linear functional on D_0 by $f \mapsto \int f\bar{h}d\theta/2\pi$.

Definition (*Sobolev space*). The space of harmonic functions

$$h(re^{i\theta}) = \sum_{n=-\infty}^{\infty} r^{|n|} b_n e^{in\theta} \qquad (4.2.30)$$

such that $h(0) = 0$ and $\sum_{n=-\infty}^{\infty} |n|^{-1}|b_n| < \infty$ is the homogeneous Sobolev space $\dot{H}^{-1/2}$, also denoted D_0'. This space has the inner product

$$\langle h, k \rangle_{\dot{H}^{-1/2}} = \sum_{n=-\infty}^{\infty} \frac{b_n \bar{c}_n}{|n|} \tag{4.2.31}$$

where $k(re^{i\theta}) = \sum_{n=-\infty}^{\infty} r^{|n|} c_n e^{i\theta}$. Evidently $\dot{H}^{-1/2}$ is the dual space of D_0 under the pairing $\langle u, h \rangle = \sum_{n=-\infty}^{\infty} a_n \bar{b}_n$. The Sobolev space $\dot{H}^{-1/2}$ should not be confused with the Hardy space H^2 of (4.2.7).

Definition (*Energy spaces*). In subsequent sections, we also require measures on subintervals of the real line, and hence some function spaces that have a similar role to D_0 and D_0'. We introduce the Green's function for the disc, namely

$$G(z, w) = \log\left|\frac{1 - z\bar{w}}{z - w}\right| \qquad (z, w \in \mathbf{D}) \tag{4.2.32}$$

and the energy space, for $[-1/2, 1/2] \subset \mathbf{D}$,

$$E^+ = \left\{\mu \in M_b^+[-1/2, 1/2] : \iint_{[-1/2,1/2]^2} G(x, y)\mu(dx)\mu(dy) < \infty\right\}; \tag{4.2.33}$$

then let

$$\langle \mu, \nu \rangle_E = \iint_{[-1/2,1/2]^2} G(x, y)\mu(dx)\nu(dy) \qquad (\mu, \nu \in E^+). \tag{4.2.34}$$

The space of differences $E = E^+ - E^+ = \{\mu - \nu : \mu, \nu \in E^+\}$ is an inner product space with inner product

$$\langle \mu_1 - \nu_1, \mu_2 - \nu_2 \rangle_E = \langle \mu_1, \mu_2 \rangle_E + \langle \nu_1, \nu_2 \rangle_E - \langle \mu_1, \nu_2 \rangle_E - \langle \nu_2, \mu_1 \rangle_E, \tag{4.2.35}$$

and has the norm $\|\mu_1 - \nu_1\|_E = \langle \mu_1 - \nu_1, \mu_1 - \nu_1 \rangle_E^{1/2}$.

The relationship between weak convergence and convergence in energy is described in the following theorem.

Theorem 4.2.8 (*Cartan [80]*). (i) E^+ *is complete for the energy norm.*
    (ii) *Let* (μ_j) *be a bounded sequence in* E^+. *Then* (μ_j) *converges weakly in* E^+ *if and only if it converges weakly in the sense of measures.*

Remark. There are two subtleties concerning this result: first, the space E may not be complete; secondly, the integral $\int G(x, y)\mu(dy)$ can be unbounded on $[-1/2, 1/2]$, even when $\mu \in E^+$.

Definition (*Equilibrium measure*). Given a compact subset S of $\mathbf{C}$, and $\mu \in Prob(S)$, the logarithmic energy is

$$\Sigma(\mu) = \iint_{[x \neq y]} \log \frac{1}{|x - y|} \mu(dx)\mu(dy) \qquad (4.2.36)$$

when the integral is absolutely convergent; here we follow the sign conventions of [137]. The energy of S is

$$E(S) = \inf\{\Sigma(\mu) : \mu \in Prob(S)\}, \qquad (4.2.37)$$

and the μ that attains the infimum is called the *equilibrium measure*.

In electrostatics, one thinks of S as a conductor and μ as the distribution of unit charge on S.

Examples 4.2.9 (i) Let $S = \{\zeta : |\zeta| \leq 1\}$. Then the equilibrium measure of S is arclength measure $d\theta/2\pi$ on the unit circle $\{\zeta : |\zeta| = 1\}$, normalized to be a probability measure. This result is intuitively natural, as the charge should tend to move towards the boundary of the conductor, and the equilibrium measure should be invariant with respect to rotations of the disc.

(ii) Now let $S = [-1, 1]$, and let ρ be the Chebyshev distribution

$$\rho(dx) = \frac{dx}{\pi\sqrt{1 - x^2}} \qquad (x \in [-1, 1]), \qquad (4.2.38)$$

otherwise known as the arcsine distribution. Then ρ is the equilibrium measure for $[-1, 1]$; as follows from the previous example, by an argument which we now sketch. The unit positive charge on $[-1, 1]$ distributes itself so that most of the unit charge accumulates near to the endpoints ± 1.

Note that the mapping $\varphi(\zeta) = 2^{-1}(\zeta + \zeta^{-1})$ takes $\{\zeta : |\zeta| > 1\}$ onto $\mathbf{C} \setminus [-1, 1]$, and takes the unit circle $\{\zeta : |\zeta| = 1\}$ onto $[-1, 1]$ twice over. One can check that φ induces the Chebyshev distribution from the normalized arclength, since

$$\int_0^{2\pi} f\left(\frac{1}{2}(e^{i\theta} + e^{-i\theta})\right) \frac{d\theta}{2\pi} = \frac{1}{\pi} \int_{-1}^1 \frac{f(x)dx}{\sqrt{1 - x^2}}. \qquad (4.2.39)$$

By the principle of subordination, ρ must be the equilibrium measure for $[-1, 1]$; see [131, p. 107].

Definition (*Chebyshev polynomials*). Let T_k and U_k be the Chebyshev polynomials of degree k of the first and second kinds, so that

$T_k(\cos\theta) = \cos k\theta$ and $U_k(\cos\theta) = \sin(k+1)\theta/\sin\theta$. Then

$$U_n(x) = 2xU_{n-1}(x) - U_{n-2}(x), \qquad (4.2.40)$$

$$T_n(x) = 2xT_{n-1}(x) - T_{n-2}(x). \qquad (4.2.41)$$

Exercise 4.2.10 Use these identities to calculate the following.

$$
\begin{aligned}
T_0(x) &= 1 & U_0(x) &= 1 \\
T_1(x) &= x & U_1(x) &= 2x \\
T_2(x) &= 2x^2 - 1 & U_2(x) &= 4x^2 - 1 \\
T_3(x) &= 4x^3 - 3x & U_3(x) &= 8x^3 - 4x \\
T_4(x) &= 8x^4 - 8x^2 + 1 & U_4(x) &= 16x^4 - 12x^2 + 1 \quad (4.2.42)
\end{aligned}
$$

(ii) Show also that

$$\int_{-1}^{1} \frac{T_j(x)T_k(x)\,dx}{\pi\sqrt{1-x^2}} = 0 \qquad (j \neq k), \qquad (4.2.43)$$

$$\int_{-1}^{1} \frac{T_j(x)^2\,dx}{\pi\sqrt{1-x^2}} = \frac{1}{2} \qquad (j = 1,2,\dots). \qquad (4.2.44)$$

Exercise 4.2.11 Let K be the space of continuous functions $f : \mathbf{T} \to \mathbf{C}$ such that $\sum_k |k| |\hat{f}(k)|^2 < \infty$. Show that K forms an algebra under pointwise multiplication of functions.

Definition (*Hardy space on the upper half-plane*). The open upper half-plane is $\mathbf{C}_+ = \{z \in \mathbf{C} : \Im z > 0\}$. Given an analytic function $f : \mathbf{C}_+ \to \mathbf{C}$, we say that f belongs to the Hardy space $H^2(\mathbf{R})$ if the expression

$$\|f\|_{H^2} = \sup_{y>0}\left(\int_{-\infty}^{\infty} |f(x+iy)|^2\,dx\right)^{1/2} \qquad (4.2.45)$$

is finite. Analogously, one can define the Hardy space for the right half plane $\{z : \Re z > 0\}$.

4.3 Free versus classical entropy on the spheres

Definition (*Free entropy on circle*). Let p and q be probability density functions with respect to $d\theta/2\pi$ on $\mathbf{T}$, and such that $p - q \in D_0'$. Then the relative free entropy of p and q is

$$\Sigma(p,q) = \iint_{\mathbf{T}^2} \log \frac{1}{|e^{i\theta} - e^{i\phi}|} \big(p(e^{i\theta}) - q(e^{i\theta})\big)\big(p(e^{i\phi}) - q(e^{i\phi})\big) \frac{d\theta}{2\pi} \frac{d\phi}{2\pi}$$

(4.3.1)

when the integral is absolutely convergent.

Theorem 4.3.1 (*Classical versus free entropy*). *Let f be a probability density function on* $\mathbf{T}$ *that has finite relative entropy with respect to $d\theta/2\pi$. Then $\Sigma(f,\mathbf{I}) \leq \mathrm{Ent}(f \mid \mathbf{I})$.*

Proof. By joint convexity of relative entropy [43], any pair of probability density functions of finite relative entropy satisfies

$$\mathrm{Ent}(f \mid u) = \int_{\mathbf{T}} P_r(e^{i\phi})\mathrm{Ent}(f_\phi \mid u_\phi)\frac{d\phi}{2\pi}$$
$$\geq \mathrm{Ent}(P_r f \mid P_r u);$$

(4.3.2)

so in particular

$$\mathrm{Ent}(f \mid \mathbf{I}) \geq \mathrm{Ent}(P_r f \mid \mathbf{I}) \quad (0 \leq r < 1).$$

(4.3.3)

Hence it suffices to prove the theorem for $P_r f$ instead of f, and then take limits as $r \to 1-$. For notational simplicity, we shall assume that f has a rapidly convergent Fourier series so that various integrals converge absolutely.

Suppose that u is a real function in the Dirichlet space D_0 that has $\int_{\mathbf{T}} u(e^{i\theta})d\theta/2\pi = -t$ and $\|u\|_{\mathrm{D}_0} = s$; by adding a constant to u if necessary, we can assume that $s^2/2 = t$. Then by (4.2.23) we have

$$\int_{\mathbf{T}} \exp u(e^{i\theta})\frac{d\theta}{2\pi} \leq \exp(s^2/2 - t) = 1$$

(4.3.4)

and consequently by the dual formula for relative entropy

$$\int_{\mathbf{T}} f(e^{i\theta}) \log f(e^{i\theta})\frac{d\theta}{2\pi} = \sup\left\{\int_{\mathbf{T}} h(e^{i\theta})f(e^{i\theta})\frac{d\theta}{2\pi} : \int_{\mathbf{T}} \exp h(e^{i\theta})\frac{d\theta}{2\pi} \leq 1\right\}$$
$$\geq \int_{\mathbf{T}} f(e^{i\theta})u(e^{i\theta})\frac{d\theta}{2\pi}.$$

(4.3.5)

Introducing the dual pairing of $\dot{H}^{-1/2}(\mathbf{T})$ with D_0, we write

$$\langle f, u \rangle = \int_{\mathbf{T}} f(e^{i\theta})u(e^{i\theta})\frac{d\theta}{2\pi} - \int_{\mathbf{T}} f(e^{i\theta})\frac{d\theta}{2\pi}\int_{\mathbf{T}} u(e^{i\theta})\frac{d\theta}{2\pi},$$

(4.3.6)

so that by (4.3.2)

$$\langle f, u \rangle \le t + \int_{\mathbf{T}} f(e^{i\theta}) \log f(e^{i\theta}) \frac{d\theta}{2\pi}. \qquad (4.3.7)$$

We choose the $\hat{u}(n)$ for $n \ne 0$ to optimize the left-hand side, and deduce that

$$\|f\|_{\dot{H}^{-1/2}(\mathbf{T})} \|u\|_{D_0} = s \|f\|_{\dot{H}^{-1/2}(\mathbf{T})}$$
$$\le s^2/2 + \int_{\mathbf{T}} f(e^{i\theta}) \log f(e^{i\theta}) \frac{d\theta}{2\pi}, \qquad (4.3.8)$$

so by choosing s we can obtain the desired result

$$2\Sigma(f, \mathbf{I}) = \|f\|^2_{\dot{H}^{-1/2}(\mathbf{T})} \le 2 \int_{\mathbf{T}} f(e^{i\theta}) \log f(e^{i\theta}) \frac{d\theta}{2\pi}. \qquad (4.3.9)$$

$\square$

Problem. Let q be a probability density function on $\mathbf{T}$ with respect to $d\theta/2\pi$. Under what further conditions does there exist $C < \infty$ such that

$$\Sigma(f, q) \le C\mathrm{Ent}(f \mid q) \qquad (4.3.10)$$

for all probability density functions f that have finite relative entropy with respect to q?

The next step is to go from the circle to the real line. Given $a + ib$ with $b > 0$, there is a linear fractional transformation $\varphi(z) = (z - a - ib)/(z - a + ib)$ that maps the upper half-plane bijectively onto the open unit disc and $\varphi(a+ib) = 0$; further, with $\varphi(x) = e^{i\theta}$ the measures satisfy

$$h_{a,b}(x)dx = \frac{bdx}{\pi((x - a)^2 + b^2)} = \frac{d\theta}{2\pi}. \qquad (4.3.11)$$

Proposition 4.3.3 *The entropy functional*

$$E(f) = \int_{-\infty}^{\infty} \int_{-\infty}^{\infty} f(x)f(t) \log |x - t| \, dx dt + \int_{-\infty}^{\infty} f(x) \log f(x) \, dx \qquad (4.3.12)$$

is defined for probability density functions with finite entropy and logarithmic energy. The probability density function $h_{a,b}$ is a stationary point of E, and $E(h_{a,b}) = -\log 2\pi$.

Proof. We introduce a Lagrange multiplier λ and calculate the functional derivative of

$$F(f) = \int_{-\infty}^{\infty} \int_{-\infty}^{\infty} f(x)f(t) \log |x - t| \, dxdt$$
$$+ \int_{-\infty}^{\infty} f(x) \log f(x) dx + \lambda \int_{-\infty}^{\infty} f(x) \, dx, \quad (4.3.13)$$

as in (6.1.1) below, obtaining

$$\frac{\delta F}{\delta f} = 2 \int_{-\infty}^{\infty} f(t) \log |x - t| \, dt + \log f(x) + 1 + \lambda. \quad (4.3.14)$$

Now $\log |z - a + ib|$ is a harmonic function in the upper half plane, so

$$\frac{\pi}{b} \int_{-\infty}^{\infty} \frac{\log |z - t|}{(t - a)^2 + b^2} dt = \log |z - a + ib|. \quad (4.3.15)$$

Hence

$$\log \frac{b}{\pi((x - a)^2 + b^2)} = \frac{2b}{\pi} \int_{-\infty}^{\infty} \frac{\log |x - t|}{(t - a)^2 + b^2} dt + \log \frac{b}{\pi} \quad (4.3.16)$$

so $h_{a,b}$ gives a solution to $\delta F/\delta f = 0$ for suitable λ. Now we use translation invariance to suppress a and calculate

$$E(h_{a,b}) = \int_{-\infty}^{\infty} \int_{-\infty}^{\infty} \frac{b/\pi}{x^2 + b^2} \log |x - t| \frac{b/\pi}{t^2 + b^2} \, dxdt + \log(b/\pi)$$
$$+ \int_{-\infty}^{\infty} \frac{b/\pi}{x^2 + b^2} \log |x + ib| \, dx \quad (4.3.17)$$
$$= \log(b/\pi) - \frac{b}{\pi} \int_{-\infty}^{\infty} \frac{\log |x + ib|}{x^2 + b^2} dx \quad (4.3.18)$$
$$= \log(b/\pi) - \log |2ib|$$
$$= -\log 2\pi. \quad (4.3.19)$$

$\square$

The functions involved in the above proof suggest the correct result for S^n and $\mathbf{R}^n$, as we now discuss. The unit sphere S^n has surface area

$$\text{vol}_n (S^n) = \frac{2\pi^{(n+1)/2}}{\Gamma((n + 1)/2)}, \quad (4.3.20)$$

and we let σ_n be the surface area normalized so as to be a probability measure. The stereographic projection $G : S^n \to \mathbf{R}^n \cup \{\infty\}$ is given by

$$G(s_0, \ldots, s_n) = \left(\frac{s_j}{1 + s_0} \right)_{j=1}^{n} \quad (4.3.21)$$

and its inverse has Jacobian $J_{G^{-1}}(x) = 2^n (1+x^2)^{-n}$; hence

$$h(x) = \text{vol}_n (S^n)^{-1} \left(\frac{2}{1 + \|x\|^2} \right)^n \qquad (4.3.22)$$

is a probability density function on $\mathbf{R}^n$. There is a linear isometric bijection $f \mapsto F = (f/h) \circ G$ from $L^1(\mathbf{R}^n, dx) \to L^1(S^n; \sigma_n)$ which takes probability density functions to probability density functions.

The stereographic projection is conformal; that is, angle preserving. Conformal maps of $\mathbf{R}^n$ include translations, rotations, dilations and inversion in the sphere.

Theorem 4.3.4 (*Carlen and Loss, [45]*).
(i) *Suppose that F is a probability density function with respect to σ_n such that $F \log F$ is integrable. Then*

$$0 \le \iint_{S^n \times S^n} (F(\omega) - 1) \log \frac{1}{\|\omega - \zeta\|} (F(\zeta) - 1)\sigma_n(d\omega)\sigma_n(d\zeta)$$

$$\le \frac{1}{n} \int_{S^n} F(\omega) \log F(\omega)\sigma_n(d\omega); \qquad (4.3.23)$$

furthermore, the optimizers of this inequality all have the form $F(\omega) = |J_T(\omega)|$ where J_T is the Jacobian of a some conformal map $T : S^n \to S^n$.

(ii) *There exist constants C_n such that the inequality*

$$\iint_{\mathbf{R}^n \times \mathbf{R}^n} f(x) \log \frac{1}{\|x - y\|} f(y) \, dx dy \le \frac{1}{n} \int_{\mathbf{R}^n} f(x) \log f(x) \, dx + C_n \qquad (4.3.24)$$

holds for all probability density functions that have finite entropy. Furthermore, all the optimizers have the form $f(x) = |J_\tau(x)| h \circ \tau(x)$ where $\tau : \mathbf{R}^n \to \mathbf{R}^n$ is conformal.

Proof. (i) We are only able to prove a small part of this result by the methods of this book, namely the left-hand inequality of (i). For unit vectors ω and ζ, we have $\|\omega - \zeta\|^2 = 2(1 - \langle \omega, \zeta \rangle)$ and hence

$$\log \frac{1}{\|\omega - \zeta\|} = -\frac{1}{2} \log 2 + \frac{1}{2} \sum_{k=1}^{\infty} \frac{\langle \omega, \zeta \rangle^k}{k}. \qquad (4.3.25)$$

Now we recall that σ_n is induced from Haar measure μ_n on $O(n)$ by the map $U \mapsto U e_n$, hence

$$\iint_{S^n \times S^n} \langle \omega, \zeta \rangle^k (F(\omega) - 1)(F(\zeta) - 1)\sigma_n(d\omega)\sigma_n(d\zeta) \qquad (4.3.26)$$

$$= \iint_{O(n) \times O(n)} \langle U e_n, V e_n \rangle^k (F(U e_n) - 1)(F(V e_n) - 1)\mu_n(dU)\mu_n(dV)$$

since

$$\langle Ue_n, Ve_n \rangle^k = \langle Ue_n \otimes \cdots \otimes Ue_n, Ve_n \otimes \cdots \otimes Ve_n \rangle \quad (4.3.27)$$

is the coefficient of a representation of $O(n)$, so the integral is nonnegative. The rest of the proof involves a delicate rearrangement argument. See [45]. □

The expressions in Theorem 4.3.4(i) give the logarithmic and classical relative entropy of F with respect to the constant function **I**.

Note that for $n > 2$, $h(x)$ is not the Poisson kernel on $\mathbf{R}^n$ and $\log \|x - y\|$ is not the Green's function for the standard Laplace operator. Hence the proof of Proposition 4.3.3 does not easily extend to the present context.

4.4 Equilibrium measures for potentials on the real line

In the remaining sections of this chapter, we consider electrostatics for charges on the real line, subject to logarithmic interaction in the presence of a force field which is represented by potential v. So we are concerned with energy expressions of the form

$$\frac{1}{n} \sum_{j=1}^{n} v(\lambda_j) - \frac{\beta}{n^2} \sum_{1 \leq j < k \leq n} \log(\lambda_k - \lambda_j) \quad (4.4.1)$$

where $\lambda_1 < \lambda_2 < \cdots < \lambda_n$, such as arise as eigenvalue distributions for:

- $\beta = 1$ generalized orthogonal ensemble:
- $\beta = 2$ generalized unitary ensemble;
- $\beta = 4$ generalized symplectic ensemble.

More generally, we can take $\beta > 0$ and contemplate an electrostatic energy associated with positive charges of strength $\beta > 0$ on a linear conductor in an electrical field that has potential v. Throughout rest of this chapter, v satisfies at least the following hypotheses.

(a) v is bounded below, so that $v(x) \geq -C$ for some $C \in \mathbf{R}$ and all $x \in \mathbf{R}$.

(b) There exist $\varepsilon > 0$ and $M < \infty$ such that

$$v(x) \geq \frac{\beta + \varepsilon}{2} \log(x^2 + 1) \quad (|x| \geq M). \quad (4.4.2)$$

(c) $v : \mathbf{R} \to \mathbf{R}$ is absolutely continuous and v' is $\alpha-$ Hölder continuous on $[-A, A]$ for some $\alpha > 0$ and each $A < \infty$.

Evidently (c) and (b) together imply (a).

Definition (*Potential and electrostatic energy*). Let $\mu \in Prob(\mathbf{R})$. Then the logarithmic energy is

$$\Sigma(\mu) = \iint_{[x\neq y]} \log \frac{1}{|x - y|} \mu(dx)\mu(dy) \qquad (4.4.3)$$

when the integral is absolutely convergent. The potential energy is

$$F_v(\mu) = \int v(x)\mu(dx) \qquad (4.4.4)$$

when the integral is absolutely convergent. If μ has compact support and belongs to Cartan's space E^+, as in Section 4.2, then the potential energy and the electrostatic energy are both finite. When both energies are defined, the electrostatic energy is

$$E_v(\mu) = \int v(x)\mu(dx) - \frac{\beta}{2} \iint_{[x\neq y]} \log |x - y|\mu(dx)\mu(dy). \qquad (4.4.5)$$

From our understanding of electrostatics, we should expect the system to seek an equilibrium configuration that minimizes the energy. The potential energy is relatively small when the measure is concentrated near to the minima of v; whereas by (b), $F_v(\mu)$ is large when μ places mass on large values of x, so v acts as a confining potential. The logarithmic energy is positive when μ is concentrated on a set of diameter less than one, reflecting the principle that charges of like sign repel one another. Hence there should be a minimizer that balances these contrary effects. The following problem is considered in detail in [137].

Problem (*Equilibrium problem with potential*). Given v that satisfies (a), (b) and (c), find $\rho \in Prob(\mathbf{R})$ that attains the infimum in

$$E_v = \inf_\nu E_v(\nu) \qquad (4.4.6)$$

where the infimum is taken over all $\nu \in Prob(\mathbf{R})$ that have finite logarithmic energy and finite potential energy with respect to v.

Theorem 4.4.1 (*Equilibrium measure given a potential*). *There exists a minimizer ρ for the equilibrium problem for a unit charge on the real line in the presence of a potential. This ρ has compact support S, and is absolutely continuous with respect to Lebesgue measure. The density*

$p = d\rho/dx$ belongs to L^2, is α-Hölder continuous and is characterized by:

$$p(x) \geq 0; \quad \int p(x)\,dx = 1; \qquad\qquad (4.4.7)$$

$$v(x) \geq \beta \int \log|x - y|p(y)dy + C \qquad (x \in \mathbf{R}); \qquad (4.4.8)$$

$$v(x) = \beta \int \log|x - y|p(y)\,dy + C \qquad (x \in S) \qquad (4.4.9)$$

where C is some constant.

Proof. We introduce the functions

$$f_\eta(x, y) = \frac{1}{2}(v(x) + v(y)) + \frac{\beta}{2}\log\frac{1}{|x - y| + \eta}, \qquad (4.4.10)$$

and observe that since $|x - y| \leq (1 + x^2)^{1/2}(1 + y^2)^{1/2}$ the functions satisfy

$$f_0(x, y) \geq \frac{1}{2}(v(x) + v(y)) - \frac{\beta}{4}\log(1 + x^2)(1 + y^2)$$

$$\geq \frac{\varepsilon}{4}\log(1 + x^2)(1 + y^2). \qquad (4.4.11)$$

Now f_0 is bounded below, the energy satisfies

$$E_v(\mu) = \iint f_0(x, y)\mu(dx)\mu(dy) \qquad (4.4.12)$$

and there exists $\nu \in Prob(\mathbf{R})$ such that $E_v(\nu) < \infty$, so the infimum is finite. Let $\mu_m \in Prob(\mathbf{R})$ satisfy

$$E_v(\mu_m) \leq E_v + 1/m$$

and observe that the set

$$\left\{\nu \in Prob(\mathbf{R}) : \int \log(1 + x^2)\nu(dx) \leq M\right\} \qquad (4.4.13)$$

is relatively compact for the weak topology by Proposition 1.3.6. Hence there exists a subsequence (μ_{n_k}) and a $\rho \in Prob(\mathbf{R})$ such that $\mu_{n_k} \to \rho$ weakly. Next we check that ρ is a minimizer, starting by using the monotone convergence theorem to obtain

$$E_v(\rho) = \lim_{\eta \to 0+} \iint f_\eta(x, y)\rho(dx)\rho(dy). \qquad (4.4.14)$$

Now $\mu_{n_k} \otimes \mu_{n_k} \to \rho \otimes \rho$ weakly and hence

$$
\begin{aligned}
E_v(\rho) &= \lim_{\eta \to 0+} \lim_{n_k} \iint f_\eta(x,y) \mu_{n_k}(dx) \mu_{n_k}(dy) \\
&= \lim_{n_k} \lim_{\eta \to 0+} \iint f_\eta(x,y) \mu_{n_k}(dx) \mu_{n_k}(dy) \\
&= \lim_{n_k} \iint f_0(x,y) \mu_{n_k}(dx) \mu_{n_k}(dy) \\
&= \lim_{n_k} E_v(\mu_{n_k}) = E_v
\end{aligned}
\tag{4.4.15}
$$

where we used monotone convergence to replace f_η by f_0.

Since $f_0(x,y) \to \infty$ as $|x| + |y| \to \infty$, the support S of ρ is compact; otherwise we could diminish $E_v(\rho)$ by moving mass towards the origin from far away.

Let $\nu = (1 + th)\rho$ where $-1 < t < 1$, $h \in C_c(\mathbf{R})$ has $\|h\|_\infty < 1$ and $\int h d\rho = 0$; so ν is also a probability measure. Then $E_v((1+th)\nu)$ has a minimum at $t = 0$, so

$$
0 = \left(\frac{d}{dt}\right)_{t=0} E_v((1+th)\nu) = \int h(x) \left(v(x) - \beta \int \log|x-y| \rho(dy) \right) \rho(dx);
\tag{4.4.16}
$$

hence there exists a constant C such that

$$
v(x) = \beta \int \log|x-y| \rho(dy) + C.
\tag{4.4.17}
$$

Now by (c), the function v is differentiable and its derivative satisfies

$$
v'(x) = \text{p.v.} \int \frac{\beta \rho(dy)}{x-y}
\tag{4.4.18}
$$

on each open subinterval S_0 of S. There are standard formulae for inverting this principal value integral equation, and here we report the conclusions. Since v' is locally an L^2 function, it follows from [137, Theorem 2.2, page 211] that ρ is absolutely continuous with respect to Lebesgue measure and that the density $p = \frac{d\rho}{dx}$ is in $L^2(S_0; dx)$ for each proper subinterval S_0 of S; furthermore, p is α-Hölder continuous on S_0 by properties of the Hilbert transform.

In principle, p could have singularities of the form $1/\sqrt{|b-x|}$ at an endpoint b of S. In the next Lemma, we show that this does not happen, so p is actually in $L^2(\mathbf{R}; dx)$.   □

A crucial feature of the equilibrium problem in the presence of a potential is that the equilibrium measure is free to choose its own support;

that is, the equilibrium measure is not constrained to lie on any particular bounded interval and S is not specified in advance. Intuitively, one sees that the energy will become large if the density p is unbounded at the end of any interval, so instead the logarithmic term will tend to force some of the mass away from the end and hence equilibrium measure should have a bounded density. To make this precise, we present a more general result based upon mass transportation.

Lemma 4.4.2 (*Endpoint regularity*). *Suppose that v satisfies (a) and (b), and that b is the supremum of the support S of the equilibrium distribution ρ_v. Suppose further that (d) v is δ-Hölder continuous on S. Then there exist $\varepsilon_0, M_v > 0$ such that*

$$\int_{[b-\varepsilon,b]} \rho_v(dx) \leq M_v \varepsilon^\delta \qquad (\varepsilon \in (0,\varepsilon_0)). \qquad (4.4.19)$$

In particular, when v satisfies (c) and hence v' is locally bounded, this inequality holds with $\delta = 1$ and the equilibrium density is bounded at the endpoint b.

Proof. We introduce the Lipschitz function $\varphi : \mathbf{R} \to \mathbf{R}$ that satisfies $\varphi(x) = x$ for $x \leq b - \varepsilon$ and $\varphi'(x) = 2$ for $x > b - \varepsilon$. The effect of φ is to shift some of the mass from $[b - \varepsilon, b]$ to outside S; clearly, $\varphi \sharp \rho_v$ is an admissible probability measure only because b is a free end in the equilibrium problem. By the mean value theorem, the function φ satisfies the simple properties $\varphi(x) - x \leq 2\varepsilon$ for $x \in [b - \varepsilon, b]$;

$$\frac{\varphi(x) - \varphi(y)}{x - y} \begin{cases} \geq 1, & \text{for all } x, y \in \mathbf{R}; \\ = 2, & \text{for all } x, y \in [b - \varepsilon, b]). \end{cases} \qquad (4.4.20)$$

Comparing the energies associated with the minimizer ρ_v and $\varphi \sharp \rho_v$, we have

$$0 \leq E_v(\varphi \sharp \rho_v) - E_v(\rho_v) = \int \big(v(\varphi(x)) - v(x) \big) \rho_v(dx)$$
$$+ \frac{\beta}{2} \iint \log \left| \frac{x - y}{\varphi(x) - \varphi(y)} \right| \rho_v(dx)\rho_v(dy) \qquad (4.4.21)$$

which we can rearrange to give

$$\frac{\beta}{2} \iint_{[b-\varepsilon,b]^2} \log \left| \frac{\varphi(x) - \varphi(y)}{x - y} \right| \rho_v(dx)\rho_v(dy)$$
$$\leq \int_{[b-\varepsilon,b]} \big(v(\varphi(x)) - v(x) \big) \rho_v(dx), \qquad (4.4.22)$$

which gives

$$\frac{\beta}{2}(\log 2)\left(\int_{[b-\varepsilon,b]}\rho(dx)\right)^2 \le C_v \int_{[b-\varepsilon,b]}(\varphi(x)-x)^\delta\rho_v(dx)$$

$$\le C_v(2\varepsilon)^\delta\int_{[b-\varepsilon,b]}\rho(dx). \qquad (4.4.23)$$

By cancelling the final integral, we obtain the stated result. □

Proposition 4.4.3 (*Support of the equilibrium measure*).

(i) *Suppose that v is strictly convex. Then ρ is supported on a single interval $[a,b]$ and $p > 0$ on (a,b).*

(ii) *Suppose that v is analytic. Then there exist finitely many intervals $[a_j, b_j]$ $(j = 1,\ldots,m)$ such that the support of ρ equals*

$$S = \bigcup_{j=1}^{m}[a_j, b_j] \qquad (4.4.24)$$

and $p > 0$ inside S.

(iii) *If v is a polynomial with degree $2k > 0$ and positive leading term, then $m \le k+1$.*

Proof. (i) Suppose on the contrary that the support of ρ has a gap $[c_1, c_2]$ and observe that

$$\varphi(s) = \beta\int_{c_2}^{\infty}\frac{\rho(dt)}{t-s} + \beta\int_{-\infty}^{c_1}\frac{\rho(dt)}{t-s} \qquad (4.4.25)$$

is an increasing function on (c_1, c_2). But

$$\varphi(c_1) = -v'(c_1) > -v'(c_2) = \varphi(c_2) \qquad (4.4.26)$$

holds by strict convexity of v. Hence the support of ρ is a compact and connected set, so equals $[a,b]$ for some a, b.

(ii) Deift *et al.* [57] show that the number of intervals in the support of ρ is bounded by a quantity that depends only upon the number of local minima of v. The details involve mainly potential theory, so are omitted here.

(iii) See [57, Theorem 1.46 and p. 408]. □

Proposition 4.4.4 (*Integral formula for equilibrium density*). *For S as in Proposition 4.4.3(ii), let*

$$R(z) = \prod_{j=1}^{m}(z-a_j)(z-b_j) \qquad (4.4.27)$$

and choose branches of square roots so that $\sqrt{R(z)}$ is analytic on $\mathbf{C} \setminus S$, and $\sqrt{R(z)}/z^m = 1 + O(1/z)$ as $|z| \to \infty$. Then the density of the equilibrium measure satisfies

$$p_v(x) = \text{p.v.} \frac{\sqrt{R(x)}}{\beta \pi^2} \int_S \frac{v'(t)\, dt}{\sqrt{R(t)}(x-t)} \qquad (x \in S). \qquad (4.4.28)$$

Proof. This follows from [119, p. 252; 57, Theorem 1.34]. □

Exercise 4.4.5 (i) Let v be a polynomial of degree $2k$ with positive leading coefficient, and let m be the number of local minima of v on $[-1, 1]$. Show that $m \le k$.

(ii) Show that if v has degree $2k + 1$, then $m \le k + 1$.

Remarks. (i) As in Proposition 4.4.4, we introduce the algebraic equation

$$y^2 = \prod_{j=1}^{m} (x - a_j)(x - b_j)$$

so that $S = \{x : y^2 \le 0\}$. The natural coordinates for such points are $\cos \theta_j$ since $(a_j + b_j)/2 + ((b_j - a_j)/2) \cos \theta_j$ gives an element of S. In particular, the quartics $y^2 = (1 - x^2)(1 - k^2 x^2)$ with $k^2 \ne 0, 1$ are associated with the elliptic integrals

$$\int_0^x \left((1 - t^2)(1 - k^2 t^2) \right)^{-1/2} dt.$$

(ii) The Riemann surface for $\sqrt{R(z)}$ consists of a handle-body with $m - 1$ handles, which is obtained by taking two copies of $\mathbf{C} \cup \{\infty\}$ and joining them crosswise along the cuts $[a_j, b_j]$.

(iii) Starting from the equilibrium problem in the presence of a potential with $\nu \in Prob(\mathbf{R})$, we have shown that the minimizer is supported on some compact subset S. Generally it is difficult to determine S, even when one has a simple formula for v; as an illustration, in Section 4.7 we find S for $v(x) = x^2/2 + gx^4/4$. In the standard theory of singular integral equations, one specifies S in advance and then solves the integral equation in terms of functions that are determined by S and the index data; whereas in our problem we need to find S. We contrast this equilibrium problem from the more standard 'constrained' equilibrium problem which we consider in Section 4.7.

(iv) The Cauchy transform of a measure uniquely determines the measure by Propositions 1.8.6 and 1.8.9. In the next section we show how to solve the singular integral equation on a single interval.

Example 4.4.6 (Johansson, [94]). The minimization problem

$$J = \inf_{\nu} \left\{ \iint_{\mathbf{T} \times \mathbf{T}} \log \frac{1}{|e^{i\theta} - e^{i\phi}|} \nu(d\theta)\nu(d\phi) : \nu \in Prob(\mathbf{T}) \right\}$$

has solution given by normalized arclength measure $\nu(d\theta) = d\theta/(2\pi)$. By making the change of variables $t = \tan\theta/2$ and $s = \tan\phi/2$, one can easily convert this to a minimization problem for probability measures on $\mathbf{R}$, namely

$$J = \inf_{\mu} \left\{ \int_{\mathbf{R}} \log(1 + t^2)\mu(dt) \right.$$

$$\left. + \iint_{\mathbf{R} \times \mathbf{R}} \log \frac{1}{2|t - s|} \mu(dt)\mu(ds) : \mu \in Prob(\mathbf{R}) \right\}. \quad (4.4.29)$$

This is essentially the map that was used in Section 4.3 to go from the line to the circle. Here the potential $v(t) = \log(1 + t^2)$ is of slower growth at infinity than is required by (b), so Theorem 4.4.1 does not apply. By the transformation, the minimizer is the Cauchy distribution $\mu(dt) = (\pi(1 + t^2))^{-1} dt$, which is not of compact support.

In subsequent sections we introduce the standard equilibrium distributions, namely:

- the semicircular distribution in Section 4.5;
- quartic potentials in Sections 4.6 and 4.7;
- the Marchenko–Pastur distribution in Section 5.5;
- Vershik's Ω distribution in Section 7.3.

4.5 Equilibrium densities for convex potentials

- The equilibrium measure for a convex potential can be found from a singular integral equation on an interval.
- The semicircular distribution arises thus.
- Tricomi's method gives the equilibrium density for convex polynomial potentials.

If the support of the equilibrium measure is a single interval S, and the density is strictly positive inside S, then the integral equation (4.4.9) simplifies considerably. Without further loss of generality, we can reduce the integral equation to

$$w(x) = \frac{1}{\pi} \int_{-1}^{1} \frac{q(y)dy}{x - y} \qquad (x \in (-1, 1)) \qquad (4.5.1)$$

where q is a probability density function on $[-1, 1]$, and w is the derivative of the potential. Here the Hilbert transform of a function is evaluated on $[-1, 1]$, so the term *finite Hilbert transform* is often used; this is distinct from the Hilbert transform on the circle, as considered in Section 4.2.

First we show how to obtain an orthogonal polynomial basis for $L^2([-1, 1]; q)$. Let $W(\theta) = 2\pi q(\cos \theta)|\sin \theta|$ so that $W(\theta)$ is a probability density function with respect to $d\theta/(2\pi)$ on $\mathbf{T}$. We let $(\varphi_n)_{n=0}^{\infty}$ be the orthogonal basis of $L^2(\mathbf{T}; W)$ that is given by 4.2.5, and introduce $x = \cos \theta$ and

$$p_n(x) = e^{-in\theta} \varphi_{2n}(e^{i\theta}) + e^{in\theta} \varphi_{2n}(e^{-i\theta}). \tag{4.5.2}$$

Proposition 4.5.1 *The* $(p_n)_{n=0}^{\infty}$ *are polynomials of degree n which are are orthogonal with respect to the weight q and have real coefficients.*

Proof. See [145, p. 292]. □

By Proposition 4.4.4, we should expect q to have the form $q(x) = \sqrt{1 - x^2} h(x)$ for some smooth function h that has $h(x) > 0$ for all $x \in (-1, 1)$. This suggests that the Chebyshev functions of Example 4.2.10 should be useful in this context.

Lemma 4.5.2 (The Aerofoil equations). *The finite Hilbert transforms of the Chebyshev polynomials satisfy*

$$T_k(x) = \text{p.v.} \frac{1}{\pi} \int_{-1}^{1} \frac{U_{k-1}(y)\sqrt{1 - y^2}}{x - y} dy, \quad (x \in (-1, 1); k = 1, 2, 3, \dots),$$

$$U_{k-1}(x) = -\text{p.v.} \frac{1}{\pi} \int_{-1}^{1} \frac{T_k(y)}{x - y} \frac{dy}{\sqrt{1 - y^2}}, \quad (x \in (-1, 1), k = 1, 2, \dots),$$

$$0 = \text{p.v.} \frac{1}{\pi} \int_{-1}^{1} \frac{dy}{(x - y)\sqrt{1 - y^2}} \quad (x \in (-1, 1)). \tag{4.5.3}$$

Proof. The basic idea behind the proof is that for $u \in L^2(\mathbf{R})$, the function $f = u + i\mathcal{H}u$ extends to define an analytic function on the upper half plane. The idea is to select f such that u is supported on $[-1, 1]$. The basic idea is that $\sqrt{1 - z^2}$ is real for $z \in [-1, 1]$ and purely imaginary for $x \in \mathbf{R} \setminus [-1, 1]$, so this is useful for constructing suitable functions u.

The function

$$\Phi(z) = \left(z - i\sqrt{1 - z^2}\right)^k \tag{4.5.4}$$

is analytic on $\mathbf{C} \setminus [-1,1]$ and is real on $\mathbf{R} \setminus [-1,1]$, where we choose the root so that $\sqrt{1 - z^2} \asymp -iz$ as $|z| \to \infty$. On $[-1,1]$, we let $x = \cos\theta$ and

$$\begin{aligned}
\Phi(x) &= \cos k\theta - i \sin k\theta \\
&= T_k(\cos\theta) - iU_{k-1}(\cos\theta)\sin\theta \\
&= T_k(x) - iU_{k-1}(x)\sqrt{1 - x^2}.
\end{aligned} \tag{4.5.5}$$

Hence $\Im\Phi(x) = -U_{k-1}(x)\sqrt{1-x^2}\mathbf{I}_{[-1,1]}(x)$, so

$$\mathcal{H} : U_{k-1}(x)\sqrt{1-x^2} \mapsto T_k(x)\mathbf{I}_{[-1,1]}(x) + \frac{1}{(x + \sqrt{x^2-1})^k}\mathbf{I}_{\mathbf{R}\setminus[-1,1]}(x), \tag{4.5.6}$$

and hence the first identity holds.

Now let

$$\Psi(z) = \frac{(z - \sqrt{1 - z^2})^k}{\sqrt{1 - z^2}} \tag{4.5.7}$$

is analytic on $\mathbf{C} \setminus [-1,1]$ and is purely imaginary for $x \in \mathbf{R} \setminus [-1,1]$. On $[-1,1]$, we let $x = \cos\theta$ and obtain

$$\begin{aligned}
\Psi(x) &= \frac{\cos k\theta - i\sin k\theta}{\sin\theta} \\
&= \frac{T_k(x)}{\sqrt{1 - x^2}} - iU_{k-1}(x).
\end{aligned} \tag{4.5.8}$$

Hence $\Re\Psi(x) = T_k(x)/\sqrt{(1-x^2)}\mathbf{I}_{[-1,1]}(x)$, so $T_k = \mathcal{H}U_{k-1}$ on $[-1,1]$ and the second identity holds. The sign difference between the expressions is ultimately due to the property $\mathcal{H}^2 = -I$ of the Hilbert transform.

Finally, the function $R(z) = 1/\sqrt{1 - z^2}$ is analytic on $\mathbf{C} \setminus [-1,1]$ and

$$R(x) = \frac{1}{\sqrt{1-x^2}}\mathbf{I}_{[-1,1]}(x) - i\frac{\operatorname{sgn}(x)}{\sqrt{x^2-1}}\mathbf{I}_{\mathbf{R}\setminus[-1,1]}(x), \tag{4.5.9}$$

so $\mathcal{H}R(x)$ vanishes for $x \in (-1,1)$.   □

Proposition 4.5.3 (*Tricomi [160]*). *Suppose that w is a polynomial with expansion $w(x) = \sum_{k=0}^{\infty} \alpha_k T_{k+1}(x)$ as a finite sum of Chebyshev polynomials. Then in $L^p(-1,1)$ with $p > 1$ the general solution of the integral equation*

$$w(x) = \text{p.v.} \frac{1}{\pi} \int_{-1}^{1} \frac{q(y)dy}{x - y} \qquad (x \in (-1,1)) \tag{4.5.10}$$

is given by

$$q(x) = \sqrt{1-x^2} \sum_{k=0}^{\infty} \alpha_k U_k(x) + \frac{\gamma}{\sqrt{1-x^2}}; \qquad (4.5.11)$$

further, if q belongs to $L^2(-1,1)$, then $\gamma = 0$.

Proof. The fact that q satisfies the integral equation is immediate from Lemma 4.5.2. The unique solution in $L^p(-1,1)$ with $p > 1$ of the corresponding homogeneous equation

$$\text{p.v.} \int_{-1}^{1} \frac{q(y)dy}{x-y} = 0 \qquad (x \in (-1,1)) \qquad (4.5.12)$$

is given by $q(x) = \gamma(1-x^2)^{-1/2}$. When $\gamma \neq 0$, this q belongs to $L^p(-1,1)$ for $p < 2$, but does not belong to $L^2(-1,1)$. □

Proposition 4.5.3 gives an effective means for determining the equilibrium density for convex polynomial potentials w. We shall use this result for quadratics in the remainder of this section then for quartics in Section 4.6.

Example 4.5.4 (*Semicircle law*). Let $v(x) = \beta x^2$, and note that the integral equation for the equilibrium distribution p and v' is

$$v'(x) = 2\beta x = \text{p.v.} \int_{-1}^{1} \frac{\beta p(t)dt}{x-t} \qquad (x \in (-1,1)). \qquad (4.5.13)$$

Since $p \in L^2(-1,1)$, the unique solution is the semicircular law

$$\sigma_{0,1}(dx) = \frac{2}{\pi}\sqrt{1-x^2}\, \mathbf{I}_{[-1,1]}(x)dx. \qquad (4.5.14)$$

Definition (*Semicircular distribution*). The probability distribution that has probability density function

$$p(x) = \frac{2}{\pi r^2}\sqrt{r^2 - (x-a)^2}\,\mathbf{I}_{[a-r,a+r]}(x) \qquad (4.5.15)$$

is the semicircular $S(a,r)$ distribution.

Proposition 4.5.5 *The moment sequence for the $S(0,2)$ distribution is*

$$m_n = \begin{cases} 0, & \text{for odd } n; \\ \dfrac{2}{n+2}\dbinom{n}{n/2}, & \text{for even } n. \end{cases} \qquad (4.5.16)$$

Proof. This following computation dates back to Wallis, and is a familiar calculus exercise. The odd case is clear. In the even case, we substitute $x = 2\sin\theta$ and obtain

$$m_n = \frac{2^{n+1}}{\pi} \int_0^{\pi/2} \sin^n \theta \cos^2 \theta \, d\theta$$

$$= \frac{2^n \Gamma(3/2)\Gamma((n+1)/2)}{\Gamma(n/2+2)\pi}, \qquad (4.5.17)$$

and one can simplify the gamma functions. □

Definition (*Catalan numbers*). The Catalan numbers

$$C_n = \frac{1}{n+1}\binom{2n}{n} \qquad (4.5.18)$$

arise in various problems in enumerative geometry. See [83].

Proposition 4.5.6 *The Cauchy transform of the $S(a,r)$ distribution is*

$$G(z) = \frac{2}{r}\left(\frac{z-a}{r} - \sqrt{\left(\frac{z-a}{r}\right)^2 - 1}\right), \qquad (4.5.19)$$

where the branch of the square root is chosen so that $G(z) \sim 1/z$ as $|z| \to \infty$. Furthermore

$$G(z) = \frac{2}{r} \sum_{k=0}^{\infty} \left(\frac{2(z-a)}{r}\right)^{-2k-1} C_k. \qquad (4.5.20)$$

Proof. By expanding the integrand as a geometric series, we have

$$G(z) = \frac{2}{\pi r^2} \int_{a-r}^{a+r} \frac{\sqrt{r^2 - (x-a)^2}}{z-x}\, dx$$

$$= \frac{2}{\pi r} \sum_{n=0}^{\infty} \left(\frac{z-a}{r}\right)^{-n-1} \int_{-1}^{1} x^n \sqrt{1-x^2}\, dx \qquad (|z-a| > r).$$

$$(4.5.21)$$

Plainly only the even terms contribute, and as in Proposition 4.5.4 we obtain

$$G(z) = \frac{1}{r} \sum_{k=0}^{\infty} \left(\frac{z-a}{r}\right)^{-2k-1} \binom{2k}{k} \frac{1}{2^{2k}(k+1)}. \qquad (4.5.22)$$

By comparing with the binomial expansion, we obtain

$$G(z) = -\frac{2}{r}\sum_{k=0}^{\infty}(-1)^k\binom{1/2}{k+1}\left(\frac{z-a}{r}\right)^{-2k-1}$$

$$= \frac{2}{r}\left(\frac{z-a}{r}\right)\sum_{k=0}^{\infty}(-1)^{k+1}\binom{1/2}{k+1}\left(\frac{z-a}{r}\right)^{-2(k+1)}$$

$$= \frac{2}{r}\left(\frac{z-a}{r}\right)\left(1-\left(1-\left(\frac{z-a}{r}\right)^{-2}\right)^{1/2}\right) \qquad (4.5.23)$$

We choose $\sqrt{(z-a)^2-r^2}$ to be asymptotic to $z-a$ as $z\to\infty$; hence the statement of the Proposition follows from an identity of the form $1/(\sqrt{c}+\sqrt{b}) = (\sqrt{c}-\sqrt{b})/(c-b)$. □

Corollary 4.5.7 *The Hilbert transform of the $S(a,r)$ distribution is*

$$\frac{2}{\pi^2 r^2}\text{p.v.}\int_{a-r}^{a+r}\frac{\sqrt{r^2-(t-a)^2}}{x-t}dt$$

$$= \frac{(2/\pi)\text{sgn}(x-a)}{|x-a|+\sqrt{(x-a)^2-r^2}} \qquad (|x-a|>r)$$

$$= \frac{2}{\pi r^2}(x-a) \qquad (|x-a|<r). \qquad (4.5.24)$$

□

Proof. The integrand $f(x)$ is Hölder continuous inside $(a-r,a+r)$ and hence we have

$$\text{p.v.}\int\frac{f(t)}{x-t}\frac{dt}{\pi} = \frac{G^+(x)+G^-(x)}{2\pi}, \qquad (4.5.25)$$

where by (4.5.21), we have

$$G(z) = \frac{2}{z-a+\sqrt{(z-a)^2-r^2}} \qquad (|z-a|>r). \qquad (4.5.26)$$

Note that the root has opposite signs on opposite sides of the cut $[a-r,a+r]$, but is positive across $[a+r,\infty)$ and negative across $(-\infty,a-r)$. □

4.6 The quartic model with positive leading term

In this section we consider an historically important example which involves several important ideas. The even quartic potential

- $gx^4/4 \pm x^2/2$ with $g > 0$ satisfies (a), (b) and (c) of Section 4.4;

- $gx^4/4 + x^2/2$ with $g \geq 0$ has an equilibrium measure supported on a single interval;
- $gx^4/4 - x^2/2$ with $g > 0$ has an equilibrium measure supported on one or two intervals.

For $g \geq 0$, we introduce the symmetric and uniformly convex potential

$$v(x) = \frac{g}{4}x^4 + \frac{1}{2}x^2 \qquad (x \in \mathbf{R}). \tag{4.6.1}$$

We shall determine the equilibrium measure by using the technique of Section 4.4. Most of the results appear in [39, 83], and here we fill in the details of the calculations that were reported there. Such potentials have also been considered in free probability, as in [15].

Theorem 4.6.1 *Let $a, g > 0$ satisfy $3ga^4 + a^2 = 1$. Then the density of the equilibrium distribution for the potential v is supported on $[-2a, 2a]$ and has the form*

$$p(x) = \frac{1}{2\pi}(gx^2 + 1 + 2ga^2)\sqrt{4a^2 - x^2}. \tag{4.6.2}$$

Proof. By Proposition 4.4.3, the equilibrium distribution of v is supported on a single interval which is symmetrical about zero, so we take the support to be $[-2a, 2a]$ and write

$$v'(x) = \text{p.v.} \int_{-2a}^{2a} \frac{2p(y)dy}{x - y}, \tag{4.6.3}$$

where a is to be determined. Next we scale the support to $[-1, 1]$ by writing $x = 2au$

$$8a^3gu^3 + 2au = \text{p.v.} \frac{2}{\pi} \int_{-1}^{1} \frac{\pi p(2at)\, dt}{u - t}, \tag{4.6.4}$$

which we express in Chebyshev polynomials as

$$a^3gT_3(u) + (a + 3a^3g)T_1(u) = \frac{1}{\pi} \int_{-1}^{1} \frac{\pi p(2at)dt}{u - t}. \tag{4.6.5}$$

By Theorem 4.4.1, the equilibrium distribution of v belongs to $L^2(\mathbf{R})$, so in Lemma 4.5.2, we must have $\gamma = 0$. By Lemma 4.5.2, the required solution is

$$\pi p(2au) = \left(a^3gU_2(u) + (a + 3a^3g)U_0(u)\right)\sqrt{1 - u^2}, \tag{4.6.6}$$

which reduces to

$$p(t) = \frac{1}{2\pi}\left(1 + 2a^2g + gt^2\right)\sqrt{4a^2 - t^2}. \tag{4.6.7}$$

To ensure that p is a probability density, we need to substitute $t = 2a\cos\theta$ and impose the condition

$$1 = \int_{-2a}^{2a} p(x)\,dx = a^2 + 3a^4 g. \tag{4.6.8}$$

We note that $p(x) \geq 0$ provided that $1 + 6a^2 g \geq 0$, where $1 + 6a^2 g = \sqrt{1 + 12g}$. When these conditions are satisfied, $\rho_v(dx) = p(x)dx$ gives the equilibrium distribution, since one can show that there exists C such that

$$v(x) \geq \int_{-2a}^{2a} \log|x - y|p(y)dy + C \qquad (x \in \mathbf{R}) \tag{4.6.9}$$

with equality if and only if $x \in [-2a, 2a]$, so ρ_v satisfies the uniqueness conditions of Theorem 4.4.1. $\qquad\square$

Corollary 4.6.2 *The Cauchy transform of p is*

$$G(z) = \frac{1}{2}(z + gz^3) - \frac{1}{2}(1 + 2ga^2 + gz^2)\sqrt{z^2 - 4a^2} \qquad (z \in \mathbf{C} \setminus [-2a, 2a]) \tag{4.6.10}$$

where the root is so chosen that $\sqrt{z^2 - 4a^2} \asymp z$ as $|z| \to \infty$. The odd moments m_{2k-1} are all zero, whereas the even moments are

$$m_n = \binom{n}{n/2}\frac{2(1 + 2a^2 g)a^{n+2}}{n+2} + \binom{n+2}{(n+2)/2}\frac{2ga^{n+4}}{n+4}. \tag{4.6.11}$$

Proof. By the calculation of the moments for the semicircular law, we have

$$m_n = \frac{1}{2\pi}\int_{-2a}^{2a} t^n(1 + 2a^2 g + gt^2)\sqrt{4a^2 - t^2}\,dt$$

$$= (1 + 2a^2 g)a^{n+2}\int_{-2}^{2} t^n\sqrt{4 - t^2}\,\frac{dt}{2\pi} + ga^{n+4}\int_{-2}^{2} t^{n+4}\sqrt{4 - t^2}\,\frac{dt}{2\pi}. \tag{4.6.12}$$

The given $G(z)$ satisfies:

(i) $G(z) = 1/z + O(1/z^2)$ as $|z| \to \infty$;
(ii) $G(z)$ is analytic on $\mathbf{C} \setminus [-2a, 2a]$;
(iii)

$$p(x) = \lim_{h \to 0+} \frac{-1}{2\pi i}(G(x + ih) - G(x - ih)). \tag{4.6.13}$$

We choose the square root so that $\sqrt{z^2 - 4a^2} = z(1 - 4a^2/z^2)^{1/2}$ and one can use the binomial theorem to verify (i). Then (ii) is clear, and (iii) follows immediately. The stated result on moments follows from the expansion of $G(z)$ by the binomial theorem.   □

Now consider the potential $w(x) = gx^4/4 - x^2/2$ with $g > 0$. When x is large, the quartic term predominates; whereas when $|x|$ is small, $w(x)$ is concave and has a local maximum at 0. When $g > 1/4$, the local maximum at 0 has little effect, so the equilibrium measure is supported on a single interval and has a density that is given by a formula resembling that in Theorem 4.6.1. When $0 < g < 1/4$, the equilibrium measure is supported on two disjoint intervals which surround the local minima of w, as we prove in the following result.

Proposition 4.6.3 *Suppose that* $0 < g < 1/4$. *Then the equilibrium distribution for* $w(x) = gx^4/4 - x^2/2$ *is an even function* p *that is supported on the pair of disjoint intervals given by* $\{x \in \mathbf{R} : 1 - 2\sqrt{g} < gx^2 < 1 + 2\sqrt{g}\}$, *and on these intervals*

$$p(x) = \frac{g|x|}{2\pi}\left[\left(\frac{1 + 2\sqrt{g}}{g} - x^2\right)\left(x^2 - \frac{1 - 2\sqrt{g}}{g}\right)\right]^{1/2}. \qquad (4.6.14)$$

Proof. By means of a trick whereby we exploit the symmetry of the problem, we construct a solution to the integral equation by using Lemma 4.5.2, and then invoke uniqueness to show that this gives the density of the equilibrium measure. Suppose that p is an even probability density function that is supported on $[-b, -c] \cup [c, b]$; then

$$w'(x) = \text{p.v.} \int_c^b \frac{4xp(y)dy}{x^2 - y^2} \qquad (x \in (c, b)) \qquad (4.6.15)$$

and with the substitutions $x = \sqrt{\xi}$ and $y = \sqrt{\eta}$ we obtain

$$g\xi - 1 = \text{p.v.} \int_{c^2}^{b^2} \frac{2p(\sqrt{\eta})\, d\eta}{(\xi - \eta)\sqrt{\eta}}. \qquad (4.6.16)$$

We introduce the variables X and Y by

$$\xi = 2^{-1}(c^2 + b^2) + 2^{-1}(b^2 - c^2)X, \quad \eta = 2^{-1}(c^2 + b^2) + 2^{-1}(b^2 - c^2)Y \qquad (4.6.17)$$

so that

$$\frac{b^2 - c^2}{8\pi}\left((b^2 - c^2)gX + g(c^2 + b^2) - 2\right) = \text{p.v.}\frac{1}{\pi}\int_{-1}^1 \frac{q(Y)dY}{X - Y}, \qquad (4.6.18)$$

where q is a probability density function in $L^2(-1, 1)$. By comparing this with Lemma 4.5.1, we deduce that the constant term on the left-hand side must vanish, and

$$q(Y) = \frac{g(b^2 - c^2)^2}{8\pi} \sqrt{1 - Y^2} \qquad (-1 < Y < 1), \qquad (4.6.19)$$

so

$$g(b^2 - c^2)^2 = 16, \quad g(c^2 + b^2) = 2. \qquad (4.6.20)$$

Writing $\eta = y^2$, from the equation

$$\frac{p(y)}{|y|} = \frac{2}{(b^2 - c^2)} q(Y) = \frac{4}{\pi(b^2 - c^2)} \left((1 - Y)(1 + Y)\right)^{1/2} \qquad (4.6.21)$$

we recover

$$p(y) = \frac{8|y|}{\pi(b^2 - c^2)^2} \left((b^2 - y^2)(y^2 - c^2)\right)^{1/2} \qquad (c^2 < y^2 < b^2), \quad (4.6.22)$$

which is equivalent to the stated result. ☐

Exercise 4.6.4 Let $w(x) = gx^4/4 - x^2/2$ where $g > 1/4$. Show that the equilibrium measure is supported on $[-b, b]$ where

$$\frac{3}{16} g b^4 - \frac{b^2}{4} = 1 \qquad (4.6.23)$$

and that the equilibrium density is

$$p(x) = \frac{1}{2\pi} \left(gx^2 + \frac{gb^2}{2} - 1\right) \sqrt{b^2 - x^2}. \qquad (4.6.24)$$

Show further that as $g \to 1/4+$, $b^2 \to 8-$ and $p(0) \to 0+$. Hence there is a phase transition at $g = 1/4$.

Example 4.6.5 The double-well potential $v(x) = 4^{-1}(x^2 - 1)^2$ has minima at ± 1 and is even. By Exercise 4.6.4, the equilibrium distribution is supported on $[-b, b]$, where

$$b = \left(\frac{2}{3}(1 + \sqrt{13})\right)^{1/2}. \qquad (4.6.25)$$

Carlen *et al.* [46] have considered displacement convexity for this potential in a model for binary fluids.

Exercise 4.6.6 Let v be an even polynomial of degree 6 with positive leading term; so that $v(x) = w(x^2)$ for some cubic w. Let p_v be the equilibrium measure for v. Show that there exist $0 \leq a \leq b \leq c$ such that

$$w'(\xi) = \text{p.v.} \int_{[0,a]\cup[b,c]} \frac{p_v(\sqrt{\eta})d\eta}{(\xi - \eta)\sqrt{\eta}} \qquad (\xi \in [0,a] \cup [b,c]). \qquad (4.6.26)$$

4.7 Quartic models with negative leading term

In this section we consider even quartic potentials with negative leading term. In particular the potential $v(x) = gx^4/+x^2/2$ with $g < 0$ violates hypotheses (a) and (b) of Section 4.4 and does not have an equilibrium measure in the sense of Section 4.4; so in this section we constrain the support of the measures in the equilibrium problem to be contained in a compact set S. Thus we obtain an equilibrium measure which depends upon v and S.

We can regard $v(x) = x^2/2 + gx^4/4$ as a perturbation of the quadratic potential $x^2/2$, and seek to expand various quantities as power series in g for small $|g|$. Later, we consider whether the energy thus obtained relates to a meaningful variational problem. We continue with the notation and conditions from Section 4.6.

Proposition 4.7.1 *(i) For $|g| < 1/12$, a^2 is an analytic function of g with*

$$a^2 = \frac{1}{6g}\left(-1 + \sqrt{1+12g}\right). \qquad (4.7.1)$$

(ii) With $p(x) = (1 + 2a^2g + gx^2)\sqrt{4a^2 - x^2}/(2\pi)$ the energy function

$$E(g) = \int_{-2a}^{2a} v(x)p(x)\,dx + \iint \log\frac{1}{|x-y|}p(x)p(y)\,dxdy \qquad (4.7.2)$$

satisfies $E(0) = 3/4$ and

$$E(g) = E(0) + \frac{1}{24}(a^2-1)(9-a^2) - \frac{1}{2}\log a^2. \qquad (4.7.3)$$

Proof. (i) From the condition that $\int_{-2a}^{2a} p(x)dx = 1$, we have $3ga^4 + a^2 = 1$ and as a consequence we have $p(x) \geq 0$. Then we recover a^2 from this quadratic equation, choosing the root such that $a^2 \to 1$ as $g \to 0$.

(ii) Since $v(0) = 0$, we can integrate (4.4.18) and obtain the identity

$$v(x) = 2\int_{-2a}^{2a} (\log|x-y| - \log|y|)p(y)\,dy, \qquad (4.7.4)$$

and hence

$$E(g) = \frac{1}{2} \int_{-2a}^{2a} v(x)p(x)\,dx - \int_{-2a}^{2a} (\log|x|)p(x)\,dx. \qquad (4.7.5)$$

In the first of these integrals we substitute $x = 2a\sin\theta$ and obtain after some elementary integration

$$\frac{1}{2} \int_{-2a}^{2a} v(x)p(x)\,dx = \frac{1}{4}a^4 + \frac{5}{12}a^2(1 - a^2) + \frac{1}{8}(1 - a^2)^2. \qquad (4.7.6)$$

In the other integral we need an identity due to Euler [35, p. 245]

$$\int_0^{\pi/2} \log|\sin\theta|\,d\theta = -\frac{\pi}{2}\log 2 \qquad (4.7.7)$$

and two consequences of integrating this logsine integral by parts, namely

$$\int_0^{\pi/2} \sin^2\theta \log|\sin\theta|\,d\theta = \frac{-\pi}{4}\log 2 + \frac{\pi}{8}, \qquad (4.7.8)$$

$$\int_0^{\pi/2} \sin^4\theta \log|\sin\theta|\,d\theta = \frac{-3\pi}{16}\log 2 + \frac{7\pi}{64}. \qquad (4.7.9)$$

Hence with $x = 2a\sin\theta$ we have

$$\int_{-2a}^{2a} \log|x|p(x)\,dx = \log|2a|$$

$$+ \frac{4a^2}{\pi} \int_0^{\pi/2} \big((1 + 2a^2 g) + (2a^2 g - 1)\sin^2\theta$$

$$- 4a^2\sin^4\theta\big)\log|\sin\theta|\,d\theta$$

$$= \log a - \frac{1}{4}(1 + a^2). \qquad (4.7.10)$$

When $g = 0$, we have $a = 1$ and $E(0) = 3/4$, Hence

$$E(g) - E(0) = \frac{1}{4}a^4 + \frac{5}{12}a^2(1 - a^2) + \frac{1}{8}(1 - a^2)^2 + \frac{1}{4} + \frac{a^2}{4} - \frac{3}{4} - \log a \qquad (4.7.11)$$

which reduces to the formula

$$E(g) - E(0) = \frac{1}{24}(a^2 - 1)(9 - a^2) - \frac{1}{2}\log a^2. \qquad (4.7.12)$$

$\square$

When $g < 0$, the potential does does not satisfy the condition (b) stated in Section 4.4, so Theorem 4.4.1 does not apply. We consider v

to represent an electrical field and ρ to be the distribution of electrical charge on a conductor S.

Problem. (*Constrained equilibrium problem with potential*). We choose S to be some interval and consider the variational problem of finding $\rho \in Prob(S)$ that attains the infimum

$$\inf_{\rho}\Big\{ E_{v,S}(\rho) : E_{v,S}(\rho) = \int_S (x^2/2 + gx^4/4)\rho(dx)$$
$$+ \iint_{S \times S} \log \frac{1}{|x-y|}\rho(dx)\rho(dy)\Big\}. \qquad (4.7.13)$$

The constrained variational problem is considered in detail by Saff and Totik in [137], and the paper [57] contains some subtle refinements; here we summarize the conclusions without attempting to fit together the proofs needed to address the different cases. Clearly we can make $E_{v,S}(\rho)$ tend to $-\infty$ when $S = \mathbf{R}$ and $g < 0$ by moving mass towards $\pm\infty$; so we choose $S = [-\lambda, \lambda]$ to make the problem nontrivial. There are a few intuitive observations concerning the solution of this constrained minimization problem.

(1) The minimizer ρ_v will be symmetrical about 0.
(2) The support S_0 of ρ_v can consist of several intervals

$$S_0 = [-\lambda, -a_j] \cup [-b_{j-1}, -a_{j-1}] \cup \cdots \cup [a_j, \lambda]. \qquad (4.7.14)$$

(3) On S_0, the integral equation

$$gx^3 + x = \text{p.v.} \int_{S_0} \frac{\rho(dy)}{x-y} \qquad (4.7.15)$$

holds, so ρ must be absolutely continuous on S_0 and have a L^2 density on compact subsets inside S_0.
(4) When we replace the potential by $v = 0$, leaving only the logarithmic energy, by [137, page 25] the equilibrium distribution is the Chebyshev (inverse sine) distribution

$$\rho_0(dx) = \frac{1}{\pi\sqrt{\lambda^2 - x^2}}\mathbf{I}_{(-\lambda,\lambda)}(x)dx \qquad (4.7.16)$$

by [137, page 25]. Thus Lemma 4.4.2 does not apply to $\pm\lambda$ in constrained variational problems on $[-\lambda, \lambda]$.

Proposition 4.7.2 *Let S_0 be the support of ρ_v. There exist critical values $g_2 < g_1 < g_0$ such that:*

(i) for $g > g_0$, there exists $b < \lambda$ such that $[-b, b] = S_0$;
(ii) for $g_1 < g < g_0$, $S_0 = [-\lambda, \lambda]$;

(iii) for $g_2 < g < g_1$, there exists $0 < b < c < \lambda$ such that $S_0 = [-\lambda, -c] \cup [-b, b] \cup [c, \lambda]$;

(iv) for $g < g_2$, there exists $0 < c < \lambda$ such that $S_0 = [-\lambda, -c] \cup [c, \lambda]$.

Definition. We choose the radicals so that $(z^2 - \lambda^2)^{1/2} > 0$ for real $z > \lambda$, then we write $(x^2 - \lambda^2)^{1/2}_+$ for the value of this radical on the top half of the cut $[-\lambda, \lambda]$. To deal with case (ii), we introduce the function

$$F(z) = -\frac{(z^2 - \lambda^2)^{1/2}}{2\pi^2} \int_{-\lambda}^{\lambda} \frac{v'(y)}{(y^2 - \lambda^2)^{1/2}_+} \frac{dy}{z - y} + \frac{i\gamma}{(z^2 - \lambda^2)^{1/2}}.$$

(4.7.17)

In the notation of case (iii), let

$$G(z) = \frac{R(z)^{1/2}}{2\pi^2} \int_{S_0} \frac{v'(y)}{R(y)^{1/2}} \frac{dy}{z - y}$$

(4.7.18)

where

$$R(z) = \frac{(z + c)(z + b)(z - b)(z - c)}{z^2 - \lambda^2},$$

(4.7.19)

and we choose the radicals so that $R(z)^{1/2} > 0$ for real $z > \lambda$.
In the notation of case (iv), let

$$H(z) = \frac{T(z)^{1/2}}{2\pi^2} \int_{S_0} \frac{v'(y)}{T(y)^{1/2}} \frac{dy}{z - y}$$

(4.7.20)

where

$$T(z) = \frac{z^2 - c^2}{z^2 - \lambda^2}.$$

(4.7.21)

We choose the radicals so that $T(z)^{1/2} > 0$ for real $z > \lambda$. Given a radical $K(z)$ on a neighbourhood of (a, b), we let $K_+(x) = \lim_{h \to 0+} K(x + ih)$ for $x \in (a, b)$.

Theorem 4.7.3 *In case (ii), the minimizer ρ_v is absolutely continuous with respect to Lebesgue measure and its density p satisfies $p(x) = \Re F(x)_+$, so*

$$p(x) = \frac{\sqrt{\lambda^2 - x^2}}{2\pi} \left(g(16x^2 - 7\lambda^2) + 1\right) + \frac{\gamma}{\sqrt{\lambda^2 - x^2}} \quad (-\lambda < x < \lambda).$$

(4.7.22)

Furthermore, the interior of the support of ρ_v equals $\{x : \Re F(x)_+ > 0\}$. Corresponding results hold in the cases (iii) with F replaced by G and (iv) with F replaced by H.

Instead of providing a proof of Theorem 4.7.3, we consider how the various cases arise and give some indication of how the various parameters are related in the specific cases.

Remarks. (i) Suppose the $\lambda > 2$, and let g decrease from $g = 0$. The constrained variational problem on $[-\lambda, \lambda]$ is essentially equivalent to the free variational problem since the potential is either convex or sufficiently close to convex that the equilibrium distribution is supported on the single interval $[-2a, 2a]$ which is contained in $[-\lambda, \lambda]$. Hence the equilibrium distribution is given by Theorem 4.6.1, so

$$p(x) = \frac{1}{2\pi}(gx^2 + 1 + gb^2/2)\sqrt{b^2 - x^2}.$$

Then in the notation of Section 4.6, $b = 2a$ where $a = 1$ at $g = 0$ and $2a = 2\sqrt{2}$ at $g = -1/12$.

(ii) As g decreases, b increases so that the support of p fills out $(-\lambda, \lambda)$ and $\pm\lambda$ become endpoints. To find the equilibrium distribution p, we substitute $z = \lambda \cos \theta$ and $y = \lambda \cos \phi$ in the formula for F so that

$$v'(\lambda \cos \phi) = g\lambda^3(4\cos 3\phi - 3\cos \phi) + \lambda \cos \phi \qquad (4.7.23)$$

then

$$F(\lambda \cos \theta) = \frac{-\sin \theta}{2\pi^2} \int_0^\pi \frac{v'(\lambda \cos \phi)d\phi}{\cos \theta - \cos \phi} + \frac{\gamma}{\lambda \sin \theta}, \qquad (4.7.24)$$

which we evaluate with the help of Lemma 4.7.1 to obtain

$$F(\lambda \cos \theta) = \frac{1}{2\pi^2}\left(4g\lambda^3 \sin 3\theta - 3g\lambda^3 \sin \theta + \lambda \sin \theta\right) + \frac{\gamma}{\lambda \sin \theta}, \qquad (4.7.25)$$

which in the original variables is

$$F(x) = \frac{\sqrt{\lambda^2 - x^2}}{2\pi}\left(g(16x^2 - 7\lambda^2) + 1\right) + \frac{\gamma}{\sqrt{\lambda^2 - x^2}}. \qquad (4.7.26)$$

Now F is a probability density function on $[-\lambda, \lambda]$, only if

$$1 = \int_{-\lambda}^{\lambda} F(x)\, dx$$

$$= \int_0^\pi F(\lambda \cos \theta)\lambda \sin \theta\, d\theta$$

$$= \frac{1}{2\pi} \int_0^\pi \left(4g\lambda^4 \sin 3\theta \sin \theta + (\lambda^2 - 3g\lambda^4)\sin^2 \theta\right) d\theta + \pi\gamma$$

$$= \frac{\lambda^2 - 3g\lambda^4}{4} + \pi\gamma, \qquad (4.7.27)$$

and $F(x) \geq 0$ only if $\gamma \geq 0$. When $\gamma > 0$, $F(x)$ behaves like a Chebyshev (inverse sine) distribution as $x \to \pm\lambda$, indicating that the equilibrium distribution tends to accumulate towards the ends of the interval $S = [-\lambda, \lambda]$.

Suppose now that $\lambda < 2$. Then $\gamma > 0$ for $|g|$ sufficiently small and one can check that F does define a probability density function on $[-\lambda, \lambda]$.

(iii) Note that $v(x)$ is convex near to $x = 0$, so we would still expect the equilibrium distribution to accumulate near to the local minimum of v at $x = 0$. As g decreases further, two gaps open up in the support of p on either side of the local minimum of v, so there are Chebyshev type singularities at $\pm\lambda$, and semicircle type endpoints at $\pm c$ and $\pm b$ inside $(-\lambda, \lambda)$.

(iv) As g decreases further still, the central interval $[-b, b]$ disappears from the support of p, leaving only the intervals $[-\lambda, -c]$ and $[c, \lambda]$ at the ends. When $|g|$ is large, ρ will have mass accumulating at $\pm\lambda$, so $d\rho/dx$ is unbounded near to $\pm\lambda$, indeed the equilibrium density grows like $C/\sqrt{\lambda - |x|}$.

- The notion of an equilibrium measure can depend upon the potential and on the constraints imposed upon the support.

4.8 Displacement convexity and relative free entropy

In this section, we use a special kind of affine structure on the cone of probability density functions, which was introduced by McCann in the context of gas dynamics. See [114, 162, 174].

Example 4.8.1 Let Φ and Ψ be strictly increasing and continuous cumulative distribution functions, and let $\phi : (0, 1) \to \mathbf{R}$ be the inverse function of Φ and $\psi : (0, 1) \to \mathbf{R}$ be the inverse function of Ψ. Then for each $0 < t < 1$, the function $R_t(x) = (1 - t)\Phi(x) + t\Psi(x)$ is likewise a strictly increasing and continuous cumulative distribution function. The corresponding Riemann–Stieltjes integrals give the standard notion of convexity on the space of probability measures on $\mathbf{R}$; so that

$$\int v(x)dR_t(x) = (1 - t) \int v(x)d\Phi(x) + t \int v(x)d\Psi(x)$$

$$= (1 - t) \int_0^1 v(\phi(y))dy + t \int_0^1 v(\psi(y))dy \quad (4.8.1)$$

for all bounded and continuous real functions v.

In contrast, $\rho_t(y) = (1-t)\phi(y) + t\psi(y)$ is a strictly increasing and continuous function from $(0,1)$ onto $\mathbf{R}$, and hence is the inverse of some cumulative distribution function S_t; but the corresponding integrals satisfy

$$\int v\, dS_t = \int_0^1 v(\rho_t(y))\, dy = \int_0^1 v\big((1-t)\phi(y) + t\psi(y)\big)\, dy. \qquad (4.8.2)$$

In each case, we obtain a one-parameter family of measures which interpolates between $d\Phi$ and $d\Psi$, but the properties of the intermediate measures dR_t and dS_t are different. Whereas (4.8.1) is an affine function of t, one can easily choose example of f such that (4.8.2) is not a convex function of t.

Definition (*Displacement interpolation*). Let μ and ν be probability measures on $[a,b]$ that have no atoms. Then by the proof of Proposition 1.6.1, there exists an increasing $\varphi : [a,b] \to [a,b]$ such that $\varphi \sharp \mu = \nu$; then we let

$$\varphi_t(x) = (1-t)x + t\varphi(x) \qquad (t \in [a,b], x \in [a,b]) \qquad (4.8.3)$$

be increasing functions and let $\nu_t = \varphi_t \sharp \mu$. We say that the family $(\nu_t)_{0 \le t \le 1}$ interpolates between μ and ν. In Section 6.1 we shall return to this concept and formulate it more generally for probability measures on $\mathbf{R}^n$, where we need a less obvious choice of inducing map φ.

Proposition 4.8.2 (*Displacement convexity of energy*). *Let v be twice continuously differentiable on $[a,b]$ with $v''(x) \ge \alpha$ for all $x \in (a,b)$ for some $\alpha > 0$. Then $t \mapsto E_v(\nu_t)$ is convex on $[a,b]$ with*

$$\frac{d^2}{dt^2} E_v(\nu_t) \ge \alpha W_2(\nu, \mu)^2. \qquad (4.8.4)$$

Proof. We use the definition of induced measure to write

$$E_v(\nu_t) = \int v(x)\, \nu_t(dx) - \frac{\beta}{2} \iint \log|x-y| \nu_t(dx)\nu_t(dy)$$

$$= \int v(\varphi_t(x))\mu(dx) - \frac{\beta}{2} \iint \log|\varphi_t(x) - \varphi_t(y)|\, \mu(dx)\mu(dy). \qquad (4.8.5)$$

Now $(d/dt)\varphi_t(x) = \varphi(x) - x$, hence

$$\frac{d}{dt} E_v(\nu_t) = \int v'(\varphi_t(x))(\varphi(x) - x)\, \mu(dx)$$

$$- \frac{\beta}{2} \iint \frac{\varphi(x) - x - \varphi(y) + y}{\varphi_t(x) - \varphi_t(y)} \mu(dx)\mu(dy) \qquad (4.8.6)$$

and so

$$\frac{d^2}{dt^2} E_v(\nu_t) = \int v''(\varphi_t(x))(\varphi(x) - x)^2 \,\mu(dx)$$
$$+ \frac{\beta}{2} \iint \left(\frac{\varphi(x) - x - \varphi(y) + y}{\varphi_t(x) - \varphi_t(y)} \right)^2 \mu(dx)\mu(dy)$$
$$\geq \alpha \int (\varphi(x) - x)^2 \mu(dx)$$
$$\geq \alpha W_2(\nu, \mu)^2. \tag{4.8.7}$$

$\square$

Definition (*Relative free entropy*). Let v be a potential with equilibrium measure ρ as in Theorem 4.4.1. Then for an arbitrary ν of finite energy, we define the relative free entropy with respect to ρ by

$$\Sigma(\nu \mid \rho) = E_v(\nu) - E_v(\rho). \tag{4.8.8}$$

By definition of ρ, we have $0 \leq \Sigma(\nu \mid \rho)$; note that some authors reverse the signs.

Theorem 4.8.3 (*Free transportation inequality*). *Let v be twice continuously differentiable on $[a, b]$ with $v''(x) \geq \alpha$ for all $x \in (a, b)$ for some $\alpha > 0$. Then ρ satisfies the free transportation inequality*

$$W_2(\nu, \rho)^2 \leq \frac{2}{\alpha} \Sigma(\nu \mid \rho). \tag{4.8.10}$$

Proof. There exists an increasing function that induces ν from ρ, so we can introduce a one-parameter family of measures ν_t with $\nu_0 = \rho$ and $\nu_1 = \nu$. By Proposition 4.8.2, the function $E_v(\nu_t)$ is convex and takes its minimum value at 0, hence by the mean value theorem

$$E_v(\nu_1) = E_v(\nu_0) + \frac{1}{2} \frac{d^2}{dt^2} E_v(\nu_t) \tag{4.8.11}$$

for some t, so

$$E_v(\nu_1) - E_v(\nu_0) \geq \frac{\alpha}{2} W_2(\nu_1, \nu_0)^2 \tag{4.8.12}$$

and hence

$$\Sigma(\nu \mid \rho) \geq \frac{\alpha}{2} W_2(\nu, \rho)^2. \tag{4.8.13}$$

$\square$

We can now recover a formula for the relative free entropy which resembles more closely the definition of relative free entropy over the circle.

Proposition 4.8.4 *Suppose that ν has finite energy and that the support of ν is contained in the support of ρ. Then*

$$\Sigma(\nu \mid \rho) = \frac{\beta}{2} \iint \log \frac{1}{|x-y|} (\nu(dx) - \rho(dx))(\nu(dy) - \rho(dy)). \quad (4.8.14)$$

Proof. On the support of ρ we have

$$v(x) = \beta \int \log |x-y| \rho(dy) + C \quad (4.8.15)$$

for some constant C; hence

$$\int v(x)(\nu(dx) - \rho(dx)) = \beta \iint \log |x-y| \rho(dy)(\nu(dx) - \rho(dx)). \quad (4.8.16)$$

We deduce that

$$\begin{aligned}
E_v(\nu) - E_v(\rho) &= \int v(x)(\nu(dx) - \rho(dx)) \\
&\quad + \frac{\beta}{2} \iint \log \frac{1}{|x-y|} \nu(dx)\nu(dy) \\
&\quad - \frac{\beta}{2} \iint \log \frac{1}{|x-y|} \rho(dx)\rho(dy) \\
&= \frac{\beta}{2} \iint \log \frac{1}{|x-y|} (\nu(dx) - \rho(dx))(\nu(dy) - \rho(dy)).
\end{aligned}$$
$$(4.8.17)$$

$\square$

4.9 Toeplitz determinants

The subspaces of H^2 are traditionally described by prediction theory. At each epoch n, we consider $0, 1, \ldots, n-1$ as the past and n as the future and, for a given bounded and positive Radon measure ω on with infinite support on $\mathbf{T}$, we wish to find

$$\alpha_n = \inf \left\{ \int_{\mathbf{T}} \left| e^{in\theta} - \sum_{j=0}^{n-1} a_j e^{ij\theta} \right|^2 \omega(d\theta) : a_j \in \mathbf{C} \right\}. \quad (4.9.1)$$

Our presentation is based upon [145], where one can find much more detail.

Definition (*Toeplitz determinant*). A $n \times n$ Toeplitz matrix has the form

$$T^{(n)} = [c_{k-j}]_{j,k=0,\ldots,n-1} \quad (4.9.2)$$

for some complex coefficients c_j so that the leading diagonals are constant; the corresponding Toeplitz determinant is $D_{n-1} = \det T^{(n)}$. In particular, given a measure ω on $\mathbf{T}$, we introduce the Fourier coefficients $c_j = \int_{\mathbf{T}} e^{-ij\theta}\omega(d\theta)$ and form the corresponding Toeplitz matrices $T^{(n)}$ and determinants D_{n-1}. Note that $D_n^{1/(n+1)}$ is the geometric mean of the eigenvalues of $T^{(n+1)}$.

We observe that $T^{(n)}$ is self-adjoint, and hence that its eigenvalues are real. Let ω be absolutely continuous, so $\omega(d\theta) = W(e^{i\theta})d\theta/(2\pi)$ where $W \geq 0$ is integrable.

Exercise. Verify that T^n as a positive definite matrix.

Proposition 4.9.1 (*Heine*). (*i*) *There exists a unique monic polynomial* Φ_n *of degree n such that*

$$\alpha_n = \int_{\mathbf{T}} |\Phi_n(e^{i\theta})|^2 W(e^{i\theta})\frac{d\theta}{2\pi}. \tag{4.9.3}$$

(*ii*) *This* Φ_n *is given by*

$$\Phi_n(z) = \frac{1}{D_{n-1}}\begin{vmatrix} c_0 & \bar{c}_1 & c_2 & \cdots & \bar{c}_n \\ c_1 & c_0 & \bar{c}_1 & \cdots & \bar{c}_{n-1} \\ \vdots & \vdots & \vdots & \ddots & \vdots \\ 1 & z & z^2 & \cdots & z^n \end{vmatrix}. \tag{4.9.4}$$

(*iii*) *The* Φ_n *are orthogonal in* $L^2(W)$.

Proof. (i) Since the measure has infinite support, the $e^{ij\theta}$ give linearly independent vectors in $L^2(W)$. By elementary Hilbert space theory, there is a unique element in span$\{1, e^{i\theta}, \ldots, e^{i(n-1)\theta}\}$ that is closest in $L^2(W)$ to $e^{in\theta}$, and this closest element is orthogonal to the span. See [171].

(ii) The given Φ_n is monic by the definition of the D_{n-1}. Furthermore, by substitution we obtain

$$\int_{\mathbf{T}} \Phi_n(e^{i\theta})e^{-ij\theta}W(e^{i\theta})\frac{d\theta}{2\pi} = \frac{1}{D_{n-1}}\begin{vmatrix} c_0 & \bar{c}_1 & c_2 & \cdots & \bar{c}_n \\ c_1 & c_0 & \bar{c}_1 & \cdots & \bar{c}_{n-1} \\ \vdots & \vdots & \vdots & \ddots & \vdots \\ c_j & c_{j-1} & c_{j-2} & \cdots & c_{j-n} \end{vmatrix} \tag{4.9.5}$$

where $c_{-k} = \bar{c}_k$, and since a determinant with two equal rows vanishes, we have

$$\int_{\mathbf{T}} \Phi_n(e^{i\theta})e^{-ij\theta}\frac{d\theta}{2\pi} = \begin{cases} 0 & \text{for } j = 0, 1, \ldots, n-1; \\ D_n/D_{n-1} & \text{for } j = n. \end{cases} \tag{4.9.6}$$

(iii) Given (i) and (ii), this is part of the Gram–Schmidt process. $\square$

It is convenient to write

$$\alpha_\infty = \exp\Big(\frac{1}{2\pi}\int_{\mathbf{T}} \log W(e^{i\theta})\, d\theta\Big). \qquad (4.9.7)$$

Theorem 4.9.2 (*Helson–Szegö, [102]*). *Suppose that ω is absolutely continuous with $d\omega = W(e^{i\theta})\frac{d\theta}{2\pi}$ and that $\log W$ is integrable. Then*

$$\alpha_n \to \alpha_\infty \qquad (n \to \infty). \qquad (4.9.8)$$

Proof. We introduce $P_n = \{\sum_{j=1}^n a_j z^j : a_j \in \mathbf{C}\}$ and observe that

$$\alpha_n = \inf\Big\{\int_{\mathbf{T}} |1 + p_n(e^{i\theta})|^2 W(e^{i\theta})\frac{d\theta}{2\pi} : p_n \in P_n \Big\}. \qquad (4.9.9)$$

As in Proposition 4.2.5(ii), we can introduce the harmonic conjugate v of $u = \log W$ and then form the function

$$h(z) = \exp\Big(\frac{1}{2}P_z u + \frac{i}{2}P_z v\Big) \qquad (4.9.10)$$

which belongs to H^2 and satisfies

$$|h(e^{i\theta})|^2 = W(e^{i\theta}) \qquad (4.9.11)$$

almost everywhere, and

$$h(0) = \exp\Big(\frac{1}{2}\int_{\mathbf{T}} \log W(e^{i\theta})\frac{d\theta}{2\pi}\Big). \qquad (4.9.12)$$

Hence for any $p_n \in P_n$, the function $(1 + p_n)h$ belongs to H^2 and so we have

$$\alpha_n \geq |(1 + p_n(0))h(0)|^2 = \exp\Big(\int_{\mathbf{T}} \log W(e^{i\theta})\frac{d\theta}{2\pi}\Big). \qquad (4.9.13)$$

The sequence (α_n) is evidently decreasing, and to show that the right-hand side of (4.9.7) is the limit, we exploit the fact that h is an outer function. By Beurling's theorem [102], there exist polynomials q_n of degree n such that

$$q_n(e^{i\theta})h(e^{i\theta}) \to h(0) \qquad (4.9.14)$$

in H^2 as $n \to \infty$, and hence $q_n(0) \to 1$ as $n \to \infty$. We now choose

$$p_n(z) = \frac{q_n(z)}{q_n(0)} - 1 \in P_n \qquad (4.9.15)$$

such that

$$\int_{\mathbf{T}} |1 + p_n(e^{i\theta})|^2 |h(e^{i\theta})|^2 \frac{d\theta}{2\pi} \to |h(0)|^2 \qquad (n \to \infty). \qquad (4.9.16)$$

$\square$

- The geometric mean of the eigenvalues of $T^{(n)}$ converges as $n \to \infty$ to the geometric mean of the weight W.

Corollary 4.9.3 (*Szegö*). *Let W be as in Theorem 4.9.2, and let D_n be the corresponding Toeplitz determinants. Then*

$$D_n^{1/n} \to \alpha_\infty \qquad (n \to \infty). \qquad (4.9.17)$$

Proof. Using orthogonality as in (4.9.5), one can check that

$$
\begin{aligned}
\alpha_n &= \int_{\mathbf{T}} \Phi_n(e^{i\theta}) \bar{\Phi}_n(e^{i\theta}) W(e^{i\theta}) \frac{d\theta}{2\pi} \\
&= \int_{\mathbf{T}} \Phi_n(e^{i\theta}) e^{-in\theta} W(e^{i\theta}) \frac{d\theta}{2\pi} \\
&= \frac{D_n}{D_{n-1}}
\end{aligned}
\qquad (4.9.18)
$$

by Proposition 4.9.1(ii). Hence we have

$$D_n = (D_n/D_{n-1})(D_{n-1}/D_{n-2})\ldots(D_2/D_1)D_1 = \alpha_n \alpha_{n-1} \ldots \alpha_2 D_1, \qquad (4.9.19)$$

where $\alpha_n \to \alpha_\infty$ as $n \to \infty$ by Theorem 4.9.2; the result now follows by elementary analysis. $\qquad \square$

Lemma 4.9.4 *Let ω be as in Theorem 4.9.2. Then*

$$D_{n-1} = \frac{1}{n!} \int \cdots \int_{\mathbf{T}^n} \prod_{0 \le j < k \le n-1} |e^{i\theta_j} - e^{i\theta_k}|^2 \omega(d\theta_0) \ldots \omega(d\theta_{n-1}). \qquad (4.9.20)$$

Proof. The right-hand side involves the Vandermonde determinant

$$\prod_{0 \le j < k \le n-1} (e^{i\theta_k} - e^{i\theta_j}) = \det[e^{ik\theta_j}]_{j,k=0,\ldots,n-1}; \qquad (4.9.21)$$

so the expression on the right-hand side of (4.9.20) is

$$\frac{1}{n!} \int_{\mathbf{T}} \det[W(e^{i\theta_k})e^{ij\theta_k}]_{j,k=0,\ldots,n-1} \det[e^{-i\ell\theta_k}]_{k,\ell=0,\ldots,n-1} \frac{d\theta_0}{2\pi} \cdots \frac{d\theta_{n-1}}{2\pi}; \qquad (4.9.22)$$

so by Andréief's Lemma 2.2.2, this expression reduces to

$$D_{n-1} = \det\left[\int_{\mathbf{T}} e^{i(j-k)\theta} W(e^{i\theta}) \frac{d\theta}{2\pi}\right]_{j,k=0,\ldots,n-1}. \qquad (4.9.23)$$

$\qquad \square$

Definition (*Linear statistics*). Let $f : \mathbf{T} \to \mathbf{R}$ be a continuous function. A linear statistic on $SU(n)$ is the function $F : SU(n) \to \mathbf{R}$

$$F_n(U) = \operatorname{trace} f(U) = \sum_{j=1}^{n} f(e^{i\psi_j}) \qquad (U \in U(n)) \tag{4.9.24}$$

where $(e^{i\psi_j})$ are the eigenvalues of U.

Proposition 4.9.5 *(i) The exponential integral over $U(n)$ of a linear statistic of f satisfies*

$$\int_{U(n)} e^{F_n(U)} \mu_{U(n)}(dU) = \det\left[\int_0^{2\pi} \exp\bigl(f(e^{i\psi}) + i(j-k)\psi\bigr) \frac{d\psi}{2\pi} \right]_{1 \le j,k \le n}. \tag{4.9.25}$$

(ii) The linear statistic satisfies a law of large numbers in the sense that

$$\frac{1}{n} \log\left(\int_{U(n)} e^{F_n(U)} \mu_{U(n)}(dU) \right) \to \int_{\mathbf{T}} f(e^{i\theta}) \frac{d\theta}{2\pi} \qquad (n \to \infty). \tag{4.9.26}$$

Proof. (i) We can regard F_n as a function on the maximal torus $\mathcal{T}$ and replace the Haar measure by the expression in Proposition 2.7.1. Then we have

$$\int_{U(n)} e^{F_n(U)} \mu_{U(n)}(dU)$$

$$= \frac{1}{n!} \int_{\mathcal{T}} \exp\left(\sum_{j=1}^{n} f(e^{i\psi_j})\right) \prod_{1 \le j < k \le n} |e^{i\psi_j} - e^{i\psi_k}|^2 \frac{d\psi_1}{2\pi} \cdots \frac{d\psi_n}{2\pi}. \tag{4.9.27}$$

Now we can apply Lemma 4.9.4 to obtain the stated identity.

(ii) Given (i), this follows directly by Corollary 4.9.3 with $W = e^f$. $\square$

In Section 10.6, we shall state a stronger version of Szegö's theorem which gives more precise information about the fluctuation of linear statistics; effectively we obtain a central limit theorem.

5

Convergence to equilibrium

Abstract

In this chapter we combine the results from Chapter 3 about concentration of measure with the notion of equilibrium from Chapter 4, and prove convergence to equilibrium of empirical eigenvalue distributions of $n \times n$ matrices from suitable ensembles as $n \to \infty$. We introduce various notions of convergence for eigenvalue ensembles from generalized orthogonal, unitary and symplectic ensembles. Using concentration inequalities from Chapter 3, we prove that the empirical eigenvalue distributions, from ensembles that have uniformly convex potentials, converge almost surely to their equilibrium distribution as the number of charges increases to infinity. Furthermore, we obtain the Marchenko–Pastur distribution as the limit of singular numbers of rectangular Gaussian matrices. To illustrate how concentration implies convergence, the chapter starts with the case of compact groups, where the equilibrium measure is simply normalized arclength on the circle.

5.1 Convergence to arclength

Suppose that n unit positive charges of strength $\beta > 0$ are placed upon a circular conductor of unit radius, and that the angles of the charges are $0 \leq \theta_1 < \theta_2 < \cdots < \theta_n < 2\pi$. Suppose that the θ_j are random, subject to the joint distribution

$$\sigma_n^{(\beta)}(d\Theta) = Z^{-1} \prod_{1 \leq j < k \leq n} |e^{i\theta_j} - e^{i\theta_k}|^\beta d\theta_1 d\theta_2 \ldots d\theta_n. \qquad (5.1.1)$$

Then we would expect that the θ_j would tend to form a uniform distribution round the circle as $n \to \infty$ since the uniform distribution appears to minimize the energy. We prove this for $\beta = 2$.

177

We introduce the product of maximal tori for the circular unitary ensembles, on which lives the probability measures given by the Weyl integration formula as in Proposition 2.7.1:

$$(T, \sigma) = \left(\prod_{n=1}^{\infty} \mathbf{T}^n, \bigotimes_{n=1}^{\infty} \sigma_{SU(n+1)} \right). \tag{5.1.2}$$

The empirical eigenvalue distribution for the nth factor is

$$\mu_n^\Theta = \frac{1}{n} \sum_{j=1}^{n} \delta_{\exp(i\theta_j)}, \tag{5.1.3}$$

so $\mu_n^\Theta \in Prob(\mathbf{T})$, with $\Theta = (e^{i\theta_j})_{j=1}^n$ is a random point of $\mathbf{T}^n$ which is subject to $\sigma_{SU(n+1)}$.

Theorem 5.1.1 (*Metric convergence of empirical distribution*). *The empirical distribution of the eigenvalues from the circular ensembles $SU(n)$ converges to normalized arclength measure almost surely; so that*

$$W_1(\mu_n^\Theta, d\theta/(2\pi)) \to 0 \qquad (n \to \infty) \tag{5.1.4}$$

almost surely on (T, σ).

Proof. Let $f : \mathbf{T} \to \mathbf{R}$ be a 1-Lipschitz function and let $\varepsilon > 0$. Then $F_n : (\mathbf{T}^n, \ell^2(n)) \to \mathbf{R}$ defined by

$$F_n(\Theta) = \frac{1}{n} \sum_{j=1}^{n} f(e^{i\theta_j}) - \int_{\mathbf{T}^n} \left(\frac{1}{n} \sum_{j=1}^{n} f(e^{i\theta_j'}) \right) \sigma_{SU(n+1)}(d\Theta') \tag{5.1.5}$$

is a K–Lipschitz function for some K independent of n as in Lemma 3.9.4, and

$$\int F_n(\Theta) \sigma_{SU(n+1)}(d\Theta) = 0. \tag{5.1.6}$$

By the concentration inequality Corollary 3.9.3, there exists a universal constant κ such that

$$\sigma_{SU(n+1)} \big[|F_n(\Theta)| > \varepsilon \big] \le 2 \exp(-\varepsilon^2 \kappa n) \tag{5.1.7}$$

hence

$$\sum_{n=1}^{\infty} \sigma_{SU(n+1)} \big[|F_n(\Theta)| > \varepsilon \big] < \infty. \tag{5.1.8}$$

So by the first Borel–Cantelli lemma, the event

$$\big[|F_n(\Theta)| > \varepsilon \quad \text{for infinitely many distinct} \quad n \big] \tag{5.1.9}$$

has zero probability; hence $F_n(\Theta) \to 0$ almost surely as $n \to \infty$.

With the convention that $e^{i\theta_{n+1}} = e^{-i(\theta_1 + \cdots + \theta_n)}$, the maximal torus of $SU(n+1)$ and the measure $\sigma_{SU(n+1)}$ are invariant under the transformation $\theta_j \mapsto \theta_j + 2\pi/(n+1)$ for $j = 1, \ldots, n+1$, which we can apply repeatedly to generate points on the maximal torus. So one can check by a simple averaging argument that

$$\left| \frac{1}{2\pi} \int_{\mathbf{T}} f(e^{i\theta}) \, d\theta - \int_{\mathbf{T}^n} \frac{1}{n} \sum_{j=1}^{n} f(e^{i\theta_j}) \sigma_{SU(n+1)}(d\Theta) \right| \leq \frac{1}{n}; \qquad (5.1.10)$$

hence the equilibrium measure is indeed normalized arclength measure on the circle, in the sense that

$$\frac{1}{n} \sum_{j=1}^{n} f(e^{i\theta_j}) \rightarrow \int_{\mathbf{T}} f(e^{i\theta}) \frac{d\theta}{2\pi} \qquad (5.1.11)$$

almost surely as $n \rightarrow$ on (T, σ).

The Arzelà–Ascoli theorem [141] shows that $\{g : \mathbf{T} \rightarrow \mathbf{R} : g(1) = 0, \|g\|_{Lip} \leq 1\}$ is a relatively compact subset of $(C(\mathbf{T}; \mathbf{R}), \|\cdot\|_\infty)$, and hence for each $\varepsilon > 0$ has a finite ε-net $f_1, \ldots, f_{M(\varepsilon)}$ such that $f_j(1) = 0$ and $\|f_j\|_{Lip} \leq 1$. We shall spell out details of how to choose ε-nets in the proof of Theorem 5.5.3, so leave the details until then.

Suppose that $W_1(\mu_n^\Theta, d\theta/2\pi) > 2\varepsilon$. Then by the Kantorovich duality Theorem 3.3.2, there exists g such that $g(1) = 0$, $\|g\|_{Lip} \leq 1$ and $\int g(\mu_n^\Theta(d\theta) - \frac{d\theta}{2\pi}) > 2\varepsilon$, so we can choose a j such that $\|g - f_j\|_\infty < \varepsilon$ and so $\int f_j(\mu_n^\Theta(d\theta) - \frac{d\theta}{2\pi}) > \varepsilon$. Hence

$$\left[W_1(\mu_n^\Theta, d\theta/2\pi) > 2\varepsilon \right] \subseteq \bigcup_{j=1}^{M(\varepsilon)} \left[\left| \int_{\mathbf{T}} f_j(e^{i\theta})(\mu_n^\Theta(d\theta) - \frac{d\theta}{2\pi}) \right| > \varepsilon \right].$$

$$(5.1.12)$$

Repeating the preceding proof with f_j instead of f, we deduce that

$$\left[W_1(\mu_n^\Theta, d\theta/2\pi) > 2\varepsilon \quad \text{for infinitely many distinct} \quad n \right] \qquad (5.1.13)$$

has probability zero. $\qquad \square$

Likewise, one can formulate almost sure convergence results for the circular orthogonal ensemble and the circular symplectic ensemble.

5.2 Convergence of ensembles

The discussion in this section is carried out at three levels:

- the joint distribution $\sigma_n^{(\beta)}(d\lambda)$ of $\lambda \in \Delta^n$;
- the empirical eigenvalue distribution $\mu_n^\lambda(dx) = \frac{1}{n} \sum_{j=1}^{n} \delta_{\lambda_j}(dx)$;
- the integrated density of states.

In the typical applications to random matrix theory, $\sigma_n^{(\beta)}(d\lambda)$ is the probability measure on Δ^n that describes the joint distribution of eigenvalues from the orthogonal, unitary or symplectic ensembles; but without worrying about the source, we simply assume here that

$$\sigma_n^{(\beta)} = Z_n^{-1} \exp\left(-n\sum_{j=1}^{n} v(\lambda_j)\right) \prod_{j<k}(\lambda_k - \lambda_j)^\beta d\lambda_1 \ldots d\lambda_n \qquad (5.2.1)$$

where $\beta > 0$ and $\lambda = (\lambda_1, \ldots, \lambda_n)$ has $\lambda_1 \leq \cdots \leq \lambda_n$. Then we introduce the probability measure μ_n^λ, where λ is random and subject to the distribution $\sigma_n^{(\beta)}$.

Let μ_n^λ be the empirical eigenvalue distribution of a self–adjoint matrix X, and let $|zI - X| = ((zI - X)(\bar{z}I - X))^{1/2}$. Then by functional calculus,

$$\text{trace}_n \log|zI - X| = \int \log|z - w|\, \mu_n^\lambda(dw) \qquad (5.2.2)$$

since $\text{trace}_n \log|zI - X| = n^{-1}\log\det|zI - X|$.

Definition (*IDS*). The integrated density of states ρ_n is the probability measure on $\mathbf{R}$ that satisfies

$$\int f(x)\rho_n(dx) = \iint f(x)\mu_n^\lambda(dx)\sigma_n^{(\beta)}(d\lambda); \qquad (5.2.3)$$

we use the term density although ρ_n need not *a priori* be absolutely continuous.

Definition (*Modes of convergence*). The principal modes of convergence to equilibrium that are discussed in the literature are:

(1°) weak convergence of the *IDS*, so $\rho_n \to \rho$ as $n \to \infty$ weakly for some $\rho \in Prob(\mathbf{R})$;

(2°) weak convergence in probability of the empirical distributions to the equilibrium measure, so

$$\sigma_n^{(\beta)}\left[\lambda \in \Delta^n : \left|\int f d\mu_n^\lambda - \int f d\rho\right| > \varepsilon\right] \to 0 \qquad (n \to \infty) \quad (5.2.4)$$

for all $\varepsilon > 0$ and $f \in C_b(\mathbf{R})$;

(3°) almost sure weak convergence. Let $(\Delta, \sigma) = (\prod_{n=1}^{\infty}\Delta^n, \otimes_{n=1}^{\infty}\sigma_n^{(\beta)}(d\lambda))$. Say that μ_n^λ converges weakly almost surely to ρ as $n \to \infty$ if, for each $f \in C_b(\mathbf{R})$, the sequence of random variables $\int f(x)\mu_n^\lambda(dx)$ on (Δ, σ) converges almost surely to $\int f(x)\rho(dx)$.

(4°) Metric almost sure weak convergence occurs when $W_1(\mu_n^\lambda, \rho) \to 0$ almost surely on (Δ, σ) as $n \to \infty$.

Lemma 5.2.1 *Suppose that all the measures are supported on $[-A, A]$ for some $A < \infty$. Then the modes of convergence satisfy*

$$(4^o) \Rightarrow (3^o) \Rightarrow (2^o) \Rightarrow (1^o). \qquad (5.2.5)$$

Proof. $(4^o) \Rightarrow (3^o)$. This follows from Proposition 3.3.4.

$(3^o) \Rightarrow (2^o)$. Almost sure convergence implies convergence in probability by the bounded convergence theorem or [88].

$(2^o) \Rightarrow (1^o)$. Suppose that $\|f\|_\infty \le K$. Then $|\int f(x)\mu_n^\lambda(dx) - \int f(x)\rho(dx)|$ is bounded by $2K$ always, and is less than 2ε on a set with $\sigma_n^{(\beta)}$ measure that converges to one as $n \to \infty$. Hence the integral with respect to $\sigma_n^{(\beta)}(d\lambda)$ converges to zero as $n \to \infty$, as in [88]. $\square$

Remarks. (i) The probability measures in condition (1°) are not random. In $(2^o), (3^o)$ and (4°), the μ_n^λ are random; whereas their limit ρ is not random. In Section 5.4 we introduce another condition $(3')$ to deal with this anomaly.

(ii) Conditions such as (4°) feature in Bolley's thesis [29], and give a metric notion of weak convergence. Bolley shows that for certain interacting particle systems, these can improve upon the Sanov-type theorems concerning large deviations.

(iii) On the probability space (Δ, σ), the ordered lists of eigenvalues $\lambda_n \in \Delta^n$ and $\lambda_m \in \Delta^m$ are statistically independent when $n \ne m$. This construction is artificial, but in practice one seldom considers eigenvalues from distinct n and m simultaneously; more often, one is interested in the μ_n^λ as $n \to \infty$.

Conclusion (1°) holds under very general hypotheses. First we identify ρ, after restating the main hypotheses of Section 4.4.

(a) v is bounded below, so that $v(x) \ge -C$ for some $C \in \mathbf{R}$ and all $x \in \mathbf{R}$.

(b) There exist $\varepsilon > 0$ and $M < \infty$ such that

$$v(x) \ge \frac{\beta + \varepsilon}{2} \log(x^2 + 1) \qquad (|x| \ge M). \qquad (5.2.6)$$

(c) $v : \mathbf{R} \to \mathbf{R}$ is absolutely continuous and v' is $\alpha-$ Hölder continuous on $[-A, A]$ for some $\alpha > 0$ and each $A > 0$.

Proposition 5.2.2 *Suppose that v satisfies (a), (b) and (c), and let ρ be as in Theorem 4.4.1. Then the ground state energy of the statistical mechanical system equals the minimum value of the electrostatic* *energy, so*

$$- \lim_{n \to \infty} \frac{\log Z_n}{n^2} = E_v = \int v(x)\rho(dx) - \frac{\beta}{2} \iint \log |x - y| \, \rho(dx)\rho(dy).$$
$$(5.2.7)$$

Proof. See [37] and Section 5.3. □

Theorem 5.2.3 (*Convergence of IDS*). *Suppose that v satisfies the hypotheses (a), (b) and (c), and let ρ be the equilibrium measure. Then $\rho_n \to \rho$ weakly as $n \to \infty$.*

Boutet de Monvel, Pastur and Shcherbina [37] proved this theorem using operator theory, and went on to establish convergence in probability (2°) for the ensembles. In [37] the result is stated for a hypothesis weaker than (c), but Totik [155] showed that the stronger hypothesis (c) really is needed. In Section 5.3, we shall prove some of the key steps towards Theorem 5.2.3. Theorem 5.2.3 has equivalent statements in terms of the Cauchy transforms and the eigenvalue counting function

$$N_n(t) = \sharp\{j : \lambda_j \leq t\} \qquad (t \in \mathbf{R}). \qquad (5.2.8)$$

Corollary 5.2.4 *Suppose that the hypotheses of Theorem 5.2.3 hold. Then*

$$\int \frac{\rho_n(dx)}{z - x} \to \int \frac{\rho(dx)}{z - x} \qquad (5.2.9)$$

uniformly on compact subsets of $\mathbf{C} \setminus \mathbf{R}$ as $n \to \infty$, and

$$\frac{1}{n}\mathbf{E}N_n(t) \to \int_{(-\infty,t]} \rho(dx) \qquad (5.2.10)$$

as $n \to \infty$ at all points of continuity of the right-hand side.

Proof. By Lemma 1.8.7, weak convergence of the ρ_n to the compactly supported ρ is equivalent to uniform convergence of the Cauchy transforms on compact sets. In another language, the normalized counting function $N_n(t)/n$ is the cumulative distribution function for the empirical distribution, so $\mathbf{E}N_n(t)/n$ is the cumulative distribution function for the integrated density of states ρ_n. Hence the result follows from Proposition 3.3.4. □

5.3 Mean field convergence

In mean field theory, we select a probability density function f and replace

$$\frac{1}{n}\sum_{j=1}^{n} v(\lambda_j) + \frac{\beta}{2n^2} \sum_{j,k:j\neq k} \log \frac{1}{|\lambda_j - \lambda_k|} \qquad (5.3.1)$$

by

$$\frac{1}{n}\sum_{j=1}^{n} \left(v(\lambda_j) + \frac{\beta}{2} \int \log \frac{1}{|\lambda_j - x|} f(x)\,dx \right) \qquad (5.3.2)$$

so that the eigenvalues decouple. This idea was previously mentioned in Exercise 3.1.5, where the joint eigenvalue distribution $\sigma_n^{(\beta)}$ was compared with a product measure. With a suitable choice of f, this gives a sufficiently good approximation to the true potential since the interaction between eigenvalues is controlled by a logarithmic term which has a relatively weak effect.

Using this idea, we now sketch the main stages in the proof of Theorem 5.2.3. Since by Corollary 3.6.5 and Theorem 4.4.1, the equilibrium measure is supported on $[-A, A]$ for some $A < \infty$, we can rescale the variables to $x_j = \lambda_j/(2A)$ and the potential to $u(x) = v(Ax)$ so that $x \in [-1/2, 1/2]$ and $u : [-1/2, 1/2] \to [M_0, \infty)$ for some $M_0 > -\infty$. We consider probability density functions f such that: $f(x) \geq 0$ for all $x \in [-1/2, 1/2]$;

$$\int_{-1/2}^{1/2} f(x)\,dx = 1; \qquad (5.3.3)$$

$$\iint \log \frac{1}{|x-y|} f(x)f(y)dxdy < \infty. \qquad (5.3.4)$$

To formulate these in a Hilbert space context, we observe that each nonnegative $g \in C([-1/2, 1/2])$ gives a bounded linear functional on $L^2[-1/2, 1/2]$, so

$$P \cap L^2 = \left\{ f \in L^2[-1/2, 1/2] : \int_{-1/2}^{1/2} f(x)\,dx = 1; f(x) \geq 0 \right\}$$

$$= \bigcap_g \left\{ f \in L^2[-1/2, 1/2] : \int_{-1/2}^{1/2} f(x)\,dx = 1; \right.$$

$$\left. \times \int_{-1/2}^{1/2} f(x)g(x)\,dx \geq 0 \right\} \qquad (5.3.5)$$

is a closed and convex subset of $L^2[-1/2, 1/2]$.

Lemma 5.3.1 *Let* $\mathcal{L} : L^2[-1/2, 1/2] \rightarrow L^2[-1/2, 1/2]$ *be the linear operator*

$$\mathcal{L}f(x) = \int_{-1/2}^{1/2} \log |x - y| \, f(y) \, dy. \tag{5.3.6}$$

Then:

(i) $\mathcal{L}$ *is a Hilbert–Schmidt and self-adjoint operator;*

(ii) $\mathcal{L}$ *is negative definite; so that for* $f_1, f_2 \in P \cap L^2$, *the quadratic form associated with* $\mathcal{L}$ *satisfies*

$$-\langle \mathcal{L}(f_1 - f_2), (f_1 - f_2) \rangle_{L^2} \geq 0. \tag{5.3.7}$$

Proof. (i) The kernel of $\mathcal{L}$ is real and symmetric, so $\mathcal{L}$ is self-adjoint. Furthermore, one can check by direct calculation that

$$\iint_{[-1/2, 1/2]^2} \left(\log |x - y| \right)^2 dx dy < \infty. \tag{5.3.8}$$

Hence $\mathcal{L}$ is a Hilbert–Schmidt operator, so is completely continuous and maps weakly convergent sequences into norm convergent sequences.

(ii) Passing to the Fourier transform, we have the identity

$$\iint_{[-1/2, 1/2]^2} \frac{(f_1(x) - f_2(x))(f_1(y) - f_2(y))}{|x - y|^s} dx dy$$
$$= \frac{C(s)}{(2\pi)^2} \int_{-\infty}^{\infty} \frac{|\hat{f}_1(\xi) - \hat{f}_2(\xi)|^2}{|\xi|^{1-s}} d\xi, \tag{5.3.9}$$

where $0 < s < 1$ and

$$C(s) = \int_{-\infty}^{\infty} \frac{e^{-ix}}{|x|^s} dx = 2\Gamma(1-s) \sin \frac{\pi s}{2} \geq 0, \tag{5.3.10}$$

so both sides of this formula are nonnegative. Since f_1 and f_2 are both probability density functions, we have

$$\iint_{[-1/2, 1/2]^2} \left(f_1(x) - f_2(x) \right) \left(f_1(y) - f_2(y) \right) dx dy = 0, \tag{5.3.11}$$

so the right-hand derivative of (5.3.10) at $s = 0+$ must be nonnegative; that is

$$-\iint_{[-1/2, 1/2]^2} \log |x - y| \left(f_1(x) - f_2(x) \right) \left(f_1(y) - f_2(y) \right) dx dy \geq 0. \tag{5.3.12}$$

$\square$

Definition (*Energy norm* [80]). Let D_0' be the completion of

$$\left\{ f \in L^2[-1/2, 1/2] : \int f(x)dx = 0 \right\} \qquad (5.3.13)$$

for the energy norm $\| \ \|_E$ associated with the inner product

$$\langle f, g \rangle_E = \iint_{[-1/2, 1/2]^2} \log \frac{1}{|x-y|} f(x) \bar{g}(x) \, dx dy. \qquad (5.3.14)$$

Say that a sequence (f_j) in D_0' converges in energy norm when $\| f_j - f_k \|_E \to 0$ as $j, k \to \infty$. There is a natural map $P \cap L^2 \to D_0' : f \mapsto f - \mathbf{I}$; so we say that (f_j) in $P \cap L^2$ converges in energy when $(f_j - \mathbf{I})$ converges in energy norm D_0'. [We do not assert that the limit is an L^2 function.]

For each $n \geq 1$, and $f \in P \cap L^2$ we introduce the approximating Hamiltonian

$$H_n(x_1, \ldots, x_n; f) = \frac{1}{n} \sum_{j=1}^{n} (u(x_j) - \beta \mathcal{L}f(x_j)) + \frac{\beta}{2} \int f(x)\mathcal{L}f(x) \, dx;$$

$$(5.3.10)$$

this has properties that there is no interaction between each x_j and the other variables and $H_n(; f)$ is a symmetrical function. In effect, we have replaced one of the copies of μ_n by a mean field f, and allowed the other to remain as an empirical distribution. The partition function $Z_n(f)$ for the Hamiltonian $n^2 H_n(; f)$ satisfies

$$\begin{aligned} \Phi_n(f) &= \frac{1}{n^2} \log Z_n(f) \\ &= \frac{1}{n^2} \log \int_{-1/2}^{1/2} \exp\left(-n^2 H_n(x_1, \ldots, x_n; f)\right) dx_1 \ldots dx_n \\ &= \frac{-\beta}{2} \int f(x)\mathcal{L}f(x) \, dx \\ &\quad + \frac{1}{n} \log \int_{-1/2}^{1/2} \exp\left(-nu(x) + n\beta \mathcal{L}f(x)\right) dx. \end{aligned} \qquad (5.3.16)$$

In particular, suppose that p_n is a probability density function that satisfies the molecular field equation

$$p_n(x) = Z_n(p_n)^{-1} \exp(-nu(x) + \beta n \mathcal{L} p_n(x)); \qquad (5.3.17)$$

then

$$\int f(x)p_n(x)dx = Z_n(p_n)^{-1} \int_{\mathbf{R}^n} \left(\frac{1}{n} \sum_{j=1}^{n} f(x_j)\right)$$

$$\times \exp(-n^2 H_n(x_1, \ldots, x_n; p_n)dx_1 \ldots dx_n, \qquad (5.3.18)$$

so p_n gives an approximation to the integrated density of states ρ_n.

Lemma 5.3.2 *The functional* Φ_n *has a unique minimizer* p_n *such that* $p_n - \mathbf{I} \in D_0'$.

Proof. First we prove that Φ_n is bounded below. By Jensen's inequality

$$\Phi_n(f) \geq \int_{-1/2}^{1/2} v(x)f(x)dx + \beta \int_{-1/2}^{1/2} \mathcal{L}f(x)\,dx - \frac{\beta}{2}\int_{-1/2}^{1/2} f(x)\mathcal{L}f(x)\,dx$$

$$= \int_{-1/2}^{1/2} v(x)f(x)\,dx - \frac{\beta}{2}\int_{-1/2}^{1/2}(f(x)-1)\mathcal{L}(f-1)(x)\,dx + \frac{\beta}{2}$$

$$\geq M_0 + \frac{\beta}{2}, \tag{5.3.19}$$

where we have used (ii) of the preceding Lemma 5.3.1.

Now we show that Φ_n is convex. By Hölder's inequality, we have

$$\log \int_{-1/2}^{1/2} \exp\big(-nv(x) + nt\beta\mathcal{L}f_1(x) + n(1-t)\beta\mathcal{L}f_2(x)\big)dx$$

$$\leq t\log \int_{-1/2}^{1/2} \exp\big(-nv(x) + n\beta\mathcal{L}f_1(x)\big)dx$$

$$+ (1-t)\log\int_{-1/2}^{1/2}\exp\big(-nv(x)\big) + n\beta\mathcal{L}f_2(x)\big)dx,$$

$$\tag{5.3.20}$$

so by Lemma 5.3.1

$$t\Phi_n(f_1) + (1-t)\Phi_n(f_2) - \Phi_n(tf_1 + (1-t)f_2)$$

$$\geq -\frac{t(1-t)}{2}\int\big(f_1(x) - f_2(x)\big)\mathcal{L}(f_1 - f_2)(x)\,dx \geq 0. \tag{5.3.21}$$

We let f_j be a minimizing sequence in $P \cap L^2$, for Φ_n so that $\Phi_n(f_j) \to \inf_f \Phi_n(f)$. Then (f_j) is a Cauchy sequence in D_0'; hence $(f_j - \mathbf{I})$ is likewise a Cauchy sequence and hence converges in energy to a unique $p_n - \mathbf{I} \in D_0'$. □

There are in principle two cases to consider.

(1) If $\|f_j\|_{L^2} \to \infty$ as $j \to \infty$, then p_n may be a singular measure.
(2) There exists a subsequence (f_{m_j}) such that $f_{m_j} \to p_n$ weakly in L^2 as $m_j \to \infty$.

Then the infimum is attained at $p_n \in P \cap L^2$ since $\mathcal{L}f_{m_j} \to \mathcal{L}p_n$ in norm by Lemma 5.3.1(i), so $\Phi_n(f_{m_j}) \to \Phi_n(p_n)$ as $m_j \to \infty$.

In fact (2) occurs, so the minimizer p_n belongs to $P \cap L^2$.

Theorem 5.3.3 (*Mean field convergence*). *The p_n converge weakly in L^2 to ρ, the equilibrium measure that corresponds to the potential v as in Theorem 4.4.1.*

The proof involves a few more Lemmas.

Lemma 5.3.4 *There exists ρ such that $p_n \to \rho$ in energy as $n \to \infty$.*

Proof. By Hölder's inequality, we have $\Phi_n(f) \geq \Phi_{n+1}(f)$ for all f, and $\Phi_n(f) \geq \Phi_{n+1}(p_{n+1})$ by the choice of p_{n+1}, so

$$\Phi_n(p_n) \geq \Phi_{n+1}(p_{n+1}) \geq M_0 + \frac{1}{2}. \tag{5.3.22}$$

Hence $(\Phi_n(p_n))$ is a decreasing sequence which is bounded below, and so converges to some limit m_0 as $n \to \infty$. For $m \geq n$ we have

$$\Phi_n(p_n) - \Phi_m(p_m) \geq \Phi_m(p_n) - \Phi_m(p_m)$$
$$\geq -\beta \int (p_m(x) - p_n(x))\mathcal{L}(p_m - p_n)(x)dx$$
$$- 2\Phi_m(p_m) + 2\Phi_m((p_n + p_m)/2)$$
$$\geq -\beta \int (p_m(x) - p_n(x))\mathcal{L}(p_m - p_n)(x)dx. \tag{5.3.23}$$

The left-hand side converges to zero as $n \to \infty$, hence $(p_n - \mathbf{I})$ gives a Cauchy sequence in D_0', so there is a limit $\rho - \mathbf{I}$ in D_0'. $\qquad \square$

Lemma 5.3.5 (*Molecular field equation*). *Suppose that (2) holds. Then p_n satisfies the molecular field equation*

$$p_n(x) = \frac{\exp(-nu(x) + n\beta\mathcal{L}p_n(x))}{\int_{-1/2}^{1/2} \exp(-nu(y) + n\beta\mathcal{L}p_n(y))dy}. \tag{5.3.24}$$

Proof. Let $p \in P \cap L^2$ and introduce $q = p - p_n$. We have

$$\Phi_n(p_n) \leq \Phi((1-t)p_n + tp) = \Phi_n(p_n + tq) \tag{5.3.25}$$

for $0 \leq t \leq 1$, so

$$0 = \left(\frac{d}{dt}\right)_{t=0} \Phi_n(p_n + tq)$$

hence

$$0 = \beta \int p_n(x)\mathcal{L}(q)(x)dx - \frac{\beta}{Z_n(p_n)} \int \exp(-nu(x) + n\beta\mathcal{L}p_n(x))\mathcal{L}(q)(x)\,dx, \tag{5.3.26}$$

hence the result. $\qquad \square$

When we consider the limits as $n \to \infty$, there are again two possibilities, either:

(3) $\|p_n\|_{L^2} \to \infty$; or
(4) there exists $\rho \in P \cap L^2$ and a subsequence (p_{m_k}) such that $p_{m_k} \to \rho$ weakly.

When v satisfies (c), conditions (1) and (3) cannot hold, since by Theorem 4.4.1, the equilibrium measure ρ belongs to L^2.

Proposition 5.3.6 *Condition (4) holds, and there exists a constant M^* such that*

$$u(x) \geq \frac{\beta}{2} \int \log |x - y| \rho(y) dy - M^* \qquad (5.3.27)$$

with equality on the support of ρ.

Proof. By Theorem 4.4.1, (4) holds, and by Lemma 5.3.4 $\mathcal{L}p_{n_k} \to \mathcal{L}\rho$ in L^2 as $n_k \to \infty$. Then $\mathcal{L}\rho$ is continuous since $x \mapsto \log |x - y|$ is continuous $[-1/2, 1/2] \to L^2[-1/2, 1/2]$; hence we can introduce $M^* = \sup_x\{-u(x) + \mathcal{L}\rho(x)\}$ and, for each $\varepsilon > 0$, also introduce the set

$$S^{(\varepsilon)} = \{x : -u(x) + \beta\mathcal{L}\rho(x) \geq M^* - \varepsilon\}, \qquad (5.3.28)$$

which has positive Lebesgue measure $m(S^{(\varepsilon)}) > 0$. Likewise we introduce

$$S_n^{(\varepsilon)} = \{x : -u(x) + \beta\mathcal{L}p_n(x) \geq M^* - \varepsilon\}, \qquad (5.3.29)$$

which satisfies $m(S_{n_j}^{(\varepsilon)} \setminus S^{(\varepsilon)}) + m(S^{(\varepsilon)} \setminus S_{n_j}^{(\varepsilon)}) \to 0$ as $n_j \to \infty$ since $\mathcal{L}p_{n_j} \to \mathcal{L}\rho$ in L^2. By the molecular field equation, any x outside of $S_{n_j}^{(2\varepsilon)}$ satisfies

$$p_{n_j}(x) \leq \frac{\exp(nM^* - 2n\varepsilon)}{\int_{S_{n_j}^{(\varepsilon)}} \exp(-nu(y) + \beta\mathcal{L}p_n(y))dy} \leq \frac{\exp(-n\varepsilon)}{m(S_{n_j}^{(\varepsilon)})} \qquad (5.3.30)$$

so $p_{n_j}(x) \to 0$ as $n_j \to \infty$. Finally, we have

$$\Phi_n(p_n) \to M^* - \frac{\beta}{2} \int \rho(x)\mathcal{L}\rho(x)\, dx \qquad (n \to \infty). \qquad (5.3.31)$$

Hence we obtain the result, consistent with Proposition 5.3.2,

$$-\lim_n \Phi_n(p_n) = E_u = u(x) - \frac{\beta}{2} \int \rho(x)\mathcal{L}\rho(x)\, dx. \qquad (5.3.32)$$

$\square$

5.4 Almost sure weak convergence for uniformly convex potentials

This section contains the main result of this chapter, which refines Theorem 6.1 of [17]. In Section 5.3, we established convergence in the sense of condition (1°) for the integrated densities of states, and we wish to proceed to establish convergence in the sense of (4°) by using the results of Section 3.6. We first introduce another notion of convergence, which does not refer to the limit distribution ρ and which is well suited to analysis by concentration techniques.

Definition (*weak almost sure convergence*). Let μ_n^λ and ρ_n on (Δ, σ) be as in Section 5.2. We say that $\mu_n^\lambda - \rho_n \to 0$ weakly almost surely when

$$(3') \qquad \int f(x)(\mu_n^\lambda(dx) - \rho_n(dx)) \to 0 \qquad (5.4.1)$$

almost surely on (Δ, σ) for all $f \in C_b(\mathbf{R})$ as $n \to \infty$.

Proposition 5.4.1 *Suppose that the $\sigma_n^{(\beta)}$ on (Δ, σ) satisfy the concentration of measure phenomenon $C(\alpha_n)$ of 3.2 with constants $\alpha_n \geq Cn^\alpha$ for some $C, \alpha > 0$. Then $\mu_n^\lambda - \rho_n \to 0$ weakly almost surely as $n \to \infty$.*

Proof. Let $f \in C_b(\mathbf{R}; \mathbf{R})$ be uniformly continuous, and let $\varepsilon > 0$. Then there exists $L < \infty$ and an L-Lipschitz function g such that $\|f - g\|_\infty < \varepsilon/4$; so

$$\left[\left|\int f d\mu_n^\lambda - \int f d\rho_n\right| > \varepsilon\right] \subseteq \left[\left|\int g d\mu_n^\lambda - \int g d\rho_n\right| > \varepsilon/2\right]. \qquad (5.4.2)$$

Now let $G_n : (\Delta^n, \ell^2(n)) \to \mathbf{R}$ be

$$G_n(\lambda) = \frac{1}{n} \sum_{j=1}^n g(\lambda_j) = \int g d\mu_n^\lambda, \qquad (5.4.3)$$

which is L-Lipschitz since

$$|G_n(\lambda) - G_n(\lambda')| \leq \frac{1}{n} \sum_{j=1}^n |g(\lambda_j) - g(\lambda'_j)|$$

$$\leq \frac{L}{n} \sum_{j=1}^n |\lambda_j - \lambda'_j|$$

$$\leq L\left(\frac{1}{n} \sum_{j=1}^n |\lambda_j - \lambda'_j|^2\right)^{1/2}. \qquad (5.4.4)$$

Now $\int G_n(\lambda)\sigma_n^{(\beta)}(d\lambda) = \int g(x)\rho_n(dx)$ by definition of the integrated density of states, so

$$\left[\left| G_n - \int G_n d\sigma_n^{(\beta)} \right| \geq \frac{\varepsilon}{2}\right] = \left[\left| \int g d\mu_n^\lambda - \int g d\rho_n \right| \geq \frac{\varepsilon}{2}\right]. \qquad (5.4.5)$$

By the concentration of measure Corollary 3.6.2, and Lemma 3.2.3

$$\sigma_n^{(\beta)}\left[\left| G_n - \int G_n d\sigma_n^{(\beta)} \right| \geq \frac{\varepsilon}{2}\right] \leq 2\exp\left(-\varepsilon^2 \alpha_n /(8L^2)\right) \qquad (5.4.6)$$

and hence

$$\sum_{n=1}^{\infty} \sigma_n^{(\beta)}\left[\left| G_n - \int G_n d\sigma_n^{(\beta)} \right| \geq \frac{\varepsilon}{2}\right] \leq \sum_{n=1}^{\infty} 2\exp\left(-\varepsilon^2 \alpha_n /(8L^2)\right) < \infty.$$
$$(5.4.7)$$

By the Borel–Cantelli lemma, the event

$$\left[\left| G_n - \int G_n d\sigma_n^{(\beta)} \right| \geq \frac{\varepsilon}{2} \quad \text{for infinitely many} \quad n\right] \qquad (5.4.8)$$

occurs with probability zero with respect to $\otimes_{n=1}^{\infty} \sigma_n^{(\beta)}$; hence

$$\int g d\mu_n^\lambda - \int g d\rho_n \to 0 \qquad (5.4.9)$$

almost surely as $n \to \infty$. Likewise

$$\int f d\mu_n^\lambda - \int f d\rho_n \to 0 \qquad (5.4.10)$$

almost surely as $n \to \infty$. $\qquad \square$

Corollary 5.4.2 (*Almost sure weak convergence of the empirical distribution*). *Suppose that v is twice continuously differentiable and that there exists $\kappa > 0$ such that $v''(x) \geq \kappa$ for all $x \in \mathbf{R}$. Then (3°) holds, so that the empirical eigenvalue distribution μ_n^λ converges to the equilibrium distribution ρ weakly almost surely as $n \to \infty$.*

Proof. The hypotheses of Theorem 5.2.3 are clearly satisfied, so (1°) holds. Further, by Theorem 3.6.1, the $\sigma_n^{(\beta)}$ satisfy the concentration of measure phenomenon with constants κn^2, so $(3')$ holds by Proposition 5.4.1. Clearly, we have

$$(3')\&(1^\circ) \Rightarrow (3^\circ), \qquad (5.4.11)$$

hence the result. $\qquad \square$

By refining the proof, we can obtain a stronger theorem, namely convergence in the sense of (4°).

Theorem 5.4.3 (*Metric almost sure weak convergence of the empirical distribution*). *Suppose that v is twice continuously differentiable and that there exists $\kappa > 0$ such that $v''(x) \geq \kappa$ for all $x \in \mathbf{R}$. Then*

$$W_1(\mu_n^\lambda, \rho) \to 0 \qquad (5.4.12)$$

almost surely as $n \to \infty$.

Proof. By Theorem 5.2.3 we have $\rho_n \to \rho$ weakly as $n \to \infty$, and $W_1(\rho_n, \rho) \to 0$ as $n \to \infty$, so it suffices to prove that $W_1(\mu_n^\lambda, \rho_n) \to 0$ almost surely.

By Corollary 3.6.5, the event that μ_n^λ has support inside $[-nA, nA]$ for all sufficiently large n occurs with probability one. Hence we suppose that μ_n and ρ_n are probability measures on $[-nA, nA]$, and let $\varepsilon > 0$.

The Arzelà–Ascoli theorem [141] asserts that a uniformly bounded and uniformly equicontinuous subset of $C([-nA, nA])$ is relatively compact for the $\| \cdot \|_\infty$ norm. In particular, the inclusion map $g \mapsto g$:

$$\{g : [-nA, nA] \to \mathbf{R} : g(0) = 0, \|g\|_{Lip} \leq 1\} \to C([-nA, nA]) \quad (5.4.13)$$

has relatively compact range K_n, hence K_n is totally bounded for the $\| \cdot \|_\infty$ norm. Further, empirical distributions define bounded linear functionals on $C([-nA, nA])$, so we can regard the Kantorovich formula as a variational problem on the relatively compact set K_n.

Recall that

$$W_1(\mu_n^\lambda, \rho_n) = \sup_g \left\{ \int_{-nA}^{nA} g(d\mu_n^\lambda - d\rho_n) : g(0) = 0, \|g\|_{Lip} \leq 1 \right\}, \quad (5.4.14)$$

and observe that for each n and ε we can choose some such g satisfying

$$\left[W_1(\mu_n^\lambda, \rho_n) > 4\varepsilon \right] \subseteq \left[\left| \int_{-nA}^{nA} g(d\mu_n^\lambda - d\rho_n) \right| > 4\varepsilon \right]. \quad (5.4.15)$$

The next idea is to replace g at stage n by some f_j which is to be chosen from a net that depends upon ε, and involves more terms as n increases.

Let $\lfloor x \rfloor = \max\{m \in \mathbf{Z} : m \leq x\}$. We introduce the grid points $(k\varepsilon, \ell\varepsilon)$ for

$$k, \ell = -\left(\left\lfloor \frac{nA}{\varepsilon} \right\rfloor + 1 \right), -\left\lfloor \frac{nA}{\varepsilon} \right\rfloor, \dots, \left(\left\lfloor \frac{nA}{\varepsilon} \right\rfloor + 1 \right); \quad (5.4.16)$$

then we form the graphs of 1-Lipschitz functions f_j such that $f_j(0) = 0$ by joining grid points with straight line segments. When we move one step along the x-axis away from the origin, we have only three choices available when constructing the graph of such a function: either a horizontal line segment, the segment that slopes upwards with gradient $+1$, or the segment that slopes downwards with gradient -1. Thus one constructs functions $f_j : [-nA, nA] \to \mathbf{R}$ where $j = 1, \ldots, 3^{M(n)}$, where $M(n) = 2\lfloor nA/\varepsilon \rfloor + 3$.

The (f_j) form an ε-net for the supremum norm on $\{g \in C[-nA, nA] : g(0) = 0\}$, so there exists j such that

$$\left| \int_{-nA}^{nA} g(d\mu_n^\lambda - d\rho_n) - \int_{-nA}^{nA} f_j(d\mu_n^\lambda - d\rho_n) \right| \leq 2\|g - f_j\|_\infty \leq 2\varepsilon; \quad (5.4.17)$$

hence we have the inclusion

$$\left[\left| \int_{-nA}^{nA} g(d\mu_n^\lambda - d\rho_n) \right| > 4\varepsilon \right] \subseteq \bigcup_{j=1}^{3^{M(n)}} \left[\left| \int_{-nA}^{nA} f_j(d\mu_n^\lambda - d\rho_n) \right| > 2\varepsilon \right],$$

$$(5.4.18)$$

with the probabilistic consequence

$$\sigma_n^{(\beta)} \left[W_1(\mu_n^\lambda, \rho_n) > 4\varepsilon \right] \leq \sum_{j=1}^{3^{M(n)}} \sigma_n^{(\beta)} \left[\left| \int_{-nA}^{nA} f_j(d\mu_n^\lambda - d\rho_n) \right| > 2\varepsilon \right].$$

$$(5.4.19)$$

By the concentration inequality Theorem 3.6.1, this gives

$$\sigma_n^{(\beta)} \left[W_1(\mu_n^\lambda, \rho_n) > 4\varepsilon \right] \leq 3^{M(n)} 2\exp\left(-4\varepsilon^2 \kappa n^2/8\right), \quad (5.4.20)$$

where $M(n) \leq (2nA/\varepsilon) + 4$, hence

$$\sum_{n=1}^\infty \sigma_n^{(\beta)} \left[W_1(\mu_n^\lambda, \rho_n) > 4\varepsilon \right] < \infty. \quad (5.4.21)$$

By the first Borel–Cantelli lemma, the event

$$\left[W_1(\mu_n^\lambda, \rho_n) > 4\varepsilon \quad \text{for infinitely many distinct} \quad n \right] \quad (5.4.22)$$

occurs with probability zero. $\qquad \square$

Problem. Is there an analogue of Theorem 5.4.3 for the generalized orthogonal ensemble that has potential w as in Proposition 4.6.3?

The technical problem here is that the concentration inequalities such as Theorem 3.6.3 that are used in the proof of Theorem 5.4.3 depend upon convexity of the potential. When the support of the equilibrium distribution splits into two disjoint intervals, it is difficult to see how this method of proof can be adapted to prove almost sure convergence of the empirical eigenvalue distribution.

5.5 Convergence for the singular numbers from the Wishart distribution

- The singular numbers of rectangular matrices with Gaussian IID entries satisfy the Wishart distribution.
- The corresponding equilibrium distribution is the Pastur–Marchenko distribution.

Let Y be a $m \times n$ matrix with mutually independent standard Gaussian $N(0,1)$ random variables as entries, and let the real symmetric matrix $S = Y^t Y$ have nonnegative eigenvalues $s_1 \leq s_2 \leq \cdots \leq s_n$. We let $m = \lfloor \lambda n \rfloor$ for $\lambda > 0$, scale the eigenvalues to $x_j = s_j / \min\{n, m\}$, and then form the empirical distribution $\mu_n = \frac{1}{n} \sum_{j=1}^n \delta_{x_j}$. Evidently, S has rank$(S) \leq \min\{n, m\}$, so S is singular when $\lambda < 1$. Wishart considered such random matrices in [170].

Definition (*Marchenko–Pastur distribution* [83]). The Marchenko–Pastur distribution is

$$\rho_\lambda(dx) = \frac{\sqrt{4\lambda - (x - 1 - \lambda)^2}}{2\pi x} \mathbf{I}_\Lambda(x)dx \qquad (\lambda \geq 1),$$

$$\rho_\lambda(dx) = (1 - \lambda)\delta_0 + \frac{\sqrt{4\lambda - (x - 1 - \lambda)^2}}{2\pi x} \mathbf{I}_\Lambda(x)dx \qquad (0 < \lambda < 1),$$

$$(5.5.1)$$

where $\Lambda = [(1 - \sqrt{\lambda})^2, (1 + \sqrt{\lambda})^2]$.

Theorem 5.5.1 (*Singular numbers of Wishart matrices*).

(i) *Suppose that $\lambda > 1$. Then the empirical eigenvalue distribution μ_n of $Y^t Y/n$ converges weakly almost surely as $n \to \infty$ to the equilibrium distribution ρ_λ on the bounded interval $\Lambda \subset (0, \infty)$.*

(ii) *Suppose that $0 < \lambda < 1$. Then the empirical eigenvalue distribution μ_n associated with $Y^t Y/(\lambda n)$ converges weakly almost surely as $n \to \infty$ to*

$$(1 - \lambda)\delta_0 + \lambda\rho_{1/\lambda}. \qquad (5.5.2)$$

Proof. (i) By Theorem 3.7.5, the potential function is

$$v(x) = \frac{1}{2}\big(x - (\lambda - 1)\log x\big) \qquad (5.5.3)$$

which is uniformly convex on $(0, K]$ for each K. Since $v(x) \to \infty$ as $x \to 0+$, and since Proposition 3.7.2 shows that the eigenvalue distributions are concentrated on bounded intervals, there is no loss in supposing that the equilibrium distribution is supported on $[\delta, K]$ for some small $\delta > 0$ and some large $K < \infty$. Then v is uniformly convex on $[\delta, K]$, and we can follow through the proofs in Sections 4.4–4.7. In particular, Proposition 4.4.4 shows that the equilibrium measure is supported on a single interval Λ, and Theorem 4.4.1 characterizes ρ_λ in terms of the integral equation

$$\frac{1}{2}(x - (\lambda - 1)\log x) = \int_\Lambda \log|x - y|\rho_\lambda(dy) + C \qquad (x \in \Lambda) \quad (5.5.4)$$

for some constant C. We shall verify in Lemma 5.5.2 that ρ_λ furnishes a solution.

(ii) Now suppose that $\lambda < 1$, so $m < n$. Then $\mathrm{rank}(Y^t Y) \leq m$, so the spectrum of S contains at least $n - m$ zero eigenvalues, hence μ_n assigns at least $1 - \lambda$ of its mass to zero. Now by elementary spectral theory, each nonzero eigenvalue of S is also an eigenvalue of YY^t where Y^t is a $n \times \lfloor \lambda n \rfloor$ matrix with mutually independent $N(0, 1)$ entries. By (i), we can obtain the limiting eigenvalue distribution of $YY^t/(\lambda n)$. $\qquad \square$

We now complete the proof of Theorem 5.5.1 by verifying that the Marchenko–Pastur distribution satisfies the integral equation.

Lemma 5.5.2 *For $\lambda \geq 1$, the Hilbert transform of the Marchenko–Pastur probability density function is*

$$\mathcal{H}\rho_\lambda(x) = \mathrm{p.v.} \int_{(\sqrt{\lambda}-1)^2}^{(\sqrt{\lambda}+1)^2} \frac{\sqrt{4\lambda - (t - 1 - \lambda)^2}}{(x - t)t}\, \frac{dt}{2\pi^2}$$

and this is equal to

$$\mathcal{H}\rho_\lambda(x) = \frac{x + 1 - \lambda}{2\pi x} \qquad (x \in \Lambda),$$

$$= \frac{x + 1 - \lambda - \sqrt{(x - 1 - \lambda)^2 - 4\lambda}}{2\pi x} \qquad (x \in \mathbf{R} \setminus \Lambda). \quad (5.5.5)$$

Proof. Consider the function

$$G(z) = \frac{z + 1 - \lambda - \sqrt{(z - 1 - \lambda)^2 - 4\lambda}}{2z} \qquad (5.5.6)$$

with square root such that

$$\sqrt{(z - 1 - \lambda)^2 - 4\lambda} \asymp z - 1 - \lambda \qquad (|z| \to \infty) \qquad (5.5.7)$$

so that $G(z)$ is analytic on $\mathbf{C} \setminus \Lambda$ and $G(z) \asymp 1/z$ as $z \to \infty$, due to the power series. Then $G(x)$ is real for $x \in \mathbf{R} \setminus \Lambda$, whereas for $x \in \Lambda$ we have

$$G(x) = \frac{x + 1 - \lambda - i\sqrt{4\lambda - (x - 1 - \lambda)^2}}{2x} \qquad (x \in \Lambda), \qquad (5.5.8)$$

hence G is the Cauchy transform of the Marchenko–Pastur distribution. So with $f(x) = \sqrt{4\lambda - (x - 1 - \lambda)^2}/(2x)$ we have

$$\mathcal{H}f(x) = \frac{x + 1 - \lambda}{2\pi x} \qquad (x \in \Lambda);$$

$$= \frac{1}{\pi} \frac{2\,\mathrm{sgn}(x + 1 - \lambda)}{|x + 1 - \lambda| + \sqrt{(x - 1 - \lambda)^2 - 4\lambda}} \qquad (x \in \mathbf{R} \setminus \Lambda). \quad (5.5.9)$$

$\square$

Remarks. (i) The Marchenko–Pastur distribution is sometimes referred to as the free Poisson law due to its representation as a limit of stable laws; see [83].

(ii) Our analysis does not cover the case of square matrices, when $\lambda = 1$, and we are not able to prove almost sure weak convergence by this type of argument. But the formula for ρ_1 still makes good sense in this case.

Exercise 5.5.3 Suppose that ξ is a random variable with probability density function

$$\frac{d\sigma_{0,4}}{dx} = \sqrt{4 - x^2}\,\mathbf{I}_{[-2,2]}(x)/(2\pi). \qquad (5.5.10)$$

Show that ξ^2 has distribution given by the Marchenko–Pastur ρ_1 law.

6

Gradient flows and functional inequalities

Abstract

In this chapter we introduce various functionals such as entropy and free entropy that are defined for suitable probability density functions on $\mathbf{R}^n$. Then we introduce the derivatives of such functionals in the style of the calculus of variations. This leads us to the gradient flows of probability density functions associated with a given functional; thus we recover the famous Fokker–Planck equation and the Ornstein–Uhlenbeck equation. A significant advantage of this approach is that the free analogues of the classical diffusion equations arise from the corresponding free functionals. We also prove logarithmic Sobolev inequalities, and use them to prove convergence to equilibrium of the solutions to gradient flows of suitable energy functionals. Positive curvature is a latent theme in this chapter; for recent progress in metric geometry has recovered analogous results on metric spaces with uniformly positive Ricci curvature, as we mention in the final section.

6.1 Variation of functionals and gradient flows

In this chapter we are concerned with evolutions of families of probability distributions under partial differential equations. We use ρ for a probability density function on $\mathbf{R}^n$ and impose various smoothness conditions as required. For simplicity, the reader may suppose that ρ is C^∞ and of compact support so that various functionals are defined. The fundamental examples of functionals are:

- Shannon's entropy $S(\rho) = -\int \rho(x) \log \rho(x)\, dx$;
- Potential energy $F(\rho) = \int v(x)\rho(x)\, dx$ with respect to a potential function v;
- Fisher's information $I(\rho) = \int \|\nabla \log \rho\|^2 \rho(x)\, dx$;

and in the special case of $n = 1$, Voiculescu introduced [163, 164]:

- free entropy $\chi(\rho) = \iint \log |x - y| \, \rho(x)\rho(y) \, dx dy$;
- free information $\Phi(\rho) = \int \rho(x)^3 \, dx$.

We wish to calculate variations of these functionals by differentiating with respect to ρ along the direction of some test function. The following definition is suggested by the standard notion of Gateaux derivative, but we adapt this since the probability density functions do not form a linear space, hence do not have a dual in the classical sense.

Definition (*Weak derivative*). Let P be a subset of $L^2(\mathbf{R}^n)$, let D be a Banach space that is continuously included as a linear subspace of $L^2(\mathbf{R}^n)$, and let D' be the dual space of D so that $D \subseteq L^2(\mathbf{R}^n) \subseteq D'$. (We do not require D to be closed as a subspace of L^2.) Suppose that for each $\rho \in P$ and $h \in D$, a real functional $F(\rho + \varepsilon h)$ is defined for sufficiently small $\varepsilon > 0$ and that there exists $\delta F / \delta \rho \in D'$ such that

$$\lim_{\varepsilon \to 0+} \frac{F(\rho + \varepsilon h) - F(\rho)}{\varepsilon} = \left\langle \frac{\delta F}{\delta \rho}, h \right\rangle_{L^2}. \qquad (6.1.1)$$

Then F is weakly differentiable on P and the domain of the weak derivative $\delta F / \delta \rho$ contains D.

Examples 6.1.1 There exist suitable P and D such that the above functionals are weakly differentiable with:

- $\dfrac{\delta}{\delta \rho} S = -\log \rho - 1$;

- $\dfrac{\delta}{\delta \rho} F = v(x)$;

- $\dfrac{\delta}{\delta \rho} \chi = 2 \displaystyle\int \log |x - y| \, \rho(y) \, dy$;

- $\dfrac{\delta}{\delta \rho} \Phi = 3\rho(x)^2$.

We shall use these facts in calculations to follow. Before then, we introduce a special metric structure on the probability densities.

Definition (*Flows* [162]). Let P_2 denote the space of probability density functions ρ on $\mathbf{R}^n$ such that $\int \|x\|^2 \rho(x) \, dx < \infty$. Then P_2 has the metric associated with W_2.

In this section we shall see that (P_2, W_2) is associated with a Riemannian structure whereby suitable $\rho_0, \rho_1 \in P_2$ are connected by a path

$\{(\rho_t)_{0 \le t \le 1}\}$ in P_2 that has minimal length with respect to W_2; so

$$W_2(\rho_0, \rho_1) = \inf_{\rho_t} \sup_{\pi} \left\{ \sum_{j=0}^{m-1} W_2(\rho_{t_j}, \rho_{t_{j+1}}) : \pi = \{0 = t_0 < \cdots < t_m = 1\} \right\}.$$

We shall construct such a path by means of a differential equation. Let $\xi_t : \mathbf{R}^n \to \mathbf{R}^n$ be a C^1 vector field for each $t > 0$, and let φ_t be the unique solution of the initial value problem

$$\frac{d}{dt}\varphi_t(x) = \xi_t(\varphi_t(x)), \qquad \varphi_0(x) = x_0. \tag{6.1.2}$$

It helps to think of ρ as the density of a fluid that moves with velocity ξ, where $\xi = \xi_t(x)$ depends upon position x and time t.

Proposition 6.1.2 (*Continuity equation*). *For each bounded open subset Ω of $\mathbf{R}^n$ there exists $t_0 > 0$ such that $\varphi_t : \Omega \to \Omega_t$ is a diffeomorphism to some open bounded set Ω_t for $0 < t < t_0$. For ρ a probability density function on Ω, the probability densities $\rho_t = \varphi_t \sharp \rho$ satisfy the continuity equation*

$$\frac{\partial}{\partial t}\rho_t = -\mathrm{div}(\xi_t \rho_t) \tag{6.1.3}$$

in the weak sense.

Proof. For a smooth vector field $\varphi : \mathbf{R}^n \to \mathbf{R}^n$, let $[\nabla \varphi]$ denote the matrix of derivatives of the components of φ. By the existence theory for ordinary differential equations, the solution $\varphi_t(x)$ is a differentiable function of the initial value x, and

$$\frac{\partial}{\partial t}\nabla \varphi_t = \left[\nabla \xi_t(\varphi_t(x))\right]\left[\nabla \varphi_t(x)\right], \tag{6.1.4}$$

$$\nabla \varphi_0(x) = I.$$

Hence $\det[\nabla \varphi_t] > 0$ for small $t > 0$ and by the open mapping theorem φ_t is a diffeomorphism.

Probability is conserved under transportation, so for any $f \in C^\infty(\Omega)$ we have

$$\frac{\partial}{\partial t} \int f(x)\rho_t(x)\, dx = \frac{\partial}{\partial t} \int f(\varphi_t(x))\rho(x)\, dx$$

$$= \int \langle \nabla f(\varphi_t(x)), \xi_t(\varphi_t(x))\rangle \rho(x)\, dx$$

$$= \int \langle \nabla f(x), \xi_t(x)\rangle \rho_t(x)\, dx \tag{6.1.5}$$

since $\rho_t = \varphi_t \sharp \rho$, so the continuity equation holds in the weak sense. Furthermore, when $\rho \in C^1$ the continuity equation holds in the classical sense since all the terms in the expression

$$\rho_t(\varphi_t(x)) \det[\nabla \varphi_t(x)] = \rho(x) \qquad (6.1.6)$$

are then differentiable.

□

The total kinetic energy associated with the flow (6.1.2) is

$$T = \int_0^1 \int_{\mathbf{R}^n} \|\xi_t(x)\|^2 \rho_t(x) \, dx dt. \qquad (6.1.7)$$

Given ρ_0 and ρ_1, we wish to find a flow that takes ρ_0 to ρ_1 and that minimizes the total kinetic energy.

Theorem 6.1.3 (*Optimal transportation*). *Let* $\rho_0, \rho_1 \in P_2$. *Then there exists a locally* L^2 *vector field* ξ_t *that generates by (6.1.2) a flow of probability densities* $(\rho_t)_{0 \le t \le 1}$ *such that*

$$W_2(\rho_t, \rho_0)^2 = t^2 \int \|\varphi_1(x) - x\|^2 \rho_0(x) \, dx. \qquad (6.1.8)$$

Remarks. A rigorous proof of this result involves subtle points concerning the regularity of solutions of the transportation partial differential equations, so we give only a sketch that emphasizes the main ideas [44]. In earlier transportation arguments such as Proposition 3.3.1, the optimal transportation strategy was given by a monotonic inducing map, and for $\mathbf{R}^n$ the correct notion of a monotone function is given by the gradient of a convex function. By a theorem of Brenier and McCann [162], there exists a convex function $\Phi : \mathbf{R}^n \to \mathbf{R}$ such that $\varphi_1(x) = \nabla \Phi(x)$ induces ρ_1 from ρ_0; this φ is defined except on a set of Lebesgue measure zero. Now we introduce an interpolating family of bijective maps $\varphi_t : \mathbf{R}^n \to \mathbf{R}^n$ by

$$\varphi_t(x) = (1 - t)x + t\varphi_1(x) \qquad (6.1.9)$$

and a vector field ξ_t such that

$$\xi_t(\varphi_t(x)) = \varphi_1(x) - x, \qquad (6.1.10)$$

hence the differential equation (6.1.2) holds. In the special case in which Φ is C^2, we have a positive definite matrix

$$\nabla \varphi_t(x) = (1 - t)I + t\text{Hess } \Phi(x); \qquad (6.1.11)$$

so φ_t is clearly bijective for $0 \le t < 1$. In general, one can recover this formula by interpreting the Hessian in Alexsandrov's sense [162].

The merit of this construction is that the flow $\rho_t = \varphi_t \sharp \rho$ gives the path of shortest length in (P_2, W_2) from ρ_0 to ρ_1, and φ_t gives an optimal transportation strategy, hence

$$W_2(\rho_0, \rho_t)^2 = \int \|\varphi_t(x) - x\|^2 \rho_0(x)\, dx = t^2 \int \|\varphi_1(x) - x\|^2 \rho_0(x)\, dx.$$
(6.1.12)

The reader wishing to see how this can be made precise may refer to Carlen and Gangbo's paper [44], or Villani's book [162]. Our subsequent analysis does not require the technical details of the proof.

We think of the paths $\{\rho_t : 0 \le t \le 1\}$ as geodesics in (P_2, W_2), and investigate other functionals defined along these paths.

Definition (*Displacement convexity*). Let (ρ_t) be as in Theorem 6.1.3, and let $\Psi : P_2 \to \mathbf{R} \cup \{\infty\}$ be a functional such that $t \mapsto \Psi(\rho_t)$ is continuous $[0, 1] \to \mathbf{R}$. We say that Ψ is κ-uniformly displacement convex for some $\kappa > 0$ if

$$\Psi(\rho_{t+h}) + \Psi(\rho_{t-h}) - 2\Psi(\rho_t) \ge \kappa h^2 W_2(\rho_0, \rho_1)^2$$

for all $(0 \le t - h \le t \le t + h \le 1)$.   (6.1.13)

Example 6.1.4 (*Uniform displacement convexity of electrostatic energy and relative free entropy*). Let v be twice continuously differentiable on $[a, b]$ with $v''(x) \ge \kappa > 0$. Then we saw in Proposition 4.8.1 that the energy

$$E_v(\mu) = \int_{[a,b]} v(x)\mu(dx) - \frac{\beta}{2} \iint_{[x \ne y]} \log|x - y|\mu(dx)\mu(dy) \quad (6.1.14)$$

is κ-uniformly displacement convex. Note that E_v is an energy in the sense of potential theory and may be considered as a free entropy in Voiculescu's theory [163, 164, 166]; but it is different from the Gibbs or Helmholtz functions of equilibrium thermodynamics [3], which are sometimes referred to as the free energy.

• If the potential is uniformly convex, then relative free entropy is uniformly displacement convex.
• Uniform displacement convexity for relative free entropy amounts to the free transportation inequality.

Exercise 6.1.5 (*Displacement convexity of Boltzmann's entropy*). Let (ρ_t) be as in Theorem 6.1.3.

(i) Show that

$$\rho_t(\varphi_t(x))J_{\varphi_t}(x) = \rho_0(x) \qquad (6.1.15)$$

where J_{ρ_t} is the Jacobian of the transformation $x \mapsto \varphi_t(x)$, which induces ρ_t from ρ_0; deduce that

$$-S(\rho_t) = \int_{\mathbf{R}^n} \rho_0(x) \log \rho_0(x) \, dx - \int_{\mathbf{R}^n} \rho_0(x) \log J_{\varphi_t}(x) \, dx.$$

$$(6.1.16)$$

(ii) By considering the eigenvalues of $\operatorname{Hess} \Phi(x)$, show that

$$\log J_{\varphi_t}(x) = \operatorname{trace} \log\big((1-t)I + t\operatorname{Hess}\Phi(x)\big)$$

and deduce that

$$-\frac{d^2}{dt^2} \log J_{\varphi_t}(x) = \operatorname{trace}\Big(R_t\big(\operatorname{Hess}\Phi(x)-I\big)R_t\big(\operatorname{Hess}\Phi(x)-I\big)\Big) \ge 0,$$

$$(6.1.17)$$

where $R_t = \big((1-t)I + t\operatorname{Hess}\Phi(x)\big)^{-1}$ is positive definite.

(iii) Deduce that $t \mapsto -S(\rho_t)$ is convex.

Exercise 6.1.6 (*Uniform displacement convexity of relative entropy*).

(i) Let ρ_1 be a probability density function on $\mathbf{R}^n$ and let v be a twice continuously differentiable function so that $\rho_0(x) = e^{-v(x)}$ is a probability density function. Show that

$$\operatorname{Ent}(\rho_1 \mid \rho_0) = \int v(x)\rho_1(x)\, dx + \int \rho_1(x) \log \rho_1(x) \, dx. \quad (6.1.18)$$

(ii) Suppose further that $v : \mathbf{R}^n \to \mathbf{R}$ is κ uniformly convex; let (ρ_t) be as in Theorem 6.1.3. Combine Exercise 6.1.5 with the proof of Proposition 4.8.2 to show that $\operatorname{Ent}(\rho_t \mid \rho_0)$ is κ uniformly convex.

(iii) Deduce that the transportation inequality holds, as in Section 3.4:

$$W_2(\rho_1, \rho_0)^2 \le \frac{2}{\kappa}\operatorname{Ent}(\rho_1 \mid \rho_0). \qquad (6.1.19)$$

- If the potential is uniformly convex, then relative entropy is uniformly displacement convex.
- Uniform displacement convexity for relative entropy amounts to the transportation inequality $T_2(\kappa)$.

In Section 6.8 we return to this point, and give further conditions under which relative entropy is indeed κ-uniformly convex.

Definition (*Gradient flow*). Let F be a functional on the geodesic $P = \{\rho_t : 0 \leq t \leq 1\}$ that is weakly differentiable and such that $\delta F/\delta\rho$ defines a C^1 function at each ρ_t. Then the gradient flow is

$$\frac{\partial \rho_t}{\partial t} = \operatorname{div}\left(\rho_t \nabla \frac{\delta F}{\delta \rho}\right). \qquad (6.1.20)$$

In other words, the gradient flow is associated with the vector field

$$\xi_t(x) = -\nabla\left(\frac{\delta F}{\delta \rho}\right)_{\rho_t}. \qquad (6.1.21)$$

Let ρ be a probability density function on $\mathbf{R}^n$ and let v be a continuously differentiable function so that $\mu(dx) = e^{-v(x)}\,dx$ and $\nu(dx) = \rho(x)\,dx$ are probability measures.

Examples 6.1.7 (*Fokker–Planck equation*, [65]). Consider the functional given by

$$F(\rho) = \int v(x)\rho(x)\,dx + \int \rho(x)\log\rho(x)\,dx, \qquad (6.1.22)$$

which is analogous to the expressions in Examples 6.1.4 and 6.1.6. Suppose that Ω is a bounded open set, that ρ is supported on $\bar{\Omega}$ and that $\rho(x) > 0$ on Ω; then for a compact subset K of Ω let

$$D = \left\{ h \in C^1(\Omega; \mathbf{R}) : \int h(x)\,dx = 0, h(x) = 0 \quad \text{on} \quad K^c \right\}. \qquad (6.1.23)$$

Then $\delta F/\delta\rho = \log\rho + 1 + v$ defines an element of $L^2(K)$, and hence a bounded linear functional on D. The gradient flow is

$$\frac{\partial \rho_t}{\partial t} = \operatorname{div}\left(\nabla \rho_t + \rho_t \nabla v\right) \qquad (6.1.24)$$

and there are two special cases of note. When $v = 0$, this is the standard heat equation $\partial \rho_t/\partial t = \sum_{j=1}^n \partial^2 \rho_t/\partial x_j^2$.

When $v(x) = \|x\|^2/2$ we have the diffusion equation

$$\frac{\partial \rho_t}{\partial t} = \operatorname{div}\nabla \rho_t + \langle \nabla \rho_t, x \rangle + n\rho_t \qquad (6.1.25)$$

that is associated to the Ornstein–Uhlenbeck process which we consider in Chapter 11.

6.2 Logarithmic Sobolev inequalities

- Logarithmic Sobolev inequalities bound relative entropy by relative information.
- Logarithmic Sobolev inequalities imply convergence to equilibrium under the gradient flow associated with relative entropy.
- The logarithmic Sobolev constant measures the rate of convergence to equilibrium.

Definition (*LSI*). Let ρ be a probability density function on $\mathbf{R}^n$. Then ρ satisfies a logarithmic Sobolev inequality with constant α when

$$LSI(\alpha): \quad \int_{\mathbf{R}^n} f^2 \log \left(f^2 \Big/ \int_{\mathbf{R}^n} f^2 \rho \right) \rho(x)\, dx \leq \frac{2}{\alpha} \int_{\mathbf{R}^n} \|\nabla f\|^2 \rho(x)\, dx$$

(6.2.1)

holds for all $f \in L^2(\rho)$ with distributional gradient $\nabla f \in L^2(\rho, \mathbf{R}^n)$.

Note that when we replace f by tf, where $t > 0$ is a constant, we scale up both sides of the inequality by t^2. Suppose that ρ is a probability density function and $\mathbf{E}$ is the corresponding expectation. Then $LSI(\alpha)$ asserts that

- $\mathbf{E}(q \log q) \leq (2/\alpha)\mathbf{E}\|\nabla \sqrt{q}\|^2$ for all probability density functions $q\rho$.

Example 6.2.1 Gross [75] showed that the standard $N(0,1)$ density γ satisfies *LSI* with $\alpha = 1$. He deduced this results from a discrete inequality by means of a central limit theorem. This approach has been revived in [71, p. 214]. In Corollary 6.3.3, we shall recover this result by a quite different argument from [27, 17] based upon the Prékopa–Leindler inequality as in Theorem 3.5.2.

The constant α measures the rate at which the system converges to equilibrium under a suitable gradient flow, in a sense made precise in the following result.

Theorem 6.2.2 (*Barron [10]*). *Suppose that ρ_∞ satisfies $LSI(\alpha)$ and that $(\rho_t)_{t\geq 0}$ is a family of probability density functions that undergoes the gradient flow associated with the relative entropy functional $F(\rho) = \mathrm{Ent}(\rho_t \mid \rho_\infty)$. Then*

$$F(\rho_t) \leq e^{-2\alpha t} F(\rho_0) \qquad (t \geq 0). \tag{6.2.2}$$

Proof. With $F(\rho_t) = \int \rho_t \log(\rho_t/\rho_\infty)dx$ we find the variation

$$\frac{\delta F}{\delta \rho} = 1 + \log \frac{\rho_t}{\rho_\infty} \tag{6.2.3}$$

and hence the gradient flow

$$\frac{\partial \rho_t}{\partial t} = \text{div}\left(\rho_t \nabla \log \frac{\rho_t}{\rho_\infty}\right). \tag{6.2.4}$$

Under this flow, the relative entropy satisfies

$$\frac{d}{dt}F(\rho_t) = \left\langle \frac{\delta F}{\delta \rho_t}, \frac{\partial \rho_t}{\partial t} \right\rangle_{L^2}$$
$$= \int \left(1 + \log(\rho_t/\rho_\infty)\right) \text{div}\left(\rho_t \nabla \log(\rho_t/\rho_\infty)\right) dx \tag{6.2.5}$$

which by the divergence theorem and Proposition 6.1.2 gives

$$\frac{d}{dt}F(\rho_t) = -\int \langle \nabla \log(\rho_t/\rho_\infty), \nabla \log(\rho_t/\rho_\infty)\rangle \rho_t(x)\, dx$$
$$= -I(\rho_t \mid \rho_\infty). \tag{6.2.6}$$

The logarithmic Sobolev inequality gives

$$\frac{d}{dt}F(\rho_t) \leq -2\alpha F(\rho_t); \tag{6.2.7}$$

hence by Gronwall's inequality we have
$$F(\rho_t) \leq e^{-2\alpha t}F(\rho_0). \tag{6.2.8}$$

$\square$

Corollary 6.2.3 *Under the hypotheses of Theorem 6.2.2,*

$$\|\rho_t - \rho_\infty\|_{L^1}^2 \leq 2e^{-2\alpha t}\text{Ent}(\rho_0 \mid \rho_\infty) \qquad (t \geq 0), \tag{6.2.9}$$

so $\rho_t \to \rho_\infty$ in $L^1(dx)$ as $t \to \infty$.

Proof. This follows directly from the theorem via Csiszár's inequality Proposition 3.1.8. $\square$

Remarks. (i) Barron [10] used this result with $\rho_\infty = \gamma_1$ to obtain a quantitative version of the central limit theorem, where ρ_t undergoes a gradient flow associated with the Ornstein–Uhlenbeck process.

(ii) For probability densities on **R**, Bobkov and Götze [25] found a relatively simple characterization of LSI in terms of cumulative distribution functions, including a formula for α. There is no such characterization known for densities on $\mathbf{R}^n$, but the following result allows us to manufacture examples of LSI on product spaces.

(iii) The following proposition is interesting because it leads to constants that are independent of dimension. The converse is part of a subsequent exercise.

Proposition 6.2.4 (*Tensorization*). *Suppose that ρ_1 on $\mathbf{R}^m$ satisfies $LSI(\alpha_1)$ and that ρ_2 on $\mathbf{R}^n$ satisfies $LSI(\alpha_1)$. Then $\rho_1 \otimes \rho_2$ satisfies $LSI(\alpha)$, where $\alpha = \min\{\alpha_1, \alpha_2\}$.*

Proof. Let ∇_x be the gradient operator with respect to the x variable. Suppose that $f(x, y)$ is a differentiable function on $\mathbf{R}^m \times \mathbf{R}^n$ and let $g(x) = (\int f(x,y)^2 \rho_2(y)\, dy)^{1/2}$; then

$$\iint_{\mathbf{R}^m \times \mathbf{R}^n} f^2 \log\left(f^2 / \int f^2 d(\rho_1 \otimes \rho_2)\right) \rho_1(x)\rho_2(y)\, dxdy \quad (6.2.10)$$

$$= \int_{\mathbf{R}^m} \left(\int_{\mathbf{R}^n} f^2(x,y) \log(f(x,y)^2/g(x)) \rho_2(y)dy\right) \rho_1(x)\, dx$$

$$+ \int_{\mathbf{R}^m} g(x)^2 \log\left(g(x)^2 / \int_{\mathbf{R}^m} g^2 \rho_1\right) \rho_1(x)\, dx$$

and by the logarithmic Sobolev inequalities, this is

$$\leq \frac{2}{\alpha_2} \int_{\mathbf{R}^m} \left(\int_{\mathbf{R}^n} \|\nabla_y f(x,y)\|^2 \rho_2(y)dy\right) \rho_1(x)dx$$

$$+ \frac{2}{\alpha_1} \int_{\mathbf{R}^m} \|\nabla g(x)\|^2 \rho_1(x)\, dx. \quad (6.2.11)$$

Now by the triangle inequality, the function g is differentiable with

$$\|\nabla g(x)\|^2 \leq \int \|\nabla_x f(x,y)\|^2 \rho_2(y)\, dy, \quad (6.2.12)$$

so the right-hand side of (6.2.9) is bounded above by

$$\frac{2}{\alpha} \iint_{\mathbf{R}^m \times \mathbf{R}^n} \left(\|\nabla_x f(x,y)\|^2 + \|\nabla_y f(x,y)\|^2\right) \rho_1(x)\rho_2(y)\, dxdy. \quad (6.2.13)$$

$\square$

Exercise 6.2.5 (i) Let $\sigma \in Prob(\Omega)$ satisfy $LSI(\alpha)$ and let $\varphi : \Omega \to \Phi$ be L-Lipschitz. Show that $\varphi \sharp \sigma$ satisfies $LSI(\alpha/L^2)$.

(ii) Let $\rho(x, y)$ be a probability density function on $\mathbf{R}^m \times \mathbf{R}^n$ that satisfies $LSI(\alpha)$. Show that the marginal density

$$\rho_1(x) = \int_{\mathbf{R}^m} \rho(x,y)\, dy \quad (6.2.14)$$

also satisfies $LSI(\alpha)$.

6.3 Logarithmic Sobolev inequalities for uniformly convex potentials

- Logarithmic Sobolev inequalities hold for Gibbs measures with uniformly convex potentials.
- This follows from the Prékopa–Leindler inequality by an argument due to Bobkov and Ledoux.
- The are logarithmic Sobolev inequalities for matrix ensembles with uniformly convex potentials.

The following theorem was important in the historical development of the theory, as Bakry and Emery used it in their theory of Dirichlet forms which linked diffusion processes with functional inequalities. Bobkov and Ledoux [27] discovered the more elementary proof below, which depends upon the Prékopa–Leindler inequality as in Theorem 1.4.2. In Section 6.8, we shall mention some more general logarithmic Sobolev inequalities and related results from metric geometry.

Definition (*Uniform convexity*). A continuous function $V : \mathbf{R}^n \to \mathbf{R}$ is uniformly convex if there exists $\alpha > 0$ such that

$$sV(x) + tV(y) - V(sx + ty) \geq \frac{st\alpha}{2}\|x - y\|^2 \qquad (6.3.1)$$

for all $x, y \in \mathbf{R}^n$ and all $0 \leq s, t \leq 1$ such that $s + t = 1$.

Example 6.3.1 The fundamental example is the quadratic function $V(x) = \|x\|^2/2$, which satisfies the condition with $\alpha = 1$. This potential gives rise to the standard Gaussian measure γ_n, and it turns out that uniformly convex potentials share several properties of the Gaussian.

Remark. The dual expression for the Wasserstein metric W_2 on $\mathbf{R}$ involves

$$g(x) - f(y) \leq \frac{1}{2}\|x - y\|^2, \qquad (6.3.2)$$

or equivalently

$$g(x) \leq \inf_y \Big\{ f(y) + \frac{1}{2}\|x - y\|^2 \Big\}. \qquad (6.3.3)$$

This expression appears in the Hamilton–Jacobi partial differential equation, an observation which has been exploited by Bobkov, Gentil and Ledoux [26] to prove various transportation and logarithmic Sobolev inequalities. We shall use the differential form of this condition in the proof of the following result, and then give a free version in Section 6.6.

Theorem 6.3.2 (*Bakry–Emery's LSI, [9]*). *Let* $V : \mathbf{R}^n \to \mathbf{R}$ *be uniformly convex with constant* $\alpha > 0$. *Then there exists* $Z < \infty$ *such that*

$$\rho(x) = Z^{-1} e^{-V(x)} \qquad (6.3.4)$$

defines a probability density function, and ρ *satisfies LSI(α).*

Proof. First we observe that $V(x)$ grows like $\alpha \|x\|^2 / 2$ as $\|x\| \to \infty$, so the existence of $Z < \infty$ follows by comparison with the Gaussian density.

Suppose that $g \in C_b^2(\mathbf{R})$. Let $x, y \in \mathbf{R}^n$ have $z = sx + ty$ on the line segment between them, where $0 \leq s, t \leq 1$ satisfy $s + t = 1$, where s will be chosen to be small; then choose $0 < \beta < \alpha - s\|\text{Hess}\,g\|$. By Taylor's theorem we have $\bar{z}$ between y and z such that

$$g(y) = g(z) - s\langle \nabla g(y), x - y \rangle + \frac{s^2}{2} \langle \text{Hess}\,g(\bar{z})(x - y), (x - y) \rangle \quad (6.3.5)$$

and so

$$g(y) - \frac{st\beta}{2} \|x - y\|^2$$
$$= g(z) - \frac{s}{2} \| \sqrt{t\beta}(x - y) + \nabla g(y)/\sqrt{\beta t}\|^2 + \frac{s}{2t\beta} \|\nabla g(z)\|^2$$
$$+ \frac{s^2}{2} \langle \text{Hess}\,g(\bar{z})(x - y), (x - y) \rangle; \qquad (6.3.6)$$

so by the choice of β

$$g(y) - \frac{st\alpha}{2} \|x - y\|^2$$
$$\leq g(z) - \frac{s}{2} \| \sqrt{t\beta}(x - y) + \nabla g(y)/\sqrt{\beta t}\|^2 + \frac{s}{2t\beta} \|\nabla g(z)\|^2. \qquad (6.3.7)$$

Now on account of this we have

$$g_t(z) = \sup \left\{ g(y) - \frac{st\alpha}{2} \|x - y\|^2 : z = sx + ty \right\}$$
$$\leq g(z) + \frac{s}{2t\beta} \|\nabla g(z)\|^2. \qquad (6.3.8)$$

The inequality

$$g(y) - \big(sV(x) + tV(y) - V(z)\big) \leq g(y) - \frac{st\alpha}{2} \|x - y\|^2 \qquad (6.3.9)$$

gives rise to the inequality

$$\exp\big(g(z) + (s/2t\beta)\|\nabla g(z)\|^2 - V(z)\big) \geq \exp\big(-sV(x)\big) \exp\big(g(y) - tV(y)\big) \qquad (6.3.10)$$

and hence by the Prékopa–Leindler inequality Theorem 1.4.2 we have

$$\int \exp\big(g(z) + (s/2t\beta)\|\nabla g(z)\|^2\big)\rho(z)\,dz$$
$$\geq \left(\int \rho(z)dz\right)^s \left(\int \exp(g(y)/t)\,\rho(y)dy\right)^t. \qquad (6.3.11)$$

We take logarithms of both sides and differentiate at $s = 0+$; thus

$$\frac{1}{2\beta}\int \|\nabla g(z)\|^2 e^{g(z)}\rho(z)dz \geq \int g(z)e^{g(z)}\rho(z)dz$$
$$-\left(\int e^{g(z)}\rho(z)dz\right)\log\left(\int e^{g(z)}\rho(z)dz\right). \qquad (6.3.12)$$

We can let $\beta \to \alpha-$ to recover the stated result. $\qquad \square$

Corollary 6.3.3 (*Gross's LSI [75]*). *The Gaussian density with covariance matrix A satisfies LSI(α), where α is the smallest eigenvalue of A^{-1}.*

Proof. Here

$$v(x) = \frac{1}{2}\langle A^{-1}x, x\rangle + \langle b, x\rangle \qquad (6.3.13)$$

for some $b \in \mathbf{R}^n$, and hence $\operatorname{Hess} v(x) = A^{-1}$. By applying Taylor's theorem to v, we deduce that α is the constant in (6.3.1). $\qquad \square$

We finally arrive at the intended application to random matrix theory. In the Definition (6.3.1), for consistency with previous scalings, we take $\|\cdot\|$ to be the $c^2(n)$ norm on $M_n^s(\mathbf{R})$. Nevertheless, it is significant that the constants in the following results genuinely improve as n increases.

Corollary 6.3.4 (*LSI for generalized ensembles*). *Suppose that $v : \mathbf{R} \to \mathbf{R}$ is twice continuously differentiable with $v''(x) \geq \alpha$ for some $\alpha > 0$, and let $V(X) = (1/n)\operatorname{trace} v(X)$ for $X \in M_n^s(\mathbf{R})$. Then for some $Z_n < \infty$ the probability measure*

$$\nu_n(dX) = Z_n^{-1}\exp\big(-n^2V(X)\big)\,dX \qquad (6.3.14)$$

satisfies the logarithmic Sobolev inequality with constant $n^2\alpha > 0$.

Proof. As in the proof of Theorem 3.6.3, n^2V is uniformly convex on $(M_n^s(\mathbf{R}), c^2(n))$ with constant $n^2\alpha$, hence we can apply Theorem 6.3.2. $\qquad \square$

Corollary 6.3.5 (*LSI for eigenvalue distributions*). *Suppose that v :* $\mathbf{R} \to \mathbf{R}$ *is twice continuously differentiable with $v''(x) \geq \alpha$ for some $\alpha > 0$, and for $\beta > 0$ let*

$$\sigma_n^{(\beta)}(d\lambda) = Z_n(\beta)^{-1} \exp\Big(-n \sum_{j=1}^{n} v(\lambda_j)\Big) \prod_{1 \leq j < k \leq n} (\lambda_k - \lambda_j)^\beta d\lambda_1 \dots d\lambda_n.$$
(6.3.15)

Then there exists $0 < Z_n(\beta) < \infty$ such that $\sigma_n^{(\beta)}$ is a probability measure that satisfies the logarithmic Sobolev inequality on $(\Delta^n, \ell^2(n))$ with constant $n^2\alpha$.

Proof. The potential

$$V(\lambda) = n \sum_{j=1}^{n} v(\lambda_j) + \beta \sum_{1 \leq j < k \leq n} \log \frac{1}{\lambda_k - \lambda_j}$$
(6.3.16)

is uniformly convex on Δ^n with constant $n^2\alpha$, so we can apply Theorem 6.3.2. □

Exercise 6.3.6 (Logarithmic Sobolev inequality for conditioned Gaussians). Let A be a positive definite $n \times n$ real matrix and let $z \in \mathbf{R}^n$; then let

$$Q(x) = 2^{-1}\langle Ax, x\rangle - \langle x, z\rangle.$$
(6.3.17)

  (i) Show that $Q(x) \geq -2^{-1}\langle A^{-1}z, z\rangle$, and the minimum value is attained at $x = A^{-1}z$.
 (ii) For $k \leq n$, let B be a $k \times n$ real matrix of rank k; let $w \in \mathbf{R}^k$. Show that the minimum of the constrained variational problem

$$\text{minimize}\{Q(x) \mid Bx = w\}$$
(6.3.18)

is attained at $x = x_0$ where

$$x_0 = A^{-1}z - A^{-1}B^t(BA^{-1}B^t)^{-1}(BA^{-1}z - w).$$
(6.3.19)

(iii) Now let

$$q(y) = 2^{-1}\langle Ay, y\rangle + \langle Ay, x_0\rangle - \langle y, z\rangle,$$

and let $V = \{v = B^t u : u \in \mathbf{R}^n\}$. Verify that $x = y + x_0$ satisfies $Bx = w$ for all $y \in V^\perp$, where $V^\perp = \{y : \langle y, v\rangle = 0, \forall v \in V\}$, and that $Q(x) = q(y) + Q(x_0)$.
(iv) Deduce that there exists a constant Z such that

$$\gamma(dy) = Z^{-1}e^{-q(y)}dy$$
(6.3.20)

defines a Gaussian probability measure on $V^\perp$, where the logarithmic Sobolev constant of γ depends only upon B and A.

6.4 Fisher's information and Shannon's entropy

- Logarithmic Sobolev inequalities bound relative entropy by relative Fisher's information.
- Gaussian densities have a special rôle in the theory.
- The logarithmic Sobolev inequality for the Gaussian measure implies Shannon's entropy power.

Logarithmic Sobolev inequalities may be expressed in many ways, and in this section we express them in the language of information theory. Indeed, some of the results were originally stated thus.

Definition (*Fisher's information*). Suppose that ρ is a probability density function on $\mathbf{R}^n$ such that $\sqrt{\rho}$ has distributional gradient $\nabla\sqrt{\rho}$ in $L^2(dx)$. Then the information of ρ is

$$I(\rho) = \int \frac{\|\nabla\rho\|^2}{\rho}\, dx. \qquad (6.4.1)$$

If X is a random variable with probability density function ρ, then we sometimes write $I(X)$ for $I(\rho)$.

Proposition 6.4.1 *Suppose that ρ_1 is a probability density function on $\mathbf{R}^m$ and that ρ_2 is a probability density function on $\mathbf{R}^n$ such that ρ_1 and ρ_2 have finite information. Then $\rho = \rho_1 \otimes \rho_2$ also has finite information and*

$$I(\rho) = I(\rho_1) + I(\rho_2). \qquad (6.4.2)$$

Proof. The reader should be able to extract the appropriate steps from the proof of Proposition 6.2.4. □

Proposition 6.4.2 (*Carlen*). *Suppose that ρ is a probability density function on $\mathbf{R}^{2n}$ that has finite information. Then the marginals ρ_1 and ρ_2 on $\mathbf{R}^n$ have finite information, and*

$$I(\rho_1) + I(\rho_2) \leq I(\rho). \qquad (6.4.3)$$

Remarks. The result asserts that $I(\rho_1 \otimes \rho_2) \leq I(\rho)$, so that ρ has more information than the tensor product of its marginals. This eminently natural result is due to Carlen, who also showed that equality occurs in (6.4.3) only when $\rho = \rho_1 \otimes \rho_2$. See [175].

In information theory, LSI is expressed differently.

Theorem 6.4.3 (*Shannon's entropy power*). *Let ρ be a probability density function on $\mathbf{R}^n$ that has finite Fisher information. Then*

$$\exp\left(\frac{-2S(\rho)}{n}\right) \le \frac{I(\rho)}{2\pi ne}, \tag{6.4.4}$$

with equality for the standard Gaussian density γ_n on $\mathbf{R}^n$.

Proof. By Corollary 6.3.3, the standard Gaussian density γ_n satisfies $LSI(1)$. Into the logarithmic Sobolev inequality

$$\int_{\mathbf{R}^n} f(x)^2 \log\left(f(x)^2 / \int f^2 \gamma_n\right) \gamma_n(x) dx \le 2 \int_{\mathbf{R}^n} \|\nabla f(x)\|^2 \gamma_n(x) dx \tag{6.4.5}$$

we substitute

$$f(x) = \rho(x)^{1/2} (2\pi)^{n/4} \exp(\|x\|^2/4) \tag{6.4.6}$$

and thus we obtain

$$\int \rho\big(\log \rho + (n/2)\log 2\pi + \|x\|^2/2\big) dx \le \frac{1}{2}\int \left\|\frac{\nabla\rho}{\sqrt{\rho}} + x\sqrt{\rho}\right\|^2 dx \tag{6.4.7}$$

hence

$$\int \rho\log\rho\, dx + (n/2)\log 2\pi \le \frac{1}{2}\int \frac{\|\nabla\rho\|^2}{\rho} dx + \int \langle x, \nabla\rho\rangle dx. \tag{6.4.8}$$

By the divergence theorem $\int \langle x, \nabla\rho\rangle\, dx = -n$, so

$$\frac{-2S(\rho)}{n} + \log 2\pi + 2 \le \frac{I(\rho)}{n}. \tag{6.4.9}$$

Now we rescale the density, replacing ρ by $t^n\rho(tx)$ and thus obtain

$$\frac{-2S(\rho)}{n} + \log 2\pi + 2 \le -2\log t + \frac{t^2 I(\rho)}{n}; \tag{6.4.10}$$

then the optimal choice $t = (n/I(\rho))^{1/2}$ gives

$$\frac{-2S(\rho)}{n} + \log 2\pi e \le \log \frac{I(\rho)}{n}, \tag{6.4.11}$$

as required. Further, we have equality in these inequalities when $f = 1$. $\square$

Definition (*Relative Fisher information*). Let f and g be probability density functions on $\mathbf{R}^n$ that are differentiable. Then the relative information of f with respect to g is

$$I(f \mid g) = \int_{\mathbf{R}^n} \left\|\nabla \log \frac{f(x)}{g(x)}\right\|^2 f(x)\, dx \tag{6.4.12}$$

whenever this integral converges.

Proposition 6.4.4 *The logarithmic Sobolev inequality holds for ρ if and only if*

$$\text{Ent}(q \mid \rho) \leq \frac{1}{2\alpha} I(q \mid \rho) \qquad (6.4.13)$$

holds for all probability density functions q that have finite information relative to ρ.

Proof. We note that $\|\nabla|f|\| \leq \|\nabla f\|$ holds in the sense of distributions, so we have $LSI(\alpha)$ for real-valued functions once we obtain the case of $f \geq 0$.

Let $f = \sqrt{q/\rho}$. Then $LSI(\alpha)$ reduces to the inequality

$$\int q \log(q/\rho)\, dx \leq \frac{1}{2\alpha} \int \left\| \frac{\nabla q}{q} - \frac{\nabla \rho}{\rho} \right\|^2 q(x)\, dx. \qquad (6.4.14)$$

$\square$

Remark. In this notation, the logarithmic Sobolev inequality $LSI(\alpha)$ for $Z \sim N(0, 1/\alpha)$ asserts that

$$\text{Ent}(X \mid Z) \leq \frac{1}{2\alpha} I(X \mid Z). \qquad (6.4.15)$$

Definition (*Score function*). Let X have probability density function f, and suppose that f is positive and differentiable. Then the score function of X is

$$\phi(x) = \nabla \log f(x). \qquad (6.4.16)$$

Let $X, Y : \Omega \to \mathbf{R}^n$ be random variables, and suppose that X has probability density function f and Y has probability density function g; let ϕ be the score function of f and ψ be the score function of g. Then we define

$$I(X \mid Y) = \mathbf{E}\|\phi(X) - \psi(X)\|^2 = I(f \mid g); \qquad (6.4.17)$$

and likewise

$$\text{Ent}(X \mid Y) = \int f(x) \log \frac{f(x)}{g(x)}\, dx, \qquad (6.4.18)$$

where this integral converges.

Example. On $(\Omega, \mathbf{P})$, let Z be a standard $N(0, I_n)$ random variable on $(\Omega, \mathbf{P})$, and let X be a random variable such that $\mathbf{E}X = 0$ and $\mathbf{E}\langle X, X \rangle = n$. Then the score function of Z is $\psi(x) = -x$, so

$$I(X \mid Z) = \mathbf{E}\|\phi(X) + X\|^2 = I(X) - n. \qquad (6.4.19)$$

Exercise. (i) Let p_0 and p_1 be probability density functions on $\mathbf{R}^n$ that have finite Fisher information, and let $p_t = (1-t)p_0 + tp_1$ for $0 \le t \le 1$. Show that

$$\frac{d^2}{dt^2} \int_{\mathbf{R}^n} \frac{\|\nabla p_t\|^2}{p_t}\,dx = 2 \int_{\mathbf{R}^n} \left\| \frac{\nabla \dot{p}_t}{p_t} - \frac{\dot{p}_t \nabla p_t}{p_t^2} \right\|^2 p_t\,dx,$$

where $\dot{p}_t = p_1 - p_0$, and hence that information is a convex functional.

(ii) Deduce that $p \mapsto I(p \mid q)$ is also convex.

6.5 Free information and entropy

- Voiculescu introduced free information and free entropy.
- Free logarithmic Sobolev inequalities bound the relative free entropy by the relative free information.
- Free logarithmic Sobolev inequalities imply convergence to equilibrium under the gradient flow of relative free entropy.

in [163, 164] Voiculescu introduced the concepts of free information and entropy in the context of non-commutative probability, and in Chapter 4 we saw how free entropy and relative free entropy arise in random matrix theory. In order to emphasize the analogy between classical and free functional equations, we introduce the concepts of free information and relative free information by simple formulae and show that relative free information arises in the gradient flow of relative free entropy. Before presenting the complex Burgers equation, we pause to consider a form of the Burgers equation that arises in fluid mechanics; see [62]. Consider the motion of a fluid in one space dimension so that the velocity at position x at time t is $u(x,t)$. When the density is ρ, and the mass of the fluid is conserved, we also have the continuity equation

$$\frac{\partial \rho}{\partial t} + \frac{\partial}{\partial x}(u\rho) = 0. \tag{6.5.1}$$

If we follow the motion of a particle and consider a quantity $w(x,t)$, then the rate of change is

$$\frac{Dw}{Dt} = \frac{\partial w}{\partial t} + u(x,t)\frac{\partial w}{\partial x}. \tag{6.5.2}$$

In particular, the acceleration satisfies an equation of motion $Du/Dt = f$, where f is the force field which includes pressure and viscosity terms.

The Burgers equation is

$$\frac{\partial u}{\partial t} + u\frac{\partial u}{\partial x} = \nu\frac{\partial^2 u}{\partial x^2},$$

(6.5.3)

where ν is a constant. The case $\nu = 0$ is already non-trivial, since shocks can occur.

In the next section, we consider a probability density function which evolves under a chosen velocity field, and we analyse the effect on certain functionals.

Proposition 6.5.1 (*The complex Burgers equation*). *Let G be a solution of the complex Burgers equation*

$$\frac{\partial G}{\partial t} + \alpha G\frac{\partial G}{\partial z} = 0 \qquad (t > 0, \Im z > 0)$$

(6.5.4)

and suppose that $\Im G_t(x) = \rho_t(x)$ is a probability density function for each $t > 0$. Then $\rho_t(x)$ satisfies

$$\frac{\partial \rho}{\partial t} = \frac{\partial}{\partial x}(\rho\mathcal{H}\rho)$$

(6.5.5)

which is the gradient flow for the scaled free entropy $\frac{\alpha}{2\pi}\chi$.

Proof. The scaled free entropy

$$\frac{\alpha}{2\pi}\chi(\rho) = \frac{\alpha}{2\pi}\iint \log|x - y|\,\rho(x)\rho(y)\,dxdy$$

(6.5.6)

has variation

$$\frac{\delta}{\delta\rho}\frac{\alpha}{2\pi}\chi(\rho) = \frac{\alpha}{\pi}\int \log|x - y|\,\rho(y)\,dy$$

(6.5.7)

and hence the gradient flow is

$$\frac{\partial}{\partial t}\rho = \alpha\frac{\partial}{\partial x}\left(\rho(x)\frac{1}{\pi}\int\frac{1}{x - y}\rho(y)\,dy\right).$$

(6.5.8)

By comparison, $G(x) = -(\mathcal{H}\rho) + i\rho$ satisfies

$$0 = -\frac{\partial}{\partial x}(\mathcal{H}\rho) + i\frac{\partial\rho}{\partial t} + \alpha\left((\mathcal{H}\rho)\frac{\partial}{\partial x}(\mathcal{H}\rho) - \rho\frac{\partial\rho}{\partial x}\right)$$

$$- \alpha i\rho\frac{\partial}{\partial x}(\mathcal{H}\rho) - \alpha i(\mathcal{H}\rho)\frac{\partial\rho}{\partial x},$$

(6.5.9)

and the imaginary part is the gradient flow as stated. □

By analogy with the case of classical logarithmic Sobolev inequalities, we introduce the free logarithmic Sobolev inequality. Then we shall prove that the analogy extends to gradient flows for the appropriate functionals.

Definition (*Free information*). Let ρ be a probability density function on $\mathbf{R}$ such that $\rho \in L^3$. Then the free information of ρ is

$$\Phi(\rho) = \int_{-\infty}^{\infty} \rho(x)^3 \, dx. \tag{6.5.10}$$

Voiculescu showed that the definition of the free information could be expressed in a style similar to that of Fisher's information, using the Hilbert transform $\mathcal{H}$ of Theorem 1.8.8; see [163, 164].

Lemma 6.5.2 *For all probability density functions $\rho \in L^3(\mathbf{R})$, the free information satisfies*

$$\Phi(\rho) = 3 \int_{-\infty}^{\infty} (\mathcal{H}\rho)^2 \rho(x) \, dx. \tag{6.5.11}$$

Proof. By Theorem 1.8.8, the function $f = \rho + i\mathcal{H}\rho$ belongs to L^3 and extends by means of the Poisson kernel to an analytic function

$$f(x+iy) = \frac{1}{\pi} \int_{-\infty}^{\infty} \frac{yf(t)}{(x-t)^2 + y^2} \, dt \qquad (y > 0) \tag{6.5.12}$$

on the upper half plane. By Cauchy's theorem, we have the identity

$$0 = \int_{-\infty}^{\infty} f(x)^3 \, dx = \int_{-\infty}^{\infty} \left(\rho(x)^3 - 3(\mathcal{H}\rho)^2 \rho \right) dx$$
$$+ i \int_{-\infty}^{\infty} \left(3\rho^2 \mathcal{H}\rho - (\mathcal{H}\rho)^3 \right) dx, \tag{6.5.13}$$

and the real part of the right hand side must be zero. $\qquad\square$

Definition (*Relative free information*). Let p and q be probability measures on $\mathbf{R}$ such that $p, q \in L^3$. Then the relative free information is

$$\Phi(q \mid p) = 3 \int_{-\infty}^{\infty} \left(\mathcal{H}p - \mathcal{H}q \right)^2 q(x) \, dx. \tag{6.5.14}$$

The integral converges on account of Hölder's inequality and M. Riesz's Theorem 1.8.8 since

$$\Phi(q \mid p) \leq 3 \left(\int_{-\infty}^{\infty} |\mathcal{H}(p-q)|^3 \, dx \right)^{2/3} \left(\int q(x)^3 \, dx \right)^{1/3}$$
$$\leq 3A_3^2 \left(\int_{-\infty}^{\infty} |q-p|^3 \, dx \right)^{2/3} \left(\int q^3 \, dx \right)^{1/3}. \tag{6.5.15}$$

For comparison, we have the relative Fisher information

$$I(q \mid p) = \int_{-\infty}^{\infty} \left(\frac{d}{dx} \log \frac{q}{p} \right)^2 q \, dx \tag{6.5.16}$$

versus the relative free information

$$\Phi(q \mid p) = \frac{3}{\pi^2} \int_{-\infty}^{\infty} \left(\frac{d}{dx} \int_{-\infty}^{\infty} \log|x-y|(q(y)-p(y))\,dy \right)^2 q(x)\,dx.$$
(6.5.17)

Pursuing this analogy with Lemma 6.4.4, we make the following definition.

Definition (*Free LSI*). A probability density function on **R** satisfies a free logarithmic Sobolev inequality when there exists $\alpha > 0$ such that

$$\Sigma(q,\rho) \le \frac{1}{2\alpha} \Phi(q,\rho).$$
(6.5.18)

Example 6.5.3 Biane and Voiculescu [16] showed that the semicircle law

$$\sigma(dx) = (2/\pi)\sqrt{1-x^2}\,dx$$
(6.5.19)

on $[-1,1]$ satisfies the free logarithmic Sobolev inequality. We shall recover this result in Section 6.6, and also show that random matrix theory links free and classical logarithmic Sobolev inequalities.

The constant α measures the rate of convergence to equilibrium under the gradient flow for the relative free entropy.

Theorem 6.5.4 (*Free convergence*). *Suppose that ρ_∞ satisfies the free logarithmic Sobolev inequality and that $(\rho_t)_{t\ge 0}$ undergoes the gradient flow associated with the relative free entropy $F(\rho_t) = \Sigma(\rho_t, \rho_\infty)$. Then ρ_t converges to ρ_∞ in relative free entropy, with*

$$F(\rho_t) \le e^{-4\alpha\pi^2 t/3} F(\rho_0).$$
(6.5.20)

Proof. With

$$F(\rho_t) = \iint (\rho_t(x)-\rho_\infty(x))(\rho_t(y)-\rho_\infty(y))\log\frac{1}{|x-y|}\,dxdy \quad (6.5.21)$$

the variation is

$$\frac{\delta F}{\delta \rho_t} = 2\int \log\frac{1}{|x-y|}(\rho_t(y)-\rho_\infty(y))\,dy \qquad (6.5.22)$$

and so the gradient flow is

$$\frac{\partial \rho_t}{\partial t} = 2\frac{\partial}{\partial x}\left(\rho_t \frac{\partial}{\partial x}\int \log\frac{1}{|x-y|}(\rho_t(y)-\rho_\infty(y))\,dy\right)$$
$$= -2\pi\frac{\partial}{\partial x}\left(\rho_t(x)\mathcal{H}(\rho_t-\rho_\infty)\right).$$
(6.5.23)

Hence the rate of change of $F(\rho_t)$ is

$$\frac{d}{dt}F(\rho_t) = \left\langle \frac{\delta F}{\delta \rho_t}, \frac{\partial \rho_t}{\partial t} \right\rangle_{L^2}$$

$$= -2\pi \iint \log \frac{1}{|x-y|}(\rho_t(y)-\rho_\infty(y))\,dy \frac{\partial}{\partial x}(\rho_t(x)\mathcal{H}(\rho_t-\rho_\infty))\,dx$$

$$= -2\pi^2 \int (\mathcal{H}(\rho_t - \rho_\infty))^2 \rho_t(x)\,dx$$

$$= -\frac{2\pi^2}{3}\Phi(\rho_t \mid \rho_\infty) \tag{6.5.24}$$

where the middle step follows by integration by parts. Hence by the free logarithmic Sobolev inequality we have

$$\frac{d}{dt}F(\rho_t) \le -(4\alpha\pi^2/3)F(\rho_t) \tag{6.5.25}$$

hence $F(\rho_t) \le \exp(-(4\alpha\pi^2 t/3))F(\rho_0)$. □

Remarks. (i) For a given energy functional E defined on probability density functions, the appropriate analogue of *LSI* is

$$E(\rho) \le \frac{1}{\alpha}\int_{\mathbf{R}^n} \rho \left\|\nabla \frac{\delta E}{\delta \rho}\right\|^2 dx. \tag{6.5.26}$$

(ii) The following table converts classical to free quantities.

Classical	Free
$\log \rho$	$\mathcal{L}\rho(x) = \int \log|x-y|\,\rho(y)dy$
$\nabla \log \rho(x)$	$\pi\mathcal{H}\rho(x)$
$S(\rho) = -\int \rho(x)\log\rho(x)dx$	$\chi(\rho) = -\int \rho(x)\mathcal{L}\rho(x)dx$
$\text{Ent}(q \mid p) = \int q\log(q/p)dx$	$\Sigma(q,p) = \int (q-p)\mathcal{L}(q-p)dx$
$I(q \mid p) = \int q(\nabla\log(q/p))^2 dx$	$\Phi(q \mid p) = 3\int (\mathcal{H}(q-p))^2 qdx$

$$(6.5.27)$$

Exercise. Let X be a random variable with probability density function ρ, and let f be a continuously differentiable and strictly increasing function.

(i) Calculate the probability density function q for $f(X)$, and hence obtain an expression for the information $I(f(X))$.
(ii) Calculate the free information of $f(X)$; that is, find $\Phi(q)$.

6.6 Free logarithmic Sobolev inequality

• The semicircle law satisfies a free logarithmic Sobolev inequality.

Whereas the statement of the free LSI involves probability density functions on $\mathbf{R}$, the proof uses the Prékopa–Leindler inequality on $\mathbf{R}^n$ as $n \to \infty$.

Theorem 6.6.1 (*Ledoux [105]*). *Suppose that v is twice continuously differentiable and that $v''(x) \geq \alpha$ for all real x for some $\alpha > 0$. Then ρ_v satisfies the free logarithmic Sobolev inequality*

$$\Sigma(q \mid \rho_v) \leq \frac{\beta\pi^2}{6\alpha}\Phi(q \mid \rho_v) \tag{6.6.1}$$

for all probability measures q that have finite energy.

Proof. Recall that there exists a constant C_v such that

$$v(x) \geq \beta \int_{-\infty}^{\infty} \log|x - y|\rho_v(y)dy + C_v \tag{6.6.2}$$

with equality on the support of v, so we introduce a real function r such that

$$r(x) \geq \beta \int_{-\infty}^{\infty} \log|x - y|q(y)\,dy \qquad (x \in \mathbf{R}) \tag{6.6.3}$$

with equality for x on the support of ρ_v, and such that $r(x) = v(x)$ whenever $|x| \geq L$, for some large L. Next we introduce, for some constant C to be chosen, the function $f(x) = v(x) - r(x) + C$ so that

$$\int_{-\infty}^{\infty} f'(x)^2 q(x)\,dx = \int_{-\infty}^{\infty} \left(v'(x) - \beta\pi\mathcal{H}q(x)\right)^2 q(x)\,dx$$
$$= \Phi(q \mid \rho_v). \tag{6.6.4}$$

Now let

$$g_t(x) = \inf_y \left\{ f(y) + \frac{(x - y)^2}{2t\alpha} \right\} \tag{6.6.5}$$

and for $z = \theta x + (1 - \theta)y$, observe that

$$g_t(x) \leq f(z) + \frac{(1 - \theta)^2}{2\alpha t}(x - y)^2 \tag{6.6.6}$$

where by continuity of v,

$$\theta v(x) + (1 - \theta)v(y) - v(z) \geq \frac{\alpha}{2}\theta(1 - \theta)(x - y)^2 \tag{6.6.7}$$

so

$$g_t(x) - \theta v(x) - (1 - \theta)v(y) \leq f(z) - v(z) \tag{6.6.8}$$

when

$$\frac{1}{2}\alpha\theta(1-\theta) \geq \frac{(1-\theta)^2}{2\alpha t} \quad \text{that is} \quad t \geq \frac{1-\theta}{\alpha^2\theta}. \tag{6.6.9}$$

Let

$$\delta(\lambda) = \prod_{1\leq j<k\leq n} (\lambda_k - \lambda_j). \tag{6.6.10}$$

By the Prékopa–Leindler inequality Theorem 1.4.2

$$\left(\int_{\Delta^n} \exp\left(n\sum_{j=1}^{n}(g(\lambda_j)-v(\lambda_j))\right)\delta(\lambda)^\beta \, d\lambda\right)^\theta$$

$$\times \left(\int_{\Delta^n} \exp\left(-n\sum_{j=1}^{n}v(\lambda_j)\right)\delta(\lambda)^\beta \, d\lambda\right)^{1-\theta}$$

$$\leq \int_{\Delta^n} \exp\left(n\sum_{j=1}^{n}(f(\lambda_j)-v(\lambda_j))\right)\delta(\lambda)^\beta \, d\lambda. \tag{6.6.11}$$

We introduce the normalizing constant $Z_n = \int_{\Delta^n} \exp(-n\sum_{j=1}^{n} v(\lambda_j)) \times \delta(\lambda)^\beta \, d\lambda$, and we recall that by the mean field theory of Proposition 5.2.2 $-n^{-2}\log Z_n \to E$ as $n \to \infty$. Similar results hold when we replace v by $g - v$ and $f - v$, so

$$\theta E_{-g/\theta+v}(\rho_{-g/\theta+v}) + (1-\theta)E_v(\rho_v) \geq E_{v-f}(\rho_{v-f}). \tag{6.6.12}$$

Now $v - f = r - C$ and this has equilibrium measure q since $f = C$ on the support of q, so

$$E_{v-f}(\rho_{v-f}) = \int (r-C)q\,dx + \frac{\beta}{2}\iint \log\frac{1}{|x-y|}q(x)q(y)\,dx\,dy; \tag{6.6.13}$$

furthermore,

$$E_v(\rho_v) = \int v\rho_v\,dx + \frac{\beta}{2}\iint \log\frac{1}{|x-y|}\rho_v(x)\rho_v(y)\,dx\,dy. \tag{6.6.14}$$

By the definition of the minimizer, we have $E_{-g/\theta+v}(q) \geq E_{-g/\theta+v}(\rho_{v-g/\theta})$, so

$$\theta E_{v-g/\theta}(q) + (1-\theta)E_v(\rho_v) \geq E_{v-f}(\rho_{v-f}), \tag{6.6.15}$$

or more explicitly

$$\int (-g+\theta v)q\,dx + (1-\theta)\int v\rho_v\,dx + \frac{\theta\beta}{2}\iint \log\frac{1}{|x-y|}q(x)q(y)\,dx\,dy$$
$$+\frac{(1-\theta)\beta}{2}\iint \log\frac{1}{|x-y|}\rho_v(dx)\rho_v(dy)$$
$$\geq \int (r-C)q\,dx + \frac{\beta}{2}\iint \log\frac{1}{|x-y|}q(x)q(y)\,dx\,dy.$$
$$(6.6.16)$$

With $g_t(x) = \inf\{f(y)+(x-y)^2/(2t\alpha)\}$, the infimum is attained where $0 = f'(y) + (y-x)/(t\alpha)$, and since f, f' and f'' are bounded

$$f(x) = f(y) + (x-y)f'(y) + O((x-y)^2)$$
$$= f(y) + (x-y)^2/(t\alpha) + O((x-y)^2). \qquad (6.6.17)$$

Hence $g(x) = f(x) - \alpha t f'(x)^2/2 + O(t^2)$ and since $f = v - r + C$,

$$-g(x) - r(x) + C + \theta v(x)$$
$$= -f(x) + \alpha t f'(x)^2/2 - r(x) + C\theta v(x) + O(t^2)$$
$$= \alpha t2 f'(x)^2/2 - (1-\theta)v(x) + O(t^2). \qquad (6.6.18)$$

When we feed this into the energy inequality (6.6.16), we obtain

$$\frac{\alpha t}{2}\int f'(x)^2 q(x)\,dx - (1-\theta)\int vq\,dx + (1-\theta)\int v\rho_v\,dx$$
$$\geq -\frac{(1-\theta)\beta}{2}\iint \log\frac{1}{|x-y|}\rho_v(dx)\rho_v(dy)$$
$$+\frac{(1-\theta)\beta}{2}\iint \log\frac{1}{|x-y|}q(x)q(y)\,dx\,dy. \qquad (6.6.19)$$

Hence

$$\frac{\alpha t}{2}\int q(x)f'(x)^2\,dx \geq (1-\theta)E_v(q) - (1-\theta)E_v(\rho_v), \qquad (6.6.20)$$

which gives by (6.6.4)

$$\frac{\alpha t}{2}\int q(x)f'(x)^2\,dx \geq (1-\theta)\Sigma(q\mid\rho_v), \qquad (6.6.21)$$

where $\alpha t/(2(1-\theta)) = 1/(2\alpha\theta)$ and $\theta\to 1-$ as $t\to 0+$, hence

$$\frac{\beta\pi^2}{6\alpha}\Phi(q\mid\rho_v) \geq \Sigma(q\mid\rho_v). \qquad (6.6.22)$$

$\square$

Corollary 6.6.2 (*Biane–Voiculescu [16]*). *The semicircular distribution satisfies the free logarithmic Sobolev inequality (6.5.18).*

Proof. The semicircular law is the equilibrium distribution for the potential $v(x) = x^2/2$, which satisfies the hypotheses of Theorem 6.6.1.

□

6.7 Logarithmic Sobolev and spectral gap inequalities

- $LSI(\alpha)$ implies the concentration inequality $C(\alpha)$.
- Uniform convexity of potential and curvature of a manifold imply concentration for the corresponding Gibbs measure.
- A spectral gap inequality has the shape $\lambda_1 \operatorname{var}(f) \leq \mathbf{E}\|\nabla f\|^2$ for all f.

In this section, we compare the logarithmic Sobolev inequality with other functional inequalities and some eigenvalue problems that are intensively studied in metric geometry. Initially, we formulate results for Euclidean space, and then mention without complete detail the corresponding results for Riemannian manifolds and other metric measure spaces.

One can show that for a probability density function on an open subset of $\mathbf{R}^n$, the following implications hold between the various inequalities with the stated constants.

$$LSI(\alpha) \Rightarrow T_2(\alpha) \Rightarrow T_1(\alpha) \Leftrightarrow C(\alpha). \qquad (6.7.1)$$

The first implication is due to Otto and Villani [162] under a technical hypothesis which was subsequently removed by Bobkov, Gentil and Ledoux [26]; the final equivalence is Theorem 3.4.4 of Bobkov and Götze [25]. The converse implications are generally not valid.

Exercise 6.7.1 ($LSI(\alpha) \Rightarrow C(\alpha)$). Suppose that $\mu \in Prob(\mathbf{R}^n)$ satisfies $LSI(\alpha)$, and let $g : \mathbf{R}^n \to \mathbf{R}$ be continuously differentiable and satisfy $\|\nabla g(x)\| \leq 1$ for all $x \in \mathbf{R}^n$, and $\int g(x)\mu(dx) = 0$. Let $f_t(x) = e^{tg(x)/2}$ and $J(t) = \int f_t(x)^2 \mu(dx)$.

(i) Show that $J(0) = 1$, $J'(0) = 0$, and that

$$tJ'(t) - J(t)\log J(t) \leq \frac{t^2}{2\alpha} J(t). \qquad (6.7.2)$$

(ii) Deduce that $J(t) \leq \exp(t^2/(2\alpha))$.

In the theory of Dirichlet forms, there is a simpler inequality which is also weaker than LSI.

Definition (*Spectral gap*). Let ρ be a probability density function on $\mathbf{R}^n$. Then ρ satisfies a *Poincaré* or spectral gap inequality with spectral gap $\lambda_1 > 0$ if

$$SG(\lambda_1) \quad \int_{\mathbf{R}^n} \left(f(x) - \int f(y)\rho(y)dy \right)^2 \rho(x)dx \leq \frac{1}{\lambda_1} \int_{\mathbf{R}^n} \|\nabla f(x)\|^2 \rho(x)\, dx$$

$$(6.7.3)$$

holds for all $f \in L^2(\rho)$ such that the distributional gradient $\nabla f \in L^2(\rho)$.

Equivalently, λ_1 is the smallest positive eigenvalue of the operator L where $Lf = -\mathrm{div}\nabla f + \langle \nabla V, \nabla f \rangle$ and $V = -\log \rho$. Hence λ_1 is given by a special case of the Rayleigh variational formula

$$\lambda_1 = \inf_f \frac{\int f(x)Lf(x)\,\rho(x)\,dx}{\int f(x)^2\,\rho(x)\,dx} \qquad (6.7.4)$$

where $f : \mathbf{R}^n \to \mathbf{R}$ is a nonconstant smooth function such that $\int f(x)\rho(x)\,dx = 0$. Here λ_1 is called the principal eigenvalue or the spectral gap.

Proposition 6.7.2 *Suppose that μ satisfies $SG(\lambda_1)$. Then any Lipschitz function $f : \mathbf{R}^n \to \mathbf{R}$ with $\|f\|_{Lip} \leq K$ satisfies*

$$\mu\left[|f - \int f d\mu| > \varepsilon\right] \leq \frac{K^2}{\lambda_1 \varepsilon^2} \qquad (\varepsilon > 0). \qquad (6.7.5)$$

Proof. This follows from the definition when one applies Chebyshev's inequality to the left-hand side. □

Proposition 6.7.3 *Suppose that ρ satisfies $LSI(\alpha)$. Then ρ also satisfies $SG(\alpha)$.*

Proof. By adding a constant to such an f we can arrange that $\int f(x)\rho(x)dx = 0$; furthermore, there in no loss is then assuming that f is bounded. Then we choose $g = 1 + \varepsilon f$ where $\varepsilon > 0$ is so small that $g \geq 0$, and apply $LSI(\alpha)$ to g. When we extract the terms of order ε^2 from the inequality,

$$\int_{\mathbf{R}^n} (1+\varepsilon f)^2 \log(1+\varepsilon f)^2\, \rho dx - \left(1 + \varepsilon^2 \int_{\mathbf{R}^n} f^2 \rho dx\right)$$
$$\times \log\left(1 + \varepsilon^2 \int_{\mathbf{R}^n} f^2 \rho dx\right) \leq \frac{2}{\alpha}\varepsilon^2 \int_{\mathbf{R}^n} \|\nabla f\|^2 \rho dx \qquad (6.7.6)$$

and we obtain the stated result. □

Exercise 6.7.4 (Bilateral exponential distribution). Let $\nu(dx) = 2^{-1}e^{-|x|}\, dx$ on $\mathbf{R}$.

(i) Show by integration by parts that ν satisfies $SG(1/4)$.

(ii) Show that ν does not satisfy $C(\alpha)$ for any $0 < \alpha < \infty$.

Hence ν does not satisfy T_1. Talagrand showed that nevertheless ν satisfies a transportation inequality with a special cost function.

Exercise 6.7.5 Let X be a random variable with probability density function q and let Z be the standard normal ransom variable with probability density function γ.

(i) By applying Poincaré's inequality with $f = (q/\gamma)^{1/2}$ and $\rho = \gamma$, show that

$$\int \left(\sqrt{q(x)} - \sqrt{\gamma(x)} \right)^2 dx \leq \frac{1}{2} I(X \mid Z). \qquad (6.7.7)$$

(ii) Use the Cauchy–Schwarz inequality to show that

$$\int |q(x) - \gamma(x)| \, dx \leq 2 \left(\int \left(\sqrt{q(x)} - \sqrt{\gamma(x)} \right)^2 dx \right)^{1/2}. \qquad (6.7.8)$$

The quantity on the right-hand side, aside from 2, is known as the Hellinger distance between the densities.

6.8 Inequalities for Gibbs measures on Riemannian manifolds

In this book we have followed an approach to the logarithmic Sobolev inequality which appears to depend upon the special properties of the Lebesgue measure, since the Prékopa–Leindler inequality is central to the proof. Indeed, the details of the proofs require modification to make them apply to other manifolds, so here we indicate the main principle and refer the reader to [51] and [58] for a full discussion.

Let M be a complete Riemannian manifold, and let ∇ be the gradient operator that is associated with the Riemannian connection; let $\sigma(dx)$ be the Riemannian volume form and let Ric be the Ricci curvature tensor. We defined this in Section 2.3 for Lie groups, and the general definition for manifolds is similar, [128]. The Ricci curvature gives the average of the Gauss curvatures of all the two-dimensional sections containing that direction, normalized by a multiplicative factor $n - 1$. See [172].

Examples 6.8.1 Let $S^n(r)$ be the sphere of radius r in $\mathbf{R}^{n+1}$ with the induced metric, so σ is the standard rotation-invariant measure. Then $\mathrm{Ric}(S^n(r)) = r^{-2}(n-1)I$.

By Exercise 2.4.2, the group $SO(n)$ has Ric $= (n-2)I/4$, so the spectral gap inequality holds with constants that improve with increasing dimension; see [48, 69].

Extension to Gibbs measures on Riemannian manifolds

In Section 6.2 we considered Gibbs measures on the flat space $\mathbf{R}^n$, and now we extend this to the context of Riemannian manifolds. Suppose that $V : M \to \mathbf{R}$ is a continuous function such that $e^{-V(x)}\sigma(dx)$ is a probability measure on M. Then there exists a second-order differential operator L such that

$$\int_M (Lf(x))\bar{g}(x)\, e^{-V(x)}\sigma(dx) = \int_M \langle \nabla f(x), \nabla g(x)\rangle\, e^{-V(x)}\sigma(dx) \quad (6.8.1)$$

for all twice continuously differentiable $f, g : M \to \mathbf{R}$ that have compact support. Bakry and Emery proposed that functional inequalities should be expressed in terms of the *carré du champ*

$$\Gamma(f,g) = \frac{1}{2}\big[fLg + (Lf)g - L(fg)\big] \quad (6.8.2)$$

and *carré du champ itéré*

$$\Gamma_2(f,g) = \frac{1}{2}\big[\Gamma(f, Lg) + \Gamma(Lf, g) - L\Gamma(f,g)\big] \quad (6.8.3)$$

which in this case simplify to $\Gamma(f,g) = \langle \nabla f, \nabla g\rangle$, and the Bochner–Weitzenböck formula is

$$\Gamma_2(f,f) = \langle \nabla f, \nabla Lf\rangle - \frac{1}{2}L\big(\|\nabla f\|^2\big)$$
$$= \|\mathrm{Hess}\, f\|_{HS}^2 + \mathrm{Hess}\, V\,(\nabla f, \nabla f) + \mathrm{Ric}\,(\nabla f, \nabla f). \quad (6.8.4)$$

The modified Ricci curvature is defined to be $\mathrm{Ric}_\infty = \mathrm{Ric} + \mathrm{Hess}\, V$.

Theorem 6.8.2 (*Bakry–Emery*). *Let M be a Riemannian manifold with no boundary, and let $\mu(dx) = Z^{-1}e^{-V(x)}\sigma(dx)$ be a probability measure on M where $\mathrm{Ric}_\infty \geq \alpha I$ for some $\alpha > 0$. Then the spectral gap inequality*

$$SG(\alpha) \quad \int_M \Big(f(x) - \int f\,d\mu\Big)^2 \mu(dx) \leq \frac{1}{\alpha}\int_M \|\nabla f(x)\|^2 \mu(dx) \quad (6.8.5)$$

holds for all $f \in L^2(\sigma)$ such that the distributional gradient $\nabla f \in L^2(\sigma)$.

Proof. Let $Lf = -\mathrm{div}\nabla f + \langle \nabla V, \nabla f\rangle$ for any twice continuously differentiable $\Psi, f : M \to \mathbf{R}$, and observe that

$$\int_M (Lf)(x)g(x)\mu(dx) = \int_M \langle \nabla f, \nabla f\rangle \mu(dx). \quad (6.8.6)$$

By the Bochner–Weitzenböck formula (6.8.4), we have

$$\langle \nabla f, \nabla L f \rangle - \frac{1}{2} L \big(\|\nabla f\|^2 \big) = \|\text{Hess}\, f\|_{HS}^2 + \text{Hess}\, V\, (\nabla f, \nabla f)$$
$$+ \text{Ric}\, (\nabla f, \nabla f); \qquad (6.8.7)$$

hence since $L\mathbf{I} = 0$ we have

$$\int_M \big(L f \big)^2 \mu(dx) = \int_M \big(\|\text{Hess}\, f\|_{HS}^2 + \text{Hess}\, V\, (\nabla f, \nabla f)$$
$$+ \text{Ric}(\nabla f, \nabla f) \big) \mu(dx) \qquad (6.8.8)$$

so by the assumption on the curvature, we have

$$\int_M \big(L f \big)^2 \mu(dx) \geq \alpha \int_M \|\nabla f\|^2 \mu(dx), \qquad (6.8.9)$$

so $\|Lf\|^2 \geq \lambda_1 \langle Lf, f \rangle_{L^2(\mu)}$. Hence the operator L has a gap containing $(0, \alpha)$ in its spectrum, so

$$\int_M \langle \nabla f, \nabla f \rangle \mu(dx) \geq \alpha \int_M \Big(f - \int f d\mu \Big)^2 \mu(dx). \qquad (6.8.10)$$

□

- The spectral gap of a Gibbs measure depends upon the lower bound of the Ricci curvature and the Hessian of the potential.

Deuschel and Stroock considered spectral gap inequalities for Gibbs measures on Riemannian manifolds in [58]. For classical groups, Rothaus considered the relationship between the constants in various functional inequalities in [135]. Bakry and Emery showed that Theorem 6.3.2 and hence Proposition 6.7.3 extend to the case in which $\text{Ric}_\infty \geq \alpha I$ for some uniform $\alpha > 0$ on M; see [108] for a discussion of the results in a more general setting, which are summarized in the following list of implications.

$$[\text{Ric}_\infty \geq \alpha I] \Rightarrow LSI(\alpha) \Rightarrow T_2(\alpha) \Rightarrow SG(\alpha). \qquad (6.8.11)$$

Conversely, Sturm has shown that Boltzmann's entropy is α-uniformly displacement convex on $P_2(M)$ if and only if M has $\text{Ric}_\infty \geq \alpha I$; see [149, 108].

Extension to measured length spaces. In the historical development of the theory, concentration inequalities were first proved for objects with a smooth structure such as Lie groups and Riemannian manifolds, and the constants in the inequalities were expressed in terms of geometrical notions such as dimension and curvature. The modern approach reverses this order, and seeks to define geometrical quantities on metric spaces without smooth structure in terms of the constants in functional inequalities. See [40, 108].

Let (Ω, d) be a metric space and $\varphi : [0,1] \to \Omega$ a continuous function; then the length of φ is $\text{length}\,(\varphi) = \sup \sum_{j=0}^{n-1} d(\varphi(t_j), \varphi(t_{j+1}))$ where the supremum is taken over all partitions $\{0 = t_0 < \cdots < t_n = 1\}$. Then (Ω, d) is called a length space if $d(x,y) = \inf_\varphi \{\text{length}\,(\varphi) : \varphi(0) = x, \varphi(1) = y\}$. Lott and Villani have introduced a notion of Ricci curvature for measured length spaces such that uniformly positive Ricci curvature implies a Poincaré inequality. In this context, one interprets $\|\nabla f(x)\| = \limsup_{y \to x} |f(x) - f(y)|/d(x,y)$ since there is no obvious notion of the Laplace operator.

7

Young tableaux

Abstract

In this chapter we consider the symmetric groups S_N and the Young diagrams which parametrize their irreducible unitary representations. The RSK correspondence identifies the Plancherel measure for S_N in terms of combinatorial formulæ from representation theory. We describe the typical shapes of scaled Young diagrams under the Plancherel measure and obtain the Vershik distribution in the limit as $N \to \infty$. We express the distribution of the longest increasing subsequence in a permutation in S_N as an integral over the unitary group with respect to Haar measure; thus we link combinatorics to random matrices via the Schur polynomials. Finally, we state various results concerning the asymptotic distribution of the length of the longest increasing subsequence in a typical permutation from S_N as $N \to \infty$. These follow from the asymptotic properties of Toeplitz determinants.

7.1 Group representations

Definition (*The group algebra of a finite group*). Let G be a finite group, and let $\mathbf{C}G = \{\sum_{g \in G} \alpha_g g : \alpha_g \in \mathbf{C}\}$ be the group algebra with the multiplication

$$\left(\sum_{g \in G} \alpha_g g\right)\left(\sum_{h \in G} \beta_h h\right) = \sum_{g,h} \alpha_g \beta_h gh. \tag{7.1.1}$$

A representation ρ is a group homomorphism $G \to GL(\mathbf{C}, n)$ for some $n < \infty$, which makes $\mathbf{C}^n$ into a unitary left $\mathbf{C}G$ module via $\mathbf{C}G \times \mathbf{C}^n \to \mathbf{C}^n : (\sum_g \alpha_g g, v) \to \sum_g \alpha_g \rho(g)v$. Conversely, for any unitary left $\mathbf{C}G$ module M that has dimension n over $\mathbf{C}$, there exists a representation

$\rho : G \to GL(\mathbf{C}, n)$ and a linear isomorphism $T : M \to \mathbf{C}^n$ such that $T^{-1}\rho(g)T = g$.

In particular, the left regular representation $\lambda_g : h \mapsto gh$ makes $\mathbf{C}G$ into a unitary left $\mathbf{C}G$ module.

A unitary left $\mathbf{C}G$ module M is irreducible if $M \neq \{0\}$ and the only $\mathbf{C}G$ submodules of M are M itself and $\{0\}$. A representation of G is said to be irreducible if the corresponding $\mathbf{C}G$ module is irreducible. By Maschke's theorem, any finite-dimensional unitary left module M can be written as a direct sum $M = M_1 \oplus M_2 \oplus \cdots \oplus M_k$ where the M_k are irreducible $\mathbf{C}G$ submodules. The group algebra $\mathbf{C}G$ is semisimple; see [91, Chapter 8].

Theorem 7.1.1 (*Dimension formula*). *There exist irreducible $\mathbf{C}G$ modules $V_{j,1}$ with $\dim(V_{j,1}) = d_j$ such that no pair of the $V_{j,1}$ are isomorphic and $\mathbf{C}G$ may be expressed as a direct sum*

$$\mathbf{C}G = \bigoplus_{j=1}^{k}(V_{j,1} \oplus \cdots \oplus V_{j,d_j})$$

where the irreducible $\mathbf{C}G$ submodule $V_{j,\ell}$ is isomorphic to $V_{j,1}$ for $\ell = 1, \ldots, d_j$. Hence the dimensions as vector spaces over $\mathbf{C}$ satisfy

$$\#(G) = \sum_{j=1}^{k} d_j^2. \tag{7.1.2}$$

Proof. See James and Liebeck [91, Chapters 10 and 11]. □

Elements g and h of G are *conjugate* if there exists $k \in G$ such that $g = khk^{-1}$. Let M be a finite-dimensional unitary left $\mathbf{C}G$ module, so each $g \in G$ gives a linear transformation $m \mapsto \rho(g)m$; then the character of M is the function $\chi : G \to \mathbf{C}$ given by $\chi(g) = \text{trace}\rho(g)$; this does not depend upon the choice of basis for M. Note that $\chi(e)$ equals the dimension of the module. Elements g and h are conjugate in G if and only if $\chi(g) = \chi(h)$ for all characters χ.

Let S_N be the symmetric group of permutations on N symbols $\{1, \ldots, N\}$. Each $\sigma \in S_N$ may be expressed as a permutation in disjoint cycles, so that the structure of the cycles is unique and is specified by integers $m_1 \geq m_2 \geq \cdots \geq m_n \geq 1$ such that $N = m_1 + \cdots + m_n$ where m_j is the length of the jth cycle. Furthermore, σ and σ' have equal cycle structures if and only if there exists $\omega \in S_N$ such that $\sigma' = \omega\sigma\omega^{-1}$, so that cycle structures correspond to conjugacy classes.

Proposition 7.1.2 *For a finite group G, there are bijective correspondences between the sets of:*

(i) inequivalent irreducible representations;
(ii) characters of irreducible representations;
(iii) conjugacy classes of G.

For $G = S_N$, these also correspond to:

(iv) partitions of N;
(v) Young diagrams with N boxes.

Proof. See [91]. We discuss (v) in Section 7.2. □

Exercise 7.1.3 (*Hamming meets Hilbert–Schmidt*). Let S_n be the symmetric group on $\{1, \ldots, n\}$ and let

$$d(\sigma, \tau) = \frac{1}{n} \#\{j : \sigma(j) \neq \tau(j)\} \qquad (\sigma, \tau \in S_n). \qquad (7.1.3)$$

For $\sigma \in S_n$, let P_σ be the permutation matrix that is given by $e_j \mapsto e_{\sigma(j)}$ for the standard basis $(e_j)_{j=1}^n$ of $\mathbf{R}^n$. Show that

$$\|P_\sigma - P_\tau\|_{c^2(n)}^2 = 2d(\sigma, \tau) \qquad (\sigma, \tau \in S_n). \qquad (7.1.4)$$

7.2 Young diagrams

• Young diagrams $\lambda \vdash N$ are used to parametrize the irreducible representations of S_N.

The symmetric group has Haar probability measure $\mu_N(A) = (\sharp A)/N!$ for $A \subseteq S_N$, so that each permutation is equally likely to occur. An increasing subsequence in $\sigma \in S_N$ of length k consists of $j_1 < j_2 < \cdots < j_k$ such that $\sigma(j_1) < \sigma(j_2) < \cdots < \sigma(j_k)$; note that we allow gaps in these subsequences. Let $\ell(\sigma)$ be the length of the longest increasing subsequence in σ, which may be realized by several subsequences.

Problem (*Ulam [5]*). If we select a permutation $\sigma \in S_N$ at random, what is the most likely length $\ell(\sigma)$ of the longest increasing subsequence in σ when N is large?

On the basis of numerical simulation, he conjectured that $\ell(\sigma)$ is close to $2\sqrt{N}$ with high probability. There are various sorting algorithms that suggest that $\sqrt{N}$ is the correct scaling factor with respect to N, but the numerical value of 2 is more subtle. See [5].

As an illustration, we consider a pack of cards, numbered $1, 2, \ldots, N$. A permutation amounts to a shuffle of the cards, so there are $N!$ possible shuffles of the pack. Clearly the identity permutation has longest increasing subsequence of length N; contrariwise the permutation that reverses the pack has longest increasing subsequence of length 1. Numerical simulations indicate that for randomly selected shuffles from a standard pack of 52 cards, the expected length of the longest increasing subsequence is about $2\sqrt{52}$. Ulam asked whether from a pack of N cards, the expected length of $\ell(\sigma)$ is $2\sqrt{N}$. Vershik and Kerov proved this as a consequence of the argument that we now sketch over the next few sections. Using random matrix theory, Johansson and others found apparently different proofs which depend upon the asymptotic properties of Toeplitz determinants, and which give detailed information about the kth longest increasing subsequence for $k = 1, 2, \ldots$.

Definition (*Young tableau, [68]*). By a partition π of N, denoted $\pi \vdash N$, we mean a list $m_1 \geq m_2 \geq \cdots \geq m_n \geq 1$ such that $m_1 + \cdots + m_n = N$. Each partition gives an array of N square boxes of unit side, arranged in left-justified rows such that the jth row down from the top has m_j boxes. Such an array is called a Young diagram λ. Numbering a λ means putting one positive integer in each of the boxes such that:

(i) the rows are weakly increasing from left to right;
(ii) the columns are strictly increasing from top to bottom.

A Young diagram, so numbered, is called a Young tableau. We use the phrase *distinctly numbered* to indicate that the numbers are used once only, and in this case the rows are strictly increasing.

Definition (*Standard Young tableau*). A standard Young tableau P is a numbering of a Young diagram π_P such that each of the integers $1, \ldots, N$ is placed in one of the boxes so that the rows are increasing from left to right, and the columns are increasing downwards. The Young diagram π_P is called the shape of P.

Theorem 7.2.1 (*Robinson–Schensted–Knuth*). *There is a bijection φ from S_N to the set of ordered pairs (P, Q) of standard Young tableaux such that $\pi_P = \pi_Q$, where π_P is any Young diagram that has N boxes.*

The length of the longest increasing subsequence in $\sigma \in S_N$ equals the length of the longest row in the Young diagram π_P where $\varphi(\sigma) = (P, Q)$.

Proof. Each $\sigma \in S_N$ produces a word; that is, the ordered list $\sigma(1), \ldots,$ $\sigma(N)$ of the numbers $\{1, \ldots, N\}$. The most profitable question to ask is:

How do we add a new entry to a Young tableau P so as to obtain another Young tableau P'? (The Young tableaux are generally not standard.)

The process involves the following steps, where the kth step is applied to $x = \sigma(k)$:

(1) If x is greater than all the entries in the first row of λ, then add x to the end of the first row;

(2) otherwise, x is less than some entry in the first row, and we let y be the smallest such entry. We replace y by x, and attempt to add y to the Young tableau ρ that arises from λ by deleting the first row. We say that y is bumped out by x. We repeat steps (1) and (2) to add y to ρ.

After N steps, the process stops, where possibly the final step may involve

(3) adding a box to the vacant row below the Young tableau.

At each stage in the construction, we have a distinctly numbered Young tableau. Ultimately, we obtain a standard Young tableau $P(\sigma)$, since we have used each of $\{1, \ldots, N\}$ exactly once.

Meanwhile, we construct another sequence of standard Young tableaux with the same shapes by placing k into the box that is added in the kth step in the construction of $P(\sigma)$. Thus we obtain a standard Young tableau $Q(\sigma)$, called the recording tableau, that has the same shape as $P(\sigma)$, and we can map $\varphi : \sigma \mapsto (P(\sigma), Q(\sigma))$.

From the construction of φ, particularly rule (1), it is evident that the length of the top row of $P(\sigma)$ is the length of the longest increasing subsequence in σ.

We show that φ is bijective by constructing the reverse *RSK* correspondence. Let Q be a standard Young tableau and P a distinctly numbered Young tableau; suppose that P and Q have equal shape. Let k be the largest entry in Q and x the entry of P that is in the corresponding position.

(4) If x is at the end of the first row, then x was the entry that was added at step k; so we let $\sigma(k) = x$.

(5) Otherwise, x was bumped out of the row R of λ that is directly above x. By condition (ii), there exists y' in R such that $y' < x$, and we let y be the largest such y'. Then x was bumped out of R by the addition of y, and either: R is the first row, in which case y was added most recently, or there exists z in the row R' directly above

R such that $z < y$. Proceeding in this way, we identify the element $\sigma(k)$ as the element that was added most recently to make P.

We now remove $\sigma(k)$ from P and restore P to what it was before $\sigma(k)$ was added; likewise we remove k from Q, and thus obtain two Young diagrams that have the same shape and one box fewer than P or Q. Then we repeat the process.

Given Young tableaux λ and ρ with N boxes and of equal shape, one can check that there exists $\sigma \in S_N$ such that $\varphi(\sigma) = (\lambda, \rho)$.   □

For S_N, the conjugacy classes are precisely the set of elements that have the same cycle structure; hence irreducible representations correspond to partitions of N. For each Young diagram λ, we now construct a complex vector space H_λ, called a Specht module, on which S_N acts irreducibly.

Definition (λ-*tableau*). A λ-tableau consists of a Young diagram λ with numbers $1, \ldots, N$ placed in the boxes, with no special condition on ordering, so a standard Young tableau is a special kind of λ-tableau.

Each $\sigma \in S_N$ operates on the λ-tableaux by taking j to $\sigma(j)$. We let C_λ be the subgroup of S_N consisting of the permutations that permute the entries in each column of λ, but leave the column fixed.

Given λ-tableaux P and Q, we say that $P \sim Q$ if there exists $\sigma \in S_N$ such that σ takes the entries in row j of P to the entries in row j of Q for all j. An equivalence class is called a λ-tabloid T, and amounts to a partition of the set $\{1, \ldots, N\}$ into unordered rows $j = 1, 2, \ldots$ of length $\lambda_1 \geq \lambda_2 \geq \ldots$. The action of S_N on the λ-tableaux gives rise to an action on the λ-tabloids where $\sigma \in S_n$ permutes the entries of T. We introduce basis vectors e_T, indexed by the λ-tabloids, and form the complex vector space $\mathrm{span}(e_T)$; then we introduce

$$\tilde{e}_T = \sum_{\sigma \in C_\lambda} \mathrm{sign}(\sigma) e_{\sigma(T)} \tag{7.2.1}$$

and $H_\lambda = \mathrm{span}\{\tilde{e}_T : \lambda - \text{tabloids } T\}$.

Definition (*Schur polynomial*). Suppose that $\lambda \vdash N$ has n rows, where $n \leq t$. A numbering P assigns $1, \ldots, t$ into the boxes of λ, and p_j is the number of occurrences of j. Then the Schur polynomial of λ in t indeterminates is

$$S_\lambda(x_1, \ldots, x_t) = \sum_P x_1^{p_1} \ldots x_t^{p_t}, \tag{7.2.2}$$

with the sum over all such numberings. We write $x^P = x_1^{p_1} \ldots x_t^{p_t}$ for short.

One can show that S_λ is symmetric, and that the S_λ give a basis for the vector space of symmetric polynomials in $(x_1, \ldots, x_t)$.

Proposition 7.2.2 (*Specht modules*). *For each $\lambda \vdash N$ there exists a finite-dimensional complex Hilbert space H_λ called a Specht module and an irreducible representation of S_N on H_λ such that distinct Young diagrams correspond to inequivalent representations. Furthermore, the H_λ satisfy*

$$N! = \sum_{\lambda \vdash N} (\dim H_\lambda)^2; \tag{7.2.3}$$

so the Specht modules give a complete set of inequivalent irreducible modules.

Proof. See [68] for a proof that the representations are well defined, inequivalent and irreducible. To prove that this gives a complete set of such representations, we count dimensions and prove (7.2.3).

One can choose a basis of H_λ consisting of standard Young tableaux ρ with shape $\pi(\rho) = \lambda$. The dimension of H_λ equals the number f^λ of standard Young tableaux ρ that have shape λ.

We expand $1/(1 - x_j y_k)$ as a geometric series, then gather terms together and find

$$\prod_{j,k=1}^{N} \frac{1}{1 - x_j y_k} = \sum_{n=0}^{N} \sum_{\lambda : \lambda \vdash n} \sum_{U : \pi(U) = \lambda} x^U \sum_{T : \pi(T) = \lambda} y^T$$

$$= \sum_{n=0}^{N} \sum_{\lambda : \lambda \vdash n} S_\lambda(x) S_\lambda(y). \tag{7.2.4}$$

Now we consider the coefficient of $x_1 \ldots x_N y_1 \ldots y_N$ on each side of this expression and we find that only standard numberings can give contribute to this coefficient, and the x_j and y_k may be permuted paired in any order, so

$$N! = \sum_{\lambda : \lambda \vdash N} (f^\lambda)^2. \tag{7.2.5}$$

This proves (7.2.3), and the dimension formula Theorem 7.1.1 from representation theory then shows that we have a complete list of inequivalent irreducible representations. $\qquad\square$

Definition (*Plancherel measure*). The Plancherel measure ν_N of S_N is the probability measure on $\Lambda_N = \{\pi : \pi \vdash N\}$ such that

$$\nu_N(\{\pi\}) = \frac{(\dim H_\pi)^2}{N!}. \tag{7.2.6}$$

Proposition 7.2.3 *Under the RSK correspondence, the map $\sigma \mapsto (P,Q) \mapsto \pi_P$ induces ν_N from the Haar measure μ_N on S_N.*

Proof. This holds since for each π there are $(\dim H_\pi)^2$ pairs (P,Q) that have shape π. □

Note that there are various natural maps from S_N to $\{\pi : \pi \vdash N\}$, such as the map considered in Exercise 7.2.9, which induce probability measures on $\{\pi : \pi \vdash N\}$ that are different from ν_N. We can regard the Plancherel measure as a probability measure on the space of inequivalent irreducible representations of S_N, and the name derives from the fact that one can prove an analogue for compact metric groups of the Plancherel formula for abelian groups; see [130, p 445].

We introduce $\ell_j = n + \lambda_j - j$, and consider the coefficient $\chi_\lambda(1^N)$ of $x_1^{\ell_1} \cdots x_n^{\ell_n}$ in

$$(x_1 + \cdots + x_n)^N \prod_{1 \le j < k \le n} (x_j - x_k). \tag{7.2.7}$$

Then $\chi_\lambda(1^N)$ equals the dimension of H_λ by a special case of Frobenius's character formula.

The dimension of H_π can otherwise be found from the hook length formula, which we state in various forms. For each box b in the Young diagram π, we form a hook consisting of b, the boxes that lie directly below b in the same column and the boxes that lie directly to the right of b in the same row; the length of this hook is h_b, the total number of boxes in the hook.

Proposition 7.2.4 (*Hook length formulæ*). *Suppose that $\lambda \vdash N$ has rows $\lambda_1 \ge \cdots \ge \lambda_n > 0$. Then with $\ell_j = \lambda_j + n - j$, the dimension of the corresponding Specht module is given by*

$$\dim H_\lambda = \frac{N!}{\prod_{b \in \lambda} h_b}$$

or equivalently

$$\dim H_\lambda = \frac{N!}{\prod_{j=1}^n \ell_j!} \prod_{1 \le j < k \le n} (\ell_j - \ell_k). \tag{7.2.8}$$

Proof. See [68]. □

Proposition 7.2.5 (*Frobenius's formula [138]*). *For $\lambda \vdash N$, the dimension of the corresponding Specht module is given by*

$$\dim H_\lambda = N! \det \left[\frac{1}{(\lambda_j - j + k)!} \right]_{j,k=1,\ldots,n}. \qquad (7.2.9)$$

Proof. In [138, p. 132] Sagan shows by induction on n that this is equivalent to the hook length formula. A basic observation is that $\lambda_j - j + n$ is the hook length for the jth entry in the first column. □

Frobenius's coordinates

For various applications in Chapter 10, we use another system of coordinates for Young diagrams. Given $\lambda \vdash N$, we introduce the diagonal of boxes (j, j) with $j = 1, \ldots, d$, and count the number P_j of boxes in row j that lie to the right of (j, j); then we count the number Q_j of boxes in column j that lie below (j, j). Evidently λ is determined by these numbers, which satisfy $P_1 > P_2 > \cdots > P_d$, $Q_1 > Q_2 > \cdots > Q_d$ and $\sum_{j=1}^{d}(P_j + Q_j) = N - d$.

We write

$$\lambda = (P_1, \ldots, P_d; Q_1, \ldots, Q_d). \qquad (7.2.10)$$

Definition (*Conjugate or transpose*). The transpose of conjugate λ^t of a Young diagram λ is the Young diagram $\lambda_1^t \geq \cdots \geq \lambda_m^t$ that has columns of height $\lambda_1, \lambda_2, \ldots, \lambda_n$. We can take $\lambda_1^t = n$ and $\lambda_1 = m$ for convenience. (The notation λ' is also in common use.)

Evidently, λ^t still has N boxes and also has d boxes on its diagonal, and one can easily see that

$$\lambda^t = (Q_1, \ldots, Q_d; P_1, \ldots, P_d). \qquad (7.2.11)$$

We now consider $\sigma = \{-1/2, -3/2, \ldots\}$ and introduce the sets of differences

$$D = \{\lambda_j + s : j = 1, \ldots, n; s \in \sigma\}, D^t = \{-\lambda_k^t - s : k = 1, \ldots, m; s \in \sigma\}. \qquad (7.2.12)$$

Proposition 7.2.6 *The sets D and D^t are complementary in $\mathbf{Z} + 1/2$.*

Proof. By some simple counting arguments, one can show that

$$\lambda_j + n - j \quad (j = 1, \ldots, n) \quad \text{and} \quad n - 1 + k - \lambda_k^t \quad (k = 1, \ldots, m) \qquad (7.2.13)$$

together give all of $\{0, 1, \ldots, m + n - 1\}$.

Suppose that $y \in \{-n - 1/2, \ldots, m - 1/2\}$. Then either there exists a unique $j \in \{0, 1, \ldots\}$ such that $y = \lambda_j - j - 1/2$, or there exists a unique $k \in \{0, 1, \ldots\}$ such that $y = k - \lambda_k^t - 1/2$.

If $y \in \{m + 1/2, m + 3/2, \ldots\}$, then $y = -\lambda_1^t + n + \ell + 1/2$ for some $\ell \in \{m, m + 1, \ldots\}$.

Likewise, if $y \in \{-n - 1/2, -n - 3/2, \ldots\}$, then $y = \lambda_1 - m - \ell - 1/2$ for some $\ell \in \{n, n + 1, \ldots\}$.   □

The Frobenius coordinates can be obtained by means of Proposition 7.2.6 from the set of differences. For $\lambda = (\lambda_1 \geq \lambda_2 \geq \ldots)$, we introduce the symmetric difference

$$\text{Fr}(\lambda) = \{\lambda_j - j : j = 1, 2, \ldots\} \triangle \{-1, -2, \ldots\} \qquad (7.2.14)$$

which satisfies

$$\text{Fr}(\lambda) = \{P_1, \ldots, P_d; -1 - Q_1, \ldots, -1 - Q_d\}. \qquad (7.2.15)$$

Thus Frobenius's coordinates are the symmetric difference of the set of differences with respect to the negative integers; this is significant in Chapter 10.

Our final version of the hook length formula is in terms of the Frobenius coordinates.

Proposition 7.2.7 (*Hook length formula in Frobenius's coordinates*). *For $\lambda \vdash N$, the dimension of the corresponding Specht module is given by:*

$$\dim H_\lambda = N! \det\left[\frac{1}{(P_j + Q_k + 1)P_j! Q_k!}\right]_{j,k=1,\ldots,d}. \qquad (7.2.16)$$

Proof. See [30, p. 489].   □

Exercise 7.2.8 Use the Cauchy determinant formula from Lemma 2.2.1(ii) to express the right-hand side similarly to Proposition 7.2.4, namely

$$\dim H_\lambda = \frac{N!}{\prod_j P_j! \prod_j Q_j!} \frac{\prod_{j<k}(P_j - P_k) \prod_{j<k}(Q_j - Q_k)}{\prod_{j,k}(1 + P_j + Q_k)}. \qquad (7.2.17)$$

Exercise 7.2.9 (i) Suppose that in the decomposition of $\sigma \in S_N$ into disjoint cycles, the distinct cycle lengths are r_j ($j = 1, \ldots, \ell$), where there are s_j disjoint cycles of length r_j. Show that the number of distinct $\tau \in S_N$ that are conjugate to σ equals

$$\frac{N!}{\prod_{j=1}^{\ell} r_j^{s_j} s_j!}. \qquad (7.2.18)$$

(ii) Let $\psi : S_N \to \{\pi : \pi \vdash N\}$ be the map that takes $\sigma \in S_N$ to the Young diagram that has shape given by the decomposition of σ into disjoint cycles. Describe the probability measure that ψ induces from Haar measure μ_N on S_N.

Exercise 7.2.10 Show that, with $\ell_j = \lambda_j + n - j$,

$$\nu_N(\{\lambda\}) = N! \det\left[\frac{1}{\ell_j!} \sum_{r=0}^{n-1} (\ell_j \ell_k)^r \frac{1}{\ell_k!}\right]_{j,k=1,\ldots,n}. \tag{7.2.19}$$

7.3 The Vershik Ω distribution

- Vershik's Ω distribution arises in various card-sorting algorithms.
- The Ω distribution describes the typical cycle structure of elements of S_N for large N.
- The Ω distribution satisfies a variational problem suggested by Plancherel measure on the symmetric group.

The Vershik Ω distribution [107, 99, 161] is important in analysis of the Plancherel measure of the symmetric group on N letters as $N \to \infty$, and hence arises in random matrix theory. The purpose of this section is to introduce the Vershik distribution by means of variational problems related to Section 4.4, and to explore the limits of what one can achieve by the general methods of this book.

Next we consider the shapes of the Young diagrams $\pi \vdash N$ as $N \to \infty$. Each Young diagram π gives a decreasing probability density function f_π on $(0, \infty)$: we reflect the diagram in the x axis to obtain a polygonal curve which resembles a descending staircase, then we scale the boxes so that they become squares of side $1/\sqrt{N}$, and then we introduce a function that has such a scaled diagram as its graph. Finally, we can smooth off the edges so that f_π is continuous and strictly decreasing.

Suppose that f is a decreasing probability density function on $(0, \infty)$. Then for $x > 0$ and $0 < y < f(x)$, the point (x, y) lies underneath the graph of f, and $f(x) - y$ represents the vertical distance to the graph, while $f^{-1}(y) - x$ represents the horizontal distance; thus $f(x) - y + f^{-1}(y) - x$ represents the *hook length*. The functional

$$\tilde{H}(f) = \int_0^\infty \int_0^{f(x)} \log(f(x) - y + f^{-1}(y) - x) \, dy \, dx \tag{7.3.1}$$

represents the integral of the hook lengths over all points under the graph.

Lemma 7.3.1 *Under the above correspondence,*

$$\nu_N(\{\pi\}) \approx (2\pi N)^{1/2} \exp\big(-N - 2N\tilde{H}(f_\pi)\big), \qquad (7.3.2)$$

for large N.

Proof. Suppose that under the above correspondence the box b is mapped to a scaled box with centre (x_b, y_b). Then we have approximately

$$\sum_{b \in \pi} \log h_b = \sum_{b \in \pi} \log\big(N^{1/2} |f_\pi(x_b) - y_b + f_\pi^{-1}(y_b) - x_b|\big)$$

$$\approx (1/2)N \log N + N \int_0^\infty \int_0^{f_\pi(x)} \log|f_\pi(x)$$

$$-y + f_\pi^{-1}(y) - x| dy dx \qquad (7.3.3)$$

and hence by Stirling's formula $\log \Gamma(x) = (x - 1/2) \log x - x + \log \sqrt{2\pi} + o(1)$, and the hook length formula

$$\nu_N(\{\pi\}) = \exp\big(\log N! - 2 \sum_{b \in \pi} \log h_b\big)$$

$$\approx (2\pi N)^{1/2} \exp\Big(-N - 2N \int_0^\infty \int_0^{f_\pi(x)} \log(f_\pi(x)$$

$$- y + f_\pi^{-1}(y) - x) dy dx\Big). \qquad (7.3.4)$$

$\square$

By Lemma 7.3.1, the representations π that are most probable with respect to ν_N are those that give f_π that make $\tilde{H}(f_\pi)$ close to its minimum. The Vershik distribution is the f that minimizes $\tilde{H}(f)$ over all decreasing probability density functions f on $(0, \infty)$. We shall show how to obtain f from the solution of an integral equation in the new variables

$$x = g(\xi) + (1/2)\xi + (1/2)|\xi|,$$
$$f(x) = g(\xi) - (1/2)\xi + (1/2)|\xi|. \qquad (7.3.5)$$

More precisely, Logan and Shepp [107] considered the problem of minimizing

$$I(g) = \int_{-\infty}^\infty g'(\xi)\xi \log \frac{e}{|\xi|} d\xi + \frac{1}{2} \iint \log \frac{1}{|\xi - \eta|} g'(\xi)g'(\eta) d\xi d\eta \qquad (7.3.6)$$

where g is a probability density function on $\mathbf{R}$ that is unimodal with mode at $\xi = 0$ and is differentiable with $|g'(\xi)| \leq 1$. By calculus, they show that minimizing $I(g)$ is equivalent to minimizing $\tilde{H}(f)$ over the given sets of functions. The functional I is bounded below on the specified set of functions and has a unique minimizer. Evidently, the minimizer satisfies

$$\xi \log \frac{e}{|\xi|} = \int_{-\infty}^{\infty} \log |\xi - \eta| \, g'(\eta) d\eta + C \qquad (7.3.7)$$

on the support of $g'(\xi)$, where C is some constant. Further, $C = 0$ when g is an even function.

For $\delta > 0$, we introduce the potential

$$v(x) = \begin{cases} (3/4)x^2 - (1/2)x^2 \log |x|, & \text{for } |x| < e; \\ (\delta/2)(|x| - e)^2 + (e^2/4), & \text{for } |x| > e; \end{cases} \qquad (7.3.8)$$

which is piecewise differentiable with

$$v'(x) = \begin{cases} x \log(e/|x|), & \text{for } |x| < e; \\ \delta \text{sgn}(x)(|x| - e), & \text{for } |x| > e; \end{cases} \qquad (7.3.9)$$

hence v satisfies (a), (b) and (c) of Chapter 4. The graph of the function v has a single well, with the minimum at $(0,0)$, and is convex on the interval $(-1, 1)$. By Theorem 4.4.1, there exists a unique minimizer ρ_v for the functional

$$\int_{-\infty}^{\infty} v(x)\rho(x)dx + \frac{1}{2} \iint \log \frac{1}{|x - y|} \rho(x)\rho(y) \, dxdy, \qquad (7.3.10)$$

and the solution satisfies

$$v'(x) = \text{p.v.} \int_{-\infty}^{\infty} \frac{\rho_v(y)dy}{x - y} \qquad (7.3.11)$$

on the support of ρ_v. Fortunately, this ρ_v is supported on $[-2, 2]$, and $\rho_v = g$ also gives the solution to the Logan–Shepp problem.

Lemma 7.3.2 *The integral equation*

$$x \log \frac{e}{|x|} = \text{p.v.} \int_{-2}^{2} \frac{g(t)}{x - t} dt \qquad (x \in (-2, 2)) \qquad (7.3.12)$$

has a unique solution in $L^2[-2, 2]$, given by

$$g_\Omega(x) = \begin{cases} \frac{2}{\pi} \left(1 - \frac{x^2}{4}\right)^{1/2} + \frac{x}{\pi} \sin^{-1} \frac{x}{2} - \frac{|x|}{2}, & \text{for } |x| < 2; \\ 0, & \text{for } |x| > 2. \end{cases} \qquad (7.3.13)$$

Proof. Logan and Shepp prove this by considering the function

$$G_\Omega(z) = \frac{2}{\pi} \left\{ \left(1 - \frac{z^2}{4}\right)^{1/2} + \frac{iz}{2} \right\} - \frac{iz}{\pi} \log \frac{z}{i} - \frac{iz}{\pi} \log \left\{ \left(1 - \frac{z^2}{4}\right)^{1/2} + \frac{iz}{2} \right\} \qquad (7.3.14)$$

which is analytic in the upper half-plane, and where the roots may be chosen so that $G_\Omega(z) = O(1/z)$ as $|z| \to \infty$. They verify that $g(x) = \Re G(x)$, and that

$$\Im G_\Omega(x) = \frac{x}{\pi} - \frac{x}{\pi}\log|x|, \qquad (|x| \le 2) \qquad (7.3.15)$$

so that the Hilbert transform is $\mathcal{H} g_\Omega(x) = v(x)/\pi$ for $|x| < 2$, as required.

$\square$

One can easily check that g_Ω has the following properties:

(i) $g_\Omega(\xi) = g_\Omega(-\xi)$, so g_Ω is even;
(ii) $0 \ge g'_\Omega(\xi) \ge -1$ for $\xi > 0$, so g_Ω is unimodal;
(iii) g_Ω maps $[0,2]$ onto $[0, \pi/2]$;
(iv) g_Ω is a probability density function since $g_\Omega(\xi) \ge 0$ and $\int_{-\infty}^{\infty} g_\Omega(\xi)d\xi = 1$.
(v) Further, g_Ω minimizes I and $I(g_\Omega) = -1/2$.

To prove (v), one writes

$$I(g_\Omega) = \int_{-2}^{2} \log|x| g_\Omega(x)\, dx + \frac{\pi}{2}\int_{-2}^{2} g_\Omega(x)\mathcal{H} g_\Omega(x)\, dx \qquad (7.3.16)$$

and then evaluates these integrals using the substitution $x = 2\sin\theta$. See Exercise 7.3.7 and Section 4.7.

Definition (*Vershik's Ω distribution*). The Vershik Ω distribution has probability density function f_Ω, where the function $f_\Omega : [0,\infty) \to [0,\infty)$ satisfies the pair of equations

$$x = g_\Omega(\xi) + (1/2)\xi + (1/2)|\xi|,$$
$$f_\Omega(x) = g_\Omega(\xi) - (1/2)\xi + (1/2)|\xi|. \qquad (7.3.17)$$

Proposition 7.3.3 *Here f_Ω satisfies the corresponding properties:*

(i) $f_\Omega(f_\Omega(x)) = x$, so that the graph of f_Ω is symmetrical about $y = x$;
(ii) f_Ω is decreasing;
(iii) f_Ω maps $[0,2]$ onto $[0,2]$;
(iv) f_Ω is a probability density function since $\int_0^\infty f_\Omega(x)dx = 1$.
(v) f_Ω minimizes $\tilde{H}$ and $\tilde{H}(f_\Omega) = -1/2$.

Proof. (i) This follows from the fact that g_Ω is even.
(ii) From the above formulae, we have

$$\frac{df_\Omega}{dx} = \begin{cases} g'_\Omega(\xi)/(1 + g'_\Omega(\xi)), & \text{for } \xi > 0; \\ (g'_\Omega(\xi) - 1)/g'_\Omega(\xi), & \text{for } \xi < 0; \end{cases} \qquad (7.3.18)$$

so $\frac{df_\Omega}{dx} \leq 0$ in all cases.

(iii) For $\xi > 2$ we have $g_\Omega(\xi) = 0$, so $x = \xi$ and $f_\Omega(x) = 0$; further, $x = 0$ at $\xi = -2$ so $f_\Omega(0) = 2$.

(iv) We introduce x_0 such that $f_\Omega(x_0) = x_0$. Then from the graph of f_Ω it is evident that

$$g_\Omega(\xi) = \frac{1}{2}\left(x + f_\Omega(x) - |x - f_\Omega(x)|\right) = \begin{cases} f_\Omega(x), & \text{for } x > x_0; \\ x, & \text{for } x < x_0; \end{cases} \quad (7.3.19)$$

so

$$\int_{-\infty}^{\infty} g_\Omega(\xi)d\xi = \int_0^{x_0} x(1 - f_\Omega'(x))dx + \int_{x_0}^{\infty} f_\Omega(x)(1 - f_\Omega'(x))dx$$
$$= x_0^2/2 + f_\Omega(x_0)^2/2 - \int_0^{x_0} x f_\Omega'(x)\,dx + \int_{x_0}^{\infty} f_\Omega(x)dx.$$
$$(7.3.20)$$

So by integrating by parts and using the choice of x_0, we obtain

$$\int_{-\infty}^{\infty} g_\Omega(\xi)d\xi = \int_0^{\infty} f_\Omega(x)\,dx. \quad (7.3.21)$$

(v) This follows from the corresponding result for g_Ω, since $\tilde{H}(f_\Omega) = I(g_\Omega)$. □

The relevance of these results in random matrix theory becomes clear with the following expression for the Plancherel measure, which resembles the expression in Proposition 2.4.3.

Proposition 7.3.5 *Let λ be a Young diagram that has n rows and let $\ell_j = \lambda_j + n - j$; so that, ℓ_j is the hook length of the box in position $(j,1)$ from the first column of λ. Then the Plancherel measure for S_N satisfies*

$$\nu_N(\lambda) = \exp\left(\log N! - 2\sum_{j=1}^n \log\Gamma(\ell_j + 1) - 2\sum_{1 \leq j < k \leq n} \log(\ell_j - \ell_k)\right).$$
$$(7.3.22)$$

Proof. This follows from the hook length formula Proposition 7.2.4. □

The function $w(x) = 2\log\Gamma(x+1)$ is uniformly convex on $[\delta, K]$ for each $0 < \delta < K < \infty$, and $w(x) \asymp \log(2\pi x) + 2x\log(x/e)$ as $x \to \infty$. For a more detailed discussion of this potential, see the author's paper [20].

Example 7.3.6 The function $v(x) = x\log(x/e)$ is uniformly convex on $[0, K]$ since $v''(x) = 1/x \geq K$. Hence by Theorem 4.8.3, the equilibrium density h_Ω satisfies

$$W_2(h, h_\Omega)^2 \leq 2K\Sigma(h \mid h_\Omega) \qquad (7.3.23)$$

for all probability density functions h that have finite relative free entropy with respect to h_Ω.

As in Section 4.5, we consider the set of absolutely continuous functions $g : [0, K] \to [0, A]$ such that g is decreasing with $g(0) = A$ and $g(K) = 0$. Then $h(x) = -g'(x)/A$ gives a probability density function on $[0, K]$. Suppose that g and g_Ω have this form, and introduce $h = -g'/A$ and $h_\Omega = -g'_\Omega/A$. Then we find

$$\begin{aligned}
E_v(h) &= \frac{1}{A}\int_0^K g'(x)x\log(e/x)dx \\
&+ \frac{\beta}{2A^2}\int_0^K\int_0^K \log\frac{1}{|x-y|}g'(x)g'(y)\,dxdy
\end{aligned} \qquad (7.3.24)$$

which resembles the functional $I(g)$ of Logan and Shepp. By Proposition 3.3.3, we have

$$W_2(h, h_\Omega) \geq W_1(h, h_\Omega) = \frac{1}{A}\int_0^K |g(x) - g_\Omega(x)|dx. \qquad (7.3.25)$$

In view of the formula for g_Ω, we should take $A = 2/\pi$ and $K = 2$. The Vershik–Kerov Theorem 7.4.6 uses sharper estimates than this; so that, the statement involves the L^∞ rather than the L^1 norm, and has an effective error bound. Transportation inequalities are not designed to give uniform bounds on probability densities.

Exercise 7.3.7 (i) Show that $I(g_\Omega) = 2^{-1}\int_{-2}^2 g_\Omega(x)\log|x|\,dx$.

(ii) Given that $\int_{-\pi/2}^{\pi/2} \log|\sin\theta|d\theta = -\pi\log 2$, by [35] show by integrating by parts that

$$\int_{-\pi/2}^{\pi/2} \cos^2\theta \log|\sin\theta|d\theta = -\frac{\pi}{2}\log 2 - \frac{\pi}{4}, \qquad (7.3.26)$$

$$\int_{-\pi/2}^{\pi/2} \theta\sin\theta\cos\theta \log|\sin\theta|d\theta = \frac{\pi}{4}\log 2 - \frac{\pi}{4}.$$

(iii) Hence show that $I(g_\Omega) = -1/2$.

Remark. In this section we have drawn analogies between Plancherel measure on the irreducible representations of S_N and certain random matrix ensembles. The sharpest results about partitions are obtained from complex analysis and appear beyond the scope of transportation methods. The distribution of lengths of rows in random Young diagrams have been thoroughly investigated by the Riemann–Hilbert technique [1, 7, 8, 56].

7.4 Distribution of the longest increasing subsequence

- The distribution of the length of the longest increasing sequence can be expressed as an integral over $U(N)$.
- The Poisson process is used to express such integrals as Toeplitz determinants.

In this section we show how to obtain the distribution of the length of the longest increasing subsequence in a permutation. The techniques essentially use random matrix theory, and the Poisson process.

When gazing at the night sky through a telescope, an observer can count the number $\rho(A)$ of bright stars within a given sector A. This $\rho(A)$ appears to be a random non-negative integer, and when A and B are disjoint sectors, $\rho(A)$ and $\rho(B)$ appear to be statistically independent. A common model for this is the Poisson process as in [101].

Definition (*Random point field*). Let (Ω_0, d) be a Polish space such that all bounded and closed subsets of Ω_0 are compact. Let ν be a measurable mapping from a probability space $(\Omega, \mathbf{P})$ to the set of Radon measures on Ω_0; the distribution is the probability measure that is induced by ν. (Hence $A \mapsto \nu_\omega(A)$ is a Radon measure for each $\omega \in \Omega$, and $\omega \mapsto \nu_\omega(A)$ is a random variable for each bounded Borel set A.) If $\nu_\omega(A)$ belongs to $\{0, 1, 2, \dots\}$ for all bounded Borel subsets A of Ω_0 and all $\omega \in \Omega$, then ν is a random point field on Ω_0.

Definition (*Poisson process* [101]). Let $(\Omega, \mathbf{P})$ be a probability space, and let m_2 be Lebesgue area measure. For $\theta > 0$, a Poisson random variable with mean θ is a measurable function $R : \Omega \to \{0, 1, \dots\}$ such that $\mathbf{P}[R = n] = e^{-\theta}\theta^n/n!$ for $n = 0, 1, \dots$. A Poisson process ρ with unit rate in the plane is a random point field on $\mathbf{R}^2$ such that:

(i) $\rho(A)$ is a Poisson random variable with mean $m_2(A)$;

(ii) $\rho(A_j)$ $(j = 1, 2, \dots)$ are mutually independent and

(iii) $\rho(A) = \sum_{j=1}^{\infty} \rho(A_j)$ whenever (A_j) $(j = 1, 2, \dots)$ are disjoint bounded Borel sets with bounded union $A = \cup_{j=1}^{\infty} A_j$.

The probability generating function g_A satisfies

$$g_A(z) = \mathbf{E}z^{\rho(A)} = \exp((z-1)m_2(A)), \qquad (7.4.1)$$

so that $A \mapsto \log g_A(z)$ determines the mean measure $m_2(A)$. Area measure is a suitable choice since it has no atoms; the Poisson random variables $\rho(A)$ can take any positive integral values, but they assign values with multiplicity one. (In this respect, the Poisson process does not provide an ideal model of stellar counts, since visual binary stars occur relatively frequently.)

For large $\theta > 0$, we consider the Poisson process on $[0, \sqrt{\theta}] \times [0, \sqrt{\theta}]$, so that $N = \rho([0, \sqrt{\theta}] \times [0, \sqrt{\theta}])$ has a $\mathrm{Po}(\theta)$ distribution. For each outcome, there exist points $(x_j, y_j) \in [0, \sqrt{\theta}] \times [0, \sqrt{\theta}]$ for $j = 1, \ldots, N$ at which the process jumps; regard (x_j, y_j) as the coordinates of the stars. Without loss of generality, we can choose the indices so that $x_1 < x_2 < \cdots < x_N$ and then we can choose a unique $\sigma \in S_N$ so that $y_{\sigma(1)} < y_{\sigma(2)} < \cdots < y_{\sigma(N)}$. For each given N, the distribution of σ is Haar measure on S_N.

We introduce a probability measure μ_∞^θ on $S_\infty = \cup_{N=0}^\infty S_N$ by selecting N according to a Poisson law with mean θ, then we select elements of S_N according to Haar measure; by convention $S_0 = \{e\}$. Later we shall consider S_∞ as an inductive limit group. The distribution of σ as given by the above Poisson process is μ_∞^θ.

To be more explicit, we write N_θ for N.

Proposition 7.4.1 $N_\theta/\theta \to 1$ *almost surely as* $\theta \to \infty$.

Proof. This is a special case of the strong law of large numbers. Indeed, for each $\varepsilon > 0$, Chebyshev's inequality gives

$$\mathbf{P}\left[\left| \frac{N_{k^2}}{k^2} - 1 \right| > \varepsilon \right] \leq \frac{1}{\varepsilon^2 k^2} \qquad (7.4.2)$$

and hence by the First Borel–Cantelli Lemma we deduce

$$\mathbf{P}\left[\left| \frac{N_{k^2}}{k^2} - 1 \right| > \varepsilon \quad \text{for infinitely many} \quad k \right] = 0. \qquad (7.4.3)$$

Taking an integer k such that $k \leq \sqrt{\theta} < k + 1$, we observe that $N_{k^2} \leq N_\theta < N_{(k+1)^2}$ and $k/\sqrt{\theta} \to 1$ as $\theta \to \infty$, hence the result. $\qquad \square$

Each increasing subsequence $\sigma(n_1) < \sigma(n_2) < \cdots < \sigma(n_\ell)$ in σ may be identified with a zig-zag path with steps $(x_{n_j}, y_{\sigma(n_j)})$ to $(x_{n_{j+1}}, y_{\sigma(n_{j+1})})$, going eastwards and northwards in the plane. For each $s < t$ we let ℓ_{st} be the number of points in the longest such path in $[s, t] \times [s, t]$. Hammersley used this device in various percolation problems.

torrg.

Theorem 7.4.2 (*Hammersley*). *There exists a constant c such that*

$$\frac{\ell_{0,t}}{t} \to c \tag{7.4.4}$$

almost surely as $t \to \infty$.

Proof. Kingman introduced the subadditive ergodic theorem to prove this result. By joining paths together we see that

$$\ell_{su} \geq \ell_{st} + \ell_{tu} \qquad (s < t < u). \tag{7.4.5}$$

The numerical sequence $-\mathbf{E}\ell_{st}$ is subadditive, in the sense that

$$-\mathbf{E}\ell_{su} \leq -\mathbf{E}\ell_{st} - \mathbf{E}\ell_{tu} \qquad (s < t < u), \tag{7.4.6}$$

and hence $\mathbf{E}\ell_{0t}/t \to c$ for some constant c by elementary results. By considering shifts of the plane along the North-East diagonal, we see that the process is strongly stationary in the sense that the joint distribution of $(-\ell_{s+1,t+1})_{s<t}$ is equal to the joint distribution of $(-\ell_{st})_{s<t}$. The invariant σ-algebra for the diagonal shift is trivial, so the process is ergodic. By the subadditive ergodic theorem [100], we have $\ell_{0,t}/t \to c$ almost surely as $t \to \infty$. Generally, c is known as the time constant. Due to Theorem 7.4.6 presented below, this constant c turns out to be 2. □

Lemma 7.4.3 (*Rains, [5]*). *Let $\ell(\sigma)$ be the length of the longest increasing sequence in σ. Then*

$$\mu_N\{\sigma \in S_N : \ell(\sigma) \leq r\} = \frac{1}{N!} \int_{U(r)} |\mathrm{trace}(U)|^{2N} \mu_{U(r)}(dU). \tag{7.4.7}$$

Proof. We choose an integer $r \leq N$. Let the eigenvalues of $U \in U(r)$ be $e^{i\theta_1}, \ldots, e^{i\theta_r}$, so that $\mathrm{trace}(U) = \sum_{j=1}^r e^{i\theta_j}$. The first elementary symmetric polynomial in r variables is $P_1(x) = \sum_{j=1}^r x_j$, and we can express its Nth power as a linear combination of the Schur functions

$$(P_1(x))^N = \sum_{\lambda \vdash N, \lambda_1^t \leq r} \chi_\lambda(1^t) s_\lambda(x) \tag{7.4.8}$$

where λ is summed over all partitions of N such that the transpose λ^t has less than or equal to k rows and χ_λ is the character that corresponds to λ. Now $\chi_\lambda(1^t) = \dim(H_\lambda)$, and the Schur functions are orthonormal with respect to $\mu_{U(r)}$. Hence we have

$$\frac{1}{N!} \int_{U(r)} |P_1(e^{i\theta_1}, \ldots, e^{i\theta_r})|^{2N} \mu_{U(r)}(dU) = \frac{1}{N!} \sum_{\lambda \vdash N, \lambda_1^t \leq r} \left(\dim(H_\lambda)\right)^2, \tag{7.4.9}$$

in which we recognise the right-hand side as

$$\nu_N\{\lambda \vdash N : \lambda_1^t \leq r\} = \nu_N\{\lambda \vdash N : \lambda_1 \leq r\}$$
$$= \mu_N\{\sigma \in S_N : \ell(\sigma) \leq r\}, \qquad (7.4.10)$$

where the final equality follows from the RSK correspondence as in Proposition 7.2.3. □

Definition (*Poissonization*). Let $(\Omega_N, \mathbf{P}_N)$ be discrete probability spaces for $N = 0, 1, 2\ldots$, and let $\Omega = \sqcup_{N=0}^{\infty}\Omega_N$ be the disjoint union of the Ω_N. Then for each $\theta > 0$ we introduce the probability measure $\mathbf{P}^\theta$ on Ω by

$$\mathbf{P}^\theta(\{\omega_N\}) = e^{-\theta}\frac{\theta^N}{N!}\mathbf{P}_N(\{\omega_N\}) \qquad (\omega_N \in \Omega_N, N = 0, 1, \ldots). \quad (7.4.11)$$

In effect, we choose the index N to be a Poisson random variable with mean θ, and then choose from Ω_N according to the law $\mathbf{P}_N$.

Definition (*Plancherel measure on partitions*). Let $\Lambda_N = \{\lambda \vdash N\}$ and $\Lambda_0 = \{1\}$; then let $\Lambda = \cup_{N=0}^{\infty}\Lambda_N$. To select a partition λ, we first choose $N = |\lambda|$ according to the Poisson law with mean θ, then we choose $\lambda \in \Lambda_N$ according to Plancherel measure; thus we define the probability measure P^θ on Λ by

$$P^\theta(\{\lambda\}) = e^{-\theta}\theta^{|\lambda|}\left(\frac{\dim(H_\lambda)}{|\lambda|!}\right)^2. \qquad (7.4.12)$$

Under the RSK correspondence, we can regard P^θ as a measure on $S_\infty = \cup_{N=1}^{\infty}S_N$.

Definition (*Bessel function of integral order [167, p. 19]*). Bessel defined J_m with integral order m by

$$J_m(x) = \int_0^\pi \cos(m\phi - x\sin\phi)\frac{d\phi}{\pi}, \qquad (m = 0, 1, \ldots; x \in \mathbf{C}) \quad (7.4.13)$$

and the relation $J_{-m}(x) = (-1)^m J_m(x)$; the general definition appears in Section 9.2.

From this definition, one can recover the J_m as the coefficients in the Laurent expansion [167, 2.11]

$$\exp\left\{\frac{x}{2}\left(z-\frac{1}{z}\right)\right\} = J_0(x) + \sum_{m=1}^{\infty}z^m J_m(x) + \sum_{m=1}^{\infty}(-z)^{-m} J_m(x). \quad (7.4.14)$$

Proposition 7.4.4 (*Gessel [5]*). *For $\theta > 0$, let*

$$\Phi_r(\theta) = P^\theta\{\lambda \in \Lambda : \lambda_1 \leq r\} \qquad (r \in \mathbf{N}) \qquad (7.4.15)$$

be the cumulative distribution function of the longest row in the partitions under P^θ. Then

$$\Phi_r(\theta) = e^{-\theta} \det\left[J_{j-k}(-2i\theta)\right]_{j,k=1,\dots,r}. \tag{7.4.16}$$

Proof. By definition of P^θ, we have

$$P^\theta\{\lambda : \lambda_1 \leq r\} = e^{-\theta} \sum_{N=0}^{\infty} \frac{\theta^N}{N!} \mu_N\{\sigma \in S_N : \ell(\sigma) \leq r\} \qquad (r \in \mathbf{N}). \tag{7.4.17}$$

Lemma 7.4.3 suggests that we can express this as an integral over the groups $U(N)$. First we observe that by the binomial theorem

$$\int_{U(r)} |\mathrm{trace}(U + U^*)|^{2N} \mu_{U(r)}(dU) = \binom{2N}{N} \int_{U(r)} |\mathrm{trace}(U)|^{2N} \mu_{U(r)}(dU), \tag{7.4.18}$$

since we need an equal numbers of factors U and U^* to produce a nonzero term, and of course $\mathrm{trace}(U + U^*) = 2\sum_{j=1}^{r} \cos\psi_j$. By Lemma 7.4.3 and the Weyl formula Theorem 2.6.2, we have

$$\frac{1}{N!} \mu_N\{\sigma \in S_N : \ell(\sigma) \leq r\}$$

$$= \frac{1}{(2N)!} \int_{U(r)} |\mathrm{trace}(U + U^*)|^{2N} \mu_{U(r)}(dU)$$

$$= \frac{1}{r!} \frac{1}{(2N)!} \int_T \left(2\sum_{j=1}^{r} \cos\psi_j\right)^{2N} \prod_{1 \leq j < k \leq r} |e^{i\psi_j} - e^{i\psi_k}|^2 \frac{d\psi_1}{2\pi} \cdots \frac{d\psi_r}{2\pi}. \tag{7.4.19}$$

When we replace $2N$ by an odd power, the resulting expression vanishes by symmetry, so when we multiply by $e^{-\theta}\theta^N$ and sum over N, we obtain

$$e^{-\theta} \sum_{N=0}^{\infty} \frac{\theta^N}{N!} \mu_N\{\sigma \in S_N : \ell(\sigma) \leq r\}$$

$$= \frac{e^{-\theta}}{r!} \int_T \exp\left(2\theta \sum_{j=1}^{r} \cos\psi_j\right) \prod_{1 \leq j < k \leq r} |e^{i\psi_j} - e^{i\psi_k}|^2 \frac{d\psi_1}{2\pi} \cdots \frac{d\psi_r}{2\pi}, \tag{7.4.20}$$

where the integral is taken over the maximal torus T. Next, we convert the right-hand side into the Toeplitz determinant by Proposition 4.9.5, and obtain

$$e^{-\theta} \det\left[\int_0^{2\pi} \exp\left(2\theta \cos\psi + i(j-k)\psi\right) \frac{d\psi}{2\pi}\right]_{j,k=1,\dots,r}. \qquad (7.4.21)$$

Next we observe that

$$\int_{-\pi}^{\pi} e^{2\theta \cos\psi + i(j-k)\psi} \frac{d\psi}{2\pi} = i^{j-k} \int_{-\pi}^{\pi} e^{-2\theta \sin\psi + i(j-k)\psi} \frac{d\psi}{2\pi}$$

$$= i^{j-k} J_{j-k}(-2i\theta), \qquad (7.4.22)$$

and by the multiplicative property of determinants

$$e^{-\theta} \det\left[\int_{-\pi}^{\pi} e^{2\theta \cos\psi + i(j-k)\psi} \frac{d\psi}{2\pi}\right]_{j,k=1,\dots,r} = e^{-\theta} \det\left[J_{j-k}(-2i\theta)\right]_{j,k=1,\dots,r}.$$

$$(7.4.23)$$

$\square$

The determinant surges in value as r passes through $2\sqrt{\theta}$; more precisely, $\Phi_r(\theta)$ satisfies the following lemma.

Lemma 7.4.5 (*Johansson [94, 8]*). *There exist constants $\delta, \varepsilon > 0$, which are independent of r and N, such that*

$$0 \le \Phi_r(\theta) \le Ce^{-\delta\theta} \qquad ((1+\varepsilon)r < 2\sqrt{\theta});$$
$$0 \le 1 - \Phi_r(\theta) \;\le C/r \qquad ((1-\varepsilon)r > 2\sqrt{\theta}). \qquad (7.4.24)$$

Proof. The proof is a lengthy deviations estimate on the determinant in Proposition 7.4.4, hence is omitted. $\square$

Theorem 7.4.6 (*Vershik–Kerov [161]*). *The scaled length of the longest increasing subsequence converges in probability to 2 as $N \to \infty$; that is, for each $\varepsilon > 0$,*

$$\mu_N\left\{\sigma_N \in S_N : \left|\frac{\ell(\sigma_N)}{\sqrt{N}} - 2\right| > \varepsilon\right\} \to 0 \qquad (N \to \infty). \qquad (7.4.25)$$

Proof. We introduce the expression

$$a_N(r) = \mu_N\{\sigma_N \in S_N : \ell(\sigma) \le r\} \qquad (7.4.26)$$

which is decreasing with respect to increasing N, and increasing with respect to increasing r. Clearly $0 \le a_N(r) \le 1$. We shall prove that $a_\lambda(r)$ jumps from a value close to 0 to a value close to 1 as r passes through $2\sqrt{\theta}$.

By a simple application of Chebyshev's inequality, we have for each $\kappa > 1$ the bounds

$$\left| \Phi_r(\theta) - \sum_{N=\theta-\kappa\sqrt{\theta}}^{\theta+\kappa\sqrt{\theta}} e^{-\theta}\frac{\theta^N}{N!} a_N(r) \right| \leq \frac{2}{\kappa^2}, \qquad (7.4.27)$$

which gives the estimates

$$a_{\theta-\kappa\sqrt{\theta}}(r) + \frac{2}{\kappa^2} \geq \Phi_r(\theta) \geq \left(1 - \frac{2}{\kappa^2}\right) a_{\theta+\kappa\sqrt{\theta}}(r) - \frac{2}{\kappa^2}. \qquad (7.4.28)$$

By Lemma 7.4.5, we have

$$Ce^{-\delta\theta} + \frac{2}{\kappa^2} \geq \left(1 - \frac{2}{\kappa^2}\right) a_{\theta+\kappa\sqrt{\theta}}\left(\frac{2\sqrt{\theta}}{1+\varepsilon}\right), \qquad (7.4.29)$$

and

$$a_{\theta-\kappa\sqrt{\theta}}\left(\frac{2\sqrt{\theta}}{1-\varepsilon}\right) \geq 1 - \frac{C(1-\varepsilon)}{2\sqrt{\theta}} - \frac{2}{\kappa^2}; \qquad (7.4.30)$$

so by taking $\kappa = \theta^{1/4}$ and letting $\theta \to \infty$, we obtain the stated result that $a_\theta(r)$ jumps near to $r = 2\sqrt{\theta}$. □

In Section 10.4 we shall revisit these ideas and give a more precise version of Theorem 7.4.6 which deals with the delicate topic of the edge distribution. The first proof of Theorem 7.4.6 did not use random matrix theory, instead it relied upon combinatorial results concerning partitions, and gives some more precise information.

Theorem 7.4.7 (*Vershik–Kerov [161]*). *Under the Plancherel measure ν_N on $\{\lambda \vdash N\}$, the scaled Young diagrams f_λ converge in probability to f_Ω as $N \to \infty$, so*

$$\nu_N\{\lambda \vdash N : \|f_\lambda - f_\Omega\|_\infty < CN^{-1/6}\} \to 1 \qquad (N \to \infty). \qquad (7.4.31)$$

Proof. See Vershik–Kerov [161]. A key ingredient of the proof is the Hardy–Ramanujan asymptotic formula [6, 79] for the number p_N of partitions of N. There exists ε_N such that

$$p_N = \sharp\{\lambda \vdash N\} = \exp\left(\pi\left(\frac{2N}{3}\right)^{1/2}(1 + \varepsilon_N)\right), \qquad (7.4.32)$$

where $\varepsilon_N \to 0$ as $N \to \infty$. The statement of the result implies convergence in Wasserstein metric due to Proposition 3.3.3(ii). □

Having established that $\ell(\sigma)$ is often close to $2\sqrt{N}$, the next step is to find the variance of $\ell(\sigma) - 2\sqrt{N}$ and the asymptotic distribution. As a result of numerical simulations, the variance was found to be of order

$N^{1/3}$ as $N \to \infty$. The nature of the asymptotic distribution is rather surprising in that it emerged in analysis of Gaussian random matrices, not in combinatorics. In Section 9.4 we introduce the Tracy–Widom distribution in the context of the GUE, and in Section 10.4 sketch how this distribution arises for measures on the space of partitions.

Exercise. Calculate the integral

$$\int_{M_n^s (\mathbf{R})} (\mathrm{trace}(X))^{2n}\, \nu_n\, (dX) \qquad (7.4.33)$$

where ν_n is the probability measure for the Gaussian orthogonal ensemble, and compare with the result for (7.4.18).

7.5 Inclusion-exclusion principle

- The inclusion-exclusion principle is a basic counting tool which is later used in the next chapter to deal with point processes.

Let $(L, \leq)$ be a finite and partially ordered set, and let $\zeta : L \times L \to \{0,1\}$ be the function

$$\zeta(\pi_1, \pi_2) = \begin{cases} 1, & \text{for } \pi_1 \leq \pi_2; \\ 0, & \text{else.} \end{cases} \qquad (7.5.1)$$

Lemma 7.5.1 (*Möbius's function*). *There exists a function $\mu : L \times L \to \mathbf{R}$ such that any pair of functions $f, g : L \to \mathbf{R}$ satisfy*

$$f(\pi) = \sum_{\omega : \omega \geq \pi} g(\omega) \qquad (\pi \in L) \qquad (7.5.2)$$

if and only if

$$g(\pi) = \sum_{\omega : \omega \geq \pi} \mu(\pi, \omega) f(\omega) \qquad (\pi \in L). \qquad (7.5.3)$$

Proof. This is by induction on $\sharp L$; see [6, p. 218]. ☐

Given a probability measure $\mathbf{P}$ on L, we can consider the correlation function

$$\rho(\pi) = \mathbf{E}\zeta(\pi, \,.\,) = \mathbf{P}\{\omega : \pi \leq \omega\}. \qquad (7.5.4)$$

By the inclusion-exclusion principle, ρ determines $\mathbf{P}$ via the equivalent formulae

$$\rho(\pi) = \sum_{\omega : \omega \geq \pi} \mathbf{P}(\{\omega\}) \qquad (7.5.5)$$

and

$$\mathbf{P}(\{\pi\}) = \sum_{\omega\,:\,\omega \geq \pi} \mu(\pi,\omega)\rho(\omega). \qquad (7.5.6)$$

Typically, the Möbius function will take both positive and negative values, which make inequalities tricky to deal with. There follow two significant cases in which μ is known.

7.5.2 Subsets of a finite set

Let L be the collection of all subsets of $\{1,\dots,n\}$, partially ordered by inclusion; see [90]. Then the Möbius function is

$$\mu(U,V) = \begin{cases} (-1)^{\sharp(V\setminus U)}, & \text{for } U \subseteq V; \\ 0, & \text{else.} \end{cases} \qquad (7.5.7)$$

7.5.3 Partitions of a finite set

By a partition of the set $\{1,\dots,n\}$ we mean a decomposition into the union of disjoint non-empty subsets; by a block we mean a nonempty subset of $\{1,\dots,n\}$. For each n, there are natural bijections between:

(i) equivalence relations on $\{1,\dots,n\}$ with r distinct equivalence classes;
(ii) partitions of $\{1,\dots,n\}$ into disjoint blocks B_k for $k = 1,\dots,r$;
(iii) polynomials in commuting indeterminates x_j that have the form

$$\sum_{k=1}^{r} \prod_{j \in B_k} x_j, \qquad (7.5.8)$$

where $\sqcup_{k=1}^{r} B_k = \{1,\dots,n\}$.

Given a partition π, with blocks of length $\lambda = (\lambda_1 \geq \cdots \geq \lambda_{r(\pi)} > 0)$, we can introduce a λ-tableau with $r(\pi)$ rows that are given by the blocks of π in some order. The lack of uniqueness is specified by the equivalence relation $\sim$ on λ-tableaux, where $\tau_1 \sim \tau_2$ if τ_2 can be obtained from τ_1 by successively interchanging numbers in each row and interchanging the rows that have equal length.

Given partitions π_1 and π_2 of $\{1,2,\dots,n\}$, we say that π_2 refines π_1 and write $\pi_2 \geq \pi_1$ if each block of π_2 is a subset of some block of π_1. Let L_n be the set of partitions of $\{1,2,\dots,n\}$ with this partial order. The maximal element is $I = [\{1\},\dots,\{n\}]$, which has n equivalence classes.

We introduce the notation $(x)_{(k)} = x(x-1)\ldots(x-k+1)$. The number of injective maps $\varphi : \{1,\ldots,w\} \to \{1,\ldots,m\}$ is $(m)_{(w)}$. Given $\pi \in L_n$, and a finite set T such that $\sharp T = t \geq r(\pi)$, we consider the maps $\varphi : \{1,\ldots,n\} \to T$ such that $\varphi(j) = \varphi(k)$ if and only if j and k both lie in the same block in the partition π. The number of such functions is $f(\pi) = (t)_{(r(\pi))}$.

Definition (*Stirling's numbers*). The Stirling numbers of the first kind $s(n,k)$ satisfy

$$(X)_{(n)} = \sum_{k=0}^{n} s(n,k)X^k; \qquad (7.5.9)$$

whereas the Stirling numbers of the second kind $S(n,k)$ satisfy

$$X^n = \sum_{k=0}^{n} S(n,k)(X)_{(k)}. \qquad (7.5.10)$$

Proposition 7.5.3 *The Stirling numbers satisfy*

$$S(n,k) = \sharp\{\pi \in L_n : r(\pi) = k\}; \qquad (7.5.11)$$

$$s(n,k) = \sum_{\pi \in L_n : r(\pi)=k} \mu(\pi; I). \qquad (7.5.12)$$

Proof. See Andrews [6].   □

Exercise 7.5.4 Find the number of partitions of $\{1,\ldots,n\}$ that have blocks of length $k_1 > \cdots > k_r$, with k_j appearing n_j times.

These ideas have been considered in detail by Borodin, Olshanski and others; see [33] and [34] for further developments.

8

Random point fields and random matrices

Abstract

The distribution of the eigenvalues of a random matrix gives a random point field. This chapter outlines Soshnikov's version of the general theory. Starting with kernels, we introduce correlation functions via determinants. Gaudin and Mehta developed a theory of correlation functions for the generalized unitary ensemble which is known as the orthogonal polynomial technique, and we show that it fits neatly into the theory of determinantal random point fields. Particular determinantal random point fields are generated by the sine kernel, the Airy kernel, and the continuous Bessel kernels; in random matrix theory these are widely held to be universal kernels in that they describe the possible asymptotic distribution of eigenvalues from large Hermitian matrices. In the final section we introduce an abstract framework in which one can describe the convergence of families of determinantal random point fields, and which we apply to fundamental examples in Chapters 9 and 10.

In Sections 8.1 and 8.2, we describe how kernels can be used to form determinantal random point fields, and in Section 8.3 we express some classical results about unitary ensembles in terms of determinantal random point fields. In Sections 9.3, 9.4 and 9.5 we look at these specific examples in more detail, and see how they arise in random matrix theory.

8.1 Determinantal random point fields

In this section, we introduce Soshnikov's theory [148] of determinantal random point fields as it applies to point fields on $\mathbf{Z}$. The idea is that, by starting with a matrix such that $0 \leq K \leq I$, one can define joint probability distributions of random arrays of points in **Z**. We begin by introducing the formalism needed to describe random arrays of points.

Definition (*Configurations*). Let (Ω, d) be a Polish space such that any closed and bounded subset A of Ω is compact. A *configuration* Λ is a function $\nu : \Omega \to \{0, 1, 2, \dots\}$ such that for each compact set A, the support of $\nu|_A = \{x \in A : \nu(x) > 0\}$ is finite. We call $\nu(x)$ the *multiplicity* of x and observe that $\{x \in \Omega : \nu(x) > 0\}$ is finite or countably infinite since Ω is the union of a sequence of closed and bounded subsets. When $\nu(x) \leq 1$ for all $x \in \Omega$, we say that Λ has *multiplicity one*; some authors use the more suggestive term *multiplicity free*. Evidently each such configuration $\nu : \Omega \to \{0, 1\}$ is the indicator function of its support $\{s \in \Omega : \nu(s) = 1\}$, so we can identify the configurations of multiplicity one with the subsets of Ω via $\mathbf{I}_S \leftrightarrow S$.

We define

$$\nu_\Lambda(A) = \sum_{x \in A} \nu(x) \tag{8.1.1}$$

to be the number of points, in A, counted according to multiplicity, and write $\nu_\Lambda(\{x\}) = \nu(x)$. We let $\mathrm{Conf}(\Omega)$ be the set of all configurations Λ on Ω, with the σ-algebra generated by the counting functions ν_Λ. Then $\Lambda \mapsto \nu_\Lambda(A)$ is a random variable for each Borel set A, while $A \mapsto \nu_\Lambda(A)$ is a measure.

Definition (*Random point field*). A random point field is a Borel probability measure on $\mathrm{Conf}(\Omega)$. There is a family of Borel measures $A \mapsto \nu_\Lambda(A)$ defined for $\Lambda \in \mathrm{Conf}(\Omega)$.

Exercise. Let ν, ω and ϖ be configurations on Ω.

(i) Show that the configurations are partially ordered by defining $\varpi \leq \nu$ if $\varpi(x) \leq \nu(x)$ for all $x \in \Omega$.

(ii) Show that $\nu + \varpi$, $\max\{\varpi, \nu\}$, $\min\{\varpi, \nu\}$ and the pointwise product $\varpi\nu$ are all configurations.

(iii) Let $\Omega = \{1, \dots, n\}$ and consider the configurations of multiplicity one. Show that the configurations of multiplicity one give a Boolean algebra with largest element $\mathbf{I}_\Omega$ for the operation $\varpi \wedge \nu = \min\{\varpi, \nu\}$ with complementation $\varpi \mapsto \mathbf{I}_\Omega - \varpi$.

Exercise. For any nonempty set Ω, let $\mathrm{Conf}_0(\Omega) = \{\nu : \Omega \to \{0, 1\} : \sum_{j \in \Omega} \nu(j) < \infty\}$, and let $\Omega_1 = \{X \subseteq \Omega : \sharp X < \infty\}$. Show that $X \mapsto \mathbf{I}_X$ gives a bijection $\Omega_1 \leftrightarrow \mathrm{Conf}_0(\Omega)$, and deduce that if Ω is countable, then $\mathrm{Conf}_0(\Omega)$ is also countable.

We now convert the above concepts into the language often used in random matrix theory.

Definition (*Linear statistics*). Suppose that $\nu : \mathbf{R} \to \{0,1\}$ is a configuration such that $\nu(\mathbf{R}) < \infty$. Letting $\{x \in \mathbf{R} : \nu(x) = 1\} = \{x_1, \ldots, x_k\}$, we observe that

$$[\max_j x_j \leq t] = [\nu(t, \infty) = 0]. \tag{8.1.2}$$

For a continuous and bounded function $f : \mathbf{R} \to \mathbf{R}$, the corresponding linear statistic is

$$\sum_{j=1}^{k} f(x_j) = \int f(x)\nu(dx). \tag{8.1.3}$$

Further, when $\nu(\mathbf{R}) = k$, the empirical distribution $\mu_k = k^{-1} \sum_{j=1}^{k} \delta_{x_j}$ satisfies $\mu_k(a, b) = k^{-1}\nu(a, b)$.

Given a probability measure on such configurations, the *characteristic function* is

$$\mathbf{E} \exp\left(i \int f d\nu\right). \tag{8.1.4}$$

Examples 8.1.1 (i) Let Y be a $n \times n$ random matrix from the Gaussian unitary ensemble, with eigenvalues $\Lambda = \{\lambda_1 \leq \cdots \leq \lambda_n\} \in \Delta^n$. We take $\Omega = \mathbf{R}$ and consider the counting function

$$\nu_\Lambda([a,b]) = \sharp\{j : a \leq \lambda_j \leq b\} \tag{8.1.5}$$

for the eigenvalues. Under the Gaussian unitary ensemble, double eigenvalues occur with probability zero and may be neglected. So ν_Λ gives a random point field on $\mathbf{R}$ with multiplicity one.

(ii) For $1 \leq n < m$, let Y be a random $n \times m$ matrix with mutually independent $N(0,1)$ entries, and let s_j be the singular numbers, so $s_j = 0$ for $j = n+1, \ldots, m$. Then $\nu_\Lambda([a,b]) = \sharp\{j : a \leq s_j \leq b\}$ has $\nu(\{0\}) \geq m - n$, so the configuration can have multiplicity greater than one. $\square$

Given a compact set A containing $\nu_\Lambda(A)$ points, the number of ways of picking n distinct points randomly from A is the random variable

$$\nu_\Lambda(A)(\nu_\Lambda(A) - 1) \ldots (\nu_\Lambda(A) - n + 1), \tag{8.1.6}$$

Definition (*Correlation measure*). For a given random point field $(\mathrm{Conf}(\Omega), \mathbf{P})$, the nth correlation measure is

$$\rho_n(A^n) = \mathbf{E}\big(\nu_\Lambda(A)(\nu_\Lambda(A) - 1) \ldots (\nu_\Lambda(A) - n + 1)\big) \tag{8.1.7}$$

with the expectation taken over configurations Λ subject to $\mathbf{P}$.

By the axioms of probability theory, $0 \le \mathbf{P}(B) \le 1$ for any event B. The correlation measures satisfy $\rho_n(A^n) \ge 0$, but need not be bounded by 1. The first correlation measure $\rho_1(A) = \mathbf{E}\nu(A)$ is the expected number of particles in A.

The probability generating functions $f_A(z) = \mathbf{E}z^{\nu(A)}$ and the cumulants

$$g_A(z) = \log \mathbf{E}\exp(z\nu(A)) \tag{8.1.8}$$

are important in the theory.

Proposition 8.1.2 *(i) If $X : \Omega \to \{0, 1, \dots\}$ is a random variable such that the probability generating function $f(z) = \mathbf{E}z^X$ defines an analytic function on some neighbourhood of $z = 1$, then*

$$f(z) = 1 + \sum_{k=1}^{\infty} \frac{(z-1)^k}{k!}\mathbf{E}(X(X-1)(X-2)\dots(X-k+1)) \tag{8.1.9}$$

and coefficients of the series determine the moments of X uniquely.

(ii) Furthermore, if $f(z)$ defines an analytic function on the annulus $\{z : 1 - \delta < |z| < 1 + \delta\}$, then $f(z)$ determines the distribution of X.

Proof. (i) We calculate the Taylor expansion of $f(z) = \mathbf{E}z^X$ about $z = 1$, and find $f^{(k)}(1) = \mathbf{E}X(X-1)\dots(X-k+1)$. Hence we can express the moments as

$$\mathbf{E}X^n = \sum_{k=0}^{n} S(n,k)f^{(k)}(1), \tag{8.1.10}$$

where $S(n, k)$ are the Stirling numbers of the second kind from (7.5.10) and thus the derivatives of f determine the moments of X.

(ii) In this case, $\mathbf{E}z^X$ determines the characteristic function $\mathbf{E}e^{itX}$, and hence the distribution. $\qquad\square$

On account of Proposition 8.1.2, the existence and uniqueness of random point fields on $\mathbf{R}$ soon reduces to the classical moment problem. Under the hypotheses of Proposition 1.8.4, the moments also determine the distribution.)

Definition (*Poisson process [101]*). Let p be a locally integrable and positive function and ν a random point field such that:

(i) $\nu(A)$ is a Poisson random variable for each bounded Borel set A;
(ii) $\mathbf{E}\nu(A) = \int_A p(x)dx$;

(iii) $\nu(A_1), \ldots, \nu(A_n)$ are mutually independent whenever $A_1, \ldots, A_n$ are mutually disjoint and bounded Borel subsets of $\mathbf{R}$.

We say that ν is a Poisson process, and p is the *intensity function*.

Example 8.1.3 Let ν be a Poisson process with intensity function $p(x) = \theta$ in the interval $[a, b]$, and let m be Lebesgue measure. Then $\mathbf{E}z^{\nu(A)} = \exp((z-1)\theta m(A \cap [a, b]))$ and hence the nth correlation measure is

$$\mathbf{E}\big(\nu(A)(\nu(A) - 1) \ldots (\nu(A) - n + 1)\big) = \left(\frac{d^n}{dz^n}\right)_{z=1} \mathbf{E}z^{\nu(A)}$$
$$= \theta^n m(A \cap [a, b])^n. \quad (8.1.11)$$

Lemma 8.1.3 *There is a bijection $K \leftrightarrow L$ between bounded linear operators K and L on Hilbert space such that*

$$L = K + KL \quad (8.1.12)$$

where $L \geq 0$ and $0 \leq K \leq I$ with $I - K$ invertible. Furthermore, K is of trace class if and only if the corresponding L is of trace class.

Proof. We have $L = K(I - K)^{-1}$ and $K = L(I + L)^{-1}$. □

The theory now divides into two basic cases:

- the discrete case, where we take $\Omega = \mathbf{Z}$ and consider matrices $K : \ell^2(\mathbf{Z}) \to \ell^2(\mathbf{Z})$;
- the continuous case as in Section 8.2, where we take $\Omega = \mathbf{R}$ and consider integral operators $K : L^2(\mathbf{R}) \to L^2(\mathbf{R})$.

Definition (*Determinantal random point field*). A determinantal random point field on $\mathbf{Z}$ consists of a matrix (or kernel) $K(x, y)$ such that $K(x, y) = \overline{K(y, x)}$, K determines a trace class operator on $\ell^2(\mathbf{Z})$ such that $0 \leq K \leq I$ and

$$\rho_n(x_1, \ldots, x_n) = \det\big[K(x_j, x_k)\big]_{j,k=1}^n. \quad (8.1.13)$$

Note that if $x_j = x_k$ for some $j \neq k$, then $\rho_n(x_1, \ldots, x_n) = 0$. Hence we define a sum over n-tuples of distinct integers by

$$\rho_n(A^n) = \sum_{x_j \in A; j=1,\ldots,n} \rho_n(x_1, \ldots, x_n). \quad (8.1.14)$$

Under the hypotheses of Lemma 8.1.3, there exists $L \geq 0$ that equivalently specifies the random point field determined by K, and some authors [30] regard L as the fundamental object when formulating definitions. In the next result we show how to introduce a probability measure on the configurations given a suitable L, and how K gives the appropriate correlation functions. In general, we follow Soshnikov's approach [148], taking K as the fundamental quantity, for reasons that will emerge in Section 8.4.

• The trace of K roughly corresponds to the number of particles.

The determinantal point field associated with a positive matrix
Let $L \geq 0$ be a finite matrix and for each finite subset X of $\mathbf{Z}$, let

$$p(X) = \frac{\det[L(x_j, x_k)]_{x_j, x_k \in X}}{\det(I + L)}, \tag{8.1.15}$$

with the convention that $p(\emptyset) = 1/\det(I + L)$; naturally $\det(I + L) = \det(\delta_{j,k} + L(j,k))_{j,k \in \mathbf{Z}}$. Then p gives a probability mass function on $\Omega = \{X \subset \mathbf{Z} : \sharp X < \infty\}$ by basic facts about determinants. Next we introduce space of finitely supported configurations with multiplicity one by $\mathrm{Conf}_0(\mathbf{Z}) = \{\nu : \mathbf{Z} \to \{0,1\} : \sum_{j=-\infty}^{\infty} \nu(j) < \infty\}$, and observe that the map $X \to I_X$ gives a bijection $\Omega \leftrightarrow \mathrm{Conf}_0(\mathbf{Z})$. Thus $(\Omega, \mathbf{P})$ determines a random point field on $\mathbf{Z}$. We define $\rho(A) = \mathbf{P}\{Y : A \subseteq Y\}$.

Furthermore, for a subset X of $\mathbf{Z}$ we write I_X for the diagonal matrix that has entry 1 for the jth diagonal entry when $j \in X$, and 0 else. We write $\det(I_X L I_X)$ for the principal minor that has $\sharp X$ rows and columns, indexed by the set X.

Theorem 8.1.4 *Let* $\nu(A) : Y \mapsto \sharp(A \cap Y)$ *where* $Y \in \Omega$ *is random, subject to* $\mathbf{P}$, *and let* $K = L(I + L)^{-1}$.

(i) Then

$$\rho(A) = \det[K(x_j, x_k)]_{x_j, x_k \in A}. \tag{8.1.16}$$

(ii) The probability generating function satisfies

$$\mathbf{E}z^{\nu(A)} = \det(I + (z-1)KI_A). \tag{8.1.17}$$

(iii) The probability that A contains exactly k points is

$$\mathbf{P}\{Y : \sharp(Y \cap A) = k\} = \frac{1}{k!}\left(\frac{d^k}{dz^k}\right)_{z=0} \det(I + (z-1)KI_A). \tag{8.1.18}$$

Proof. (i) Suppose that $\sharp A = n$, and observe that

$$[\nu(A) = n] = \{Y : \sharp(A \cap Y) = n\} = \{Y : A \subseteq Y\}, \tag{8.1.19}$$

so

$$\begin{aligned}
\rho_n(A^n) &= \mathbf{E}\big(\nu(A)(\nu(A)-1)\ldots(\nu(A)-n+1)\big)\\
&= n!\,\mathbf{P}\{Y : A \subseteq Y\}\\
&= n!\,\rho(A).
\end{aligned} \tag{8.1.20}$$

To avoid technical complications, we shall prove the result in the special case in which L has only finitely many nonzero entries. Let Z be a finite subset of $\mathbf{Z}$ which is chosen so large that L may be regarded as a matrix with indices from Z. Let Δ be any diagonal matrix on Z. Then by substituting the formula $L = K(I-K)^{-1}$, we obtain

$$\frac{\det(I+\Delta L)}{\det(I+L)} = \det(I+(\Delta-I)K), \tag{8.1.21}$$

with $I = I_Z$. With $A^c = Z \setminus A$, we choose $\Delta = \lambda I_A + I_{A^c}$, where λ is a real variable, and we expand the series to obtain

$$(\det(I+L))^{-1} \sum_{Y\,:\,Y\subseteq Z} \det(I_Y(\lambda I_A + I_{A^c})LI_Y) = \det(I+(\lambda-1)I_A K); \tag{8.1.22}$$

so the term λ^n arises from those Y such that $A \subseteq Y \subseteq Z$, and the coefficient is

$$(\det(I+L))^{-1} \sum_{Y\,:\,A\subseteq Y\subseteq Z} \det(I_Y LI_Y) = \det(I_A KI_A), \tag{8.1.23}$$

or

$$\rho(A) = \sum_{Y\,:\,A\subseteq Y\subseteq Z} p(Y) = \det(I_A KI_A). \tag{8.1.24}$$

(ii) As in the preceding computation, we have

$$\begin{aligned}
&\det(I+(z-1)KI_A)\\
&= \frac{\det(I+zLI_A + LI_{A^c})}{\det(I+L)}\\
&= (\det(I+Y))^{-1} \sum_Y \det(I_Y L(zI_A + I_{A^c})I_Y)\\
&= (\det(I+L))^{-1} \sum_Y \det(I_Y((zI_A + I_{A^c})I_Y)\det(I_Y LI_Y))\\
&= (\det(I+L))^{-1} \sum_Y z^{\sharp(A\cap Y)} \det(I_Y LI_Y)\\
&= \sum_Y p(Y) z^{\sharp(A\cap Y)}\\
&= \mathbf{E} z^{\nu(A)}.
\end{aligned} \tag{8.1.25}$$

Random Matrices: High Dimensional Phenomena

(iii) This is similar to Proposition 8.1.2 with $X = \nu(A)$. For each k, we need to extract the term in z^k from the power series from (ii), as in

$$\mathbf{P}\{Y : \sharp(Y \cap A) = k\} = \sum_{Y : \sharp(Y \cap A) = k} p(Y)$$

$$= \frac{1}{k!}\left(\frac{d^k}{dz^k}\right)_{z=0} \sum_{Y} z^{\sharp(Y \cap A)} p(Y)$$

$$= \frac{1}{k!}\left(\frac{d^k}{dz^k}\right)_{z=0} \det(I + (z-1)K\mathbf{I}_A). \quad (8.1.26)$$

$\square$

Remark. If $A \subseteq B$, then $\rho(A) \geq \rho(B)$; so ρ is not a measure. The preceding result can be paraphrased as follows.

• The principal minors of L give probabilities, while the principal minors of K give correlation measures.

We list some basic examples which arise in random matrix theory, as we shall discuss in detail in subsequent sections.

Example 8.1.5 (*Vandermonde's determinants*). Let $T_n = [x_j^{t-1}]_{j,t=1,\ldots,n}$ for $x_j \in \mathbf{R}$. Then $K_n = T_n T_n^*$ has

$$\det K_n = \prod_{1 \leq j < k \leq n} (x_j - x_k)^2 \quad (8.1.27)$$

by a familiar calculation involving the Vandermonde determinant. Suppose that $x_1 > x_2 > \cdots > x_n > 0$ are integers such that $\sum_{j=1}^n x_j = N$, and let D_n be the diagonal matrix $D_n = ((x_j!)^{-1})_{j=1}^n$. Then the positive definite matrix $D_n^* K_n D_n$ has

$$\det(D_n^* K_n D_n) = \frac{1}{\prod_{j=1}^n (x_j!)^2} \prod_{1 \leq j < k \leq n} (x_j - x_k)^2, \quad (8.1.28)$$

as in the formula for Plancherel measure on S_N.

(ii) For infinite matrices, we need to adjust the definitions, taking $T = [x_j^t]_{j,t=1,\ldots}$ where $(x_j) \in \ell^2$ has $|x_j| < 1$ for all j. Then T defines a Hilbert–Schmidt matrix such that $K = T^* T$ is of trace class with

$$K = \left[\frac{x_j x_k}{1 - x_j x_k}\right]_{j,k=1,2,\ldots}. \quad (8.1.29)$$

$\square$

We now extend these ideas to multiple sets. Given disjoint subsets E_j

for $j = 1, \ldots, k$ and integers $n_j \geq 0$ such that $\sum_{j=1}^{k} n_j = n$, we consider $\lambda \in \Lambda$ such that $\nu(E_j) \geq n_j$ for all $j = 1, \ldots, k$. Then

$$N_{E_j, n_j; j=1, \ldots, k} = \prod_{j=1}^{k} \frac{\nu_\lambda(E_j)!}{(\nu_\lambda(E_j) - n_j)!} \qquad (8.1.30)$$

gives the number of ways of choosing n_j points from the set of $\nu_\lambda(E_j)$ points from $\lambda = \{x_k\}$ that are in E_j for $j = 1, \ldots, k$. The correlation function $\rho_n : \mathbf{Z}^n \to \mathbf{R}_+$ for $\mathbf{P}$ satisfies

$$\mathbf{E}N_{E_j, n_j; j=1, \ldots, k}$$

$$= \sum_{(x_1, \ldots, x_{n_j}) \in E_1^{n_j}} \sum_{(x_{n_j+1}, \ldots, x_{n_1+n_2}) \in E_2^{n_2}} \cdots \sum_{(x_{n-n_k+1}, \ldots, x_n) \in E_k^{n_k}} \rho_n(x_1, \ldots, x_n).$$

$$(8.1.31)$$

Proposition 8.1.8 *The corresponding generating function satisfies*

$$\mathbf{E}z_1^{\nu(E_1)} \ldots z_k^{\nu(E_k)} = \det\Big(I + \sum_{j=1}^{k}(z_j - 1)\mathbf{I}_{E_j} K\Big). \qquad (8.1.32)$$

Proof. This is an extension of Theorem 8.1.4(ii). See [148]. □

8.2 Determinantal random point fields on the real line

In this section we revisit the notion of a determinantal random point field, this time with the emphasis on point fields on $\mathbf{R}$.

Definition (*Correlation measures*). The symmetric group S_n acts on Ω^n by $\sigma x^{(n)} = (x_{\sigma(1)}, \ldots, x_{\sigma(n)})$ for $x^{(n)} = (x_1, \ldots, x_n)$. For given random point field $(\Omega, \mathbf{P})$, the nth correlation measure is

$$\rho_n(A^n) = \mathbf{E}\big(\nu_\Lambda(A)(\nu_\Lambda(A) - 1) \ldots (\nu_\Lambda(A) - n + 1)\big) \qquad (8.2.1)$$

with the expectation taken over configurations Λ subject to $\mathbf{P}$. For a continuous and bounded function $F : \Omega \to \mathbf{R}$, we have

$$\int_{\Omega^n} F(x^{(n)})\rho_n(dx^{(n)}) = \mathbf{E}\Big(\sum_{\sigma x^{(n)}} F(\sigma x^{(n)}) : \quad \Lambda \quad \text{of multiplicity one}\Big)$$

$$(8.2.2)$$

where the summation is over all the orderings of the distinct points $x^{(n)} = (x_1, \ldots, x_n) \in A^n$.

Definition (*Correlation function*). Suppose that $\Omega = \mathbf{R}$ and that ρ_n is absolutely continuous with respect to Lebesgue measure; then $\rho_n = R_n(x_1, \ldots, x_n)\, dx_1 \ldots dx_n$, where $R_n : \mathbf{R}^n \to \mathbf{R}_+$ is an integrable function which is symmetrical with respect to its variables. Then R_n is the n-point *correlation function*.

Suppose that $\rho_N(\mathbf{R}^N) \leq N! C_1^N$ for all N where C_1 is some constant and let $f : \mathbf{R} \to \mathbf{C}$ be a continuous function such that $|f(x)| \leq 1/(C_1 + \delta)$ where $\delta > 0$. Then from the definition, one can easily check that

$$\mathbf{E} \prod_{j=1}^{\infty} (1 + f(x_j)) = \sum_{N=0}^{\infty} \frac{1}{N!} \int_{\mathbf{R}^N} \prod_{j=1}^{N} f(x_j)\, R_N(x_1, \ldots, x_N)\, dx_1 \ldots dx_N.$$

(8.2.3)

Suppose that we consider a small box B of sides $(dx_1, \ldots, dx_n)$ with one vertex at $(x_1, \ldots, x_n)$. Then the probability that there is exactly one particle in each interval $[x_j, x_j + dx_j]$ for $j = 1, \ldots, n$ is proportional to $R_n(x_1, \ldots x_n)\, dx_1 \ldots dx_n$.

Example. Given a symmetrical probability density function $p_N(x_1, \ldots, x_N)$, we let

$$R_{n,N}(x_1, \ldots, x_n) = \frac{N!}{(N-n)!} \int_{\mathbf{R}^{N-n}} p_N(x_1, \ldots, x_N)\, dx_{n+1} \ldots dx_N$$

(8.2.4)

be the n point correlation function, which describes the ways in which one can choose n points from N without replacement, respecting the order of selections. Evidently $R_{n,N}$ is positive and integrable, but not a probability density function for $1 \leq n \leq N - 1$ due to the initial combinatorial factor.

Definition (*Determinantal random point fields*). Let R_n be the correlation functions for a random point field on $\mathbf{R}$, and suppose that there exists a measurable kernel $K : \mathbf{R} \times \mathbf{R} \to \mathbf{C}$ such that $K(x, y) = \overline{K(y, x)}$ and

$$R_n(x_1, \ldots, x_n) = \det\left[K(x_j, x_k) \right]_{j,k=1}^{n}.$$

(8.2.5)

Then the random point field is determinantal.

The classical theory of determinants of integral operators includes the following fundamental results.

Theorem 8.2.1 (*Mercer*). *Let K be a continuous function on $[a, b] \times [a, b]$ such that $K(x, y) = \overline{K(y, x)}$ and*

$$0 \le \iint_{[a,b]^2} K(x, y) f(x) \overline{f(y)} dx dy \le \int_a^b |f(x)|^2 \, dx \qquad (f \in L^2[a, b]).$$
(8.2.6)

Then K defines a self-adjoint and trace-class linear operator on $L^2[a, b]$ such that $0 \le K \le I$, and there is a uniformly convergent expansion

$$K(x, y) = \sum_{j=1}^{\infty} \mu_j \varphi_j(x) \bar{\varphi}_j(y) \qquad (x, y \in [a, b])$$
(8.2.7)

where $(\varphi_j)_{j=1}^{\infty}$ gives an orthonormal basis in $L^2[a, b]$; furthermore

$$\text{trace } K = \int_a^b K(x, x) \, dx = \sum_{j=1}^{\infty} \mu_j.$$
(8.2.8)

Proof. See [103, p.343]. $\qquad\qquad\qquad\qquad\qquad\qquad\qquad\qquad\qquad$ □

Theorem 8.2.2 (*Fredholm*). *Let K be as in Theorem 8.2.1. Then*

$$\det(I - zK) = \prod_{j=1}^{\infty} (1 - z\mu_j)$$
(8.2.9)

defines an entire function that has all its zeros on the positive real axis at $1/\mu_j$, and has Taylor expansion

$$\det(I - zK)$$
$$= 1 + \sum_{N=1}^{\infty} \frac{(-z)^N}{N!} \int \cdots \int_{J^N} \det[K(x_j, x_\ell)]_{j,\ell=1,\dots,N} dx_1 \dots dx_N.$$
(8.2.10)

Proof. By Theorem 8.2.1, K is a trace class operator. We observe that there exists M such that $|K(x, y)| \le M$ for all $x, y \in [a, b]$ and hence that

$$0 \le \det[K(x_j, x_k)]_{j,k=1,\dots,N} \le M^N N^{N/2}$$
(8.2.11)

by Hadamard's inequality. Hence the series on the right-hand side converges absolutely and uniformly on compact sets. See [103] for the remaining details. $\qquad\qquad\qquad\qquad\qquad\qquad\qquad\qquad\qquad\qquad$ □

For a determinantal random point field, the Fredholm determinant has a simple probabilistic interpretation.

Lemma 8.2.3 *Let K be the continuous kernel of a determinantal random point field on $\mathbf{R}$, and let $K_J(x,y) = \mathbf{I}_J(x)K(x,y)\mathbf{I}_J(y)$.*

(i) Then the probability generating function satisfies, for each bounded interval J,

$$\mathbf{E}z^{\nu(J)} = \det(I + (z-1)K_J). \tag{8.2.12}$$

(ii) Hence

$$\frac{(-1)^k}{k!}\left(\frac{d^k}{dz^k}\right)_{z=1}\det(I - zK_J) \tag{8.2.13}$$

equals the probability that J contains exactly k particles.

(iii) In particular, $\det(I - K_{[x,\infty)})$ equals the probability that all the particles are in $(-\infty, x]$.

Proof. See [116], where the proof involves manipulation of generating functions and the inclusion-exclusion principle. The corresponding result in the discrete case was Theorem 8.1.4(ii). □

Exercise 8.2.4 Let K be the kernel of a determinantal random point field on $\mathbf{R}$, and suppose that K_J is as in Theorem 8.2.2. Show that the property $K_{A\cup B} = K_A + K_B$ where $K_A K_B = 0$ for disjoint A and B translates into

$$\mathbf{E}z^{\nu(A\cup B)} = \mathbf{E}z^{\nu(A)}\mathbf{E}z^{\nu(B)} \tag{8.2.14}$$

for the corresponding random point field.

In several important cases, K is given by a finite rank orthogonal projection, as described in the following result.

Proposition 8.2.5 *Let K be the kernel of a continuous function on $J \times J$ such that $K(x,y) = \overline{K(y,x)}$ and*

$$K(x,y) = \int_J K(x,z)K(z,y)\,dz \qquad (x,y \in J). \tag{8.2.15}$$

Then $R_n(x_1,\ldots,x_n) = \det[K(x_j,x_k)]_{j,k=1,\ldots,n}$ gives the n-point correlation function of a N-particle system for some $N < \infty$, where the joint probability density function is

$$p_N(x_1,\ldots,x_N) = (N!)^{-1}\det[K(x_j,x_k)]_{j,k=1,\ldots,N}. \tag{8.2.16}$$

Proof. The operator K with kernel $K(x,y)$ is self-adjoint and satisfies $K^2 = K$, hence is an orthogonal projection. Since $K(x,y)$ is continuous,

K is of trace class by Theorem 8.2.1, hence of finite rank N, where N is equal to

$$\mathrm{rank}(K) = \mathrm{trace}(K) = \int_J K(x,x)\,dx. \qquad (8.2.17)$$

Since $K = K^*K$ is positive semi-definite, $\det[K(x_j, x_k)]_{j,k=1,\dots N}$ is non-negative and integrable. Furthermore, the Fredholm determinant satisfies

$$(1-z)^N = \det(1 - zK)$$
$$= 1 + \sum_{k=1}^{N} \frac{(-z)^k}{k!} \int \cdots \int_{J^k} \det[K(x_j, x_\ell)]_{j,\ell=1,\dots,k}\,dx_1 \dots dx_k;$$

$$(8.2.18)$$

so by equating the coefficients of z^N we deduce that

$$1 = \frac{1}{N!} \int_{J^N} \det[K(x_j, x_k)]_{j,k=1,\dots,N}\,dx_1 \dots dx_N, \qquad (8.2.19)$$

and hence that $p_N(x_1, \dots x_N)$ is a probability density function.

Integrating out the last variable, we shall prove next the identity

$$\det[K(x_j, x_k)]_{j,k=1,\dots,n-1} = (N-n+1)\int_J \det[K(x_j, x_k)]_{j,k=1,\dots,n}\,dx_n.$$

$$(8.2.20)$$

By expanding the determinant on the right-hand side, we obtain

$$\sum_{\sigma \in S_n} \mathrm{sign}(\sigma) K(x_1, x_{\sigma(1)}) \dots K(x_n, x_{\sigma(n)}) \qquad (8.2.21)$$

where for each $\sigma \in S_n$ there exists a unique j such that $\sigma(j) = n$. If $\sigma(n) = n$, then we say that $\sigma \in G_n$; otherwise, σ belongs to the coset $G_n \cdot (jn)$ for the transposition (jn). Since G_n is isomorphic to S_{n-1}, the terms in G_n contribute

$$\sum_{\tau \in S_{n-1}} \mathrm{sign}(\tau) K(x_1, x_{\tau(1)}) \dots K(x_{n-1}, x_{\tau(n-1)}) \int K(x_n, x_n)\,dx_n$$
$$= N \det[K(x_j, x_k)]_{j,k=1,\dots,n-1} \qquad (8.2.22)$$

to the integral of the sum (8.2.21); whereas the terms in the coset $G_n \cdot$

(jn) typically contributes

$$-\sum_{\tau \in S_{n-1}} \text{sign}(\tau) K(x_1, x_{\tau(1)}) \ldots K(x_{n-1}, x_{\tau(n-1)})$$

$$\times \int K(x_j, x_n) K(x_n, x_{\tau(j)}) dx_n$$

$$= -\sum_{\tau \in S_{n-1}} \text{sign}(\tau) K(x_1, x_{\tau(1)}) \ldots K(x_{n-1}, x_{\tau(n-1)})$$

$$= -\det[K(x_\ell, x_k)]_{\ell, k=1, \ldots, n-1}, \tag{8.2.23}$$

since $\text{sign}(\tau(jn)) = -\text{sign}(\tau)$.

Using (8.2.21) repeatedly, we can integrate out the variables of largest index and show that

$$\det[K(x_j, x_k)]_{j,k=1,\ldots,n}$$

$$= (N-n)! \int \cdots \int_{J^{N-n}} \det[K(x_j, x_k)]_{j,k=1,\ldots,N} dx_{n+1} \ldots dx_N,$$

$$\tag{8.2.24}$$

hence we can recover the n-point correlation function as

$$\frac{N!}{(N-n)!} \int_{J^{N-n}} p_n(x_1, \ldots, x_N) dx_{n+1} \ldots dx_N = \det[K(x_j, x_k)]_{j,k=1,\ldots,n}.$$

$$\tag{8.2.25}$$

$\square$

Example 8.2.6 Let

$$Q_N(e^{i\theta}) = \frac{\sin(2N+1)\theta/2}{\sin\theta/2}. \tag{8.2.26}$$

Then Q_N gives the projection onto $\text{span}\{e^{ij\theta} : j = -N, \ldots, N\}$ in $L^2([0, 2\pi]; d\theta/2\pi)$. In example 2.7.2 we saw how $\det[Q_n(e^{i(\theta_j - \theta_k)})]$ is involved in the distribution of eigenvalues of the unitary circular ensemble.

$\square$

We state without proof the following general existence theorem for determinantal random point fields from [148].

Theorem 8.2.7 *Suppose that K is a continuous kernel such that*

(i) $K(x, y) = \overline{K(y, x)}$;

(ii) *the integral operator with kernel $K(x, y)$ on $L^2(\mathbf{R})$ satisfies $0 \leq K \leq I$;*

(iii) *the integral operator with kernel $\mathbf{I}_{[a,b]}(x) K(x, y) \mathbf{I}_{[a,b]}(y)$ on $L^2(\mathbf{R})$ is trace class for all finite $[a, b]$.*

Then there exists a determinantal random point field such that the correlation function satisfies

$$R_n(x_1, \ldots, x_n) = \det\left[K(x_j, x_k)\right]_{j,k=1}^{n}. \tag{8.2.27}$$

Furthermore, the correlation functions uniquely determine the random point field.

Proof. See [148]. □

Theorem 8.2.8 (*Bochner*). *The function* $\varphi : \mathbf{R} \to \mathbf{C}$ *is the characteristic function of a probability measure* μ *on* $\mathbf{R}$ *if and only if*

(1) $[\varphi(x_j - x_k)]_{j,k=1,\ldots,n}$ *is positive definite for all* $x_1, \ldots, x_n \in \mathbf{R}$;
(2) φ *is continuous at* $x = 0$;
(3) $\varphi(0) = 1$.

(ii) Suppose further that μ *is absolutely continuous with density* p, *where* $p(t) \leq 1/(2\pi)$ *for all* $t \in \mathbf{R}$. *Then* $K(x, y) = \varphi(x - y)$ *is the kernel of a determinantal random point field on* $[a, b]$ *for all finite* a, b.

Proof. (i) See [88, p. 91].

(ii) There exists a probability measure μ on $\mathbf{R}$ such that $\varphi(x) = \int_{-\infty}^{\infty} e^{ixt} \mu(dt)$, so we have

$$\int_{-\infty}^{\infty} \int_{-\infty}^{\infty} K(x, y) f(x) \, f(y) \, dx dy$$

$$= \int_{-\infty}^{\infty} \int_{-\infty}^{\infty} \int_{-\infty}^{\infty} e^{it(x-y)} p(t) dt f(x) f(y) \, dx dy$$

$$= \int_{-\infty}^{\infty} |\hat{f}(t)|^2 p(t) \, dt$$

$$\leq \frac{1}{2\pi} \int_{-\infty}^{\infty} |\hat{f}(t)|^2 dt; \tag{8.2.28}$$

hence by Plancherel's formula, the operators satisfy $0 \leq K \leq I$ on $L^2(\mathbf{R})$. When we cut down $K(x, y)$ to some bounded interval $[a, b]$, then we obtain a trace class operator on $L^2[a, b]$ such that $0 \leq K \leq I$ and by Theorem 8.2.1,

$$\mathrm{trace}(K) = b - a. \tag{8.2.29}$$

□

Exercise 8.2.9 (See [147]). Let K be the correlation kernel of a determinantal random point field, and for a bounded and continuous real function f, let $F(\lambda) = \sum_{j=1}^{n} f(\lambda_j)$, where the $\lambda = (\lambda_j)_{j=1}^{n}$ are random points in the field.

(i) Show that

$$\mathbf{E}F = \int K(x,x)f(x)\,dx,$$

and that

$$\mathrm{var}(F) = \int K(x,x)f(x)^2\,dx - \iint f(x)f(y)|K(x,y)|^2\,dxdy. \quad (8.2.30)$$

(ii) Suppose further that K is as in Theorem 8.2.8. Show that

$$\mathrm{var}(F) = \frac{1}{2\pi}\int_{-\infty}^{\infty}(1 - \widehat{(\varphi)^2}(t))|\hat{f}(t)|^2\,dt. \quad (8.2.31)$$

Using Theorem 8.2.2, one can express $\det(I - z\mathbf{I}_J K)$ in terms of a series of integrals involving $\det K_n$. Tracy and Widom [156] observed that there is a more compact way of expressing the determinant.

Proposition 8.2.10 *Suppose that J is a bounded interval and let K be a continuous kernel as in Proposition 8.2.5, while p_N is the corresponding joint probability density function on J^N. Let M_f be the operator on $L^2(J)$ of multiplication by $f \in C(J;\mathbf{C})$.*

(i) Then

$$\int_{\mathbf{R}^N} p_N(x_1,\ldots,x_N)\prod_{j=1}^{N}(1 + f(x_j))dx_1\ldots dx_N = \det(I + KM_f),$$

$$(8.2.32)$$

(ii) and the linear statistic associated with f satisfies

$$\int_{J^N} \exp\left(\sum_{j=1}^{N} f(x_j)\right)p_N(x_1,\ldots,x_N)dx_1\ldots dx_N = \det\left(I + K_N M_{e^f - 1}\right).$$

$$(8.2.33)$$

Proof. (i) An equivalent version of the formula is $\mathbf{E}\det(I + f(X)) = \det(I+K_N M_f)$ where the expectation is over the $X \in M_N^h(\mathbf{C})$. To justify this formula in a heuristic argument, we take $f(x) = \sum_{j=1}^{N} z_j\delta_{y_j}(x)$. By Mercer's theorem, K has an expansion $K(x,y) = \sum_{j=1}^{\infty} \mu_j\phi_j(x)\bar{\phi}_j(y)$ where (μ_j) is a positive and summable sequence and where $(\phi_j)_{j=1}^{\infty}$ gives an orthonormal basis for $L^2(J)$. Then

$$\det(I + KM_f) = \det\left[\langle (I + KM_f)\phi_j, \phi_k\rangle_{L^2(J)}\right]$$

$$= \det\left[\delta_{jk} + \sum_{\ell=1}^{N} \mu_k z_\ell\phi_j(y_\ell)\bar{\phi}_k(y_\ell)\right]. \quad (8.2.34)$$

We introduce the matrices $A = [z_\ell \phi_j(y_\ell)]$ and $B = [\mu_k \bar{\phi}_k(y_\ell)]$, and write the previous as

$$
\begin{aligned}
\det(I + KM_f) &= \det(I + AB) \\
&= \det(I + BA) \\
&= \det\left[\delta_{k\ell} + \sum_{j=1}^{\infty} \mu_j z_k \bar{\phi}_j(y_\ell)\phi_j(y_k)\right] \\
&= \det\left[\delta_{k\ell} + z_k K(y_\ell, y_k)\right]_{k,\ell=1,\ldots,N}.
\end{aligned}
$$

$$(8.2.35)$$

On the left-hand side of (8.2.32) we have

$$
\int_{J^N} p_N(x_1, \ldots, x_N) \prod_{j=1}^{N}\left(1 + \sum_{k=1}^{N} z_k \delta_{y_k}(x_j)\right) dx_1 \ldots dx_N \qquad (8.2.36)
$$

where we can express the product as a sum of 2^N terms, which are indexed by the subsets S of $\{1, \ldots, N\}$. Let $S^c = \{1, \ldots, N\} \setminus S$, and let $\sharp S$ be the cardinality of S. When we integrate with respect to all the variables x_k, the Dirac point masses contribute only when $x_k = y_k$ for $k \in S$, and the variables x_k for $k \in S^c$ are integrated out; so the expansion is

$$
\sum_{S: S \subseteq \{1,\ldots,N\}} \int_{J^{N-\sharp S}} p_N(y_j; x_k) \prod_{k \in S^c} dx_k \prod_{k \in S} z_k, \qquad (8.2.37)
$$

where in particular the coefficients of $z_1 z_2 \ldots z_N$ is

$$
N! p_N(y_1, \ldots, y_N) = \det[K(y_j, y_k)]_{j,k=1,\ldots,N} \qquad (8.2.38)
$$

as in Proposition 8.2.5. Now one can repeat the analysis of Proposition 8.2.5 to identify the remaining coefficients of $\prod_{k \in S} z_k$ for various S and establish equality of the coefficients of $\prod_{j \in S} z_j$ between (8.2.36) and (8.2.38), and thus deduce (8.2.32).

(ii) An equivalent version of the formula is $\mathbf{E} \exp(\operatorname{trace} f(X)) = \det(I + K_N M_{e^f - 1})$. We take $e^f - 1$ in place of f in (i) and simplify the left-hand side. $\square$

Example 8.2.11 Let H be a reproducing kernel Hilbert space on a domain Ω, as in Section 4.2, and suppose that $K(z, w)$ is the kernel such that $k_w(z) = K(z, w)$ satisfies $f(w) = \langle f, k_w \rangle_H$. Then for $w_1, \ldots, w_n \in \Omega$, the matrix $[K(w_j, w_\ell)]_{j,\ell=1,\ldots,n}$ is positive definite since $K(w_j, w_\ell) = \langle k_{w_j}, k_{w_\ell} \rangle$ and Proposition 2.1.2 applies.

Example 8.2.12 Let $f_j : \mathbf{R} \to \mathbf{C}$ be continuous and let $F(x^{(n)}) = [f_j(x_k)]_{j,k=1,\ldots,n}$ for $x^{(n)} = (x_1, \ldots, x_n) \in \mathbf{R}^n$. Then $K(x^{(n)}) = F(x^{(n)}) F(x^{(n)})^* \geq 0$ and

$$\det[K(x_j, x_k)]_{j,k=1,\ldots,n} = \det\left[\sum_{\ell=1}^{n} f_\ell(x_j)\bar{f}_\ell(x_k)\right]_{j,k=1,\ldots,n}. \qquad (8.2.39)$$

As we shall see, this case often arises in random matrix theory.

8.3 Determinantal random point fields and orthogonal polynomials

- The generalized unitary ensemble gives a determinantal random point field.

By reformulating results of Gaudin and Mehta [116, 153], we show that the generalized unitary ensemble is given by a determinantal random point field. The best way to see this is to consider some classical results concerning orthogonal polynomials.

Example 8.3.1 (*Orthogonal polynomials*). Let μ be a probability measure with infinite support on $\mathbf{R}$ such that $\int e^{\varepsilon|x|}\mu(dx)$ is finite for some $\varepsilon > 0$. Then the natural inner product on $L^2(\mu)$ is

$$\langle f, g \rangle = \int f(x)\bar{g}(x)\,\mu(dx) \qquad (f, g \in L^2(\mu)). \qquad (8.3.1)$$

By the Gram–Schmidt process, there exists a unique sequence of orthonormal polynomials p_j such that $p_j(x)$ has degree j, and p_j has leading coefficient $k_j > 0$; further, the p_j give a complete orthonormal basis for $L^2(\mu)$ as in Proposition 1.8.4. Heine showed that with

$$D_n = \det\left[\int_{-\infty}^{\infty} x^{j+k}\mu(dx)\right]_{j,k=0,\ldots,n} \qquad (8.3.2)$$

the polynomials satisfy $p_0 = 1$ and

$$p_n(x) = \frac{1}{(D_{n-1}D_n)^{1/2}} \det\left[\int_{-\infty}^{\infty} (x-y)y^{j+k}\mu(dy)\right]_{j,k=0,\ldots,n-1} \qquad (8.3.3)$$

and the leading coefficient of $p_n(x)$ is $k_n = (D_{n-1}/D_n)^{1/2}$; see [153, (2.2.9)].

Lemma 8.3.2 *There exist constants A_n, B_n and C_n such that the p_n satisfy the three-term recurrence relation*

$$p_n(x) = (A_n x + B_n)p_{n-1}(x) - C_n p_{n-2}(x) \qquad (n = 2, 3, \ldots). \qquad (8.3.4)$$

Moreover, the constants satisfy

$$A_n = \frac{k_n}{k_{n-1}}, \quad C_n = \frac{A_n}{A_{n-1}}. \tag{8.3.5}$$

Proof. See [153, p. 42]. □

Proposition 8.3.3 (*Christoffel–Darboux formula*). *The kernel K_n of the orthogonal projection of $L^2(\mu)$ onto*

$$H_n = \left\{ \sum_{j=0}^{n-1} a_j x^j; a_j \in \mathbf{C} \right\} \tag{8.3.6}$$

is given by

$$\sum_{j=0}^{n-1} p_j(x)p_j(y) = \left(\frac{k_{n-1}}{k_n}\right) \frac{p_n(x)p_{n-1}(y) - p_{n-1}(x)p_n(y)}{x-y}. \tag{8.3.7}$$

Further, H_n is a reproducing kernel Hilbert space and K_n gives the reproducing kernel.

Proof. This follows from the three-term recurrence relation by an induction argument; see [153, page 42]. □

Let w be a weight on an interval I such that $\int (\cosh tx)w(x)dx < \infty$ for some $t > 0$. Then by the Gram–Schmidt process, there exists a sequence of orthonormal polynomials $(p_j)_{j=0}^{\infty}$ in $L^2(I; w)$ where $p_j(x)$ has leading term $k_j x^j$.

Theorem 8.3.4 *The function*

$$(n!)^{-1} \prod_{j=0}^{n-1} k_j^2 \prod_{1 \le j < k \le n} (x_j - x_k)^2 \prod_{j=1}^{n} w(x_j) \tag{8.3.8}$$

is a probability density on I^n.

Proof. We introduce $\phi_j(x) = p_j(x)w(x)^{1/2}$ and let $K_n(x,y) = \sum_{j=0}^{n-1} \phi_j(x)\phi_j(y)$. Then by orthonormality, we have

$$\int_I K_n(x,y)K_n(y,z)dy = K_n(x,z) \tag{8.3.9}$$

so by Proposition 8.2.5

$$\int_{I^n} \det\left[K_n(x_j, x_k)\right]_{j,k=1}^{n} dx_1 \dots dx_n = n!. \tag{8.3.10}$$

Now introduce the matrix

$$A = A(x_1, \ldots, x_n) = \left[x_j^{m-1} w(x_j)^{1/2}\right]_{j,m=1}^n \qquad (8.3.11)$$

so that by elementary row operations

$$k_0 \ldots k_{n-1} \det A = \det[\phi_{m-1}(x_j)]_{j,m=1}^n \qquad (8.3.12)$$

and

$$k_0^2 \ldots k_{n-1}^2 \det A^t A(x_1, \ldots, x_n) = \det\left[\sum_{m=1}^n \phi_{m-1}(x_j)\phi_{m-1}(x_\ell)\right]_{j,\ell=1,\ldots,n}$$

$$= \det[K_n(x_j, x_\ell)]_{j,\ell=1}^n. \qquad (8.3.13)$$

We deduce from Lemma 2.2.1 that

$$(n!)^{-1} \det[K_n(x_j, x_\ell)]_{j,\ell=1}^n = (n!)^{-1} \prod_{j=0}^{n-1} k_j^2 \prod_{1\leq j<k\leq n} (x_j - x_k)^2 \prod_{j=1}^n w(x_j)$$

$$(8.3.14)$$

is a probability density function.   □

There are two other significant cases in random matrix theory in which one can calculate the determinants in such an expansion: the expansion of $\log \det(I - zK\mathbf{I}_J)$ in Lemma 10.1.2, and Gaudin's Theorem, which we now discuss.

We recall that for any positive measure μ on $\mathbf{R}$ such that $\int \cosh(st)\mu(dt) < \infty$ for some $s > 0$, the polynomials form a dense linear subspace of $L^2(\mu)$. The Gram–Schmidt process gives a unique system of orthonormal polynomials of degree $n = 0, 1, 2, \ldots$ with positive leading coefficients. We can carry out this construction with the measures $\mu_n(dt) = e^{-nv(t)} dt$ when $v(t) > s|t|$ for all sufficiently large $|t|$.

Next we introduce the orthonormal polynomials $(p_{j,n})_{j=0}^\infty$ of degree j for the weight $e^{-nv(x)}$, so that $p_{j,n}(x)$ has leading term $k_{j,n}x^j$. In Section 10.6 we shall consider the growth of $k_{j,n}$ as $n \to \infty$.

Corollary 8.3.5 (*Gaudin*). *The joint eigenvalue distribution of a $n \times n$ generalized unitary ensemble is given by a determinantal point field with kernel*

$$K_n(x,y) = Z_n^{-1} \exp\left(-\frac{(nv(x) + nv(y))}{2}\right) \sum_{j=0}^{n-1} p_{j,n}(x)p_{j,n}(y). \quad (8.3.15)$$

Proof. See also [116]. The joint probability density function for eigenvalues from the $n \times n$ generalized unitary ensemble has the form

$$Z_n^{-1} \exp\left(-n \sum_{j=1}^{n} v(\lambda_j)\right) \prod_{1 \le j < \ell \le n} (\lambda_j - \lambda_\ell)^2 \qquad (8.3.16)$$

by Theorem 2.5.3. Note that we scale the potential to $nv(x)$ where n is the number of eigenvalues of the matrix. By Theorem 8.3.4, K_n gives a determinantal random point field with probability density function

$$(n!)^{-1} \prod_{j=0}^{n-1} k_{j,n}^2 \prod_{1 \le j < k \le n} (\lambda_j - \lambda_k)^2 \prod_{j=1}^{n} e^{-nv(\lambda_j)} \qquad (8.3.17)$$

on $\mathbf{R}^n$, and so by symmetry of the variables we do not need the factor $1/n!$ when we integrate over $\{\lambda_1 \le \lambda_2 \le \cdots \le \lambda_n\}$. $\square$

One can use Proposition 8.3.3 to obtain an equivalent expression for K_n.

Corollary 8.3.5 is the basis for the orthogonal polynomial technique in random matrix theory, as developed in [116, 56]. Note that it works most effectively in the context of unitary ensembles, since the proof of Theorem 8.3.4 involves squaring the determinant; less familiar variants are known for orthogonal and symplectic ensembles. Many results concerning generalized unitary ensembles have natural generalizations to determinantal random point fields since Theorem 8.3.5 captures sufficient information concerning the joint eigenvalue distribution, and the origin of the problem in random matrices is inconsequential.

Exercise 8.3.6 In the notation of Corollary 8.3.5, let

$$k_w(z) = \frac{p_{n,n}(z)\overline{p_{n-1,n}(w)} - p_{n-1,n}(z)\overline{p_{n,n}(w)}}{(z - \bar{w})}. \qquad (8.3.18)$$

Show that for an entire function f, there exists C such that $|f(z)|^2 \le C k_z(z)$ for all $z \in \mathbf{C}$, if and only if f is a polynomial of degree less than or equal to $n - 1$.

Exercise 8.3.7 Let $\mathbf{P}$ be a random point field on Ω, and let $f = \sum_{j=1}^{n} t_j \mathbf{I}_{A_j}$ be a step function with disjoint Borel sets A_j. Then the stochastic integral of f with respect to the configurations ν_Λ is the random variable

$$\int f d\nu_\Lambda = \sum_{j=1}^{n} t_j \nu_\Lambda(A_j). \qquad (8.3.19)$$

(i) Use Kac's theorem to show that the $\nu(A_j)$ are mutually independent under $\mathbf{P}$ if and only if

$$\mathbf{E}\exp\left(i\int f\,d\nu\right) = \prod_{j=1}^{n}\mathbf{E}\exp\bigl(it_j\nu(A_j)\bigr) \qquad (t_j \in \mathbf{R}). \qquad (8.3.20)$$

(ii) The generating function of the random point field is

$$\Phi(f) = \mathbf{E}\exp\int f\,d\nu \qquad (8.3.21)$$

for any continuous and compactly supported real function. Let ν be a Poisson process with intensity function p on the real line. By considering first the case in which f is a step function, show that

$$\Phi(f) = \exp\int (e^{f(x)} - 1)p(x)dx. \qquad (8.3.22)$$

8.4 De Branges's spaces

- De Branges [38] introduced axioms for some reproducing kernel Hilbert spaces that satisfy a type of spectral theorem.
- In random matrix theory, some important kernels are reproducing kernels in Hilbert spaces that satisfy de Branges's axioms.

Examples 8.4.1 (*Universal ensembles*). The eigenvalues distributions from generalized unitary ensembles are described in terms of the following fundamental examples. In Chapter 9, we shall discuss how they arise from fundamental examples in the theory of unitary ensembles; now we introduce them in a common style.

(i) (*Spectral bulk*) [116] The sine kernel

$$D_b(x,y) = \frac{\sin\pi b(x-y)}{\pi b(x-y)} = \int_{-1}^{1} e^{i\pi btx}e^{-i\pi bty}\frac{dt}{2} \qquad (8.4.1)$$

defines a trace-class operator on $L^2([-1,1],dt/2)$ such that $0 \le D_b \le I$.

(ii) (*Soft edge*) [157] The Airy kernel (see (9.2.18))

$$W(x,y) = \frac{\mathrm{Ai}(x)\mathrm{Ai}'(y) - \mathrm{Ai}'(x)\mathrm{Ai}(y)}{x-y} \qquad (8.4.2)$$

defines a trace-class operator on $L^2(0,\infty)$ such that $0 \le W \le I$.

(iii) (*Hard edge*) [158] The Bessel kernel

$$F(x,y) = \frac{J_\alpha\sqrt{x})\sqrt{y}J_\alpha'(\sqrt{y}) - \sqrt{x}J_\alpha'(\sqrt{x})J_\alpha(\sqrt{y})}{2(x-y)} \qquad (8.4.3)$$

defines a trace-class operator on $L^2[0,1]$ such that $0 \le F \le I$.

Definition (De Branges's Axioms [38]).

Let H be a reproducing kernel Hilbert space such that the elements of H are entire functions $f : \mathbf{C} \to \mathbf{C}$, and such that H obeys de Branges's axioms, as follows.

(H1) If $f \in H$ satisfies $f(w) = 0$ for some $w \in \mathbf{C} \setminus \mathbf{R}$, then $(z - \bar{w})f(z)/(z - w)$ also belongs to H and has the same norm as f.

(H2) $f \mapsto f(w)$ is a bounded linear functional for all $w \in \mathbf{C} \setminus \mathbf{R}$.

(H3) $f^*(z) = \overline{f(\bar{z})}$ belongs to H for all $f \in H$, and $\|f^*\|_H = \|f\|_H$.

Condition (H2) asserts that H is a reproducing kernel Hilbert space on $\mathbf{C} \setminus \mathbf{R}$, while (H1) asserts that one can divide out zeros of functions in H. The condition (H3) is related to the Schwarz reflection principle, as applied to the real axis.

Definition (*De Branges's spaces [38]*). Let E be an entire function such that $|E(x - iy)| < |E(x + iy)|$ for all $x \in \mathbf{R}$ and $y > 0$ and introduce entire functions A and B, which are real on the real axis, such that $E(z) = A(z) - iB(z)$. In particular, $E(z)$ has no zeros in the upper half plane. Then there is a reproducing kernel Hilbert space H_E with reproducing kernel

$$K(z, w) = \frac{\overline{A(w)}B(z) - \overline{B(w)}A(z)}{\pi(z - \bar{w})} \qquad (8.4.4)$$

and with norm given by

$$\|f\|_{H_E}^2 = \int_{-\infty}^{\infty} |f(x)/E(x)|^2 \, dx, \qquad (8.4.5)$$

such that an entire function f belongs to H_E if and only if $\|f\|_{H_E}^2 < \infty$ and $|f(z)|^2 \leq \|f\|_{H_E}^2 K(z, z)$ for all $z \in \mathbf{C}$. De Branges showed that for any non-zero Hilbert space H that satisfies (H1), (H2) and (H3), there exists such an E so that H is isometric to H_E; see [38]. We do not describe de Branges's result in detail here, since we are mainly concerned with particular examples of kernels that arise in random matrix theory, rather than the general theory of entire functions.

In the table below, we list how Examples 8.4.1 and 8.3.6 fit into de Branges's framework and mention the names commonly used in random matrix theory. The following reproducing kernels K operate on L^2 spaces which contain closed linear subspaces H_E such that $K(z, w)$ is the reproducing kernel on H_E. The final column gives the association with random matrix theory.

Name	$B(z)$	$A(w)$	L^2	RMT
Sine	$\sin \pi b z$	$\cos \pi b w$	$L^2(-1,1)$	spectral bulk
Airy	$\mathrm{Ai}(z)$	$\mathrm{Ai}'(w)$	$L^2(0,\infty)$	soft spectral edge
Bessel	$J_\nu(\sqrt{z})$	$\sqrt{w}J_\nu'(\sqrt{w})$	$L^2(0,1)$	hard spectral edge
Christoffel –Darboux	$k_{n-1}p_n(z)/k_n$	$p_{n-1}(w)$	$L^2(\mu)$	orthogonal polynomials

Exercise 8.4.2 Let $E(z) = e^{-i\pi b z}$ for $b > 0$. Find $K(z,w)$.

Exercise 8.4.3 See Section 12.1 for the case of the Hermite polynomials in Corollary 8.3.5.

Definition (*Canonical system*) [133]. Let

$$J = \begin{bmatrix} 0 & -1 \\ 1 & 0 \end{bmatrix} \tag{8.4.6}$$

and let $\Omega(x)$ be a 2×2 real symmetric matrix for $x \in [0,b]$ such that $\Omega(x) \geq 0$ and the entries of Ω are in $C([0,b]; \mathbf{R})$. Then a canonical system is the differential equation

$$J\frac{d}{dx}u(x;z) = z\Omega(x)u(x;z). \tag{8.4.7}$$

Suppose further that

$$u(x;z) = \begin{bmatrix} u_1(x;z) \\ u_2(x;z) \end{bmatrix}, \qquad u(0;z) = \begin{bmatrix} 1 \\ 0 \end{bmatrix}, \tag{8.4.8}$$

and let $E_b(z) = u_1(b;z) + iu_2(b;z)$. Then under various technical conditions $E_b(z)$ gives an entire function as in the definition of a de Branges space. We can use canonical systems to construct kernels that satisfy Theorem 8.2.7.

Proposition 8.4.4 *The continuous kernel*

$$K_b(z,\zeta) = \frac{u(b;\zeta)^* J u(b;z)}{z - \zeta} \tag{8.4.9}$$

gives a positive operator on $L^2(0,b)$.

Proof. First suppose that $\Im z, \Im \zeta > 0$. From the differential equation, we have

$$(z - \bar\zeta)\int_0^b u(x;\zeta)^*\Omega(x)u(x;z)\,dx = \int_a^b \frac{d}{dx}\big(u(x;\zeta)^* J u(x;z)\big)\,dx$$
$$= \big[u(x;\zeta)^* J u(x;z)\big]_0^b \tag{8.4.10}$$

where $u(0;\zeta)^*Ju(0;z) = 0$ by the initial condition. Hence $K_b(z,\zeta)$ is given by

$$\frac{u(b;\zeta)^*Ju(b;z)}{z-\bar{\zeta}} = \int_0^b u(x;\zeta)^*\Omega(x)u(x;z)\,dx \qquad (8.4.11)$$

where $z \mapsto u(x;z)$ is analytic and such that $u(x;\bar{z}) = \overline{u(x;z)}$. We can therefore extend both sides of this identity by continuity to all $z, \zeta \in \mathbf{C}$, and observe that $K_b(z,\zeta)$ is a continuous function. Evidently the right-hand side of (8.4.11) gives a positive kernel. $\qquad\square$

A particularly important source of examples arises from Schrödinger's equation.

Let $V \in C([0,b];\mathbf{R})$ be the potential, let y_1 and y_2 be solutions of the equation $-y'' + Vy = 0$ such that $y_1(0) = y_2'(0) = 1$, and $y_1'(0) = y_2(0) = 0$; then we let

$$T(x) = \begin{bmatrix} y_1(x) & y_2(x) \\ y_1'(x) & y_2'(x) \end{bmatrix}, \qquad (8.4.12)$$

so $T(0) = I_2$, and

$$\Omega(x) = \begin{bmatrix} y_1(x)^2 & y_1(x)y_2(x) \\ y_1(x)y_2(x) & y_2(x)^2 \end{bmatrix}. \qquad (8.4.13)$$

Let $f(x;z)$ be a solution of Schrödinger's equation $-f'' + Vf = zf$, and let

$$u(x;z) = T(x)^{-1}\begin{bmatrix} f(x;z) \\ f'(x;z) \end{bmatrix}. \qquad (8.4.14)$$

Proposition 8.4.5 *The vector function $u(x;z)$ satisfies the canonical system*

$$J\frac{d}{dx}u(x;z) = z\Omega(x)u(x;z). \qquad (8.4.15)$$

Proof. By Proposition 2.1.2, $\Omega(x) \geq 0$. We have

$$\frac{d}{dx}u(x;z) = -T(x)^{-1}\Big(\frac{d}{dx}T(x)\Big)T(x)^{-1}\begin{bmatrix} f(x;z) \\ f'(x;z) \end{bmatrix}$$

$$+ T(x)^{-1}\frac{d}{dx}\begin{bmatrix} f(x;z) \\ f'(x;z) \end{bmatrix}; \qquad (8.4.16)$$

so Schrödinger's equation gives

$$\frac{d}{dx}u(x;z) = -T(x)^{-1}\begin{bmatrix} 0 & 0 \\ z & 0 \end{bmatrix}T(x)u(x;z). \qquad (8.4.17)$$

The Wronskian $\det T(x)$ is a constant, namely $\det T(x) = 1$, so this reduces to the differential equation

$$J\frac{d}{dx}u(x;z) = z\begin{bmatrix} y_1(x)^2 & y_1(x)y_2(x) \\ y_1(x)y_2(x) & y_2(x)^2 \end{bmatrix} u(x;z). \qquad (8.4.18)$$

$\square$

8.5 Limits of kernels

- Pairs of self-adjoint projections gives rise to K with $0 \leq K \leq I$.
- The compression of the projection Q to the subspace PH is $K = PQP$, where $0 \leq K \leq I$.

In particular, Theorem 8.2.7 shows that we can introduce determinantal random point fields by taking pairs of orthogonal projections P and Q, which do not commute, and forming $K = PQP$. The operator K can sometimes be of trace class and of infinite rank. The next result shows that this is in some sense the typical form of the kernel for a determinantal random point field.

Proposition 8.5.1 *Suppose that K is an operator on Hilbert space H such that $0 \leq K \leq I$. Then there exists a projection Q on $H \oplus H$ such that $K = P_1 Q P_1$ where $P_1 : H \oplus H \to H \oplus 0 \subset H \oplus H$ is the projection onto the first summand.*

Then K is trace class if and only if $Q P_1$ is Hilbert–Schmidt. In this case, the spectra satisfy

$$\operatorname{Spec}(K) = \operatorname{Spec}(QP_1). \qquad (8.5.1)$$

Proof. We choose

$$Q = \begin{bmatrix} K & \sqrt{K(I-K)} \\ \sqrt{K(I-K)} & K \end{bmatrix} \qquad (8.5.2)$$

which is an orthogonal projection with the required properties.

We evidently have $K = (QP_1)^*(QP_1)$, so that K is trace class if and only if QP_1 is Hilbert–Schmidt. Then QP_1 is compact, so its spectrum contains zero, and so the spectrum of QP_1^2 is equal to the spectrum of $P_1 Q P_1$, namely K. $\square$

Kernels associated with families of projections

Any separable and infinite-dimensional complex Hilbert space is unitarily equivalent to $L^2(\mathbf{R})$. The eigenvalues of random Hermitian matrices give random point fields on $\mathbf{R}$, so it is natural to work with kernels that reflect the structure of subintervals of the real line. Let R_t be the orthogonal projection onto the subspace $L^2(-\infty, t)$ of $L^2(\mathbf{R})$, and then let $P_{(a,b)} = R_b - R_a$ be the projection onto $L^2(a, b)$. Then for a suitable unitary U, we take $Q_{(a,b)} = U^* P_{(a,b)} U$, and then consider kernels such as

$$K_{s,t} = P_{(t,\infty)} Q_{(-s,s)} P_{(t,\infty)}, \qquad (8.5.3)$$

which satisfies $0 \leq K_{s,t} \leq I$. If $K_{s,t}$ is of trace class, then we can form the determinant $\det(I - K_{s,t})$, which is increasing as a function of t and decreasing as a function of s. In the next few sections, we use t to identify the interval (t, ∞) where the eigenvalues of the random matrix lie; whereas we use s to describe scaling properties of the eigenvalue distribution. Common choices for U are:

- the Fourier transform $f \mapsto \int_{-\infty}^{\infty} e^{-i\xi x} f(x)/\sqrt{2\pi}$ for the sine kernel;
- the Airy transform $\int_0^{\infty} \mathrm{Ai}(x + y) f(y) dy$ for soft edges as in [157];
- the Hankel transform $\int_0^{\infty} J_\nu(xy) f(y) y dy$ for hard edges as in [158].

The importance of Theorem 8.2.7 is that the operator K can have infinite rank, as in Section 9.1 we shall present a systematic means for generating examples. We include hypothesis (iii) so that Theorem 8.3.1 gives useful probabilistic information.

Definition (*Finite model for a random point field*). Let K be a self-adjoint operator on a Hilbert space H such that $0 \leq K \leq I$. Then a finite model for K is a sequence (K_n) of self-adjoint operators on H such that:

(i) $0 \leq K_n \leq I$;
(ii) $\mathrm{rank}(K_n) \leq n$;
(iii) $\|K - K_n\|_{op} \to 0$ as $n \to \infty$.

In particular, the Airy kernel arises as the pointwise limit of finite-rank kernels associated with the Gaussian unitary ensemble. The Bessel kernel arises as the pointwise limit of kernels associated with the Laguerre ensemble. Often it is necessary to scale operators and restrict them to specially chosen intervals to display the required convergence.

The following result gives realistic conditions under which we can convert pointwise convergence of kernels into convergence in trace norm of the associated operators.

Proposition 8.5.2 *Suppose that K_N $(N = 1, 2, \dots)$ and K are continuous kernels on a closed and bounded interval J and that $K_N \geq 0$ as an operator on $L^2(J; dx)$ for $N = 1, 2, \dots$. Suppose further that*

$$\int_J K_N(x, x)\, dx \to \int_J K(x, x)\, dx \qquad (N \to \infty) \qquad (8.5.4)$$

and $\langle K_N f, g \rangle_{L^2(J)} \to \langle Kf, g \rangle_{L^2(J)}$ as $N \to \infty$ for all $f, g \in L^2(J)$.

(i) Then

$$\|K_N - K\|_{c^1} \to 0 \qquad (N \to \infty) \qquad (8.5.5)$$

and the ordered sequences of eigenvalues satisfy

$$\|(\mu_j(K_N) - \mu_j(K))_{j=1}^\infty\|_{\ell^1} \to 0 \qquad (N \to \infty). \qquad (8.5.6)$$

(ii) The sequence of entire functions $(\det(I - zK_N))_{N=1}^\infty$ converges uniformly on compact subsets to the function $\det(I - zK)$.

Proof. (i) First we observe that $\langle Kf, f \rangle \geq 0$ for all f by weak convergence, so K is also positive semidefinite. Since $K_N \geq 0$ and the kernels are continuous, we have by Mercer's theorem

$$\|K_N\|_{c^1} = \operatorname{trace}(K_N) = \int_J K_N(x, x)\, dx, \qquad (8.5.7)$$

and likewise with K in place of K_N.

Hence $K_N \to K$ in the weak operator topology and $\|K_N\|_{c^1} \to \|K\|_{c^1}$ as $N \to \infty$. By a result from [143], we have convergence in trace norm, so $\|K_N - K\|_{c^1} \to 0$ as $N \to \infty$. Finally, we can use Lidskii's Lemma 2.1.5 to deduce the last part.

(ii) The map $T \mapsto \det(I - T)$ is continuous for the norm on c^1, and hence we have uniform convergence under the stated conditions. $\quad\square$

9

Integrable operators and differential equations

Abstract

The most important examples of kernels in random matrix theory involve integrable operators, as we describe here. Tracy and Widom considered a particular class of integrable operators and found a method for computing the correlation functions in the determinantal random point field. Here we extend their method and make systematic use of Hankel operators in the theory. We apply these methods to:

- the sine kernel, which represents the bulk of the spectrum;
- the Airy kernel, which represents the soft edge of the spectrum of GUE;
- the Bessel kernel, which represents the hard edge of the spectrum of the Jacobi and Laguerre ensembles.

We emphasize differential equations as a unifying theme, since the above kernels arise from some second order linear differential equations with rational coefficients. In particular, Weber's differential equations can be scaled so that suitable limiting forms give the bulk and soft edge cases. The chapter begins with integrable operators, as in [55].

9.1 Integrable operators and Hankel integral operators

- We give a sufficient condition for an integrable operator to generate a determinantal random point field.
- The Airy, sine, Bessel and Laguerre kernels arise from integrable operators.

Definition (*Integrable operators*). Let S be a subinterval of $\mathbf{R}$. An integrable operator on $L^2(S; dx)$ is a bounded linear operator W with kernel

$$W(x, y) = 2 \sum_{j=1}^{n} \frac{f_j(x) g_j(y)}{x - y} \qquad (x, y \in S; x \neq y) \qquad (9.1.1)$$

where f_j, g_j are bounded and measurable functions and such that $\sum_{j=1}^{n} f_j(x) g_j(x) = 0$ almost everywhere on S.

Thus an integrable operator is a particular kind of integral operator that has a special choice of kernel. In most examples below, W is a continuous function, and we refer to $W(x, x)$ as the diagonal of the kernel.

Example 9.1.1 (i) Of particular importance is the case when $n = 1$ and

$$W(x, y) = \frac{A(x)B(y) - A(y)B(x)}{x - y}. \qquad (9.1.2)$$

(ii) The reproducing kernels mentioned in Section 8.4 are integrable operators.

(iii) The Christoffel–Darboux kernel of Proposition 8.3.3 gives an integrable kernel, which has the additional special property of giving a finite-rank orthogonal projection on $L^2(\mathbf{R})$.

(iv) Let K be the integral operator that has kernel $\sum_{j=1}^{n} f_j(x) g_j(y)$. Then K has finite rank, and the diagonal of K is $\sum_{j=1}^{n} f_j(x) g_j(x)$.

Proposition 9.1.2 *Suppose that K is an integrable operator on $L^2(S)$ that is of trace class. Then the set Z of zeros of $\det(I - \zeta K)$ is countable with no limit points in the finite plane. For $\zeta \in \mathbf{C} \setminus Z$, there exists a trace-class integrable operator L_ζ such that $I + L_\zeta = (I - \zeta K)^{-1}$.*

Proof. Mostly, this follows as in Lemma 8.1.3, but we need to check that L_ζ is integrable.

Let M be the operator on $L^2(S)$ of multiplication by x, so $M : f(x) \mapsto xf(x)$; so that an operator T with kernel $T(x, y)$ has $MT - TM$ with kernel $(x - y)T(x, y)$. Hence T is integrable if and only if $TM - MT$ is of finite rank with a kernel that is zero on the diagonal $x = y$. For $\zeta \in \mathbf{C} \setminus Z$, the operator $L_\zeta M - ML_\zeta$ has a kernel that is zero on the diagonal; furthermore, we observe that

$$L_\zeta M - ML_\zeta = (I - \zeta K)^{-1} \zeta(KM - MK)(I - \zeta K)^{-1} \qquad (9.1.3)$$

is of finite rank, as required. □

Lemma 9.1.3 *Suppose further that the f_j and g_j are real-valued. Then W is self-adjoint if and only if W may be written as*

$$W(x,y) = \frac{\langle Jv(x), v(y)\rangle}{x-y} \qquad (x,y \in S; x \neq y) \qquad (9.1.4)$$

where $v(x) = \mathrm{col}[f_1(x), \ldots, f_n(x); g_1(x), \ldots, g_n(x)]$ and

$$J = \begin{bmatrix} 0 & -I_n \\ I_n & 0 \end{bmatrix} \qquad (9.1.5)$$

with the identity matrix $I_n \in M_n(\mathbf{R})$.

Proof. Clearly the kernel of W is symmetric if and only if the numerator of $W(x,y)$ is skew-symmetric, in which case we can write

$$2\sum_{j=1}^n f_j(x)g_j(y) = \sum_{j=1}^n f_j(x)g_j(y) - \sum_{j=1}^n f_j(y)g_j(x), \qquad (9.1.6)$$

and the matrix expression follows directly. $\square$

Given an integrable operator, it can be difficult to determine whether or not it satisfies Theorem 8.2.7 and thus gives a determinantal random point field. Tracy and Widom [157, 158] observed that some important kernels could be expressed as products of Hankel operators which are easier to deal with.

Definition (*Hankel integral operator [127]*). Let H be a separable Hilbert space, and let $\phi \in L^2((0,\infty); H)$. The Hankel operator with symbol ϕ is the integral operator

$$\Gamma_\phi f(s) = \int_0^\infty \phi(s+t)f(t)\, dt \qquad (9.1.7)$$

from a subspace of $L^2(0,\infty)$ into $L^2((0,\infty); H)$. We sometimes refer to $\phi(x+y)$ as the *kernel* of Γ_ϕ; whereas the Fourier transform $\hat{\phi}$ is known as the *symbol*.

Proposition 9.1.4 *Suppose that there exists a Hankel operator $\Gamma_\phi : L^2(0,\infty) \to L^2((0,\infty); H)$ such that $\|\Gamma_\phi\| \leq 1$ and*

$$\|\Gamma_\phi\|_{HS}^2 = \int_0^\infty s\|\phi(s)\|_H^2\, ds < \infty. \qquad (9.1.8)$$

Then the hypotheses of Theorem 8.2.7 are satisfied by

$$K(x,y) = \int_0^\infty \langle \phi(x+t), \phi(y+t)\rangle_H\, dt, \qquad (9.1.9)$$

so $0 \leq K \leq I$ and $\mathbf{I}_{[a,b]}(x)K(x,y)\mathbf{I}_{[a,b]}(y)$ is a trace-class kernel.

Proof. Clearly $K = \Gamma_\phi^* \Gamma_\phi$ satisfies $0 \le K \le I$ as an operator on $L^2(0, \infty)$. The Hilbert–Schmidt norm of Γ_ϕ satisfies

$$\|\Gamma_\phi\|_{HS}^2 = \int_0^\infty \int_0^\infty \|\phi(s+t)\|_H^2 \, dt ds < \infty, \qquad (9.1.10)$$

so K is of trace class. $\square$

Tracy and Widom [159] observed that most integrable operators in random matrix theory can be transformed into the form

$$K(x,y) = \frac{A(x)B(y) - A(y)B(x)}{x - y} \qquad (9.1.11)$$

where A and B satisfy

$$m(x) \frac{d}{dx} \begin{bmatrix} A(x) \\ B(x) \end{bmatrix} = \begin{bmatrix} \alpha(x) & \beta(x) \\ -\gamma(x) & -\alpha(x) \end{bmatrix} \begin{bmatrix} A(x) \\ B(x) \end{bmatrix} \qquad (9.1.12)$$

for some real polynomials $\alpha(x), \beta(x), \gamma(x)$ and $m(x)$.

Definition. We call a differential equation of the form (9.1.12) a *Tracy–Widom system* and the associated kernel of the form (9.1.11) a *Tracy–Widom kernel*.

We associate these kernels with the names Tracy and Widom since they identified the importance of the kernels in random matrix theory. Several familiar differential equations can be cast in this form. Tracy–Widom systems of (9.1.11) are a generalization of the classical theory of orthogonal polynomials with respect to weights

$$w(x) = \prod_{j=1}^n |x - x_j|^{\alpha_j} e^{-v(x)}. \qquad (9.1.13)$$

Suppose that $w(x)$ is a weight function on an interval S such that $w(x) \to 0$ as x tends to the ends of S from inside. Suppose further that v is a real polynomial, and that w and S are as in the table below.

name	S	$m(x)$	$w(x)$	
Hermite	$(-\infty, \infty)$	1	$e^{-x^2/2}$	
Laguerre	$(0, \infty)$	x	$x^\alpha e^{-x}$	$(\alpha > 0)$
Jacobi	$(-1, 1)$	$1 - x^2$	$(1-x)^\alpha (1+x)^\beta$	$(\alpha, \beta > 0)$
pseudo-Jacobi	$(-\infty, \infty)$	x^2	$(1+x^2)^{-N-\sigma}$	$(N \in \mathbf{N}, \sigma > 0)$

The pseudo-Jacobi ensemble does not quite fit the pattern, so we discuss it as a special case in Section 11.2. In the other cases, let $(p_n(x))_{n=0}^\infty$ be orthogonal polynomials with respect to the weight w with leading coefficients k_n, and $\varphi_k(x) = p_k(x)w(x)^{1/2}$ the corresponding orthonormal basis of $L^2(S; dx)$.

Proposition 9.1.5 *For the Hermite, Laguerre and Jacobi systems with the stated $m(x)$ and $w(x)$ on S, there exist real polynomials α_n, β_n and γ_n for $n = 1, 2, \ldots$ such that*

$$m(x)\frac{d}{dx}\begin{bmatrix} \varphi_n(x) \\ \varphi_{n-1}(x) \end{bmatrix} = \begin{bmatrix} \alpha_n(x) & \beta_n(x) \\ -\gamma_n(x) & -\alpha_n(x) \end{bmatrix}\begin{bmatrix} \varphi_n(x) \\ \varphi_{n-1}(x) \end{bmatrix} \qquad (x \in S),$$

$$(9.1.14)$$

and the corresponding normalized Tracy–Widom operator, which has kernel

$$K_n(x, y) = \frac{k_{n-1}}{k_n}\frac{\varphi_n(x)\varphi_{n-1}(y) - \varphi_{n-1}(x)\varphi_k(y)}{x - y}, \qquad (9.1.15)$$

is the orthogonal projection onto $\mathrm{span}\{\sum_{j=0}^{n-1} a_j\varphi_j\}$ *in $L^2(S; dx)$ as given by the Christoffel–Darboux formula (8.3.7).*

Proof. See [159] for the general result. The explicit formulas for the differential equations appear in (9.3.21) for the Hermite function, in (9.5.18) for the Jacobi polynomials, and (9.6.3) for the Laguerre functions. In each case, we can invoke Proposition 8.3.3 to obtain the formula for the kernel that represents the projection. □

Since orthogonal polynomials give rise to determinantal random point field as in Theorem 8.2.7, it is natural to construct other determinantal random point fields from (9.1.11) via Theorem 8.2.7.

In Chapter 4, we considered the support of the equilibrium measure of a generalized unitary ensemble, and found it to have the form $S = \cup_{j=1}^n [a_j, b_j]$; consequently, one can consider Tracy–Widom systems on S as a subset of $\mathbf{C} \cup \{\infty\}$. Here for simplicity we work on a single interval. Normally, we take $x \in (0, \infty)$, and we can transform to $u \in \mathbf{R}$ by letting $x = e^u$. To deal with circular ensembles, one can substitute $e^{i\theta}$ for x. This still allows a wide range of possibilities regarding singular points where $m(x) = 0$, so we need to restrict to special cases to make progress.

The matrix

$$\Omega(x) = -\begin{bmatrix} \gamma(x) & \alpha(x) \\ \alpha(x) & \beta(x) \end{bmatrix} \qquad (9.1.16)$$

is symmetric and satisfies

$$\Omega(y) - \Omega(x) = J\begin{bmatrix} \alpha(x) & \beta(x) \\ -\gamma(x) & -\alpha(x) \end{bmatrix} + \begin{bmatrix} \alpha(y) & \beta(y) \\ -\gamma(y) & -\alpha(y) \end{bmatrix}^t J. \quad (9.1.17)$$

Theorem 9.1.6 (*[157, 21]*). *Suppose that Ω_1 and Ω_0 are real symmetric 2×2 matrices with $\Omega_1 \geq 0$ and Ω_1 of rank one. Suppose further that A and B are bounded and continuous real functions in $L^2(0, \infty)$ such that $A(x) \to 0$ and $B(x) \to 0$ as $\to \infty$ and that*

$$\frac{d}{dx}\begin{bmatrix} A(x) \\ B(x) \end{bmatrix} = J(\Omega_1 x + \Omega_0)\begin{bmatrix} A(x) \\ B(x) \end{bmatrix} \qquad (x > 0). \qquad (9.1.18)$$

Then the integral operator K on $L^2(0, \infty)$ with kernel

$$K(x, y) = \frac{A(x)B(y) - A(y)B(x)}{x - y} \qquad (9.1.19)$$

is self-adjoint and bounded. Furthermore, there exists $\phi \in L^2(0, \infty)$ such that the Hankel operator Γ_ϕ is bounded and self-adjoint with $K = \Gamma_\phi^2$.

Proof. The integral operator with kernel $1/(x - y)$ is bounded on $L^2(\mathbf{R})$ by M. Riesz's theorem. Since A and B are real and bounded by hypothesis, it follows that K is a bounded and self-adjoint linear operator on $L^2(0, \infty)$.

We recall that J satisfies $J^* = -J$ and $J^2 = -I$. For the purposes of the following calculation, we have

$$\left(\frac{\partial}{\partial x} + \frac{\partial}{\partial y}\right)\left(\frac{1}{x - y}\right) = 0. \qquad (9.1.20)$$

Hence by the differential equation and (9.1.17), we have

$$\left(\frac{\partial}{\partial x} + \frac{\partial}{\partial y}\right) K(x, y)$$

$$= \left(\frac{\partial}{\partial x} + \frac{\partial}{\partial y}\right) \left(\left\langle J \begin{bmatrix} A(x) \\ B(x) \end{bmatrix}, \begin{bmatrix} A(y) \\ B(y) \end{bmatrix} \right\rangle / (x - y) \right)$$

$$= \left(\left\langle J^2 (\Omega_1 x + \Omega_0) \begin{bmatrix} A(x) \\ B(x) \end{bmatrix}, \begin{bmatrix} A(y) \\ B(y) \end{bmatrix} \right\rangle / (x - y) \right)$$

$$+ \left(\left\langle J \begin{bmatrix} A(x) \\ B(x) \end{bmatrix}, J(\Omega_1 y + \Omega_0) \begin{bmatrix} A(y) \\ B(y) \end{bmatrix} \right\rangle / (x - y) \right)$$

$$= \left(\left\langle (-\Omega_1 x - \Omega_0 + \Omega_1 y + \Omega_0) \begin{bmatrix} A(x) \\ B(x) \end{bmatrix}, \begin{bmatrix} A(y) \\ B(y) \end{bmatrix} \right\rangle / (x - y) \right)$$

$$= \left\langle -\Omega_1 \begin{bmatrix} A(x) \\ B(x) \end{bmatrix}, \begin{bmatrix} A(y) \\ B(y) \end{bmatrix} \right\rangle. \tag{9.1.21}$$

Since Ω_1 is real symmetric of rank one, we can choose an eigenvector $\mathrm{col}[\cos\theta, \sin\theta]$ corresponding to the eigenvalue $\lambda > 0$, then introduce the function

$$\phi(x) = \sqrt{\lambda}(A(x) \cos\theta + B(x) \sin\theta) \tag{9.1.22}$$

which by hypothesis is real and belongs to $L^2(0, \infty)$. Furthermore, we have

$$\left(\frac{\partial}{\partial x} + \frac{\partial}{\partial y}\right) K(x, y) = -\phi(x)\phi(y), \tag{9.1.23}$$

so integrating we obtain

$$K(x, y) = \int_0^\infty \phi(x + u)\phi(y + u)\, du + h(x - y) \tag{9.1.24}$$

for some real function h. Now $K(x, y) \to 0$ as $x \to \infty$ or as $y \to \infty$; likewise, the integral converges to zero as $x \to \infty$ or as $y \to \infty$ by the Cauchy–Schwarz inequality since $\phi \in L^2(0, \infty)$; hence $h = 0$.

We deduce that $K = \Gamma_\phi^2$ and that Γ_ϕ is a bounded Hankel operator. $\qquad \square$

Corollary 9.1.7 *Suppose further that $x^{1/2} A(x)$ and $x^{1/2} B(x)$ belong to $L^2(0, \infty)$. Then K is of trace class, and*

$$\mathrm{trace}(K) = \int_0^\infty u\phi(u)^2\, du. \tag{9.1.25}$$

Proof. By the proof of Theorem 9.1.6, $x^{1/2}\phi(x)$ belongs to $L^2(0,\infty)$ and hence by Proposition 9.1.4, Γ_ϕ is a Hilbert–Schmidt operator. Consequently, K is of trace class, with the stated formula for the trace.   □

The hypotheses of Corollary 9.1.7 and hence Theorem 8.2.7 hold in some significant examples, which we shall discuss. Furthermore, when $H = \mathbf{R}$ and $W = \Gamma_\phi^2$, the spectral resolution of the self-adjoint operator Γ_ϕ determines the spectral resolution of W, so we can find the eigenvalues of W from the eigenvalues of Γ_ϕ. This is the basis of the successful calculations in [21, 22, 157, 158], which also exploited the fact the eigenvectors of Γ_ϕ can be comparatively easy to analyze.

Theorem 9.1.6 enables us to use the spectral theory of Hankel operators to describe the eigenvalues of the kernels K, as in [127]. To describe the eigenvectors of K, we need to use quite different techniques. In several cases of interest, K commutes with a second order differential operator, and hence one can obtain the asymptotics of the eigenvectors of K by the WKB method. In Section 9.3 we present a familiar example of this when we introduce the prolate spheroidal functions, which are eigenvectors for the sine kernel. In Section 9.4 we present a sufficient condition for a Hankel operator to commute with a second order differential operator; this applies in particular to the Airy kernel and the Bessel kernel.

In the following sections, we shall consider some fundamental examples of determinantal random point fields. The analysis will go from differential equations to kernels, and then from kernels to determinants. In some cases, we require convergence results from the theory of orthogonal polynomials proper, but in all cases the differential equations suggest the special functions that appear in the analysis.

When K is the sine kernel, Airy kernel or Bessel kernel, we introduce:

- finite rank operators from the scaled Gaussian unitary ensemble such that $K_n \to K$ in c^1 as $n \to \infty$;
- a self-adjoint and Hilbert–Schmidt operator Γ_ϕ such that $\Gamma_\phi^2 = K$;
- a second order self-adjoint differential operator L such that $KL = LK$.

Exercise 9.1.8 Suppose that $K = \Gamma_\phi^2$, where $\phi \in L^2((0,\infty); \mathbf{R})$. Show that the kernel of K satisfies

$$\det[K(x_j, x_k)]_{j,k=1,\ldots,n} \tag{9.1.26}$$
$$= \frac{1}{n!} \int_{(0,\infty)^n} \det[\phi(x_j + y_\ell)]_{j,\ell=1,\ldots,n} \det[\phi(x_k + y_\ell)]_{k,\ell=1,\ldots,n} \, dy_1 \cdots dy_n$$

for all $(x_1, \ldots, x_n > 0)$.

9.2 Hankel integral operators that commute with second order differential operators

Theorem 9.2.1 *Let Γ_ϕ be the Hankel operator that has kernel $\phi(x+y)$, let L be the differential operator*

$$Lf(x) = -\frac{d}{dx}\left(a(x)\frac{df}{dx}\right) + b(x)f(x), \tag{9.2.1}$$

where $a'''(x) = Ca'(x)$ and $a(0) = 0$.

(i) Then there exists a real function α such that

$$\alpha(x+y) = \frac{a'(x) - a'(y)}{a(x) - a(y)}. \tag{9.2.2}$$

(ii) Suppose further that

$$\beta(x+y) = \frac{b(x) - b(y)}{a(x) - a(y)} \tag{9.2.3}$$

for some real function β and that

$$\phi''(u) + \alpha(u)\phi'(u) - \beta(u)\phi(u) = 0. \tag{9.2.4}$$

Then the operators Γ_ϕ and L commute.

Proof. We introduce the expression

$$\Phi(x,y) = (a(y) - a(x))\phi''(x+y) + (a'(y) - a'(x))$$
$$\times \phi'(x+y) - (b(y) - b(x))\phi(x+y); \tag{9.2.5}$$

then by successive integrations by parts, we obtain

$$(L\Gamma_\phi - \Gamma_\phi L)f(x) = \left[\phi(x+y)a(y)f'(y) - \phi'(x+y)a(y)f(y)\right]_0^\infty$$
$$+ \int_0^\infty \Phi(x,y)f(y)\,dy. \tag{9.2.6}$$

The term in square brackets vanishes since $a(0) = 0$; so if $\Phi = 0$, then the operators commute. The next idea is to reduce the condition $\Phi = 0$ into the differential equation

$$(a(x) - a(y))(\phi''(x + y) + \alpha(x + y)\phi'(x + y) - \beta(x + y)\phi(x + y)) = 0,$$
(9.2.7)

involving only one variable $u = x + y$. The following lemma guarantees the existence of such α and β under suitable hypotheses. The proof is by elementary calculus.

Lemma 9.2.2 *Suppose that a and g are twice continuously differentiable real functions such that*

$$a'(x)g''(y) - g''(x)a'(y) + a''(x)g'(y) - g'(x)a''(y) = 0.$$
(9.2.8)

(i) Then there exists a real function h such that

$$\left(\frac{\partial}{\partial x} - \frac{\partial}{\partial y}\right)\frac{g(x) - g(y)}{a(x) - a(y)} = \frac{h(x) - h(y)}{(a(x) - a(y))^2}.$$
(9.2.9)

(ii) When h is constant, there exists a real function f such that

$$\frac{g(x) - g(y)}{a(x) - a(y)} = f(x + y).$$
(9.2.10)

Conclusion of the proof of Theorem 9.2.1 In the special case of $g(x) = a'(x)$, the hypothesis of the Lemma reduces to the condition $a'''(x) = Ca'(x)$ for some constant C, and one can verify that both (i) and (ii) of the Lemma apply; so α exists. Having identified the suitable choices of a, one can use the hypothesis on b to obtain a suitable β, and hence use the differential equation for ϕ to make $\Phi = 0$. $\qquad\square$

Fortunately, Theorem 9.2.1 covers many cases that are of interest in random matrix theory. There are three main families of solutions of the differential equation $a'''(x) = Ca'(x)$, with $a(0) = 0$, depending upon the sign of C:

(1) quadratic: $C = 0$ and $a(x) = c_2 x^2 + c_1 x$;
(2) hyperbolic: $C > 0$ and $a(x) = c_2 \cosh tx + c_1 \sinh tx - c_2$, where $t \in \mathbf{R}$ has $t^2 = C$;
(3) trigonometric: $C < 0$ and $a(x) = c_2 \cos tx + c_1 \sin tx - c_2$, where $t \in \mathbf{R}$ has $t^2 = -C$.

In Proposition 9.4.3 we show that the Hankel operator Γ_ϕ commutes with the differential operator L_s, which involves $a(x) = x$, as in (1). In the Section 9.5, we consider a differential operator with $a(x) = e^{2x} - 1$, as in (2). Theorem 9.2.1 is most useful when Γ_ϕ is compact and L has a discrete spectrum, which happens in the main applications, but not in some of the cases in the next exercise.

Sonine [167, p. 82] considered the one-parameter families of functions Z_ν that satisfy the system

$$Z_{\nu-1} + Z_{\nu+1} = \frac{2\nu}{z} Z_\nu$$
$$Z_{\nu-1} - Z_{\nu+1} = 2Z_\nu', \qquad (9.2.11)$$

which gives rise to the Tracy–Widom system

$$z \frac{d}{dz} \begin{bmatrix} Z_\nu \\ Z_{\nu-1} \end{bmatrix} = \begin{bmatrix} -\nu & z \\ -z & \nu-1 \end{bmatrix} \begin{bmatrix} Z_\nu \\ Z_{\nu-1} \end{bmatrix}; \qquad (9.2.12)$$

see [167, p. 82]. This and related systems give many of the examples that arise in random matrix theory.

Definition (*Bessel functions*) *[167, 3.1(8)]*. The Bessel function of the first kind of order $\alpha > -1$ may be defined by the series

$$J_\alpha(z) = \sum_{n=0}^{\infty} (-1)^n \frac{z^{2n+\alpha}}{2^{2n+\alpha}\Gamma(n+1)\Gamma(\alpha+n+1)} \qquad (z \in \mathbf{C}). \quad (9.2.13)$$

There are also integral formulæ which apply for general real order α and which extend the definition in which we previously made in (7.4.13) for integral order α. The modified Bessel function, also known as MacDonald's function, may be defined by [173, 8.432]

$$K_\alpha(z) = \int_0^\infty e^{-z \cosh t} \cosh(\alpha t)\, dt \qquad (\Re z > 0). \qquad (9.2.14)$$

The Bessel function J_α satisfies [167, 2.13]

$$x^2 J_\alpha'' + x J_\alpha' + (x^2 - \alpha^2) J_\alpha = 0, \qquad (9.2.15)$$

whereas the modified Bessel function satisfies

$$x^2 K_\alpha'' + x K_\alpha' - (x^2 + \alpha^2) K_\alpha = 0. \qquad (9.2.16)$$

Bessel's equation (9.2.15) may be regarded as a transformation of a special case of the confluent hypergeometric equation.

The general solution of Airy's equation $y'' + xy = 0$ is

$$y = c_1 x^{1/2} J_{1/3}\left(\frac{2}{3} x^{3/2}\right) + c_2 x^{1/2} J_{-1/3}\left(\frac{2}{3} x^{3/2}\right). \qquad (9.2.17)$$

One solution of $y'' = xy$ is given by the oscillatory integral

$$\text{Ai}(x) = \frac{1}{2\pi} \int_{-\infty}^{\infty} e^{ixt+t^3/3} \, dt, \qquad (9.2.18)$$

which by [167, p. 190] may be expressed as

$$\text{Ai}(x) = \frac{1}{\pi} \sqrt{\frac{x}{3}} K_{1/3}\left(\frac{2}{3}x^{3/2}\right) \qquad (x > 0). \qquad (9.2.19)$$

Exercise 9.2.3 In this exercise, we present some choices of a and b such that ϕ satisfies some relatively simple differential equation.

(i) Let $a(x) = -x^2$ and $b(x) = -4^{-1}x + \gamma$. Show that $\Phi = 0$ when

$$u\phi''(u) + 2\phi'(u) + 4^{-1}\phi(u) = 0, \qquad (9.2.20)$$

and that $\phi(u) = u^{-1/2} J_1(u^{1/2})$ gives a solution.

(ii) Let $a(x) = x^2$ and $b(x) = -4^{-1}x + \gamma$. Show that $\Phi = 0$ when

$$u\phi''(u) + 2\phi'(u) - 4^{-1}\phi(u) = 0, \qquad (9.2.21)$$

and that $\phi(u) = u^{-1/2} K_1(u^{1/2})$ gives a solution where K_1 is the modified Bessel function.

(iii) Let $a(x) = -x^2$ and $b(x) = -x^2 + \gamma$. Show that $\Phi = 0$ when

$$u\phi''(u) + 2\phi'(u) + u\phi(u) = 0, \qquad (9.2.22)$$

and that $\phi(u) = u^{-1/2} J_{1/2}(u)$ gives a solution.

(iv) Let $a(x) = x^2$ and $b(x) = -x^2 + \gamma$. Show that $\Phi = 0$ when

$$u\phi''(u) + 2\phi'(u) - u\phi(u) = 0, \qquad (9.2.23)$$

and that $\phi(u) = u^{-1/2} K_{1/2}(u)$ gives a solution, where $K_{1/2}$ is the modified Bessel function.

(v) Let $a(x) = a_2 x^2 + a_1 x$ where $a_2 > 0 \; a_1 \geq 0$, and let $b = 0$. Show that $\phi(u) = C/(a_2 u + a_1)$. When $a_2 = 1$ and $a_1 = 0$, we thus obtain Carleman's Hankel operator with kernel $1/(x + y)$; see [127].

(vi) Let $a(x) = \cos 2\pi x - 1$ and $b(x) = \alpha \cos 2\pi x + \beta \sin 2\pi x + \gamma$. Show that $\Phi = 0$ when

$$\phi''(u) + 2\pi(\cot \pi u)\phi'(u) + (\beta \cot \pi u + \alpha)\phi(u) = 0. \qquad (9.2.24)$$

The following theorem constrains the possibilities for ϕ.

Theorem 9.2.4 *Suppose that a and b are as in Theorem 9.2.1. Then the differential equation (9.2.4) may be transformed by change of variables to the hypergeometric equation*

$$x(1-x)\frac{d^2\phi}{dx^2} + \{\lambda - (\mu+\nu+1)x\}\frac{d\phi}{dx} - \mu\nu\phi = 0 \qquad (9.2.25)$$

or Whittaker's form of the confluent hypergeometric equation

$$\frac{d^2 W}{dx^2} = \left(\frac{1}{4} - \frac{k}{x} + \frac{m^2 - 1/4}{x^2}\right)W. \qquad (9.2.26)$$

Proof. In [23] we consider the quadratic, hyperbolic and trigonometric cases in detail. In all cases, the differential equation reduces to a linear differential equation with rational functions as coefficients which has less than or equal to three singular points, and we determine the nature of the singularities. See [172, 9.220] for solutions of (9.2.26). □

The hypergeometric and confluent hypergeometric equations cover many classical families of special functions. Theorem 9.2.4 states that the differential equation for ϕ can be reduced by changes of variable into a standard form, and in [23] the author provides several examples of commuting pairs L and Γ_ϕ. However, the problem of deciding which kernels W factor as $W = \Gamma_\phi^* \Gamma_\phi$ involves some subtle analytical points which are not stable under changes of variable, so there is currently no simple criterion for deciding which Tracy–Widom systems lead to kernels W that factorize as $W = \Gamma^2$ where Γ commutes with L. See also [22].

9.3 Spectral bulk and the sine kernel

- The bulk of the spectrum for the Gaussian unitary ensemble is described by the sine kernel.
- The eigenfunctions for the sine kernel are the prolate spheroidal functions.
- The prolate spheroidal functions satisfy differential equations for which the limiting form is Webber's parabolic cylinder equation.

In Section 9.3, 9.4 and 9.5 we show how the bulk, soft and hard edges arise from unitary ensembles. Before introducing the integrable operators, we review the Gaussian unitary ensemble. Let $x_{j,k}$ and $y_{j,k}$ $(1 \leq j \leq k \leq n)$ be a family of mutually independent $N(0, 1/n)$ random variables. We let X_n be the $n \times n$ Hermitian matrix that has entries $[X_n]_{jk} = (x_{j,k} + iy_{j,k})/\sqrt{2}$ for $j < k$, $[X_n]_{jj} = x_{j,j}$ for

$1 \leq j \leq n$ and $[X_n]_{kj} = (x_{j,k} - iy_{j,k})/\sqrt{2}$ for $j < k$; the space of all such matrices with the probability measure $\nu_n^{(2)}$ forms the *Gaussian unitary ensemble*. The eigenvalues of X are real and may be ordered as $\lambda_1 \leq \cdots \leq \lambda_n$, so their positions are specified by the empirical distribution $\mu_n = (1/n) \sum_{j=1}^n \delta_{\lambda_j}$. As $n \to \infty$, the empirical distributions converge weakly to the Wigner semicircle law

$$\rho(dx) = \frac{1}{2\pi} \mathbf{I}_{[-2,2]}(x) \sqrt{4 - x^2} dx \tag{9.3.1}$$

for almost all sequences (X_n) of matrices chosen from these ensembles; see [116, 77, 166].

Sine kernel

The bulk of the spectrum is $[-2, 2]$. To describe the distribution of neighbouring eigenvalues within small subintervals of $[-2, 2]$, we introduce the sine kernel. This operator arises in many contexts in harmonic analysis.

Definition (*Sine kernel*). Let D_b be the operator on $L^2(\mathbf{R})$ that has the (Dirichlet) sine kernel

$$D_b(x, y) = \frac{\sin b\pi x \, \cos b\pi y - \cos b\pi x \, \sin b\pi y}{\pi(x - y)}, \tag{9.3.2}$$

where b is known as the bandwidth. An entire function f has exponential type if there exists constants $c_1, c_2 > 0$ such that $|f(z)| \leq c_1 e^{c_2 |z|}$ for all $z \in \mathbf{C}$.

Proposition 9.3.1 *The operator D_b on $L^2(\mathbf{R})$ is an orthogonal projection. The range of D_b consists of those $f \in L^2(\mathbf{R})$ that extend to entire functions of exponential type such that*

$$\lim_{y \to \pm\infty} \sup |y|^{-1} \log |f(iy)| \leq \pi b. \tag{9.3.3}$$

The range of D_b forms a reproducing kernel Hilbert space on $\mathbf{C}$ with reproducing kernel

$$k_w(z) = \frac{\sin \pi b(z - \bar{w})}{\pi(z - \bar{w})}. \tag{9.3.4}$$

Proof. The operator $\mathcal{F}^* \mathbf{I}_{[-\pi b, \pi b]} \mathcal{F}$ is an orthogonal projection, and one can easily check the identity

$$\int_{-\infty}^{\infty} \frac{\sin \pi b(x - y)}{\pi(x - y)} f(y) \, dy = \frac{1}{\sqrt{2\pi}} \int_{-\pi b}^{\pi b} e^{i\xi x} \mathcal{F} f(\xi) d\xi. \tag{9.3.5}$$

Replacing x by $z = x + iy$ on the right-hand side, we obtain an entire function $F(z)$ of exponential type that satisfies (9.3.3). The Paley–Wiener theorem [102] characterizes such F as those with Fourier transform supported on $[-\pi b, \pi b]$.

The range of D_b is known as the space of band limited functions. For a band limited function, we have

$$f(w) = \langle f, k_w \rangle_{L^2} = \int_{-\infty}^{\infty} \frac{\sin \pi b(w - y)}{\pi(w - y)} f(y) \, dy; \qquad (9.3.6)$$

hence $f \mapsto f(w)$ is continuous for $w \in \mathbf{C}$, so we have a reproducing kernel Hilbert space. □

The frequencies $\xi/2\pi$ of $D_b f$ lie in the band $[-b/2, b/2]$. Likewise, $L^2[-a, a]$ is known as the space of time limited functions. When we compress D_b to $L^2[-a, a]$, the range consists of functions that are both band and time limited; the functions that satisfy both these severe restrictions have very special properties.

Proposition 9.3.2 *The compression of D_b to $L^2[-a, a]$ gives a trace class operator such that $0 \le D_b \le I$. The eigenfunctions of D_b are the prolate spheroidal functions, namely the eigenfunctions of the differential operator L, where*

$$Lf(x) = -\frac{d}{dx}\left((a^2 - x^2)\frac{d}{dx}f(x)\right) + \pi^2 b^2 x^2 f(x) \qquad (9.3.7)$$

such that f is bounded on $[-a, a]$.

Proof. The operators Γ with kernel $(b/2a)^{1/2} e^{i\pi bxz/a}$ and Γ^* with kernel $(b/2a)^{1/2} e^{-i\pi byz/a}$ are Hilbert–Schmidt on $L^2[-a, a]$, and $\Gamma\Gamma^*$ has kernel

$$\frac{b}{2a} \int_{-a}^{a} e^{i\pi b(x-y)z/a} \, dz = \frac{\sin \pi b(x - y)}{\pi(x - y)}. \qquad (9.3.8)$$

Hence D_b is trace class, and since D_b is the compression of an orthogonal projection, we have $0 \le D_b \le I$.

The operator L commutes with D_b since by direct calculation we have

$$-\frac{d}{dx}\left((a^2 - x^2)\frac{d}{dx}\frac{\sin \pi b(x - y)}{\pi(x - y)}\right) + \frac{d}{dy}\left((a^2 - y^2)\frac{d}{dy}\frac{\sin \pi b(x - y)}{\pi(x - y)}\right)$$

$$= (x^2 - y^2)\frac{d^2}{dx^2}\frac{\sin \pi b(x - y)}{\pi(x - y)} + 2(x + y)\frac{d}{dx}\frac{\sin \pi b(x - y)}{\pi(x - y)}$$

$$= -\pi^2 b^2 (x^2 - y^2)\frac{\sin \pi b(x - y)}{\pi(x - y)}. \qquad (9.3.9)$$

296   *Random Matrices: High Dimensional Phenomena*

By compactness, each eigenspace of D_b is finite-dimensional and the eigenfunctions that correspond to non-zero eigenvalues are continuously differentiable functions on $[-a, a]$. Hence the eigenfunctions of D_b are also eigenfunctions of L and give a complete orthonormal basis of $L^2[-a, a]$. □

By a simple scaling of variables, we can reduce further analysis to the case $a = 1$. We introduce the kernels

$$K_\pm(x, y) = \frac{\sin b\pi(x - y)}{\pi(x - y)} \pm \frac{\sin b\pi(x + y)}{\pi(x + y)} \qquad (x, y \in (0, 1)); x \neq y),$$

(9.3.10)

which are clearly continuous and symmetric.

Proposition 9.3.3
Let $T_+ : L^2(0, 1) \to L^2(0, 1)$ have kernel $T_+(x, y) = \sqrt{2b}\cos b\pi xy$ and let $T_- : L^2(0, 1) \to L^2(0, 1)$ have kernel $T_-(x, y) = \sqrt{2b}\sin b\pi xy$.

(i) *Then $T_\mp$ are self-adjoint and Hilbert–Schmidt, and unitarily equivalent to Hankel integral operators on $L^2(0, \infty)$.*
(ii) *Their squares satisfy $T_+^2 = K_+$ and $T_-^2 = K_-$.*
(iii) *Let λ_n be the eigenvalues of the integral equation*

$$\lambda_n f(x) = \int_{-1}^{1} \frac{\sin b\pi(x - y)}{\pi(x - y)} f(y)dy, \qquad (9.3.11)$$

ordered so that $1 \geq \lambda_0 \geq \lambda_1 \geq \ldots$. Then K_+ has eigenvalues $\lambda_0 \geq \lambda_2 \geq \ldots$ as an integral operator on $L^2(0, 1)$, whereas K_- has eigenvalues $\lambda_1 \geq \lambda_3 \geq \ldots$ as an integral operator on $L^2(0, 1)$.

Proof. (i) Let $U : L^2(0, 1) \to L^2(0, \infty)$ be the unitary operator $Uf(x) = e^{-x/2}f(e^{-x})$, and observe that

$$\int_0^1 \int_0^1 \sqrt{2b}\cos(b\pi uv)f(u)\bar{g}(v)dudv$$

$$= \sqrt{2b}\int_0^\infty\int_0^\infty e^{-(x+y)}\cos(b\pi e^{-(x+y)})f(e^{-x})\bar{g}(e^{-y})\,dxdy \quad (9.3.12)$$

expresses T_+ in terms of a Hankel operator. By replacing cos by sin, we obtain a similar conclusion about T_-.

(ii) A simple direct calculation shows that the kernel of $T_\pm^2$ satisfies (9.3.10).

(iii) By (ii), the operators $K_\pm$ are trace class and positive semidefinite. Since $K_-(x,-y) = -K_-(x,y)$ and $K_-(-x,y) = -K_-(x,y)$, any eigenfunction $f \in L^2(0,1)$ of

$$\lambda_{2n+1} f(x) = \int_0^1 K_-(x,y)f(y)\,dy \qquad (9.3.13)$$

extends naturally to an odd eigenfunction of (9.3.11) in $L^2(-1,1)$ with the same eigenvalue. Similarly, $K_+(-x,y) = K_+(x,y)$ and $K_+(x,-y) = K_+(x,y)$, so any eigenfunction of K_+ in $L^2(0,1)$ extends naturally to an even eigenfunction of (9.3.11) in $L^2(-1,1)$ with the same eigenvalue. By results of [116, p. 244], the eigenvalues that correspond to odd and even eigenfunctions interlace. □

The prolate spheroidal functions were introduced to solve the Helmholtz equation

$$\frac{\partial^2 u}{\partial x^2} + \frac{\partial^2 u}{\partial y^2} + \frac{\partial^2 u}{\partial z^2} = -\lambda u \qquad (9.3.14)$$

in an ellipsoid that has two axes of equal length. Due to the interpretation of Proposition 9.3.2, the prolate spheroidal functions are important in signal processing; so numerical values for the eigenvalues and eigenfunctions of L have been calculated for this application. The eigenfunction equation for L is

$$-(a^2 - x^2)f''(x) + 2xf'(x) + \pi^2 b^2 x^2 f(x) = \lambda f(x). \qquad (9.3.15)$$

We choose $b = 1/(2\pi)$, then write $x = \sqrt{a}s$ and $\lambda = a(p+1/2)$, so that $g(s) = f(\sqrt{a}s)$ satisfies

$$-(a - s^2)g''(s) + 2sg'(s) + \frac{1}{4}as^2 g(s) = a\left(\frac{1}{2}+p\right)g(s); \qquad (9.3.16)$$

so the limiting form of this equation as $a \to \infty$ becomes the Weber's parabolic cylinder equation

$$-g''(s) + \frac{s^2}{4}g(s) = \left(p+\frac{1}{2}\right)g(s), \qquad (9.3.17)$$

which is also associated with the quantum harmonic oscillator and Hermite functions; see [167]. Let the monic Hermite polynomials be

$$H_n(x) = (-1)^n e^{x^2/2}\frac{d^n}{dx^n}e^{-x^2/2} \qquad (n = 0,1,2,\dots), \qquad (9.3.18)$$

and let

$$\phi_n(x) = (n!)^{-1/2}(2\pi)^{-1/4}H_n(x)e^{-x^2/4} \qquad (n=0,1,\dots,x\in\mathbf{R}) \qquad (9.3.19)$$

be the Hermite system of orthonormal functions in $L^2(\mathbf{R})$, which otherwise arises by applying the Gram–Schmidt process to $(x^k e^{-x^2/4}/(2\pi)^{1/4})$ in $L^2(\mathbf{R})$.

The Hermite functions $(\phi_n)_{n=0}^{\infty}$ satisfy $\phi_0(x) = (2\pi)^{-1/4} e^{-x^2/4}$ and the identities

$$-\phi_n'(x) + (x/2)\phi_n(x) = \sqrt{n+1}\phi_{n+1}(x),$$
$$\phi_n'(x) + (x/2)\phi_n(x) = \sqrt{n}\phi_{n-1}(x); \qquad (9.3.20)$$

these are known as the raising and lowering operations as in [112]. Equivalently, they satisfy the Tracy–Widom system

$$\frac{d}{dx}\begin{bmatrix} \phi_n(x) \\ \phi_{n-1}(x) \end{bmatrix} = \begin{bmatrix} -x/2 & \sqrt{n} \\ -\sqrt{n} & x/2 \end{bmatrix}\begin{bmatrix} \phi_n(x) \\ \phi_{n-1}(x) \end{bmatrix}. \qquad (9.3.21)$$

Combining these equations, we see that ϕ_n gives a solution of (9.3.17) with $p = n$.

We introduce

$$K_n(x,y) = \sum_{j=0}^{n-1} \phi_j(x)\phi_j(y). \qquad (9.3.22)$$

Theorem 9.3.4 *Let X_n be matrices from the $n \times n$ Gaussian unitary ensemble with potential $v(x) = x^2/2$.*

(i) Then the k-point correlation functions for the eigenvalues satisfy

$$R_{n,k}(x_1,\ldots,x_k) = \det[\sqrt{n}K_n(\sqrt{n}x_j, \sqrt{n}x_m)]_{j,m=1,\ldots,k}. \quad (9.3.23)$$

(ii) The equilibrium measure is the semicircle law on $[-2,2]$.

(iii) The k-point correlation functions for eigenvalues in the spectral bulk satisfy

$$\hat{R}_{n,k}(x_1,\ldots,x_k) = \det\left[\frac{1}{\sqrt{n}}K_n\left(\frac{x_j}{\sqrt{n}}, \frac{x_m}{\sqrt{n}}\right)\right]_{j,m=1,\ldots,k}, \qquad (9.3.24)$$

where

$$\hat{R}_{n,k}(x_1,\ldots,x_k) \rightarrow \det[D_{1/\pi}(x_j,x_m)]_{j,m=1,\ldots,k} \qquad (n \rightarrow \infty). \quad (9.3.25)$$

Proof. (i) The kernel $\sqrt{n}K_n(\sqrt{n}x, \sqrt{n}y)$ gives an orthogonal projection of rank n. As in Theorem 8.3.4, there exists a constant C_n such that

$$\det[\sqrt{n}K_n(\sqrt{n}x_j, \sqrt{n}x_m)]_{j,m=1,\ldots,n}$$

$$= C_n \prod_{1 \le j < m \le n} (x_j - x_m)^2 \exp\left(-\sum_{j=1}^{n} x_j^2/2\right) \qquad (9.3.26)$$

is proportional to the joint probability density function of the determinantal random point field associated with the kernel $\sqrt{n}K_n(\sqrt{n}x, \sqrt{n}y)$. In terms of Theorem 2.5.3, we have the joint eigenvalue distribution of generalized unitary ensemble on $M_n^h(\mathbf{C})$ with potential $v(x) = x^2/2$. The k-point correlation functions are given by the minors of the determinant.

(ii) The equation for the equilibrium measure reduces to

$$\frac{x}{2} = \text{p.v.} \frac{1}{2\pi} \int_{-2}^{2} \frac{\sqrt{4-y^2}}{x-y} dy \qquad (-2 < x < 2), \tag{9.3.27}$$

as in Theorem 4.4.1.

(iii) The bulk of the spectrum is therefore concentrated on $[-2, 2]$, and the typical separation of adjacent eigenvalues there is $O(1/n)$. To analyse the eigenvalue distribution at this level, we replace the scaling of (i) by $x \mapsto x/\sqrt{n}$, and the corresponding scaled functions satisfy

$$\frac{d}{dx}\begin{bmatrix} \phi_n(x/\sqrt{n}) \\ \phi_{n-1}(x/\sqrt{n}) \end{bmatrix} = \begin{bmatrix} -x/(2n) & 1 \\ -1 & x/(2n) \end{bmatrix}\begin{bmatrix} \phi_n(x/\sqrt{n}) \\ \phi_{n-1}(x/\sqrt{n}) \end{bmatrix} \tag{9.3.28}$$

which is comparable to the sinusoidal system

$$\frac{d}{dx}\begin{bmatrix} f \\ g \end{bmatrix} = \begin{bmatrix} 0 & 1 \\ -1 & 0 \end{bmatrix}\begin{bmatrix} f \\ g \end{bmatrix} \tag{9.3.29}$$

on compact sets as $n \to \infty$. By the Christoffel-Darboux formula in Proposition 8.3.3, we have

$$\frac{1}{\sqrt{n}}K_n\left(\frac{x}{\sqrt{n}}, \frac{y}{\sqrt{n}}\right) = \sqrt{n}\frac{\phi_n(x/\sqrt{n})\phi_{n-1}(y/\sqrt{n}) - \phi_{n-1}(x/\sqrt{n})\phi_n(y/\sqrt{n})}{x-y}; \tag{9.3.30}$$

then the Plancherel–Rotach formula [153, p. 199] and [116, A10] gives

$$\frac{1}{\sqrt{n}}K_n\left(\frac{x}{\sqrt{n}}, \frac{y}{\sqrt{n}}\right) = \frac{\sin(x-y)}{\pi(x-y)} + o(1) \qquad (n \to \infty). \tag{9.3.31}$$

Hence we have convergence of the kernels in the sense of Proposition 8.5.2. □

9.4 Soft edges and the Airy kernel

- The edge of the spectrum for the Gaussian unitary ensemble is described by the Airy kernel.
- The Hermite equation can be rescaled so as to converge to Airy's equation.

- The Airy kernel is the square of a self-adjoint Hankel operator.
- The Airy kernel commutes with a second-order self-adjoint differential operator.
- The Airy kernel is universal as a soft edge distribution for diverse unitary ensembles.

While most of the eigenvalues of a random matrix from the Gaussian Unitary Ensemble lie in $[-2, 2]$, the largest and smallest eigenvalues can lie outside; in this sense, the edges ± 2 are *soft*. We wish to describe in more detail the distribution of the largest eigenvalue, and we introduce the Airy kernel and determinantal random point field to do this.

Definition (*Airy kernel*). Let W be the integral operator on $L^2(\mathbf{R})$ defined by the *Airy kernel*

$$W(x, y) = \frac{\text{Ai}(x)\text{Ai}'(y) - \text{Ai}'(x)\text{Ai}(y)}{x - y}. \qquad (9.4.1)$$

Proposition 9.4.1 *The kernel W defines a trace-class operator on $L^2(0, \infty)$ such that $0 \le W \le I$. Hence W gives a determinantal random point field on $(0, \infty)$.*

Proof. From the asymptotic expansion [153, p. 18] the Airy function satisfies

$$\text{Ai}(x) \asymp \frac{e^{-\frac{2}{3}x^{3/2}}}{2\sqrt{\pi}x^{1/4}} \qquad (x \to \infty) \qquad (9.4.2)$$

so we deduce that the Hankel operator $\Gamma : L^2(0, \infty) \to L^2(0, \infty)$, as in

$$\Gamma f(x) = \int_0^\infty \text{Ai}(x + y)f(y)\, dy, \qquad (9.4.3)$$

is self-adjoint and Hilbert–Schmidt.

Clarkson and McLeod proved the identity

$$\frac{\text{Ai}(x)\text{Ai}'(y) - \text{Ai}'(x)\text{Ai}(y)}{x - y} = \int_0^\infty \text{Ai}(x + u)\text{Ai}(u + y)\, du, \qquad (9.4.4)$$

and Tracy and Widom [157] showed how it follows from the differential equation

$$\frac{d}{dx}\begin{bmatrix} \text{Ai}(x) \\ \text{Ai}'(x) \end{bmatrix} = \begin{bmatrix} 0 & 1 \\ x & 0 \end{bmatrix}\begin{bmatrix} \text{Ai}(x) \\ \text{Ai}'(x) \end{bmatrix} \qquad (9.4.5)$$

as in Theorem 9.1.6. Hence $W = \Gamma^2$, so W is nonnegative and of trace class with

$$\text{trace}(W) = \int_0^\infty x\text{Ai}(x)^2\, dx.$$

Further, we have

$$\mathcal{F}f(y) = \int_0^\infty \int_{-\infty}^\infty e^{i\xi^3/3 + i\xi(x+y)} \frac{d\xi}{\sqrt{2\pi}} f(x) \frac{dx}{\sqrt{2\pi}}$$
$$= \int_{-\infty}^\infty e^{i\xi^3/3 + i\xi y} \mathcal{F}^* f(\xi) \frac{d\xi}{\sqrt{2\pi}}; \tag{9.4.6}$$

so that $\Gamma f(y) = \mathcal{F}^* e^{i\xi^3/3} \mathcal{F}^* f(\xi)$. Since $\mathcal{F}$ is a unitary operator, and $e^{i\xi^3/3}$ is unimodular, this proves the inequality $\|\Gamma\| \le 1$, which shows that $W = \Gamma^2 \le I$; so $0 \le W \le I$. □

Let λ_j be eigenvalues from a $n \times n$ random matrix of the Gaussian Unitary Ensemble, and let ξ_j be the scaled eigenvalues such that

$$\lambda_j = 2 + \frac{\xi_j}{n^{2/3}} \qquad (j = 1, \dots, n). \tag{9.4.7}$$

When $\xi_j > 0$, the eigenvalue λ_j lies outside the bulk of the spectrum; whereas when $-2 < \xi_j < 0$, the eigenvalue λ_j lies inside. The edge 2 is *soft* since $\lambda_j > 2$ occurs with positive probability, and when n is large, we should expect this probability to be small, as reflected by the scaling factor $n^{-2/3}$.

Proposition 9.4.2 (*Aubrun*). *Let K_n be the kernel such that $\det[K_n(x_j, x_k)]_{j,k=1,\dots,n}$ gives the random point field that describes the distribution of the $(\xi_j)_{j=1}^n$. On compact subsets of $\mathbf{C}$, there is uniform convergence*

$$\det(I - zK_n \mathbf{I}_J) \to \det(I - zW\mathbf{I}_J) \qquad (n \to \infty) \tag{9.4.8}$$

for all intervals $J \subseteq (0, \infty)$.

Proof. The Hermite functions ϕ_n are solutions to the parabolic cylinder equation (9.3.17) with $p = n$, so

$$-\phi_n''(x) + \frac{x^2}{4}\phi_n(x) = (n + 1/2)\phi_n(x). \tag{9.4.9}$$

The scaled Hermite functions

$$\Phi_n(x) = n^{1/12}\phi_n\left(2(n+1/2)^{1/2} + n^{-1/6}x\right) \tag{9.4.10}$$

satisfy the differential equations

$$\Phi_n''(x) = \left(1 + \frac{1}{2\sqrt{n}}\right)^{1/2} x\Phi_n(x) + \frac{x^2}{4n^{2/3}}\Phi_n(x), \tag{9.4.11}$$

the coefficients of which converge to the coefficients of the Airy differential equation [153, page 18] $y'' = xy$. One can show that

$$\Phi_n(y) = \text{Ai}(y) + O(n^{-3/4}) \qquad (y \to \infty), \tag{9.4.12}$$

so $\Phi_n(y)$ converges uniformly on compact sets to $\mathrm{Ai}(y)$ as $n \to \infty$; see [153, page 99]. By following the proof of Theorem 9.1.6, one can deduce that

$$\sqrt{n} \frac{\phi_n(x)\phi_{n-1}(y) - \phi_{n-1}(x)\phi_n(y)}{x - y}$$

$$= \frac{\sqrt{n}}{2} \int_0^\infty \left(\phi_{n-1}(x+u)\phi_n(y+u) + \phi_n(x+u)\phi_{n-1}(y+u) \right) du.$$

$$(9.4.13)$$

The left-hand side represents the projection onto $\mathrm{span}\{\phi_j : j = 0, \dots, n-1\}$ as in the Christoffel–Darboux formula 8.3.3, and hence defines an operator K_n such that $0 \leq K_n \leq I$; whereas the right-hand side is a sum of products of Hankel operators. By (9.3.20), we have

$$\frac{\phi_n(x)\phi_{n-1}(y) - \phi_{n-1}(x)\phi_n(y)}{x - y} = \frac{\phi_{n-1}(x)\phi'_{n-1}(y) - \phi'_{n-1}(x)\phi_{n-1}(y)}{x - y}$$

$$- \frac{1}{2\sqrt{n}}\phi_{n-1}(x)\phi_{n-1}(y). \qquad (9.4.14)$$

Letting $\tau_n(x) = 2(n+1/2)^{1/2} + n^{-1/6}x$ in (9.4.13), we obtain after applying (9.4.14) the identity

$$K_n(x,y) = \frac{\phi_{n-1}(\tau_n(x))\phi'_{n-1}(\tau_n(y)) - \phi'_{n-1}(\tau_n(x))\phi_{n-1}(\tau_n(y))}{x - y}$$

$$+ \frac{\phi_{n-1}(\tau_n(x))\phi_{n-1}(\tau_n(y))}{2\sqrt{n}}$$

$$= \frac{n^{1/6}}{2} \int_0^\infty \left(\phi_{n-1}(\tau_n(x+v))\phi_n(\tau_n(y+v)) \right.$$

$$\left. + \phi_n(\tau_n(x+v))\phi_{n-1}(\tau_n(y+v)) \right) dv, \qquad (9.4.15)$$

and in the limit as $n \to \infty$, one obtains the Airy functions, by (9.4.12), so

$$K_n(x,y) \to W(x,y) \qquad (9.4.16)$$

where

$$W(x,y) = \int_0^\infty \mathrm{Ai}(x+v)\mathrm{Ai}(y+v)\, du. \qquad (9.4.17)$$

By means of some more detailed analysis, Aubrun showed that

$$\det(I - zK_n \mathbf{I}_J) \to \det(I - zW\mathbf{I}_J) \qquad (9.4.18)$$

uniformly on compact sets as $n \to \infty$ for each interval $J \subseteq (0, \infty)$. $\quad\square$

In particular, the distribution of the largest eigenvalue λ_n of a matrix from the GUE is described by the random variable ξ_n with $\lambda = 2 + n^{-2/3}\xi_n$, where the cumulative distribution function of ξ_n satisfies

$$\mathbf{P}[\xi_n \leq t] \to \det(I - W\mathbf{I}_{(t,\infty)}) \qquad (n \to \infty). \tag{9.4.19}$$

One can calculate the eigenvalues of the operator

$$\mathbf{I}_{(t,\infty)}W\mathbf{I}_{(t,\infty)} \longleftrightarrow \int_t^\infty \mathrm{Ai}(x+u)\mathrm{Ai}(y+u)\,du, \tag{9.4.20}$$

and hence compute the right-hand side of (9.4.19). Since $\mathrm{Ai}(t)$ decays rapidly as $t \to \infty$, the eigenvalues of $\mathbf{I}_{(t,\infty)}W\mathbf{I}_{(t,\infty)}$ decay rapidly as $t \to \infty$.

Remarkably, Tracy and Widom [157] identified this determinant in terms of a solution of the Painlevé II equation, as we shall see in Section 9.7. Next we give a method for finding the eigenfunctions and eigenvalues of W.

Proposition 9.4.3 *Let A_s be the Hankel operator on $L^2(0,\infty)$ that has kernel $\mathrm{Ai}(x+y+s)$ and let L_s be the differential operator*

$$L_s f(x) = -\frac{d}{dx}\left(x\frac{d}{dx}f\right) + x(x+s)f(x) \tag{9.4.21}$$

with boundary condition $f(0) = 0$. Then A_s and L_s commute; so eigenfunctions for A_s are eigenfunctions of L_s.

Proof. First we consider $A_s L_s f$. By integrating by parts, we obtain

$$\int_0^\infty \mathrm{Ai}(x+y+s))\left(-\frac{d}{dy}\left(y\frac{d}{dy}f\right) + y(y+s)f(y)\right)dy$$
$$= \int_0^\infty \left(-\left(y\frac{d}{dy}\mathrm{Ai}(x+y+s))\right) + y(y+s)\mathrm{Ai}(x+y+s)\right)f(y)\,dy, \tag{9.4.22}$$

which reduces by the Airy equation to

$$-\int_0^\infty \left(\mathrm{Ai}'(x+y+s) + xy\mathrm{Ai}(x+y+s)\right)f(y)\,dy, \tag{9.4.23}$$

and one can obtain the same expression by computing $L_s A_s f(x)$.

Since the Airy function is of rapid decay as $x \to \infty$, the operator A_s is Hilbert–Schmidt. In particular, A_s is compact and self-adjoint, and its eigenfunctions give a complete orthonormal basis for $L^2(0,\infty)$. Likewise, the operator $(iI + L_s)^{-1}$ is compact and normal, and its eigenfunctions give a complete orthonormal basis for $L^2(0, \infty)$. As A_s and $(iI + L_s)^{-1}$ operators commute, they have a common orthonormal basis, as required.   □

9.5 Hard edges and the Bessel kernel

- The Bessel kernel describes the hard edges of the Jacobi ensemble.
- The Jacobi polynomials satisfy a differential equation which converges to a variant of the Bessel equation.
- The Bessel kernel is the square of a Hankel operator.
- The Bessel kernel commutes with a self-adjoint second order differential operator.

Definition (*Jacobi's weight*) *[153]*. For $\alpha, \beta > -1$ we introduce the probability density function on $(-1, 1)$ by

$$w(x) = 2^{-(\alpha+\beta+1)} \frac{\Gamma(\alpha + \beta + 2)}{\Gamma(\alpha + 1)\Gamma(\beta + 1)} (1 - x)^\alpha (1 + x)^\beta \qquad (x \in (-1, 1)).$$

$$(9.5.1)$$

When $0 > \alpha, \beta > -1$ the density is unbounded as $x \to 1-$ and as $x \to (-1)+$. This weight is used to model an electrostatic system on $(-1, 1)$ in which there is a negative charge at (-1) and a negative charge at $(+1)$, so a positively charged particle is attracted to the ends of interval. When there are several positively charged particles, we have the following ensemble.

Definition (*Jacobi's Ensemble*). For n a positive integer, we introduce the simplex

$$\Delta^n = \{(x_j)_{j=1}^n \in \mathbf{R}^n : -1 \leq x_1 \leq \cdots \leq x_n \leq 1\}$$

and let $\alpha, \beta > -1$. Then there exists $Z_n < \infty$, which depends upon these constants, such that

$$\mu_n^{(2)}(dx) = \frac{1}{Z_n} \prod_{j=1}^n (1 + x_j)^\beta (1 - x_j)^\alpha \prod_{1 \leq j < k \leq n} (x_k - x_j)^2 \, dx_1 \cdots dx_n$$

$$(9.5.2)$$

determines a probability measure on Δ^n. We define the *Jacobi ensemble* of order n with parameters $\alpha, \beta > -1$ to be the probability measure $\mu_n^{(2)}$ as in Theorem 8.3.4. One can regard the $(x_j)_{j=1}^n$ as the ordered eigenvalues of some $n \times n$ Hermitian matrix which is random under a suitable probability measure. So the Jacobi unitary ensemble of order n is such a space of random matrices.

When $\alpha, \beta < 0$, the density of $\mu_n^{(2)}$ is unbounded at $1-$ and $(-1)+$; further, the eigenvalues are constrained to lie inside Δ^n. The following electrostatic interpretation of the factors is illuminating:

(i) $(1 + x_j)^\beta$ arises from the attraction between a negative charge at -1 and a positive charge at x_j;

(ii) $(1 - x_j)^\alpha$ arises from the attraction between a unit negative charge at 1 and a unit positive charge at x_j;

(iii) $(x_k - x_j)^2$ arises from the mutual repulsion between the positive charges at x_j and x_k.

Hence the charges x_j all lie in $[-1, 1]$ and are attracted towards the ends ± 1, but are prevented from all accumulating there by their mutual repulsion; so at equilibrium, the charges are sparse in the middle of $[-1, 1]$ and denser near to ± 1. For these reasons, the spectral edges ± 1 are said to be *hard*.

The polynomials that correspond to this ensemble are the Jacobi polynomials.

Definition (*Jacobi polynomials*) *[153]*. The Jacobi polynomials $P_n^{\alpha,\beta}$ are the orthogonal polynomials of degree $n = 0, 1, 2, \ldots$ with positive leading terms that are defined by the Gram–Schmidt process with respect to the weight w, and subject to the normalization

$$P_n^{\alpha,\beta}(1) = \frac{\Gamma(n+1+\alpha)}{\Gamma(n+1)\Gamma(1+\alpha)} \asymp \frac{n^\alpha}{\Gamma(\alpha+1)} \qquad (n \to \infty). \tag{9.5.3}$$

On account of this normalization at the right-hand end, some subsequent formulæ appear to be asymmetrical in α and β. In particular, when $\alpha = \beta = -1/2$, we observe that the $P_n^{-1/2,-1/2}$ are proportional to the Chebyshev polynomials T_n, which suggests that the substitution $x = \cos t$ should be helpful.

We introduce the kernel

$$K_n(x,y) = C_n(\alpha,\beta)\frac{P_n^{\alpha,\beta}(x)P_{n-1}^{\alpha,\beta}(y) - P_n^{\alpha,\beta}(y)P_{n-1}^{\alpha,\beta}(x)}{x - y}(w(x)w(y))^{1/2},$$

$$\tag{9.5.4}$$

where we have introduced the constant

$$C_n(\alpha,\beta) = \frac{2\Gamma(n+1)\Gamma(n+1+\alpha+\beta)\Gamma(\alpha+1)\Gamma(\beta+1)}{(2n+\alpha+\beta)\Gamma(n+\alpha)\Gamma(n+\beta)\Gamma(\alpha+\beta+2)}$$

from [153, p. 71] since the $P_n^{\alpha,\beta}$ need rescaling so as to become orthonormal with respect to w on $(-1, 1)$.

Proposition 9.5.1 *The kernel K_n generates the determinantal random point field associated with the eigenvalues from the Jacobi ensemble on $(-1, 1)$.*

Proof. The kernel K_n gives the orthogonal projection of rank n onto the span of $1, \ldots, x^{n-1}$ by Proposition 8.3.3. The rest follows from Theorem 8.3.4. □

We now introduce the kernel that describes eigenvalue distributions at the hard edge at $x = 1$; later in the section, we shall show how it arises for the edges of the Jacobi ensemble.

Let J_α be the Bessel function of order α; then $f(z) = z^{-\alpha/2} J_\alpha(\sqrt{z})$ satisfies

$$z\frac{d}{dz}\begin{bmatrix} f \\ f' \end{bmatrix} = \begin{bmatrix} 0 & z \\ -1/4 & -(\alpha+1) \end{bmatrix}\begin{bmatrix} f \\ f' \end{bmatrix} \qquad (9.5.5)$$

on $\mathbf{C} \setminus (-\infty, 0]$.

Definition (*Bessel kernel*). The Bessel kernel is

$$F(x, y) = \frac{J_\alpha(\sqrt{x})\sqrt{y}J_\alpha'(\sqrt{y}) - \sqrt{x}J_\alpha'(\sqrt{x})J_\alpha(\sqrt{y})}{2(x-y)}$$

for all $\qquad (0 < x, y < 1, x \neq y).$ $\qquad (9.5.6)$

In the context of Jacobi ensembles, we are principally interested in the behaviour of eigenvalues for small $x > 0$, so it is natural to restrict attention to operators on $(0, 1)$.

Theorem 9.5.2 *(i) The integral operator $T_\alpha : L^2(0, 1) \to L^2(0, 1)$ given by*

$$T_\alpha f(x) = \int_0^1 J_\alpha(2\sqrt{tx})f(t)dt \qquad (9.5.7)$$

has a square with kernel $T_\alpha^2(x, y) = 4F(4x, 4y)$.

(ii) The integral operator on $L^2((0, 1), dx)$ with kernel F defines a determinantal random point field.

Proof. (i) The following identity can be deduced from the differential equation

$$\frac{J_\alpha(2\sqrt{x})\sqrt{y}J_\alpha'(2\sqrt{y}) - J_\alpha(2\sqrt{y})\sqrt{x}J_\alpha'(2\sqrt{x})}{x - y} = \int_0^1 J_\alpha(2\sqrt{tx})J_\alpha(2\sqrt{ty})dt.$$

(9.5.8)

The left-hand side is $4F(4x, 4y)$, while the right-hand side is the kernel of T_α^2. (The author made a sign error in (5.8) of [22].)

(ii) Since $F = T_\alpha^2$ is the product of self-adjoint operators, we have $0 \le F$. Further, F is of trace class since the operators on the right-hand side are Hilbert–Schmidt.

Let $g(x) = \int_0^1 J_\alpha(2\sqrt{xt})f(t)dt$, and introduce the functions $\tilde{g}(y) = g(y^2/4)$ and $\tilde{f}(u) = 2f(u^2)\mathbf{I}_{[0,1]}(u)$. Then we have a Hankel transform

$$\tilde{g}(y) = \int_0^\infty J_\alpha(yu)\tilde{f}(u)u\,du,$$

(9.5.9)

so by Hankel's inversion theorem [146]

$$\int_0^\infty \tilde{g}(y)^2 y\,dy = \int_0^\infty \tilde{f}(u)^2 u\,du,$$

(9.5.10)

so reverting to the original variables we have

$$\int_0^1 g(v)^2 dv \le \int_0^1 f(t)^2 dt;$$

(9.5.11)

hence $\|F\| \le I$, so $0 \le F \le I$. We can now use Theorem 8.2.7 to deduce the existence of a determinantal random point field. $\square$

Proposition 9.5.3 *Let L_α be the differential operator*

$$L_\alpha f(x) = -\frac{d}{dx}\left(x(1 - x)\frac{d}{dx}f\right) + \left(x - \frac{\alpha^2}{4x} - 2\mu\right)f(x)$$

(9.5.12)

with boundary conditions $f(0) = f(1) = 0$. Then T_α commutes with L_α.

Proof. We apply a unitary transformation so that we can work on $L^2(0, \infty)$. The formula

$$e^{-u}T_\alpha f(e^{-2u}) = 2\int_0^\infty e^{-(u+v)}J_\alpha(2e^{-(u+v)})f(e^{-2v})e^{-v}dv$$

(9.5.13)

suggests that we change variable to v, where $x = e^{-2v}$. Then $\phi(v) = 2e^{-v}J_\alpha(2e^{-v})$ satisfies

$$e^v\phi''(v) + 2e^v\phi'(v) + (4e^{-v} + (1 - \alpha^2)e^v)\phi(v) = 0,$$

so we apply Theorem 9.2.1 with $a(x) = e^{2x} - 1$ and $b(x) = 4e^{-2x} - (1-\alpha^2)e^{2x} + \mu$ to obtain an operator L that commutes with the Hankel operator Γ_ϕ that has kernel $\phi(x+y)$.

The unitary transformation $S : L^2(0,1) \to L^2(0,\infty)$ is given by $Sf = g$, where $g(v) = \sqrt{2}e^{-v}f(e^{-2v})$, and the quadratic form associated to L has principal term

$$\int_0^\infty (e^{2v} - 1)g'(v)^2 \, dv = \int_0^1 \left(\frac{1}{x} - 1\right)\left(2xf'(x) + f(x)\right)^2 dx \qquad (9.5.14)$$

$$= \int_0^1 4x(1-x)f'(x)^2 \, dx + \int_0^1 \left(\frac{1}{x} + 1\right)f(x)^2 \, dx + \left[2(1-x)f(x)^2\right]_0^1,$$

as one shows by integrating by parts. The term in square brackets vanishes on account of the boundary conditions, and we recover the operator L_α by polarizing; hence the said operators commute. □

Now we return to the Jacobi ensemble. Focusing our attention on the edge at 1, we introduce the scaled eigenvalues ξ_j by $x_j = \cos \xi_j / \sqrt{n}$, to ensure that the mean spacing of the ξ_j is of order $O(1)$ near to the hard edge at $x_j \approx 1$. Then for some $C_\alpha > 0$, we rescale the kernel K_n to

$$\tilde{K}_n(x,y) = \frac{C_\alpha}{n^2} K_n\left(\cos\frac{\sqrt{x}}{n}, \cos\frac{\sqrt{y}}{n}\right). \qquad (9.5.15)$$

Theorem 9.5.4 *The kernel* $\tilde{K}_n$ *generates a determinantal random point field such that the k point correlation functions satisfy*

$$\tilde{R}_{n,k}(x_1,\ldots,x_k) = \det[\tilde{K}_n(x_j,x_m)]_{j,m=1,\ldots,k}, \qquad (9.5.16)$$

and

$$\tilde{R}_{n,k}(x_1,\ldots,x_k) \to \det[F(x_j,x_m)]_{j,m=1,\ldots,k} \qquad (n \to \infty). \qquad (9.5.17)$$

Proof. The main idea is that $\tilde{K}_n \to F$ in trace class norm on $L^2(0,1)$ as $n \to \infty$. The details of the proof are in [113]; Lemma 9.5.5 is the crucial step.

The Jacobi polynomials $P_n = P_n^{\alpha,\beta}$ satisfy the differential equation

$$(1 - x^2)P_n''(x) + (\beta - \alpha + (\alpha + \beta + 2)x)P_n'(x)$$
$$+ n(n + \alpha + \beta + 1)P_n(x) = 0. \qquad (9.5.18)$$

□

Lemma 9.5.5 *Let $f_n(x) = n^{-\alpha} P_n(\cos \frac{\sqrt{x}}{n})$. Then*

$$4x f_n''(x) + \left(2 - \frac{2\sqrt{x}}{n}(\beta - \alpha)\operatorname{cosec} \frac{\sqrt{x}}{n} \right.$$
$$+ \frac{2}{n}(\alpha + \beta + 1)\sqrt{x} \cot \frac{\sqrt{x}}{n}\Big) f_n'(x)$$
$$+ \frac{1}{n}(n + \alpha + \beta + 1) f_n(x) = 0. \tag{9.5.19}$$

As $n \to \infty$, the coefficients of this differential equation converge uniformly on compact sets, so the limiting form of this differential equation is

$$4x f''(x) + 4(\alpha + 1) f'(x) + f(x) = 0 \tag{9.5.20}$$

with $f(0) = 1/\Gamma(\alpha + 1)$.

Proof. We have

$$z \frac{d}{dz} \begin{bmatrix} f_n(z) \\ f_n'(z) \end{bmatrix} = \Omega_n(z) \begin{bmatrix} f_n(z) \\ f_n'(z) \end{bmatrix} \tag{9.5.21}$$

where

$$\Omega_n(z)$$
$$= \begin{bmatrix} 0 & z \\ -(n+\alpha+\beta+1)/(4n) & \frac{\sqrt{z}}{2n}\big((\beta-\alpha)\operatorname{cosec}\frac{\sqrt{z}}{n} - (\alpha+\beta+1)\cot\frac{\sqrt{z}}{n}\big) - \frac{1}{2} \end{bmatrix}$$

satisfies

$$\Omega_n(z) \to \begin{bmatrix} 0 & z \\ -1/4 & -(\alpha+1) \end{bmatrix} \quad (n \to \infty) \tag{9.5.22}$$

uniformly on compact sets. $\qquad\square$

Lemma 9.5.6 *The scaled Jacobi polynomials converge, so*

$$n^{-\alpha} P_n\left(\cos \frac{\sqrt{z}}{n}\right) \to 2^\alpha z^{-\alpha/2} J_\alpha(\sqrt{z}) \quad (n \to \infty) \tag{9.5.23}$$

uniformly on compact subsets of $\mathbf{C}$.

Proof. This is due to Mehler and Heine; see [153, p. 190]. By direct calculation one can show that $f(x) = x^{-\alpha/2} J_\alpha(\sqrt{x})$ satisfies the differential equation (9.5.20), or equivalently the matrix form (9.5.5). By Lemma 9.5.5, the solutions of (9.5.19) converge uniformly on compact sets to the solution of (9.5.20). When $0 \le \alpha < 1$, $f(z) = 2^\alpha z^{-\alpha/2} J_\alpha(\sqrt{z})$ is the unique solution of (9.5.20) that is bounded at $z = 0$ and that satisfies the initial condition, so the result follows from Lemma 9.5.5. $\qquad\square$

We can recover Theorem 9.5.4 by using the further identity

$$(2n + \alpha + \beta)\frac{d}{dx}P_n(x) = 2(n + \alpha)(n + \beta)P_{n-1}(x)$$
$$- n\big((2n + \alpha + \beta)x + \beta - \alpha\big)P_n(x), \quad (9.5.24)$$

which leads to

$$2(n + \alpha)(n + \beta)\frac{P_n(\cos\frac{\sqrt{y}}{n})P_{n-1}(\cos\frac{\sqrt{x}}{n}) - P_{n-1}(\cos\frac{\sqrt{y}}{n})P_n(\cos\frac{\sqrt{x}}{n})}{\cos\frac{\sqrt{y}}{n} - \cos\frac{\sqrt{x}}{n}}$$

$$= -n(2n + \alpha + \beta)P_n\left(\cos\frac{\sqrt{x}}{n}\right)P_n\left(\cos\frac{\sqrt{y}}{n}\right)$$

$$+ 2n(2n + \alpha + \beta)$$

$$\times\left(\frac{P_n(\cos\frac{\sqrt{x}}{n})\sqrt{y}\sin\frac{\sqrt{y}}{n}\frac{d}{dy}P_n(\cos\frac{\sqrt{y}}{n}) - P_n(\cos\frac{\sqrt{y}}{n})\sqrt{x}\sin\frac{\sqrt{x}}{n}\frac{d}{dx}P_n(\cos\frac{\sqrt{x}}{n})}{\cos\frac{\sqrt{y}}{n} - \cos\frac{\sqrt{x}}{n}}\right),$$

$$(9.5.25)$$

where $\cos\frac{\sqrt{y}}{n} - \cos\frac{\sqrt{x}}{n} = \frac{x-y}{2n^2} + O(\frac{1}{n^4})$ as $n \to \infty$. Hence the right-hand side is asymptotic to a multiple of the Bessel kernel

$$\frac{x^{-\alpha/2}J_\alpha(\sqrt{x})y\frac{d}{dy}(y^{-\alpha/2}J_\alpha(\sqrt{y})) - y^{-\alpha/2}J_\alpha(\sqrt{y})x\frac{d}{dx}(x^{-\alpha/2}J_\alpha(\sqrt{x}))}{x - y}.$$

$$(9.5.26)$$

These are the main steps in the proof of Theorem 9.5.4, which appears in more detail in [113].

9.6 The spectra of Hankel operators and rational approximation

- A Hankel integral operator has a natural expression in terms of the Laguerre basis.
- Hankel integral operators are unitarily equivalent to Hankel matrices.
- The spectra of Hankel operators have special properties.

Definition (*Hankel matrix [127]*). For any $(c_j) \in \ell^2(\mathbf{Z}_+)$, we can define a Hankel matrix $[c_{j+k-1}]_{j,k=1}^\infty$ which gives a densely defined linear operator in $\ell^2(\mathbf{Z}_+)$ with respect to the standard orthonormal basis.

The *symbol* of the Hankel matrix is some function $\psi \in L^2(\mathbf{T};\mathbf{C})$ such that

$$c_n = \int_{\mathbf{T}} \psi(e^{i\theta})e^{-in\theta}\frac{d\theta}{2\pi} \quad (n = 1, 2, \dots). \quad (9.6.1)$$

Note that the Hankel matrix does not determine the negative Fourier coefficients of ψ, hence one can choose them advantageously. There are different conventions in common use concerning the role of H^2 and $L^2 \ominus H^2$ as the codomain of the Hankel operator.

There is a simple relationship between Hankel integral operators and Hankel matrices.

Definition (*Laguerre functions*) Let $L_n^{(\alpha)}$ be the Laguerre polynomial of degree n, so that

$$L_n^{(\alpha)}(x) = \frac{x^{-\alpha}}{n!} e^x \frac{d^n}{dx^n}\left(e^{-x} x^{(n+\alpha)}\right) \qquad (s>0)$$

$$= \sum_{\kappa=0}^{n} (-1)^\kappa \frac{\Gamma(\alpha+n+1)}{\kappa!(n-\kappa)!\Gamma(\alpha+\kappa+1)} x^\kappa. \qquad (9.6.2)$$

and let $h_n(x) = 2^{1/2} e^{-s} L_n^{(0)}(2s)$, so that $(h_n)_{n=0}^\infty$ gives a complete orthonormal basis for $L^2(0,\infty)$.

The Laguerre polynomials satisfy the differential equation

$$x^2 \frac{d^2}{dx^2} L_n^{(\alpha)}(x) + (\alpha+1-x)\frac{d}{dx} L_n^{(\alpha)}(x) + nL_n^{(\alpha)}(x) = 0, \qquad (9.6.3)$$

hence the function $u(x) = xe^{-x/2} L_n^{(1)}(x)$ satisfies

$$\frac{d}{dx}\begin{bmatrix} u \\ u' \end{bmatrix} = \begin{bmatrix} 0 & 1 \\ 1/4 - (n+1)/x & 0 \end{bmatrix}\begin{bmatrix} u \\ u' \end{bmatrix}. \qquad (9.6.4)$$

The right half-plane is $\mathbf{C}_+ = \{z : \Re z > 0\}$. The Hardy space H^2 on $\mathbf{C}_+$ may be defined by modifying (4.2.45) in the obvious way.

Proposition 9.6.1 (*i*) *There is a unitary map* $U : H^2(\mathbf{D}; d\theta/2\pi) \to L^2((0,\infty); dx)$ *given by*

$$\sum_{n=0}^\infty a_n z^n \mapsto \sum_{n=0}^\infty a_n h_n; \qquad (9.6.5)$$

(*ii*) *the Laplace transform gives a unitary map* $\mathcal{L} : L^2((0,\infty); dx) \to H^2(\mathbf{C}_+; dy/2\pi)$;

(*iii*) *the composition* $\Phi = \mathcal{L} \circ U$ *gives a unitary map* $H^2(\mathbf{D}; d\theta/2\pi) \to H^2(\mathbf{C}_+; dy/2\pi)$ *such that*

$$\Phi f(\lambda) = \frac{\sqrt{2}}{\lambda+1} f\left(\frac{\lambda-1}{\lambda+1}\right). \qquad (9.6.6)$$

Proof. (i) The (h_n) give a complete orthonormal basis for $L^2(0,\infty)$ by Proposition 1.8.4.

(ii) This is essentially the Paley–Wiener theorem [102].

(iii) One can calculate

$$\mathcal{L}h_n(\lambda) = \int_0^\infty e^{-\lambda s} \frac{e^{s/2}}{\sqrt{2}n!} \frac{d^n}{ds^n} e^{-s} s^n \, ds$$

$$= \sqrt{2} \frac{(\lambda - 1)^n}{(\lambda + 1)^{n+1}}, \qquad (\Re\lambda > 0) \qquad (9.6.7)$$

so that

$$\Phi : \sum_{n=0}^\infty a_n z^n \mapsto \sum_{n=0}^\infty a_n \sqrt{2} \frac{(\lambda - 1)^n}{(\lambda + 1)^{n+1}}, \qquad (9.6.8)$$

from which the result is clear.   □

Exercise 9.6.2 (i) Let $\phi(x) = p(x)e^{-\alpha x}$ where $p(x)$ is a polynomial function of degree n. Show that the Hankel integral operator with kernel $\phi(x + y)$ has rank less than or equal to $n + 1$.

(ii) Let $\phi(x) = \sum_{j=0}^n a_j h_j(x)$ and consider the Hankel operator with kernel $\phi(x + y)$. Show that for a typical $f(x) = \sum_{j=0}^\infty b_j h_j(x)$ we have

$$\Gamma_\phi f(x) = \sum_{j=0}^n a_j b_j h_j(x). \qquad (9.6.9)$$

Proposition 9.6.3 *(i) A bounded Hankel integral operator is unitarily equivalent to an operator that is represented by a Hankel matrix with respect to some orthonormal basis.*
(ii) The Hankel matrix $[b_{j+k-1}]_{j,k=1}^\infty$ has finite rank if and only if the symbol $\mathcal{F}\phi$ may be chosen to be a rational function.

Proof. (i) See [127].

(ii) This is Kronecker's theorem; see [127].   □

The self-adjoint and bounded Hankel operators on Hilbert space have been characterized up to unitary equivalence by Peller, Megretskiĭ and Treil [127].

For a real sequence (γ_j) the multiplicity function $\nu : \mathbf{R} \to \{0, 1, \dots\} \cup \{\infty\}$ is

$$\nu(\lambda) = \sharp\{j : \gamma_j = \lambda\}. \qquad (9.6.10)$$

Theorem 9.6.4 *Suppose that Γ is a self-adjoint and compact Hankel operator with eigenvalues $(\gamma_j)_{j=1}^\infty$ listed according to geometric multiplicity. Then the multiplicity function satisfies*

(i) $\nu(0) \in \{0, \infty\}$, so the nullspace of Γ is zero, or infinite-dimensional;
(ii) $|\nu(\lambda) - \nu(-\lambda)| \leq 1$ for all $\lambda > 0$.

Conversely, given any real sequence $(\gamma_j)_{j=1}^{\infty}$ *that satisfies (i) and (ii), there exists* $\phi \in L^2(0, \infty)$ *such that* Γ_{ϕ} *is a bounded and self-adjoint Hankel operator with eigenvalues* (γ_j) *listed according to multiplicity.*

Suppose further that $\sum_{j=1}^{\infty} |\gamma_j| < \infty$ *and that* $\gamma_j < 1$ *for all* j. *Then* $\det(I - \Gamma_{\phi}) = \prod_{j=1}^{\infty}(1 - \gamma_j)$ *converges.*

Proof. This result is due to Megretskiĭ, Peller and Treil, as in [127]. $\square$

Exercise. Deduce that if Γ is a positive and compact Hankel operator, then the positive eigenvalues of Γ are simple.

We use the term rational approximation of the Hankel operator to mean approximation by finite-rank Hankel operators.

Proposition 9.6.5 *Given* ϕ *such that the Hankel operator* Γ_{ϕ} *is self-adjoint and Hilbert–Schmidt, there exists* ϕ_j ($j = 1, 2, \dots$) *such that* Γ_{ϕ_j} *is a finite-rank Hankel operator and such that*

$$\det(I - z\Gamma_{\phi_j}^2) \to \det(I - z\Gamma_{\phi}^2) \qquad (j \to \infty) \qquad (9.6.11)$$

uniformly on compact subsets of **C**.

Proof. Given a Hilbert–Schmidt Hankel matrix $\Gamma = [a_{j+k}]$ and $\varepsilon > 0$, the truncated Hankel matrix $\Gamma_N = [a_{j+k}\mathbf{I}_{[0,N]}(j+k)]$ has finite rank and satisfies $\|\Gamma - \Gamma_N\|_{c^2} < \varepsilon$ for all sufficiently large N. Using Proposition 9.6.3, one can pass from Hankel matrices to Hankel integral operators on $L^2(0, \infty)$. Hence for any Hilbert–Schmidt Hankel integral operator Γ_{ϕ} and $\varepsilon > 0$, there exists a finite-rank Hankel operator Γ_{ψ} such that $\|\Gamma_{\phi} - \Gamma_{\psi}\|_{c^2} < \varepsilon$. Note that

$$\Gamma_{\phi}^2 - \Gamma_{\psi}^2 = \Gamma_{\phi}^2 - \Gamma_{\psi}\Gamma_{\phi} + \Gamma_{\psi}\Gamma_{\phi} - \Gamma_{\psi}^2, \qquad (9.6.12)$$

so

$$\|\Gamma_{\phi}^2 - \Gamma_{\psi}^2\|_{c^1} \leq \|\Gamma_{\phi} - \Gamma_{\psi}\|_{c^2}\|\Gamma_{\phi}\|_{c^2} + \|\Gamma_{\psi}\|_{c^2}\|\Gamma_{\phi} - \Gamma_{\psi}\|_{c^2} \qquad (9.6.13)$$

We have $\|\Gamma_n - \Gamma\| \to 0$ as $n \to \infty$, so $\Gamma_n^2 \to \Gamma^2$ as $n \to \infty$.

By Kronecker's theorem, as in Proposition 9.6.3, Γ_{ϕ} has finite rank if and only if $\hat{\phi}$ may be chosen to be a rational function. When the rational function has the form

$$\hat{\phi}(z) = \sum_{j=1}^{N} \frac{c_j/i}{z - i\alpha_j} \qquad (9.6.14)$$

with $\Re \alpha_j > 0$ so that the poles are in the upper half plane, we have

$$\phi(x) = \int_{-\infty}^{\infty} \hat{\phi}(z) e^{izx} \frac{dz}{2\pi} = \sum_{j=1}^{N} c_j e^{-\alpha_j x} \qquad (x > 0); \qquad (9.6.15)$$

by partially differentiating with respect to α_j one can introduce higher order poles.   □

The term *exponential bases* is used somewhat loosely to describe the properties of sums $\sum_j a_j e^{-\alpha_j x}$ in $L^2((0,\infty); dx)$ for a given sequence of complex frequencies $(\alpha_j)_{j=-\infty}^{\infty}$. First we give a condition that ensures that $(e^{-\alpha_j x})_{j=-\infty}^{\infty}$ is a basic sequence, but also shows that span$\{e^{-\alpha_j x} : j \in \mathbf{Z}\}$ is not dense in $L^2((0,\infty); dx)$.

Let $\alpha = (\alpha_j)_{j=-\infty}^{\infty}$, where $\Re \alpha_j > 0$ and let

$$\psi_j(z) = \left(2\Re \alpha_j\right)^{1/2} e^{-\alpha_j z} \qquad (j \in \mathbf{Z}) \qquad (9.6.16)$$

which are unit vectors in $L^2(0,\infty)$. Then let

$$k_j(z) = \frac{i\sqrt{\Re \alpha_j}}{2\pi(z + i\bar{\alpha}_j)} \qquad (j = 1, 2, \dots) \qquad (9.6.17)$$

which belong to H^2 of the upper half plane, and which are scaled reproducing kernels in the sense that they satisfy

$$\int_{-\infty}^{\infty} f(x)\bar{k}_j(x) dx = \sqrt{\Re \alpha_j} f(i\alpha_j) \qquad (f \in H^2). \qquad (9.6.18)$$

Theorem 9.6.6 *The following conditions are equivalent.*

(i) $(\alpha_j)_{j=-\infty}^{\infty}$ *is a Carleson interpolating sequence with $\Re \alpha_j > 0$; so there exists $\delta > 0$ such that*

$$\inf_j \prod_{k:k\neq j} \left| \frac{\alpha_j - \alpha_k}{\alpha_j + \bar{\alpha}_k} \right| \geq \delta. \qquad (9.6.19)$$

(ii) $(\psi_j)_{j=-\infty}^{\infty}$ *gives a Riesz sequence in $L^2(0,\infty)$; so there exist c_α, $C_\alpha > 0$ such that*

$$c_\alpha \sum_{j=-\infty}^{\infty} |a_j|^2 \leq \int_0^{\infty} \Big| \sum_{j=-\infty}^{\infty} a_j \psi_j(x) \Big|^2 dx$$

$$\leq C_\alpha \sum_{j=-\infty}^{\infty} |a_j|^2 \qquad ((a_j) \in \ell^2). \qquad (9.6.20)$$

(iii) $(k_j)_{j=-\infty}^{\infty}$ *gives a Riesz sequence in* H^2; *so there exist* $c_\alpha, C_\alpha > 0$ *such that*

$$c_\alpha \sum_{j=-\infty}^{\infty} |a_j|^2 \le \int_{-\infty}^{\infty} \Big| \sum_{j=-\infty}^{\infty} a_j k_j(x) \Big|^2 \, dx$$

$$\le C_\alpha \sum_{j=-\infty}^{\infty} |a_j|^2 \qquad ((a_j) \in \ell^2). \qquad (9.6.21)$$

Proof. The equivalence of (ii) and (iii) follows from Plancherel's theorem. The equivalence of (i) and (iii) is an aspect of Carleson's interpolation theorem, as in [122, 72]. □

Example. The points $\alpha_j = 1 + ij$ with $j \in \mathbf{Z}$ give a sequence which satisfies (i).

9.7 The Tracy–Widom distribution

- The Tracy–Widom distribution gives the soft edge that arises from the Gaussian unitary ensemble.
- A determinant involving the Airy kernel gives the Tracy–Widom distribution.

In the preceding sections, we have emphasized the probabilistic aspects and suppressed important connections with the theory of Painlevé transcendents. A Painlevé differential equation is a second order differential equation $y'' = R(x, y, y')$ that is rational in y and y', is analytic in x, and has no movable essential singularities. One aspect of the Painlevé theory is that there are only fifty different classes of such equations, up to transformation. Many significant results concerning random matrix theory were first discovered via spectral theory and the inverse scattering method. Indeed, the term *integrable operators* refers to integrable systems of differential equations. In this section, we use an integral equation which is familiar from the theory of inverse scattering; see [62].

Suppose that $w(x)$ is a solution of the Painlevé II equation

$$w'' = 2w^3 + xw \qquad (9.7.1)$$

such that $w(x) \asymp -\mathrm{Ai}(x)$ as $x \to \infty$.

Theorem 9.7.1 *Let* $\Gamma_{(x)} : L^2(x, \infty) \to L^2(x, \infty)$ *be the Hankel operator that has kernel* $\mathrm{Ai}((y + z)/2)$. *Then*

$$w(x)^2 = -\frac{d^2}{dx^2} \log \det(I - \Gamma_{(x)}^2). \qquad (9.7.2)$$

Proof. (Sketch). We consider the differential equations $f''(x) = xf(x)$ and $g''(x) = (x + 2w(x)^2)g(x)$, and observe that $f(x) = -\mathrm{Ai}(x)$ and $g(x) = w(x)$ give solutions. This comparison suggests that we seek w by a perturbation method based upon scattering. Ablowitz and Segur [2] have shown that the pair of integral equations

$$K(x,y) - r\mathrm{Ai}\left(\frac{x+y}{2}\right) + \frac{r}{2}\int_x^\infty L(x,s)\mathrm{Ai}\left(\frac{s+y}{2}\right)ds = 0,$$

$$L(x,y) + \frac{r}{2}\int_x^\infty K(x,z)\mathrm{Ai}\left(\frac{z+y}{2}\right)dz = 0 \quad (9.7.3)$$

have unique solutions for $|r| < 1/2$ that are continuous functions of y. They deduce that the first solution satisfies

$$\left(\frac{\partial}{\partial x} + \frac{\partial}{\partial y}\right)^2 K(x,y) = \frac{x+y}{2}K(x,y) + 2K(x,x)^2 K(x,y) \quad (9.7.4)$$

and that $w(x) = K(x,x)$ satisfies Painlevé II. We wish to apply the result in the case of $r = -2$, which is not fully justified by the analysis in [2]; nevertheless, we proceed to sketch the main features of the proof. One can show that L satisfies

$$L(x,z) - \frac{r}{2}F(x,z) + \frac{r^2}{4}\int_x^\infty L(x,y)F(y,z)\,dy = 0, \quad (9.7.5)$$

where

$$F(x,y) = \int_0^\infty \mathrm{Ai}\left(\frac{x+z}{2}\right)\mathrm{Ai}\left(\frac{z+y}{2}\right)dz \quad (9.7.6)$$

satisfies

$$\frac{\partial^2}{\partial x^2}F(x,y) - \frac{\partial^2}{\partial y^2}F(x,y) = \frac{1}{8}(x-y)F(x,y). \quad (9.7.7)$$

Using standard computations from scattering theory [62] and the uniqueness of the solutions of (9.7.4), one can deduce that

$$\frac{\partial^2}{\partial x^2}L(x,y) - \frac{\partial^2}{\partial y^2}L(x,y) - \frac{1}{8}(x-y)L(x,y) = r\left(\frac{d}{dx}L(x,x)\right)L(x,y)$$

$$(9.7.8)$$

and hence that $2\frac{d}{dx}L(x,x) = w(x)^2$.

The Hankel operator with kernel $\mathrm{Ai}((x+y)/2)$ is Hilbert–Schmidt, hence is the limit in Hilbert–Schmidt norm of a sequence of finite-rank Hankel operators. We now indicate how to express the determinant from (9.7.2) in terms of these approximations. Suppose that

$((2\Re\alpha_j)^{1/2}e^{-\alpha_j x/2})$ gives a Riesz sequence in $L^2(0,\infty)$ such that $\phi(x) =$ Ai(x) where

$$\phi(x) = \sum_{j=1}^{\infty} c_j e^{-\alpha_j x},$$

$$K(x,y) = \sum_{j=1}^{\infty} \xi_j(x) e^{-\alpha_j y/2},$$

$$L(x,y) = \sum_{k=1}^{\infty} \eta_k(x) e^{-\alpha_k y/2}; \qquad (9.7.9)$$

we also introduce the matrix

$$A = \left[\frac{c_j e^{-(\alpha_j+\alpha_k)x/2}}{(\alpha_j+\alpha_k)/2}\right]_{j,k=1}^{\infty} \qquad (9.7.10)$$

and the column vectors

$$X = [\xi_j(x)]_{j=1}^{\infty}, \qquad Y = [\eta_k(x)]_{k=1}^{\infty},$$
$$C = [c_j e^{-\alpha_j x/2}]_{j=1}^{\infty}, \qquad E = [e^{-\alpha_j x/2}]_{j=1}^{\infty}. \qquad (9.7.11)$$

$\square$

Lemma 9.7.2 *Suppose that*

$$X = (I - r^2 A^2/4)^{-1} rC,$$
$$Y = -(r^2/4)(I - r^2 A^2/4)^{-1} C. \qquad (9.7.12)$$

Then K and L satisfy

$$K(x,y) - r\phi\left(\frac{x+y}{2}\right) + \frac{r}{2}\int_x^{\infty} L(x,s)\phi\left(\frac{s+y}{2}\right) ds = 0,$$

$$L(x,s) + \frac{r}{2}\int_x^{\infty} K(x,z)\phi\left(\frac{z+y}{2}\right) dz = 0. \qquad (9.7.13)$$

Proof. By substituting the given formulæ (9.7.9) and (9.7.11) into the integral equations (9.7.13), we soon derive the expressions

$$\sum_{j=1}^{\infty} \xi_j(x) e^{-\alpha_j y/2} - r\sum_{j=1}^{\infty} c_j e^{-\alpha_j(x+y)/2}$$

$$+ \frac{r}{2}\sum_{j,k=1}^{\infty} \frac{\eta_k(x) e^{-(\alpha_j+\alpha_k)x/2} e^{-\alpha_j y/2} c_j}{(\alpha_j+\alpha_k)/2} = 0,$$

$$\sum_{k=1}^{\infty} \eta_k(x) e^{-\alpha_k y/2} + \frac{r}{2}\sum_{k,\ell=1}^{\infty} \frac{\xi_\ell(x) e^{-(\alpha_\ell+\alpha_k)x/2} e^{-\alpha_k y/2} c_k}{(\alpha_k+\alpha_\ell)/2} = 0,$$

$$(9.7.14)$$

which by independence of the Riesz basis reduces to

$$\xi_j(x) - rc_j e^{-\alpha_j x/2} + \frac{r}{2}\sum_{k=1}^{\infty} \frac{\eta_k(x)e^{-(\alpha_j + \alpha_k)x/2}c_j}{(\alpha_j + \alpha_k)/2} = 0,$$

$$\eta_k(x) + \frac{r}{2}\sum_{\ell=1}^{\infty} \frac{\xi_\ell(x)e^{-(\alpha_\ell + \alpha_k)x/2}c_k}{(\alpha_k + \alpha_\ell)/2} = 0. \quad (9.7.15)$$

In terms of the vectors, these are the pair of equations

$$X - rC + \frac{r}{2}AY = 0,$$

$$Y + \frac{r}{2}AX = 0; \quad (9.7.16)$$

which reduce to

$$X - \frac{r^2}{4}A^2 X = rC,$$

$$Y = -\frac{r}{2}AX, \quad (9.7.17)$$

and the solutions are as in (9.7.12).   □

Lemma 9.7.3 *Let* $\Gamma_{(x)} : L^2(x, \infty) \to L^2(x, \infty)$ *be the Hankel operator that has kernel* $\phi((y+z)/2)$. *Then*

$$L(x, x) = -\frac{d}{dx}\frac{1}{2}\Big(\log\det(I - r\Gamma_{(x)}/2) + \log\det(I + r\Gamma_{(x)}/2)\Big). \quad (9.7.18)$$

Proof. We express the function in terms of the matrices, and write

$$L(x, x) = E^t Y$$

$$= -\frac{r^2}{4}\text{trace}\big(E^t A(I - r^2 A^2/4)^{-1}C\big)$$

$$= -\frac{r^2}{4}\text{trace}\big((I - r^2 A^2/4)^{-1}ACE^t\big). \quad (9.7.19)$$

A simple computation shows that $\frac{d}{dx}A = -CE^t$, so we can resume the chain of identities with

$$L(x, x) = \frac{r^2}{4}\text{trace}\Big((I - r^2 A^2/4)^{-1}A\frac{d}{dx}A\Big)$$

$$= -\frac{d}{dx}\frac{1}{2}\text{trace}\log(I - r^2 A^2/4)$$

$$= -\frac{d}{dx}\frac{1}{2}\log\det(I - r^2 A^2/4). \quad (9.7.20)$$

The matrix A represents the Hankel operator with kernel $\phi(y+z)/2)$ on $L^2(x, \infty)$ with respect to the Riesz basis $(e^{-\alpha_j z/2})$. Hence we can have $\det(I + rA/2) = \det(I + r\Gamma_{(x)}/2)$, and (9.7.18) follows from (9.7.20). □

Conclusion of the proof of Theorem 9.7.1 From (9.7.8) we have $2\frac{d}{dx}L(x,x) = w(x)^2$, and hence taking $r = 2$ in Lemma 9.7.3, we have

$$w(x)^2 = -\frac{d^2}{dx^2}\log\det(I - \Gamma_{(x)}^2).\tag{9.7.21}$$

□

Definition (*Tracy–Widom distribution*). Let w be the solution of Painlevé II equation that satisfies $w(x) \asymp -\text{Ai}(x)$ as $x \to \infty$. Then the Tracy–Widom distribution has cumulative distribution function

$$F(t) = \exp\left(-\int_t^\infty (x-t)w(x)^2\,dx\right).\tag{9.7.22}$$

Corollary 9.7.4 *In terms of $\Gamma_{(t)}$ as in Theorem 9.7.1, the Tracy–Widom distribution satisfies*

$$F(t) = \det(I - \Gamma_{(t)}^2).\tag{9.7.23}$$

Proof. Clearly, both sides of (9.7.23) increase to one as $t \to \infty$. From the definition of F, we have

$$\frac{d^2}{dt^2}\log F(t) = -w(t)^2,\tag{9.7.24}$$

which we can combine with (9.7.21) to give

$$\frac{d^2}{dt^2}\log F(t) = \frac{d^2}{dt^2}\log\det(I - \Gamma_{(t)}^2).\tag{9.7.25}$$

By integrating this equation, we obtain the stated result.   □

We retuen to the context of Theorem 9.3.4, but consider the edge of the spectrum.

Theorem 9.7.5 (*Tracy–Widom [157]*). *The distribution of the largest eigenvalue λ_n of a $n \times n$ matrix under the Gaussian Unitary Ensemble satisfies*

$$\nu_n^{(2)}\left[\lambda_n \le 2 + \frac{t}{n^{2/3}}\right] \to F(t) \qquad (n \to \infty).\tag{9.7.26}$$

Proof. This follows from Proposition 9.4.2 and Corollary 9.7.4.   □

In a subsequent paper, Tracy and Widom showed that the asymptotic spectral edge distributions of Gaussian orthogonal and symplectic ensembles could be described in terms of related Painlevé functions.

- The scaled largest eigenvalue of a GUE matrix has an asymptotic Tracy–Widom distribution.
- The scaled length of the longest increasing subsequence in a random permutation has an asymptotic Tracy–Widom distribution.

The precise statement of these results appear as Theorems 9.7.5 and 10.4.7, and their respective proofs start from rather different probability spaces S_N and $GUE(N)$, and involve special functions which look dissimilar. The fact that such disparate computations lead to similar results suggests that there are deeper general principles which justify the conclusions. For variances and fluctuations, the Costin–Lebowitz central limit theorem of Section 10.2 provides such a general principle.

Let X_n be $n \times n$ matrices from the generalized unitary, orthogonal or symplectic ensemble, and suppose that the equilibrium measure is supported on a single interval $[a, b]$. Dyson's universality principle [64] asserts that the spectra of the X_n within subintervals of $[a, b]$ of length $O(1/n)$ do not depend upon the particular potential of the ensemble as $n \to \infty$. This has been verified in many cases, and the eigenvalue distribution is given by the determinantal random point field from the sine kernel [125, 126].

In [148], Soshnikov observes 'There is a general belief amongst people working in random matrix theory that in the same way as the sine kernel appears to be a universal limit in the bulk of the spectrum for random Hermitian matrices, the Airy and Bessel kernels are universal limits of the soft and hard edge of the spectrum'. See [66, 67] for more on this topic.

10

Fluctuations and the Tracy–Widom distribution

Abstract

We present the Costin–Lebowitz theorem on fluctuations, which is a version of the central limit theorem for scaled eigenvalue distributions. The most striking consequence is the recent application to Young diagrams under Plancherel measure. The Tracy–Widom distribution gives the asymptotic edge distribution of the eigenvalues from GUE, and surprisingly the asymptotic distribution of the longest increasing sequence in random permutations, after suitable scalings. The discrete Bessel kernel also plays an important role in the proof, as it links the determinantal random point fields of random matrices to those of the Young diagrams.

10.1 The Costin–Lebowitz central limit theorem

This section features an important universality result, namely the Costin–Lebowitz central limit theorem from [52], in a form adapted to deal with determinantal random point fields by Soshnikov [147]. In Section 10.6, we sketch an application to fluctuations of Young tableaux, which also involves the results of Chapter 7. First we define what is meant by a fluctuation.

Definition (*Fluctuations*). Let ω_k be random point fields on $\mathbf{R}$ such that $\omega_k(\mathbf{R}) = k$, and let $\mu_k = k^{-1} \sum_{j=1}^{k} \delta_{\lambda_j}$ be the empirical distribution that corresponds to ω_k for $k = 1, 2, \ldots$. Suppose that $\mu_k \to \rho$ weakly in distribution as $k \to \infty$; so that

$$\mathbf{P}\left[\left| \frac{1}{k} \sum_{j=1}^{k} f(\lambda_j) - \int f(x)\rho(dx) \right| < \varepsilon \right] \to 1 \qquad (k \to \infty) \quad (10.1.1)$$

for all $\varepsilon > 0$ and all $f \in C_b(\mathbf{R}; \mathbf{C})$. Then *fluctuations* are the random

321

variables $\sum_{j=1}^{k} f(\lambda_j) - k \int f(x)\rho(dx)$ and the associated random measures $\sum_{j=1}^{k} \delta_{\lambda_j} - k\rho$. We sometimes replace f by the indicator function of an interval. See [147].

Before discussing the general theorem, we consider an important example of the Poisson process where we can carry out the computations explicitly.

Example 10.1.1 (*Central limit theorem for a Poisson process*). Let ν be a Poisson process with intensity function p, as in Example 8.1.3. For a step function $f(x) = \sum_{j=1}^{n} t_j \mathbf{I}_{A_j}(x)$ with disjoint bounded Borel sets A_j, we introduce the random variable $\int f d\nu = \sum_{j=1}^{n} t_j \nu(A_j)$, and we find that

$$\mathbf{E} \exp \int i f d\nu = \exp\left(\int \left(e^{if(x)} - 1 \right) p(x) dx \right). \qquad (10.1.2)$$

By Kac's theorem and the uniqueness of the characteristic functions, the right-hand side determines properties (i), (ii) and (iii) of Theorem 8.2.7 for the random point field, and hence the distribution. The probability generating function of the random point field satisfies

$$\mathbf{E} z^{\nu(A)} = \exp\left((z-1) \int_A p(x) dx \right). \qquad (10.1.3)$$

Thus the Poisson process is a degenerate type of determinantal random point field in which the kernel is $K(x_j, x_k) = \mathrm{diag}[p(x_1), p(x_2), \dots]$ so the Nth joint probability density function is $p(x_1)p(x_2)\dots p(x_N)$.

Further, the generating function of $\int f d\nu$ satisfies

$$\mathbf{E} z^{\int f d\nu} = \exp\left(\int \left(z^{f(x)} - 1 \right) p(x) \, dx \right). \qquad (10.1.4)$$

Proposition 10.1.1 *Suppose that p is locally integrable, but not integrable over $\mathbf{R}$, and choose sets A_j such that*

$$\mu_j = \int_{A_j} p(x) \, dx \to \infty \qquad (j \to \infty). \qquad (10.1.5)$$

Then the normalized random variables

$$Y_j = \frac{\nu(A_j) - \mu_j}{\mu_j^{1/2}}, \qquad (10.1.6)$$

converge in distribution to a $N(0,1)$ random variable as $j \to \infty$.

Proof. Using (10.1.2), we calculate the characteristic functions

$$\mathbf{E}\exp isY_j = \exp\left(\mu_j\left(e^{is\mu_j^{-1/2}} - 1\right) - is\mu_j^{1/2}\right)$$
$$\to \exp\left(-s^2/2\right) \qquad (j \to \infty). \qquad (10.1.7)$$

Hence by Theorem 1.8.2, (Y_j) converges in law to a $N(0,1)$ random variable as $j \to \infty$. □

We now consider a typical determinantal random point field on $\mathbf{R}$. As our version of the central limit theorem is proved using cumulants and probability generating functions, we start with a few results concerning the series that were introduced in Proposition 8.2.5. The probability generating function of a determinantal random point field

$$\mathbf{E}z^{\nu(J)} = \det(I + (z-1)K\mathbf{I}_J) \qquad (10.1.8)$$

involves a determinant which is hard to deal with, so we replace it by a simpler expression $\log\det(I - zK_J)$ which we can compute more easily.

Lemma 10.1.2 *Suppose that Γ is a Hilbert–Schmidt operator such that $K = \Gamma^*\Gamma$ is a trace–class kernel on $L^2(J; dx)$ and that $0 \le K \le I$. Then the series*

$$-\log\det(I - zK)$$
$$= \sum_{k=1}^{\infty} \frac{z^k}{k} \int \ldots \int_{J^k} K(x_1, x_2)K(x_2, x_3)\ldots K(x_k, x_1)dx_1 \ldots dx_k$$

$$(10.1.9)$$

converges for $|z| < 1$.

Proof. The trace norm satisfies $\|K^k\|_{c^1} \le \|K\|_{c^1}$ since $\|K\|_{op} \le 1$; so the expansion

$$-\log(I - zK) = \sum_{k=1}^{\infty} \frac{z^k}{k}K^k \qquad (10.1.10)$$

converges in the norm of c^1 for $|z| < 1$, and we can take the trace to get

$$-\text{trace}\log(I - zK) = \sum_{k=1}^{\infty} \frac{z^k}{k}\text{trace}(K^k). \qquad (10.1.11)$$

The stated result follows directly since by the Hilbert–Schmidt theorem

$$\text{trace}(K) = \int_J \int_J \Gamma^*(y,x)\Gamma(x,y)\,dxdy = \int_J K(x,x)dx, \quad (10.1.12)$$

$$\text{trace}(K^2) = \int_J \int_J K(x,y)K(y,x)\,dxdy \qquad (10.1.13)$$

and similarly for the powers K^k. □

Let I_t be a family of intervals in $\mathbf{R}$, parametrized by $t > 0$, and let $K_t : L^2(I_t) \to L^2(I_t)$ be a family of integral operators that satisfy the hypotheses of Lemma 10.1.2. For a determinantal random point field $\mathbf{P}$, we let ν_t be the number of particles in I_t. The next step is to find the distribution of the fluctuations, which we normalize to

$$\frac{\nu_n - \mathbf{E}(\nu_n)}{\operatorname{var}(\nu_n)^{1/2}}. \qquad (10.1.14)$$

The following theorem was proved for the sine kernel by Costin and Lebowitz, who noted an observation of Widom that a similar result should hold for more general kernels; see [52]. This hope was realised in [147] by Soshnikov, who made the applications to the Airy and Bessel kernels and proved some more subtle results concerning joint distributions for families of disjoint intervals.

Theorem 10.1.3 (*Costin–Lebowitz–Soshnikov*). *Suppose that K_t are trace-class operators associated with $\mathbf{P}$ on I_t such that*

$$\operatorname{var}(\nu_t) = \operatorname{trace}(K_t - K_t^2) \qquad (10.1.15)$$

diverges to infinity as $t \to \infty$. Then

$$\frac{\nu_t - \mathbf{E}(\nu_t)}{(\operatorname{var}(\nu_t))^{1/2}} \qquad (10.1.16)$$

converges in distribution to a Gaussian $N(0,1)$ random variable as $t \to \infty$.

Proof. The logarithmic moment generating function of a Gaussian $N(0,1)$ random variable satisfies $g_\gamma(iz) = -z^2/2$, and the cumulants of ν_t satisfy $\kappa_1 = \mathbf{E}(\nu_t)$, and $\kappa_2 = \operatorname{var}(\nu_t)$. We wish to show that κ_j is comparatively small for $j \geq 3$.

Lemma 10.1.4 *There exist constants C_j independent of t such the cumulants satisfy*

$$|\kappa_j| \leq C_j \operatorname{trace}(K_t - K_t^2). \qquad (10.1.17)$$

Proof. The probability generating function satisfies

$$\psi(z) = \mathbf{E} z^{\nu_t} = \det\bigl(I + (z-1)K_t\bigr), \qquad (10.1.18)$$

and hence by Lemma 10.1.2 the logarithmic moment generating function satisfies

$$
\begin{aligned}
g(iz) &= \log \mathbf{E}(e^{i\nu_t z}) \\
&= \log \det\big(I + (e^{iz} - 1)K_t\big) \\
&= \operatorname{trace} \log\big(I + (e^{iz} - 1)K_t\big),
\end{aligned} \tag{10.1.19}
$$

and expanding terms in series, we obtain

$$
\sum_{k=1}^{\infty} \frac{\kappa_k (iz)^k}{k!} = \sum_{k=1}^{\infty} \frac{(-1)^{k-1}}{k}(e^{iz} - 1)^k \operatorname{trace}(K_t^k) \tag{10.1.20}
$$

where the κ_k are the cumulants. We next observe that $(e^{iz} - 1)^j$ has a Maclaurin series beginning with $(iz)^j$, and so by Cauchy's integral formula

$$
\frac{\kappa_j i^j}{j!} = \sum_{k=1}^{j} \frac{(-1)^{k-1}}{2\pi i k} \int_{C(0,\delta)} \frac{(e^{iz} - 1)^k}{z^{j+1}} \operatorname{trace}(K_t^k)\, dz; \tag{10.1.21}
$$

thus in particular we obtain the formulas

$$
\kappa_1 = \operatorname{trace}(K_t), \tag{10.1.22}
$$

$$
\kappa_2 = \operatorname{trace}(K_t) - \operatorname{trace}(K_t^2). \tag{10.1.23}
$$

More generally, we obtain the formula

$$
\begin{aligned}
\kappa_j = \operatorname{trace}(K_t) &+ \sum_{k=2}^{j-1} \frac{(-1)^{k-1} j!}{2\pi i^{j+1} k} \int_{C(0,\delta)} \frac{(e^{iz} - 1)^k}{z^{j+1}} \operatorname{trace}(K_t^k)\, dz \\
&+ (-1)^{j-1}(j-1)!\, \operatorname{trace}(K_t^j),
\end{aligned} \tag{10.1.24}
$$

which leads to the identity

$$
\kappa_j = (-1)^{j-1}(j-1)!\, \operatorname{trace}(K_t^j - K_t) + \sum_{k=2}^{j-1} C_{k,j}\kappa_k \tag{10.1.25}
$$

for some constants $C_{k,j}$ which are independent of t.

Now we have

$$
\begin{aligned}
\operatorname{trace}(K_t - K_t^j) \\
= \operatorname{trace}(K_t - K_t^2) + \operatorname{trace}(K_t^2 - K_t^3) + \cdots + \operatorname{trace}(K_t^{j-1} - K_t^j) \\
\leq (j-1)\operatorname{trace}(K_t - K_t^2)
\end{aligned} \tag{10.1.26}
$$

since $0 \leq K_t \leq I$, so

$$
\operatorname{trace}(K_t - K_t^j) \leq (j-1)\operatorname{trace}(K_t - K_t^2). \tag{10.1.27}
$$

On combining (10.1.26) with (10.1.15) and a simple recursion argument, we obtain the Lemma. □

Proof of Theorem 10.1.3 We choose $a_t = \text{var}(\nu_t)^{-1/2}$ and $b_t = -\mathbf{E}(\nu_t)\text{var}(\nu_t)^{-1/2}$, then scale ν_t to $a_t\nu_t + b_t$, so that the logarithmic moment generating function satisfies

$$g_{a_t\nu_t+b_t}(iz) = izb_t + g_{\nu_t}(ia_t z) = izb_t + \sum_{j=1}^{\infty} \frac{\kappa_j}{j!}(ia_t z)^j. \quad (10.1.28)$$

By Lemma 10.1.4 and (10.1.15), we have

$$a_t^j|\kappa_j| \le C_j(\text{trace}(K_t - K_t^2))^{1-j/2}; \quad (10.1.29)$$

so the cumulants of $a_t\nu_t + b_t$ have $a_t^j|\kappa_j| \to 0$ as $t \to \infty$ for $j = 3, 4, \dots$. By Proposition 1.8.5, $a_t\nu_t + b_t$ converges in distribution to a $N(0,1)$ random variable. □

The Costin–Lebowitz–Soshnikov fluctuation theorem holds for

- the sine kernel, representing the spectral bulk;
- the Airy kernel, representing soft edges;
- the Bessel kernel, representing hard edges;
- the Plancherel measure on partitions.

The quantity $\text{trace}(K - K^2)$ measures the extent to which K deviates from being an orthogonal projection. It can be large only when the eigenvalues of K decay slowly to zero. To be more precise, we introduce the eigenvalues κ_j of K and recall that $0 \le \kappa_j \le 1$ may be chosen so that κ_j decreases to 0 as $j \to \infty$; then $\text{trace}(K - K^2) = \sum_{j=1}^{\infty}(\kappa_j - \kappa_j^2)$. We also introduce the counting function $n(s) = \#\{j : s\kappa_j \ge 1\}$. Then $n(s)$ is an increasing function with $n(s) = 0$ for $0 < s < 1$, and

$$\text{trace}(K - K^2) = \int_1^{\infty}\left(1 - \frac{2}{s}\right)\frac{n(s)}{s^2}ds. \quad (10.1.30)$$

Proposition 10.1.5 *Suppose that $n(s) \asymp Cs^r$ for some $C > 0$ and $0 < r < 1$ as $s \to \infty$. Then*

$$\log\det(1 + xK) \asymp \frac{C\pi x^r}{\sin \pi r} \quad (x \to \infty). \quad (10.1.31)$$

The converse also holds.

Proof. Here $n(s)$ gives the number of zeros of the entire function $\det(I + zK)$ that lie inside $\{z : |z| \le s\}$, and evidently the zeros lie on the negative real axis. The stated result follows by a standard asymptotic formula, with the converse due to Valiron; see [154]. □

10.2 Discrete Tracy–Widom systems

In several cases one wishes to apply the central limit theorem to a kernel as in Lemma 10.1.2. In this section we introduce linear systems in discrete time that give rise to such kernels, and are analogous to the systems of differential equations considered in Section 9.1.

The notion of a Tracy–Widom differential equation has a natural analogue for sequences, and special cases are important in mathematical physics, especially the theory of discrete Schrödinger operators. See [30, 113]. In this section we look briefly at three important examples, after motivating the definition.

Suppose that we have a Tracy–Widom differential equation

$$\frac{d}{dt} f(t) = J\Omega(t) f(t) \qquad (10.2.1)$$

in the sense of (9.1.12); then

$$f(t+h) = (I + hJ\Omega(t)) f(t) + o(h) \qquad (h \to 0) \qquad (10.2.2)$$

where $J\Omega(t)$ has trace zero, so

$$\det(I + hJ\Omega(t)) = 1 + h\,\mathrm{trace}(J\Omega(t)) + O(h^2) = 1 + O(h^2) \qquad (h \to 0). \qquad (10.2.3)$$

This suggests the following definition.

Definition (*Discrete Tracy–Widom system [24]*). Suppose that S_x is a 2×2 matrix such that $\det S_x = 1$, and $x \mapsto S_x$ is a rational function, then let (a_n) be an ℓ^2 sequence such that

$$\begin{bmatrix} a_{x+1} \\ a_x \end{bmatrix} = S_x \begin{bmatrix} a_x \\ a_{x-1} \end{bmatrix}. \qquad (10.2.4)$$

We call S_x the *one-step transition matrix*. Then the discrete Tracy–Widom kernel is

$$K(x,y) = \frac{a_{x+1}a_y - a_x a_{y+1}}{x - y} \qquad (x \neq y). \qquad (10.2.5)$$

With $A_x = \mathrm{col}\,[a_x, a_{x-1}]$, we have

$$\frac{\langle JA_x, A_y \rangle}{x - y} = \frac{a_{x-1}a_y - a_x a_{y-1}}{x - y}, \qquad (10.2.6)$$

and by analogy with (9.1.17) we have

$$\langle JA_{x+1}, A_{y+1} \rangle - \langle JA_x, A_y \rangle = \langle (S_y^t J S_x - J) A_x, A_y \rangle. \qquad (10.2.7)$$

Example 10.2.1 (*Mathieu's equation*). Mathieu's equation [111] gives an historically important example. If x is a 2π periodic solution of the equation $x''(t) + (\alpha + \beta \cos t)x(t) = 0$, then x has Fourier series $x(t) = \sum_{n=-\infty}^{\infty} a_n e^{int}$, where (a_n) satisfies the recurrence relation with

$$S_n = \begin{bmatrix} 2(n^2 - \alpha)/\beta & -1 \\ 1 & 0 \end{bmatrix}. \tag{10.2.8}$$

The values of α and β such that a nonzero $(a_n) \in \ell^2(\mathbf{Z})$ exists are given by Hill's determinantal equation.

Example 10.2.2 (*Almost Mathieu's equation*). The almost Mathieu's equation

$$a_{n+1} + a_{n-1} + g \cos 2\pi(\omega n + \theta)\, a_n = E a_n \tag{10.2.9}$$

can be expressed as a recurrence relation involving the one step transition matrix

$$S_n = \begin{bmatrix} E - g \cos 2\pi(\omega n + \theta) & -1 \\ 1 & 0 \end{bmatrix}. \tag{10.2.10}$$

Jitomirskaya [93] has determined the nature of the point spectrum of the almost Mathieu operator, and shown that the solutions are of exponential decay as $|n| \to \infty$. When $g > 2$ and ω satisfies a Diophantine condition, there exists a set of θ of full Lebesgue measure such that (10.2.4) has a solution with $(a_n) \in \ell^2(\mathbf{Z})$. This is an instance of the famous 'Anderson localization' phenomenon.

Example 10.2.3 (*The discrete Bessel system*). A fundamentally important example is the discrete Bessel system. Let $\theta > 0$ and abbreviate $J_m = J_m(2\sqrt{\theta})$; the order of the Bessel function is the main variable in the following discussion. Also, let $\mathbf{N}_0 = \{0, 1, 2, \dots\}$. The three–term recurrence relation [167, p.17] for the Bessel sequence can be written in matrix form as

$$\begin{bmatrix} J_{x+1} \\ J_x \end{bmatrix} = \begin{bmatrix} 2x/t & -1 \\ 1 & 0 \end{bmatrix} \begin{bmatrix} J_x \\ J_{x-1} \end{bmatrix}. \tag{10.2.11}$$

10.3 The discrete Bessel kernel

In this section we investigate the kernel that arises from the example at the end of the previous section. As we shall see, the Airy kernel makes a surprise return to the theory. In Section 10.4, we shall apply the results to the asymptotic shapes of Young diagrams. Most of the calculations in this section are extracted from [8, 30], and we abbreviate some of the proofs to sketches for the benefit of the general reader.

Definition (*Discrete Bessel kernel*). With the abbreviation $J_m = J_m(2\sqrt{\theta})$, we define the discrete Bessel kernel by

$$J(x, y; \theta) = \sqrt{\theta}\frac{J_x J_{y+1} - J_{x+1}J_y}{x - y} \qquad (x \neq y; x, y = 0, 1, 2, \dots);$$

$$= \sqrt{\theta}(J'_x J_{x+1} - J'_{x+1}J_x) \qquad (x = y; x = 0, 1, 2, \dots) \quad (10.3.1)$$

where $J'_x = \frac{\partial}{\partial x}J_x(2\sqrt{\theta})$, and so by formulae 6 in 8.486(1) and 1 in 8.477 of [173]

$$J(x, x; \theta) = \frac{1}{2\sqrt{\theta}} - \frac{x+1}{2\sqrt{\theta}}J_x(2\sqrt{\theta})^2 + \frac{x!}{2}\sum_{k=0}^{x-1}\frac{\theta^{(k-1-x)/2}}{k!}J_k(2\sqrt{\theta})$$

$$\times \left(\frac{\sqrt{\theta}J_{x+1}(2\sqrt{\theta})}{x-k} - \frac{(x+1)J_x(2\sqrt{\theta})}{x+1-k}\right). \quad (10.3.2)$$

Theorem 10.3.1 (*i*) *The matrix* $J(x, y; \theta)$ *is the square of a Hankel matrix and satisfies*

$$J(x, y; \theta) = \sum_{k=0}^{\infty} J_{x+k+1}J_{y+k+1}. \quad (10.3.3)$$

(*ii*) *The matrix* $J(x, y; \theta)$ *defines a trace-class linear operator on* $\ell^2(\mathbf{N}_0)$ *such that* $0 \leq J \leq I$.

(*iii*) *The matrix* $[J_{x+y}]_{x,y=0}^{\infty}$ *has eigenvalue one.*

Proof. (i) We first prove that $J(x, y; \theta)$ is the square of the Hankel matrix $[J_{x+y+1}]_{x,y=0}^{\infty}$. We recall the definition of the Bessel function $J_m(t)$ of integral order and the resulting Laurent series from (7.4.14). Writing $t = 2\sqrt{\theta}$ and taking $z = e^{i\phi}$ there, we have as in [167, 2.22]

$$e^{it\sin\phi} = J_0(t) + 2\sum_{m=1}^{\infty}J_{2m}(t)\cos(2m\phi) + 2i\sum_{m=1}^{\infty}J_{2m-1}(t)\sin(2m-1)\phi,$$

$$(10.3.4)$$

since

$$2\int_0^{\pi}\cos(t\sin\phi)\cos m\phi\frac{d\phi}{\pi}$$

$$= \int_0^{\pi}\cos(t\sin\phi + m\phi)\frac{d\phi}{\pi} + \int_0^{\pi}\cos(t\sin\phi - m\phi)\frac{d\phi}{\pi}$$

$$= J_m(t) + J_{-m}(t) \quad (10.3.5)$$

and

$$2 \int_0^\pi \sin(t \sin \phi) \sin m\phi \frac{d\phi}{\pi} = \int_0^\pi \cos(t \sin \phi - m\phi) \frac{d\phi}{\pi}$$
$$- \int_0^\pi \cos(t \sin \phi - m\phi) \frac{d\phi}{\pi}$$
$$= J_m(t) - J_{-m}(t). \qquad (10.3.6)$$

Hence by Parseval's identity, we have

$$1 = J_0(t)^2 + 2 \sum_{m=1}^\infty J_m(t)^2, \qquad (10.3.7)$$

so the sequence $(J_m)_{m=0}^\infty$ is square summable. Further, this fact and the three-term recurrence relation

$$J_{x+2} - \frac{2(x+1)}{t} J_{x+1} + J_x = 0 \qquad (10.3.8)$$

together imply that $(m J_m)$ is square summable, so $[J_{x+y-1}]$ defines a self-adjoint operator on $\ell^2(\mathbf{N}_0)$ of Hilbert–Schmidt class.

From the definition of J and the recurrence relation, we have

$$J(x+1, y+1; \theta) - J(x, y; \theta)$$
$$= \sqrt{\theta} \frac{J_{x+1} J_{y+2} - J_{x+2} J_{y+1} - J_x J_{y+1} + J_{x+1} J_y}{x - y}$$
$$= \frac{\sqrt{\theta}}{x - y} \left[\left(\frac{y+1}{\sqrt{\theta}} - \frac{x+1}{\sqrt{\theta}} \right) J_{x+1} J_{y+1} \right]$$
$$= -J_{x+1} J_{y+1}. \qquad (10.3.9)$$

which is analogous to (9.1.21). By summing this identity down the diagonals of the matrix $J(x + k, y + k; \theta)$, we obtain

$$J(x, y; \theta) = \sum_{k=0}^\infty J_{x+k+1} J_{y+k+1} \qquad (10.3.10)$$

since $J(x, y; \theta) \to 0$ as $x \to \infty$ or $y \to \infty$. Hence $J(x, y; \theta)$ is the square of a Hankel matrix $[J_{x+k+1}]$.

(ii) By (i), $J(x, y; \theta)$ determines a positive semidefinite operator on $\ell^2(\mathbf{N}_0)$ of trace class.

Now let (a_n) and (b_n) belong to ℓ^2. By the Cauchy–Schwarz inequality, we have

$$
\left| \sum_{n,m} J_{n+m} a_n \bar{b}_m \right| = \left| \int_{-\pi}^{\pi} \sum_{n,m} a_n \bar{b}_m e^{i(n+m)\phi - it\sin\phi} \frac{d\phi}{2\pi} \right|
$$

$$
= \left| \int_{-\pi}^{\pi} \sum_n a_n e^{in\phi} \sum_m \bar{b}_m e^{im\phi} e^{-it\sin\phi} \frac{d\phi}{2\pi} \right|
$$

$$
\leq \left(\int_{-\pi}^{\pi} |\sum_n a_n e^{in\phi}|^2 \frac{d\phi}{2\pi} \right)^{1/2} \left(\int_{-\pi}^{\pi} |\sum_n \bar{b}_m e^{im\phi}|^2 \frac{d\phi}{2\pi} \right)^{1/2}
$$

$$
= \left(\sum_n |a_n|^2 \right)^{1/2} \left(\sum_m |b_m|^2 \right)^{1/2}; \tag{10.3.11}
$$

hence $\|J\| \leq 1$.

(iii) Lommel's identity [167, 5.22(7)] gives

$$
\sum_{m=0}^{\infty} \frac{\theta^{m/2}}{m!} J_{n+m}(2\sqrt{\theta}) = \frac{\theta^{n/2}}{n!}, \tag{10.3.12}
$$

and thus displays the eigenvector that corresponds to eigenvalue one. $\square$

The success of the calculation in the proof of Theorem 10.3.1 appears to depend on a strange assortment of identities with little connection to the original question; in some respects, this is true. One can make the computations more systematic, and formulate sufficient conditions for a given operator to be the square of a self-adjoint Hankel operator, but all cases known to the author involve a special form of recurrence relation and delicate estimates. McCafferty [113] has a version of Theorem 10.3.1 which applies to matrices associated with first-order difference equations.

Corollary 10.3.2 *There exists a determinantal random point field* P^θ *on* $\mathbf{Z}$ *with correlation functions*

$$
\rho_k^\theta(x_1, \ldots, x_k) = \det[J(x_j, x_\ell; \theta)]_{j,\ell=1,\ldots,k}. \tag{10.3.13}
$$

Proof. This follows from the Proposition 10.3.1 and Theorem 8.1.4. $\square$

Remarkably, one can identify this P^θ with a natural probability measure on the space of partitions, namely the Plancherel measure, as we discuss in the next section.

The following results will be used in Section 10.4. We use Pochhammer's notation and write $(\alpha)_m = \alpha(\alpha+1)\ldots(\alpha+m-1)$ and $(\alpha)_0 = 1$.

Lemma 10.3.3 *The Bessel kernel satisfies*

$$J(x, y; \theta) = \sum_{m=0}^{\infty} \frac{(-1)^m}{m!} \frac{\theta^{(x+y+2+2m)/2}(x+y+m+2)_m}{\Gamma(x+m+2)\Gamma(y+m+2)}. \quad (10.3.14)$$

Proof. This follows directly from the formula [167, 5.41(1)]

$$J_\mu(z)J_\nu(z) = \sum_{m=0}^{\infty} \frac{(-1)^m (z/2)^{\mu+\nu+2m}(\mu+\nu+m+1)_m}{m!\Gamma(\mu+m+1)\Gamma(\nu+m+1)}, \quad (10.3.15)$$

which in turn follows from multiplying the series together. The crucial identity in the calculation is

$$(x+a)_n = \sum_{k=0}^{n} \binom{n}{k} (x)_k (a)_{n-k}. \quad (10.3.16)$$

□

Lemma 10.3.4 (*Nicholson's approximation*).

$$J_n(t) \asymp \frac{2^{1/3}}{t^{1/3}} \operatorname{Ai}\left(\frac{2^{1/3}(n-t)}{t^{1/3}}\right) \qquad (0 < t < n; n \to \infty). \quad (10.3.17)$$

Proof. The following heuristic argument motivates the formula, and is made precise in [167, 8.43]. Expanding the sine in a Maclaurin series, then letting $\phi = (t/2)^{1/3}\theta$, we have

$$J_n(t) = \frac{1}{\pi} \int_0^\pi \cos\left(n\theta - t\sin\theta\right) d\theta$$

$$\approx \frac{1}{\pi} \int_0^\pi \cos\left((n-t)\theta + \frac{t\theta^3}{6}\right) d\theta$$

$$\approx \frac{2^{1/3}}{t^{1/3}\pi} \int_0^\infty \cos\left(\frac{2^{1/3}(n-t)}{t^{1/3}}\phi + \frac{\phi^3}{3}\right) d\phi \quad (10.3.18)$$

since, by Kelvin's stationary phase principle [167, p. 229], the greatest contribution to the integral arises from the range of integration near to where the phase is stationary. By the choice of $0 < t < n$, the derivative of the phase is

$$\frac{d}{d\theta}(n\theta - t\theta + 6^{-1}t\theta^3) = n - t + \frac{t}{2}\theta^2 \geq n - t. \quad (10.3.19)$$

Recalling the definition of the Airy function in (9.2.18), we recover the stated result. □

The continuous Tracy–Widom system (9.4.5) for the Airy kernel resembles formally the system (10.2.11) for the discrete Bessel kernel. The discrete Bessel kernel gives a determinantal random point field on **Z** such that the scaled right edge is soft in the sense of Section 9.4. Recall that the cumulative distribution function of the Tracy–Widom distribution satisfies

$$F(t) = \det(I - \mathbf{I}_{[t,\infty]} W \mathbf{I}_{[t,\infty)}) \qquad (t > 0). \qquad (10.3.20)$$

Proposition 10.3.5 *Let P^θ be the determinantal random point field on* **Z** *associated with the discrete Bessel kernel* $J(\,,\,;\theta)$, *and* $\nu(a,b)$ *be the number of points in* $(a,b) \cap$ **Z**. *Then as* $\theta \to \infty$,

$$P^\theta[\nu(2\sqrt{\theta} + t\theta^{1/6}, \infty) = 0] \to F(t) \qquad (t > 0). \qquad (10.3.21)$$

Proof. (*Sketch*). We now approximate the discrete Bessel kernel by the Airy kernel as $\theta \to \infty$. By Lemma 10.3.4, we have

$$J(x,y;\theta)$$

$$\approx \frac{2^{2/3}\theta^{1/2}}{(2\sqrt{\theta})^{2/3}} \left(\frac{\text{Ai}\left(\frac{x-2\sqrt{\theta}}{\theta^{1/6}}\right)\text{Ai}\left(\frac{y+1-2\sqrt{\theta}}{\theta^{1/6}}\right) - \text{Ai}\left(\frac{x+1-2\sqrt{\theta}}{\theta^{1/6}}\right)\text{Ai}\left(\frac{y-2\sqrt{\theta}}{\theta^{1/6}}\right)}{x-y} \right),$$

$$(10.3.22)$$

in which we apply the mean value theorem

$$\text{Ai}\left(\frac{y+1-2\sqrt{\theta}}{\theta^{1/6}}\right) \approx \text{Ai}\left(\frac{y-2\sqrt{\theta}}{\theta^{1/6}}\right) + \theta^{-1/6}\,\text{Ai}'\left(\frac{y-2\sqrt{\theta}}{\theta^{1/6}}\right)$$

and similarly with the term involving $x+1$ in the numerator. As $\theta \to \infty$ with $x,y > 2\sqrt{\theta}$, we deduce that

$$J(x,y;\theta) \approx \theta^{-1/6} W\left(\frac{x-2\sqrt{\theta}}{\theta^{1/6}}, \frac{y-2\sqrt{\theta}}{\theta^{1/6}}\right). \qquad (10.3.23)$$

where W is as in (9.4.1).

We now replace the determinantal random point field on **Z** by a determinantal random point field on **R**. Now there is a linear isometry $\varphi_\theta : \ell^2(\mathbf{Z}) \to L^2(\mathbf{R})$ given by

$$\varphi_\theta : (a_n)_{n=-\infty}^\infty \mapsto \sum_n \theta^{1/12} a_n \mathbf{I}_{[\theta^{-1/6}n,\theta^{-1/6}(n+1)]}, \qquad (10.3.24)$$

thus $J(x,y;\theta)$ with $x,y \in \mathbf{Z}_+$ gives rise to an operator $\tilde{J}(\,;\theta)$: $L^2(0,\infty) \to L^2(0,\infty)$ such that $0 \leq \tilde{J}(\,;\theta) \leq I$.

The authors of [30] adopt an alternative approach via contour integration and make the above approximation arguments rigorous; thus they show that $\tilde{J}(\,;\theta) \to W$ as trace-class operators on $L^2(0,\infty)$, so

$$\det(I + (z-1)\mathbf{I}_{(t,\infty)}\tilde{J}(\,,\,;\theta)) \to \det(I + (z-1)\mathbf{I}_{(t,\infty)}W) \qquad (\theta \to \infty)$$
$$(10.3.25)$$

and in particular

$$\det(I - \mathbf{I}_{(t,\infty)}\tilde{J}(\ ,\ ;\theta)) \to \det(I - \mathbf{I}_{(t,\infty)}W) = F(t) \qquad (\theta \to \infty).$$
$$(10.3.26)$$

$\square$

Exercise 10.3.6 (i) Substitute Nicholson's approximate formula into the series

$$\sum_{k=0}^{\infty} J_{x+k+1}(t)J_{y+k+1}(t) \qquad (10.3.27)$$

and hence obtain an approximate formula for $\mathrm{J}(x,y;t)$ in terms of the Airy kernel W. (The lack of effective error terms in Lemma 10.3.4 means that this is not rigorous.)

(ii) Similarly give a non rigorous argument to justify the statement: factorization of the discrete Bessel kernel as the square of a self-adjoint Hankel matrix implies factorization of the Airy kernel as the square of a self-adjoint Hankel operator on $L^2(0,\infty)$.

10.4 Plancherel measure on the partitions

Consider a random Young diagram with N boxes.

- The discrete Bessel kernel gives the correlation functions for the Plancherel measure on the partitions of order N, where N is subject to Poisson randomization.
- The length of the longest row of a Young diagram with N boxes is typically about $2\sqrt{N}$,
- The fluctuations in a row length are typically about $\sqrt{\log N}$.

In this section we consider in more detail the Plancherel measure on the set of all partitions, as introduced in Section 7.4, and thus obtain a combinatorial interpretation of the discrete Bessel kernel.

Borodin and Olshanski have introduced a theory of z-measures on the space of partitions which unifies many probability density functions into a consistently defined family, and gives rise to the hypergeometric kernel [34]. One can obtain various kernels from the hypergeometric kernel by making suitable choices of parameters and taking limits. In our presentation, we prefer to express results in terms of representations of S_N and the discrete Bessel kernel.

For this application we need a version of the inclusion-exclusion principle as it applies to random point fields.

Definition (*Correlation measures*). Let $\mathbf{P}$ be a random point field on $\mathbf{Z}$ which is given by a Borel probability measure on $\mathrm{Conf}_0(\mathbf{Z})$ or equivalently on $\Omega = \{X \subset \mathbf{Z} : \sharp X < \infty\}$. For $Z \in \Omega$, there is an map $\Omega \to \Omega : X \mapsto X \triangle Z$, which we can associate with a map on the configurations. By elementary set theory, we have

$$(X \triangle Z)\triangle Z = X \triangle (Z \triangle Z) = X \triangle(\emptyset) = X, \qquad (10.4.1)$$

so the map is an involution. Now for each $X \in \Omega$, let $X_1 = X \setminus Z \in \Omega$ and $X_2 = Z \setminus X \in \Omega$; then introduce the first correlation measures as in Theorem 8.1.4 by

$$\rho(X) = \mathbf{P}\{Y : X \subseteq Y\}, \qquad (10.4.2)$$
$$\rho^{\triangle}(X) = \mathbf{P}\{Y : X \triangle Z \subseteq Y \triangle Z\}. \qquad (10.4.3)$$

Lemma 10.4.1 (*Complement principle*) *[30].*

$$\rho^{\triangle}(X) = \sum_{S:S \subseteq X_2} (-1)^{\sharp S} \rho(X_1 \cup S). \qquad (10.4.4)$$

Proof. We introduce the functions $g(S) = \rho(X_1 \cup S)$ and

$$f(T) = \mathbf{P}\{Y : Y \cap (X_1 \cup X_2) = X_1 \cup T\} \qquad (10.4.5)$$

for $S, T \subseteq X_2$, and we identify

$$\rho^{\triangle}(X) = \mathbf{P}\{Y : X \triangle Z \subseteq Y \triangle Z\} = \mathbf{P}\{Y : X_1 \subset Y, Y \cap X_2 = \emptyset\} = f(\emptyset). \qquad (10.4.6)$$

Next observe that by disjointness

$$g(S) = \sum_{T:S \subseteq T \subseteq X_2} f(T) = \mathbf{P}\{Y : Y \supseteq X_1 \cup S\} \qquad (10.4.7)$$

for all $S \subseteq X_2$. Now by the inclusion-exclusion principle of Section 7.5, we can invert this relation to obtain

$$f(T) = \sum_{S:T \subseteq S \subseteq X_2} (-1)^{\sharp S} g(S) \qquad (10.4.8)$$

and in particular

$$f(\emptyset) = \sum_{S:S \subseteq X_2} (-1)^{\sharp S} g(S), \qquad (10.4.9)$$

as required. $\qquad \square$

Suppose further that $\mathbf{P}$ is given by a determinantal random point field; so that there exists a kernel $K : \mathbf{Z} \times \mathbf{Z} \to \mathbf{C}$ such that $0 \leq K \leq I$ on

$\ell^2(\mathbf{Z})$ and

$$\rho(X) = \det[K(x_j, x_k)]_{x_j, x_k \in X} \qquad (X \in \Omega). \qquad (10.4.10)$$

We partition $\mathbf{Z}$ into $\mathbf{Z} \setminus Z$ and Z, and partition matrices accordingly, so

$$K = \begin{bmatrix} A & B \\ C & D \end{bmatrix} \begin{matrix} \mathbf{Z} \setminus Z \\ Z \end{matrix}. \qquad (10.4.11)$$

Proposition 10.4.2 *Let $K^\triangle$ be the matrix that has block form*

$$K^\triangle = \begin{bmatrix} A & B \\ -C & I - D \end{bmatrix} \begin{matrix} \mathbf{Z} \setminus Z \\ Z \end{matrix} \qquad (10.4.12)$$

for Z and X finite. Then

$$\rho^\triangle(X) = \det[K^\triangle(x_j, x_k)]_{x_j, x_k \in X_1 \cup X_2}. \qquad (10.4.13)$$

Proof. Let $N = \sharp(X_1 \cup X_2)$, and $n = \sharp X_2$, and introduce commuting indeterminates $z_{s_1}, \ldots, z_{s_n}$ for $s_j \in X_2$. We replace I by $\Delta(z) = \mathrm{diag}(z_{s_1}, \ldots, z_{s_n})$ and expand the determinant

$$\det \begin{bmatrix} A & B \\ -C & \Delta(z) - D \end{bmatrix} = z_{s_1} \ldots z_{s_n} \det \begin{bmatrix} A & B \\ 0 & I \end{bmatrix} + \cdots + \det \begin{bmatrix} A & B \\ -C & -D \end{bmatrix}$$

$$= z_{s_1} \ldots z_{s_n} \det[A] + \cdots + (-1)^n \det \begin{bmatrix} A & B \\ C & D \end{bmatrix}$$

$$= z_{s_1} \ldots z_{s_n} \rho(X_1) + \cdots + (-1)^n \rho(X_1 \cup X_2)$$

$$(10.4.14)$$

as a polynomial in the indeterminates. In the expansion of a determinant by its diagonal elements as in [4], with z_{s_j} corresponding to $S = \{s_j\} \subseteq X_2$ taken together, there is a complementary principal minor of order $N - \sharp S$. By considering the contribution of each subset $S \subseteq X_2$ and applying Lemma 10.4.1, we obtain by setting $z_{s_j} = 1$ the required identity

$$\det K^\triangle = \rho(X_1) - \cdots + (-1)^n \rho(X_1 \cup X_2) = \rho^\triangle(X). \quad (10.4.15)$$

$\square$

We intend to apply these results to a particular determinantal random point field on the integers. Let $\lambda \vdash N$ be a partition with rows $\lambda_1 \geq \cdots \geq \lambda_n > 0$, and introduce the hook lengths $\ell_j = \lambda_j + n - j$, as in Proposition 7.2.4. Then the Plancherel measure of this partition λ satisfies

$$\nu_N(\{\lambda\}) = \frac{N!}{\prod_{j=1}^n (\ell_j!)^2} \prod_{1 \leq j < k \leq n} (\ell_j - \ell_k)^2. \qquad (10.4.16)$$

The quantities λ_j are coupled due to the determinantal expression

$$\prod_{1 \le j < k \le n} (\ell_j - \ell_k)^2 = \det[\ell_j^{k-1}]_{j,k=1,\ldots,n}^2 \qquad (10.4.17)$$

and satisfy the constraint

$$\sum_{j=1}^{n} \lambda_j = N. \qquad (10.4.18)$$

We can remove the constraint by Poisson randomization as in Section 7.4.

The choice of Plancherel probability measure ν_N on $\Lambda_N = \{\lambda \in \Lambda : \lambda \vdash N\}$ is canonical by Proposition 7.2.3, but we choose Plancherel measure P^θ on $\Lambda = \cup_{N=0}^\infty \Lambda_N$ as in Section 7.4 to take advantage of special properties of the Poisson process.

Using Frobenius's coordinates, we can replace Λ by $\Omega = \{X \subset \mathbf{Z} : \sharp X < \infty\}$.

As is standard, we index matrices of $\mathbf{Z} \times \mathbf{Z}$ by $[a_{jk}]$ where the signs of (j, k) are

$$\begin{bmatrix} (-,-) & (-,+) \\ (+,-) & (+,+) \end{bmatrix}. \qquad (10.4.19)$$

Lemma 10.4.3 *Suppose that partitions are specified by their Frobenius coordinates. Then there exists a positive kernel $L(\,,\,;\theta)$ on the set $(\mathbf{Z}+1/2) \times (\mathbf{Z}+1/2)$ such that*

$$P^\theta(\{\lambda\}) = e^{-\theta} \det[L(x_j, x_k; \theta)]_{1 \le j,k \le 2d}, \qquad (10.4.20)$$

where $(x_j)_{j=1}^{2d}$ corresponds to λ.

Proof. For each partition λ, we have Frobenius coordinates

$$\mathrm{Fr}(\lambda) = (P_1, \ldots, P_d; -Q_1 - 1, \ldots, -Q_d - 1) \qquad (10.4.21)$$

with d positive and d negative terms, and such that $\sum_{j=1}^d (P_j + Q_j) = |\lambda| - d$; so $\mathrm{Fr}(\lambda) \in \Omega$. In view of Theorem 8.1.4, it suffices to introduce a determinantal random point field on $\mathbf{Z}+1/2$ given by a kernel L such that L has principal minors satisfying (10.4.20).

We introduce the modified Frobenius coordinates

$$\mathrm{Fr}^*(\lambda) = (P_1 + 1/2, \ldots, P_d + 1/2; -Q_1 - 1/2, \ldots, -Q_d - 1/2) \quad (10.4.22)$$

where each entry x_j belongs to $\mathbf{Z} + 1/2$, then we introduce for each θ the kernel

$$L(x, y; \theta) = \begin{bmatrix} 0 & L_\theta(x, y) \\ L_\theta(x, y) & 0 \end{bmatrix} \qquad (10.4.23)$$

where the off-diagonal blocks are

$$L_\theta(x, y) = \frac{\theta^{(|x|+|y|)/2}}{(x - y)\Gamma(|x| + 1/2)\Gamma(|y| + 1/2)} \qquad (xy < 0) \quad (10.4.24)$$

on $\mathbf{Z} + 1/2$. To prove (10.4.20), we introduce

$$A = \left[\frac{\theta^{(1+P_j+Q_k)/2}}{(P_j + Q_k + 1)P_j!Q_k!} \right]_{1 \le j, k \le d} \qquad (10.4.25)$$

and recall that by Proposition 7.2.7

$$\det A = \left(\theta^{\sum_{j=1}^d (P_j + Q_j + 1)/2} \right) \frac{\dim(H_\lambda)}{|\lambda|!}$$

$$= \theta^{|\lambda|/2} \frac{\dim(H_\lambda)}{|\lambda|!}. \qquad (10.4.26)$$

By considering the positive and negative entries of the modified Frobenius coordinates, and recalling the sign conventions, we see that

$$\det[L(x_j, x_k; \theta)]_{1 \le j, k \le 2d} = \det \begin{bmatrix} 0 & A \\ -A & 0 \end{bmatrix} = (\det A)^2, \qquad (10.4.27)$$

so that by (10.4.26)

$$e^{-\theta} \det[L(x_j, x_k; \theta)]_{1 \le j, k \le 2d} = \frac{e^{-\theta} \theta^{|\lambda|} (\dim(H_\lambda))^2}{(|\lambda|!)^2}, \qquad (10.4.28)$$

as in the formula for P^θ in Section 7.4. $\qquad \square$

Lemma 10.4.4 *The following identities hold:*

$$(i) \qquad \sum_{m=0}^\infty \frac{z^{m+\nu}}{m+\nu} \frac{J_m(2z)}{m!(\nu - 1)!} = J_\nu(2z), \qquad (10.4.29)$$

$$(ii) \qquad \sum_{m=0}^\infty \frac{z^{m+\nu+1}}{m+\nu+1} \frac{J_{m+1}(2z)}{m!\nu!} = \frac{z^\nu}{\nu!} - J_\nu(2z). \quad (10.4.30)$$

Proof. (i) This follows from (10.3.12) by an argument presented in [167, 5.23].

(ii) The matrix A is a submatrix of

$$B = [B_{p,q}] = \left[\frac{\theta^{(1+p+q)/2}}{(p + q + 1)p!q!} \right]_{p,q=0,1,\ldots}, \qquad (10.4.31)$$

where B has the special property that

$$[B_{p+1,q}] + [B_{p,q+1}] = \left[\frac{\theta^{(2+p+q)/2}}{(p+1)!(q+1)!}\right]$$

(10.4.32)

is of rank one. This translates into an identity for the power series of the Bessel functions.

Using the partial fractions

$$\frac{1}{(m+\nu+1)\nu} = \frac{1}{(m+1)\nu} - \frac{1}{(m+\nu+1)(m+1)}$$

(10.4.33)

on the mth terms, one can easily check that

$$\sum_{m=0}^{\infty} \frac{z^{1+m+\nu} J_{m+1}(2z)}{(m+\nu+1)m!\nu!} = \sum_{m=0}^{\infty} \frac{z^{1+m+\nu} J_{m+1}(2z)}{(m+1)!\nu!}$$

$$- \sum_{m=0}^{\infty} \frac{z^{1+m+\nu} J_{m+1}(2z)}{(1+m+\nu)(m+1)!(\nu-1)!}$$

$$= \sum_{m=0}^{\infty} \frac{z^{m+\nu} J_m(2z)}{m!\nu!} - \sum_{m=0}^{\infty} \frac{z^{m+\nu} J_{m+1}(2z)}{(m+\nu)m!(\nu-1)!}$$

$$= \frac{z^{\nu}}{\nu!} - J_{\nu}(2z),$$

(10.4.34)

where we have used (i) and (10.3.12) to simplify the sums. □

In Theorem 8.1.4, we saw that a determinantal random point field could be described equivalently by a matrix L, with principal minors that gives probabilities, or a matrix K with minors that give the correlation functions. The next step is to identify the K for our L.

We introduce a new variable s so that $\theta = s^2$. Let K be the symmetric kernel on $(\mathbf{Z}+1/2) \times (\mathbf{Z}+1/2)$ such that

$$K(x,y;s)$$
$$= s\frac{J_{|x|-1/2}(2s)J_{|y|+1/2}(2s) - J_{|x|-1/2}(2s)J_{|y|+1/2}(2s)}{|x|-|y|} \qquad (xy > 0),$$

$$K(x,y;s)$$
$$= s\frac{J_{|x|-1/2}(2s)J_{|y|-1/2}(2s) + J_{|x|+1/2}(2s)J_{|y|+1/2}(2s)}{x-y} \qquad (xy < 0),$$

(10.4.35)

or in block matrix form

$$
\begin{bmatrix}
\dfrac{J_{-x-1/2}J_{-y+1/2} - J_{-x+1/2}J_{-y-1/2}}{x-y} & \dfrac{J_{-x-1/2}J_{y-1/2} + J_{-x+1/2}J_{y+1/2}}{-x-y} \\[2ex]
\dfrac{J_{x-1/2}J_{-y-1/2} + J_{x+1/2}J_{-y+1/2}}{x+y} & \dfrac{J_{x-1/2}J_{y+1/2} - J_{x+1/2}J_{y-1/2}}{x-y}
\end{bmatrix}.
$$

$$(10.4.36)$$

On the diagonal $x = y$, we use l'Hôpital's rule to take the limit and apply (10.3.1).

We observe that the block of the $K(x,y)$ matrix that corresponds to $x, y > 0$ is given by the discrete Bessel matrix. Following [30], we now verify that K and L are related as in Lemma 8.1.3.

Lemma 10.4.5 *The operator L is nonnegative, $I + L$ is invertible and*

$$(I - K)(I + L) = I. \qquad (10.4.37)$$

Proof. By Proposition 2.1.2, we have $L \geq 0$ since the principal minors are positive by Proposition 7.2.7. Hence $I + L$ is invertible. The identity holds at $s = 0$, so it suffices to show that

$$-\frac{\partial K}{\partial s} - \frac{\partial K}{\partial s}L + \frac{\partial L}{\partial s} - K\frac{\partial L}{\partial s} = 0. \qquad (10.4.38)$$

By the previous formulae, we have

$$\frac{\partial L}{\partial s} = \begin{bmatrix} 0 & -B \\ B & 0 \end{bmatrix} \qquad (10.4.39)$$

where the off-diagonal blocks are

$$B(x,y) = \frac{s^{|x|+|y|-1}}{\Gamma(|x| + 1/2)\Gamma(|y| + 1/2)} \qquad (xy < 0). \qquad (10.4.40)$$

The Bessel functions satisfy a system of differential equations quite analogous to those considered in Section 9.1, namely

$$\frac{d}{ds}\begin{bmatrix} J_{x+1}(2s) \\ J_x(2s) \end{bmatrix} = \begin{bmatrix} -(x+1)/s & 2 \\ -2 & x/s \end{bmatrix}\begin{bmatrix} J_{x+1}(2s) \\ J_x(2s) \end{bmatrix}; \qquad (10.4.41)$$

hence one can verify the identity

$$
\frac{\partial K(x,y;s)}{\partial s} = J_{|x|-1/2}(2s)J_{|y|+1/2}(2s)
$$
$$
+ J_{|x|+1/2}(2s)J_{|y|-1/2}(2s) \qquad (xy > 0)
$$

$$
\frac{\partial K(x,y;s)}{\partial s} = \mathrm{sgn}(x)\big(J_{|x|-1/2}(2s)J_{|y|-1/2}(2s)
$$
$$
- J_{|x|+1/2}(2s)J_{|y|+1/2}(2s)\big) \qquad (xy < 0). \qquad (10.4.42)
$$

By using Lemma 10.4.4 to simplify the terms $\frac{\partial K}{\partial s}L$ and $K\frac{\partial L}{\partial s}$, one can verify (10.4.38). □

For a partition $\lambda = (\lambda_1 \geq \lambda_2 \geq \dots)$, the sequence $\lambda_j - j$ is clearly strictly decreasing, and we form the set $D(\lambda) = \cup_{j=1}^\infty \{\lambda_j - j\}$ of such differences. Since $\lambda_j = 0$ for all sufficiently large j, the symmetric difference between $D(\lambda)$ and $\{-j : j \in \mathbf{N}\}$ is finite. For each $\theta > 0$, the k-point correlation function is

$$\rho_k^\theta(x_1, \dots, x_k) = P^\theta\{\lambda \in \Lambda : x_1, \dots, x_k \in D(\lambda)\}. \qquad (10.4.43)$$

Theorem 10.4.6 (*Borodin, Okounkov, Olshanskii, [30]*). *Let* $J : \mathbf{Z} \times \mathbf{Z} \to \mathbf{R}$ *be the discrete Bessel kernel as in (10.3.1). Then*

$$\rho_\ell^\theta(x_1, \dots, x_\ell) = \det[J(x_j, x_k; \theta)]_{j,k=1,\dots,\ell}. \qquad (10.4.44)$$

Proof. By the preceding calculations, P^θ is given by the determinantal random point field that is equivalently specified by K or L. Now we wish to return to determinantal random point fields on $\mathbf{Z}$.

With $\mathbf{Z} + 1/2$ and $Z = \{-1/2, -3/2, \dots\}$ we have $(\mathbf{Z} + 1/2) \setminus Z = \{1/2, 3/2, \dots\}$, and with $x = m + 1/2$ we have $x \in (\mathbf{Z} + 1/2) \setminus Z$ if and only if $m \in \{0, 1, 2, \dots\}$. Now the set of differences $D(\lambda) = \{\lambda_j - j\}$ gives the set of modified Frobenius coordinates as in $\mathrm{Fr}^*(\lambda) = (D(\lambda) + 1/2)\triangle Z$ by Proposition 7.2.6. To recover the determinantal random point field on $D(\lambda)$, we need to apply the operation $K \mapsto K^\triangle$, and by Proposition 10.4.2, check the identity

$$K^\triangle(x, y) = \mathrm{sgn}(m)^{m+1}\mathrm{sgn}(n)^{n+1}J(m, n; \theta), \qquad (10.4.45)$$

where $x = m + 1/2$ and $y = n + 1/2$. The unimodular factors do not affect the determinants. For $m \neq n$, the identity follows readily from the identity $J_n = (-1)^n J_{-n}$ and the definitions of the kernels.

On the diagonal we need to use the definition of the kernel involving l'Hôpital's rule. In particular, for $x = -k - 1/2 = y$, we have

$$K^\triangle(x, x) = 1 - K(x, x) = 1 - J(k, k; \theta); \qquad (10.4.46)$$

so we show by induction that

$$1 - J(k, k; \theta) = J(-k - 1, -k - 1; \theta) \qquad (k = 0, 1, \dots). \qquad (10.4.47)$$

The basis of induction is the identity $1 - \mathrm{J}(0,0;\theta) = \mathrm{J}(-1,-1;\theta)$, which follows from Lemma 10.3.3 since

$$\mathrm{J}(-1,-1;\theta) = \sum_{n=0}^{\infty} \frac{(-1)^n \theta^n (n)_n}{n!\Gamma(n+1)\Gamma(n+1)}$$

$$= 1 - \sum_{m=0}^{\infty} \frac{(-1)^m \theta^{m+1}(m+2)_m}{m!\Gamma(m+2)\Gamma(m+2)}$$

$$= 1 - \mathrm{J}(0,0;\theta). \tag{10.4.48}$$

The induction step follows from the identity

$$\mathrm{J}(k+1,k+1;\theta) - \mathrm{J}(k,k;\theta) = -J_{k+1}J_{k+1}. \tag{10.4.49}$$

as in (10.3.9).   □

The preceding result is mainly interesting when $\theta = \mathbf{E}|\lambda|$ is large. The arguments sketched at the end of Section 10.3 can be made precise, and lead to the following result.

Theorem 10.4.7 (*Baik, Deift, Johansson [8]*). *The distribution of the length of the longest increasing subsequence in* $\sigma \in S_N$ *scales to a random variable* $N^{-1/6}(\ell(\sigma) - 2\sqrt{N})$ *that converges in distribution to the Tracy-Widom distribution as* $N \to \infty$; *so that*

$$\mu_N\left\{\sigma \in S_N : \frac{\ell(\sigma) - 2\sqrt{N}}{N^{1/6}} \leq t\right\} \to F(t) \quad (N \to \infty). \tag{10.4.50}$$

Proof. By the RSK correspondence in Proposition 7.2.3, we have

$$\mu_N\left\{\sigma \in S_N : \frac{\ell(\sigma) - 2\sqrt{N}}{N^{1/6}} \leq t\right\} = \nu_N\left\{\lambda \vdash N : \frac{\lambda_1 - 2\sqrt{N}}{N^{1/6}} \leq t\right\}, \tag{10.4.51}$$

where ν_N here denotes Plancherel measure on the partitions of N. We can replace this by the Poisson version of Plancherel measure, namely P^θ where $\theta = N$, and obtain approximately

$$P^\theta\left\{\lambda : \frac{\lambda_1 - 2\sqrt{\theta}}{\theta^{1/6}} \leq t\right\}. \tag{10.4.52}$$

Converting this to Frobenius coordinates and using Theorem 10.4.6, we can express this as the determinant

$$\det(I - \mathbf{I}_{(t,\infty)}\tilde{\mathrm{J}}(\ ,\ ;\theta)), \tag{10.4.53}$$

and by Proposition 10.3.5,

$$\det(I - \mathbf{I}_{(t,\infty)}\tilde{\mathrm{J}}(\ ,\ ;\theta)) \to F(t) \quad (\theta \to \infty). \tag{10.4.54}$$

Subsequently, Borodin, Okounkov and Olshanski [30] refined this result and obtained the asymptotic distribution of the kth longest row of the Young tableaux $\lambda \vdash N$ as $N \to \infty$.   □

10.5 Fluctuations of the longest increasing subsequence

The Costin–Lebowitz theorem [52, 147] has found several significant applications in random matrix theory. Recently, Bogachev and Su [28] found the fluctuations of Young diagrams about the Vershik distribution.

In this section we combine the ideas of Sections 10.1 and 10.4, and thus present a result concerning the fluctuations of the length of the longest increasing subsequence in a randomly chosen partition. We state the following results without their proofs, which depend upon asymptotic analysis of Bessel functions. Given a partition $\lambda \vdash N$ with Young diagram $\lambda_1 \geq \lambda_2 \geq \lambda_3 \geq \ldots$, we extend the definition of λ_j to $\lambda(x) = \lambda_{\lceil x \rceil}$ where $\lceil x \rceil = \min\{j \in \mathbf{N} : j \geq x\}$. Then we introduce the scaled Young diagram

$$f_\lambda(x) = \frac{1}{\sqrt{N}}\lambda(x\sqrt{N}) \qquad (x > 0) \qquad (10.5.1)$$

as in 7.3, and the Vershik distribution f_Ω; then we introduce

$$\Delta^{(\lambda)}(x) = \sqrt{N}(f_\lambda(x) - f_\Omega(x)); \qquad (10.5.2)$$

we regard $\Delta(x)/\sqrt{\log N}$ as a stochastic process with time parameter x. It is convenient to scale the variance according to the time by introducing

$$\psi(x) = \cos^{-1}\big((1/2)(f_\Omega(x) - x)\big) \qquad (10.5.3)$$

and then introducing the scaled random variables

$$\lambda \mapsto Y^{(\lambda)}(x) = \frac{2\psi(x)\Delta^{(\lambda)}(x)}{\sqrt{\log N}} \qquad (\lambda \in \Lambda) \qquad (10.5.4)$$

where $\lambda \vdash N$. We briefly return to the notation of (7.3.5), and observe that with $\xi = 0$, we have $g(0) = 2/\pi$ and $f_\Omega(x) = x = 2/\pi$. Hence

$$\psi(0) = 0, \quad \psi(2) = \pi, \quad \psi(2/\pi) = \pi/2. \qquad (10.5.5)$$

Theorem 10.5.1 (*Bogachev and Su, [28]*). *For $x \in (0,2)$, the distribution of $Y(x)$ with respect to the measure P^θ converges to the $N(0,1)$ distribution as $\theta \to \infty$.*

Proof. We wish to show that

$$P^{\theta}\{\lambda \in \Lambda : Y^{(\lambda)}(x) > z\} \to \int_{z}^{\infty} e^{-u^2/2} \frac{du}{\sqrt{2\pi}} \qquad (\theta \to \infty), \qquad (10.5.6)$$

where the rescaling gives

$$\{\lambda \in \Lambda : Y^{(\lambda)}(x) > z\}$$
$$= \Big\{\lambda \in \Lambda : \lambda(x\sqrt{\theta}) - \sqrt{\theta} f_{\Omega}(x) > 2\sqrt{\theta}\cos\psi(x) + \frac{z\sqrt{\log\theta}}{2\psi(x)}\Big\}.$$

$$(10.5.7)$$

We introduce the interval

$$I_{\theta,x} = \Big[2\sqrt{\theta}\cos\psi(x) + \frac{z\sqrt{\log\theta}}{2\psi(x)}, \infty\Big). \qquad (10.5.8)$$

We introduce the matrix $K_{\theta,x}$, which is given by the kernel $J(m,n;\theta)$ with $m, n \in I_{\theta,x}$. Now we let $\nu_{\theta,x}$ be the random variable that gives the number of points in $I_{\theta,x}$, subject to the probability field P^{θ}. Then by (10.1.22) and (10.1.23), we have

$$\mathbf{E}(\nu_{\theta,x}) = \operatorname{trace} K_{\theta,x}, \qquad (10.5.9)$$

$$\operatorname{var}(\nu_{\theta,x}) = \operatorname{trace}(K_{\theta,x} - K_{\theta,x}^2). \qquad (10.5.10)$$

By estimation on $K_{\theta,x}$, one can prove the following Lemma.   □

Lemma 10.5.2 *As $\theta \to \infty$, the following hold:*

$$\mathbf{E}(\nu_{\theta,x}) = x\sqrt{\theta} - \frac{z\log\theta}{2\pi} + O(1), \qquad (10.5.11)$$

$$\operatorname{var}(\nu_{\theta,x}) = \frac{\log\theta}{4\pi^2}(1 + o(1)). \qquad (10.5.12)$$

Proof. This follows by estimating the determinant. See [28].   □

Conclusion of the proof of Theorem 10.5.1 The variance of ν_{θ} is much smaller than its mean value. One can think of θ as being approximately equal to N, so that the variance is of order only $\log N$, whereas the mean is of order $\sqrt{N}$. Given these Lemmas, Theorem 10.5.1 follows from Theorem 10.1.3.   □

For each $0 < x < 1$ and N, we introduce the scaled random variables

$$\lambda \mapsto Y^{(\lambda)}(x) = \frac{2\psi(x)\Delta_N^{(\lambda)}(x)}{\sqrt{\log N}} \qquad (\lambda \vdash N). \qquad (10.5.13)$$

Theorem 10.5.3 *For $x \in (0,2)$, the distribution of $Y_N(x)$ with respect to the Plancherel measure converges to the $N(0,1)$ distribution as $N \to \infty$.*

The proof of Theorem 10.5.3 follows from Theorem 10.5.1 by a technique known as 'de-Poissonization', which is an aspect of the classical theory of Borel summability Suppose that $\sum_j a_j$ is a real series with partial sums s_n. If $\sum_{j=0}^{\infty} a_j$ converges to sum A, then it is summable in Borel's sense so that

$$\lim_{\theta \to \infty} e^{-\theta} \sum_{j=0}^{\infty} s_j \theta^j / j! = A. \qquad (10.5.14)$$

This is an easy exercise in real analysis; the converse requires further hypotheses as in [78].

Theorem 10.5.4 (*The principal Tauberian theorem for Borel summability*). *Suppose that $\sum_{j=0}^{\infty} a_j$ is summable in Borel's sense to A, so that (10.5.14) holds, and that $a_n = O(n^{-1/2})$ as $n \to \infty$. Then $\sum_{j=1}^{\infty} a_j$ converges with sum A.*

Proof. See [78]. ☐

10.6 Fluctuations of linear statistics over unitary ensembles

As in Section 4.9, let $f : \mathbf{T} \to \mathbf{R}$ be a continuous function. The corresponding linear statistic on $SU(n)$ is the function $F : SU(n) \to \mathbf{R}$

$$F_n(U) = \operatorname{trace} f(U) = \sum_{j=1}^{n} f(e^{i\theta_j}) \qquad (U \in U(n)) \qquad (10.6.1)$$

where $(e^{i\theta_j})$ are the eigenvalues of U. In this section, we consider the difference

$$\sum_{j=1}^{n} f(e^{i\theta_j}) - n \int_{\mathbf{T}} f(e^{i\theta}) \frac{d\theta}{2\pi}; \qquad (10.6.2)$$

and seek bounds that are independent of n, and have a simple dependence upon f.

- The exponential integral for a linear statistic with respect to the circular ensemble is given by a Toeplitz determinant as in Section 4.9.

Remark (*Concentration and fluctuations*). Suppose that $f : \mathbf{T} \to \mathbf{R}$ is a 1-Lipschitz function that gives rise to linear statistics $F_n : (U(n), c^2(n)) \to \mathbf{R}$ which are n-Lipschitz. Then by Theorem 3.9.2,

$(U(n), c^2(n), \sigma_{U(n)})$ exhibit concentration of measure with $\alpha_n = n^2/c$, so that

$$\log \int_{U(n)} e^{tF_n}\, d\mu_{U(n)} - t \int_{U(n)} F_n\, d\mu_{U(n)} \le \frac{ct^2}{2} \qquad (t \in \mathbf{R}; n = 1, 2, \dots).$$

$$(10.6.3)$$

Definition (*Fluctuation theorem*). We say that the F_n satisfy a fluctuation theorem when there exist a linear functional $\phi : \mathrm{Lip} \to \mathbf{R}$ and a bilinear form $A : \mathrm{Lip} \times \mathrm{Lip} \to \mathbf{R}$ such that

$$\log \int_{U(n)} e^{tF_n}\, d\mu_{U(n)} - t \int_{U(n)} F_n\, d\mu_{U(n)} \to t\phi(f) + t^2 A(f, f) \quad (n \to \infty).$$

$$(10.6.4)$$

In both (10.6.3) and (10.6.4), there is a quadratic term on the right-hand side which is independent of n; but the concentration inequality is a uniform bound, whereas the fluctuation theorem gives a limit. Such a fluctuation theorem resembles both the central limit theorem and the law of large numbers of classical probability.

Exercise. Assuming that a fluctuation theorem holds, we can partially determine the form of ϕ and A. Let $f_\psi(e^{i\theta}) = f(e^{i(\theta - \psi)})$ and observe that the Haar measure is invariant under the translation $u \mapsto e^{-i\psi}u$; so $\phi(f_\psi) = \phi(f)$ and $A(f_\psi, f_\psi) = A(f, f)$ for all f and ψ. Deduce that there exist constants δ_0 and γ_j for $j \in \mathbf{Z}$ such that for $f(e^{i\theta}) = \sum_{k=-\infty}^{\infty} a_k e^{ik\theta}$ we have $\phi(f) = \delta_0 a_0$ and $A(f, f) = \sum_{k=-\infty}^{\infty} \gamma_k a_k a_{-k}$.

The following theorem refines Corollary 4.9.3, where the hypotheses on f are as in the definition of Dirichlet space in Section 4.2.

Theorem 10.6.1 (*Strong Szegö limit theorem*). *Suppose that* $f(e^{i\theta}) = \sum_{n=-\infty}^{\infty} a_n e^{in\theta}$ *has* $\sum_{n=-\infty}^{\infty} |n||a_n|^2$ *finite and that* $\sum_{n=1}^{\infty} a_n e^{in\theta}$ *and* $\sum_{n=-\infty}^{0} a_n e^{in\theta}$ *are bounded on* $\mathbf{T}$. *Then the Toeplitz determinants associated with* e^f *satisfy*

$$\log \det \left[\int_0^{2\pi} \exp\big(f(e^{i\theta}) + i(j - \ell)\theta\big) \frac{d\theta}{2\pi} \right]_{1 \le j, \ell \le n} - na_0$$

$$\to \sum_{k=1}^{\infty} k a_k a_{-k} \qquad (n \to \infty). \tag{10.6.5}$$

Proof. See [153, 145]. There are several proofs of this due to Kac, Bump and Diaconis, Johansson and others; all of them involve difficult steps. □

Corollary 10.6.2 *Let $F_n(U) = \mathrm{trace} f(U)$ be the real linear statistic on $U(n)$ corresponding to $f : \mathbf{T} \to \mathbf{R}$ as in Proposition 4.9.5.*

(i) Then

$$\log \int_{U(n)} \exp\big(F_n(U)\big)\mu_{U(n)}(dU)$$

$$- \int_{U(n)} F_n(V)\mu_{U(n)}(dV) \to \sum_{k=1}^{\infty} k|a_k|^2 \qquad (t \in \mathbf{R}) \quad (10.6.6)$$

as $n \to \infty$.

(ii) Suppose further that $f : \mathbf{T} \to \mathbf{R}$ is L-Lipschitz. Then $\sum_{k=1}^{\infty} k|a_k|^2 \leq L^2$.

Proof. (i) First we observe that

$$\int_{U(n)} F_n(U)\mu_{U(n)}(dU) = \int_{U(n)} \mathrm{trace} f(U)\mu_{U(n)}(dU) = na_0, \quad (10.6.7)$$

and we also note that $a_{-k} = \bar{a}_k$ since f is real. Then we can apply Theorem 10.6.1 to obtain the limit of the formula in Proposition 4.9.5.

(ii) By the proof of Proposition 4.2.6, we have

$$\sum_k |k||a_k|^2 = \iint_{\mathbf{T}^2} \left| \frac{f(e^{i\theta}) - f(e^{i\phi})}{e^{i\theta} - e^{i\phi}} \right|^2 \frac{d\theta}{2\pi} \frac{d\phi}{2\pi} \leq L. \quad (10.6.8)$$

$\square$

Remarks 10.6.3 (i) The linear statistic F_n in Corollary 10.6.2 is nL-Lipschitz on $(U(n), c^2(n))$, so Theorem 3.9.2 gives the correct dependence of the concentration constant upon n.

(ii) In [11], Basor considers variants of the strong Szegö limit formula for symbols that do not satisfy the smoothness assumptions that we impose in this section.

Example 10.6.4 In Section 7.4 we considered $f(e^{i\theta}) = 2\lambda \cos\theta$ so that $F_n(U) = \lambda \mathrm{trace}(U + U^*)$ and

$$\int_0^{2\pi} e^{2\lambda \cos\theta + im\theta} \frac{d\theta}{2\pi} = i^m J_m(-i2\lambda). \quad (10.6.9)$$

Applying Corollary 10.6.2, we recover a known identity that

$$\int_{U(n)} \exp\big(\lambda \mathrm{trace}\,(U + U^*)\big)\mu_{U(n)}(dU) \to e^{\lambda^2} \qquad (n \to \infty). \quad (10.6.10)$$

We now consider fluctuations of linear statistics for the generalized unitary ensemble with potential function v. Let $(\lambda_j)_{j=1}^n$ be the eigenvalues of $X \in M_n^h(\mathbf{C})$, let ρ_v be the equilibrium measure and recall that in Chapter 5 we showed that $\frac{1}{n} \sum_{j=1}^n \delta_{\lambda_j}$ converges weakly almost surely to ρ_v as $n \to \infty$. To analyse the fluctuations, we consider the measure $\sum_{j=1}^n \delta_{\lambda_j} - n\rho_v$ which is associated with the normalized linear statistic

$$F_n(\lambda_1, \ldots, \lambda_n) = \sum_{j=1}^n f(\lambda_j) - n \int f(\lambda)\rho_v(d\lambda) \qquad (10.6.11)$$

for a suitably differentiable function $f : \mathbf{R} \to \mathbf{R}$ of compact support. We wish to show that the variance of F_n converges to $Q(f, f)$ as $n \to \infty$ for some bilinear form Q.

When ρ_v is supported on a single interval $[a, b]$, we can rescale so that the measures are on $[-1, 1]$; then we can use Chebyshev polynomials to give a useful basis. First we introduce the appropriate form of determinant.

Definition (*Hankel determinant*). A (finite) Hankel matrix has the form $[a_{j+k}]_{0 \leq j,k \leq n-1}$ so that the cross diagonals are constant; the corresponding Hankel determinant is $\det[a_{j+k}]_{0 \leq j,k \leq n-1}$.

Exercise. Let μ be a positive and bounded measure of compact support on $\mathbf{R}$, and let $c_j = \int x^j \mu(dx)$ be the moments. Show that the matrix $\Gamma_n = [c_{j+k}]_{j,k=0,\ldots,n}$ is positive semidefinite.

Proposition 10.6.5 *Let μ be a probability measure on $\mathbf{R}$ such that $\int \cosh(\delta t)\mu(dt) < \infty$ for some $\delta > 0$, and let $(p_j)_{j=0}^\infty$ be the sequence of monic orthogonal polynomials in $L^2(\mu)$. Then*

$$\det\left[\int_{\mathbf{R}} x^{j+k} \mu(dx)\right]_{j,k=0,\ldots,n} = \prod_{j=0}^n \int_{\mathbf{R}} p_j(x)^2 \mu(dx). \quad (10.6.12)$$

Proof. Let $(X_j)_{j=0,\ldots,n}$ be vectors in $\mathbf{R}^{n+1}$ with the standard inner product, and let

$$S = \left\{\sum_{j=0}^n t_j X_j : 0 \leq t_j \leq 1, j = 0, \ldots, n\right\} \qquad (10.6.13)$$

be the parallelepiped that they generate. Then $\mathrm{vol}(S) = \det[X_0 \ldots X_n]$, and hence

$$\mathrm{vol}(S)^2 = \det[X_0 \ldots X_n]\det[X_0 \ldots X_n]^t = \det[\langle X_j, X_k \rangle]_{j,k=0,\ldots,n}. \qquad (10.6.14)$$

As in the Gram–Schmidt process [171], we can replace X_j by

$$\hat{X}_j = X_j - \sum_{k=0}^{j-1} \langle X_j, X_k \rangle X_k \|X_k\|^{-2}, \qquad (10.6.15)$$

so that $\det[X_0 \ldots X_n] = \det[\hat{X}_0 \ldots \hat{X}_n]$.

We can now take $x^j \leftrightarrow X_j$ and $p_j \leftrightarrow \hat{X}_j$ to deduce that

$$\det[\langle x^j, x^k \rangle_{L^2(\mu)}] = \det[\langle p_j, p_k \rangle_{L^2(\mu)}], \qquad (10.6.16)$$

as stated in the Proposition. □

- The exponential integral for a linear statistic with respect to the generalized unitary ensemble is given by a Hankel determinant.

Exercise 10.6.6 (i) The following gives the analogue for unitary ensembles of Proposition 4.9.5. Instead of D_n, we introduce the Hankel determinants

$$Z_n = \det\left[\int_{-\infty}^{\infty} e^{-nv(\lambda)} \lambda^{j+k} \, d\lambda \right]_{j,k=0,\ldots,n-1}, \qquad (10.6.17)$$

as normalizing constants. Now use Lemma 2.2.2 to show that

$$\int_{M_n^h(\mathbf{C})} \exp(F_n(X)) \nu_n^{(2)}(dX)$$

$$= Z_n^{-1} \det\left[\int_{-\infty}^{\infty} e^{f(\lambda)-nv(\lambda)} \lambda^{j+k} \, d\lambda \right]_{j,k=0,\ldots,n-1}. \qquad (10.6.18)$$

(ii) Now use Lemma 2.2.2(ii) to obtain an analogous result for generalized orthogonal ensembles.

Lemma 10.6.7 *Let* $f \in C^1(\mathbf{R}; \mathbf{R})$ *have compact support, and let* f *have Chebyshev coefficients*

$$a_k = \frac{2}{\pi} \int_{-1}^{1} f(x) \frac{T_k(x)}{\sqrt{1-x^2}} dx. \qquad (10.6.19)$$

Then

$$\sum_{k=1}^{\infty} k a_k^2 = \frac{2}{\pi^2} \int_{-1}^{1} \frac{f(y)}{\sqrt{1-y^2}} \left(\text{p.v.} \int_{-1}^{1} \frac{f'(x)\sqrt{1-x^2}}{y-x} dx \right) dy. \qquad (10.6.20)$$

Hence the function $f(\cos\theta) = \sum_{k=1}^{\infty} a_k \cos k\theta$ *belongs to Dirichlet space as in Example 4.2.6.*

Proof. The key observation is that the Chebyshev polynomials satisfy $T_k'(x) = kU_{k-1}(x)$, and hence by Lemma 4.5.2

$$\frac{1}{\pi}\int_{-1}^1 \frac{f'(x)\sqrt{1-x^2}}{y-x}dx = \sum_{k=1}^\infty \frac{ka_k}{\pi}\int_{-1}^1 \frac{U_{k-1}(x)\sqrt{1-x^2}}{y-x}dx$$

$$= \sum_{k=1}^\infty ka_k T_k(y); \qquad (10.6.21)$$

so by orthogonality

$$\frac{2}{\pi^2}\int_{-1}^1 \frac{f(y)}{\sqrt{1-y^2}}\left(\text{p.v.}\int_{-1}^1 \frac{f'(x)\sqrt{1-x^2}}{y-x}dx\right)dy$$

$$= \sum_{k=1}^\infty \frac{2ka_k}{\pi}\int_{-1}^1 f(y)\frac{T_k(y)}{\sqrt{1-y^2}}dy$$

$$= \sum_{k=1}^\infty ka_k^2. \qquad (10.6.22)$$

$\square$

We can express the series in (10.6.19) otherwise by Proposition 4.2.5 and obtain

$$\frac{1}{2}\sum_{k=1}^\infty ka_k^2 = \iint_{\mathbf{T}^2}\left|\frac{f(\cos\theta)-f(\cos\psi)}{e^{i\theta}-e^{i\psi}}\right|^2 \frac{d\theta}{2\pi}\frac{d\psi}{2\pi}. \qquad (10.6.23)$$

The following fluctuation theorem gives the analogue for generalized unitary ensembles of Corollary 10.6.2.

Theorem 10.6.8 (*Johansson, [94]*). *Suppose that v is a uniformly convex potential with equilibrium distribution ρ_v supported on $[-1,1]$; let f be as in Lemma 10.6.7. Then*

$$\log\int_{M_n^h(\mathbf{C})}\exp(F_n(X))\nu_n^{(2)}(dX) - n\int_{-1}^1 f(x)\rho_v(dx)$$

$$\to \frac{1}{8}\sum_{k=1}^\infty ka_k^2 \qquad (n\to\infty). \qquad (10.6.24)$$

Proof. See [94]. $\square$

This result is sometimes described as a central limit theorem since the asymptotic distribution of $F_n(X)$ is like a Gaussian random variable. In Chapter 11, we introduce stochastic integrals and make this interpretation more explicit. A remarkable feature of this result is that the variance term on the right-hand side does not depend upon the special choice of the potential v, whereas ρ_v depends upon v as in Proposition 4.4.4.

To emphasize the analogy between Toeplitz and Hankel determinants, we have stated Theorem 10.6.8 for unitary ensembles. Johansson proved a more general fluctuation theorem for $\sigma_n^{(\beta)}$ as in (5.2.1), which includes cases of the generalized orthogonal ensemble for $\beta = 1$ and generalized symplectic ensembles for $\beta = 4$. The proof uses mean field theory as in Section 5.3, and some delicate potential theory.

Remark 10.6.9 If $f(\cos\theta)$ is Lipschitz, then $\sum_k ka_k^2$ converges. Theorem 10.6.8 shows that the constant c/n^2 in Theorem 3.6.3 has optimal dependence upon n.

11
Limit groups and Gaussian measures

Abstract
We introduce some groups that arise as inductive limits of classical compact metric groups and which have natural representations on Hilbert space. Classical results on characters extend to limit groups of compact groups. Some groups are not locally compact, and hence do not have Haar probability measures. As substitutes, we construct the Hua–Pickrell measure on the infinite unitary group, and Gaussian measure on Hilbert space.

11.1 Some inductive limit groups

In applications to physics, there arise some groups that are not locally compact, but are the inductive limits of the classical compact groups. In this section we introduce:

- the infinite symmetric group S_∞, which is countable;
- the infinite torus $\mathbf{T}^\infty$, a compact abelian group with Haar measure;
- the infinite orthogonal group $O(\infty)$ and unitary group $U(\infty)$.

First we define the groups and their invariant measures, then we revisit the examples and state results concerning their representations on Hilbert space. Most of the results appear in Borodin and Olshanski's papers [31, 32]. Remarkably, the irreducible unitary representations of $U(\infty)$ and S_∞ arise from limits of representations of $U(N)$ and S_N.

Definition (*Inductive limits*) *[87]*. Given a sequence of compact metric groups with continuous inclusions $G_j \leq G_{j+1}$, let $G_\infty = \cup_{j=1}^\infty G_j$ have the finest topology such that all the inclusion maps $G_j \leq G_\infty$ are continuous; equivalently, a function $f : G_\infty \to Z$ to an arbitrary topological

352

space Z is continuous if and only if the restriction $f : G_j \to Z$ is continuous for each j. A continuous function $\chi : G_\infty \to \mathbf{C}$ is *approximable* if there exists a sequence of continuous functions $\chi_j : G_j \to \mathbf{C}$ such that $\chi_j \to \chi$ uniformly on G_N for each N as $j \to \infty$.

Definition (*Haar probability measure*). Let G_∞ be an inductive limit group that is a Polish group. A Haar measure μ_G for G_∞ is a Radon probability measure such that

$$\int_G f(gx)\mu_G(dx) = \int_G f(x)\mu_G(dx) \qquad (11.1.1)$$

for all continuous $f : G \to \mathbf{C}$ and $\mu_G(E) > 0$ for all non-empty open sets E.

When G_∞ is a compact metric group, this is equivalent to the definition of Section 1.2. However, not all inductive limit groups have a Haar probability measure.

Lemma 11.1.1 *Suppose that (G, d) is a Polish group with left invariant metric d. If G has an infinite discrete closed subgroup S_∞, then G does not have a Haar probability measure.*

Proof. Since S_∞ is a discrete subgroup, we have $\delta = \inf\{g(s, e) : s \in S_\infty s \neq e\} > 0$. Now let $B(g, r) = \{h \in G : d(g, h) < r\}$, and observe that since the metric is invariant, we can choose an infinite sequence $B(s_j, \delta/2)$ of such disjoint sets for $s_j \in S_\infty$ and $j = 1, 2, \ldots$. Suppose that μ is a Haar probability measure; then

$$1 = \mu_G(G) \geq \sum_{j=1}^{\infty} \mu(B(s_j, \delta/2)). \qquad (11.1.2)$$

By invariance of the measure, we must have $\mu(B(s_j, \delta/2)) = 0$ for all j, so there exist non-empty open sets with zero measure, and μ_G is not a Haar measure. $\qquad\square$

11.1.2 Infinite torus

Let $(\mathbf{T}^n, c^2(n))$ be the n-dimensional torus with Haar measure $\otimes_{j=1}^{n} d\theta_j/(2\pi)$, and let $\mathbf{T}^\infty = \{(e^{i\theta_j})_{j=1}^{\infty} : \theta_j \in \mathbf{R}\}$ be the infinite-dimensional counterpart with Haar measure $d\Theta = \otimes_{j=1}^{\infty} d\theta_j/(2\pi)$. Then $\mathbf{T}^\infty$ is a compact metric space, and there is a natural inclusion of groups

$$\mathbf{T}^1 \leq \mathbf{T}^2 \leq \cdots \leq \mathbf{T}^\infty \qquad (11.1.3)$$

where the maps $\mathbf{T}^n \to \mathbf{T}^{n+1} : (e^{i\theta_1}, e^{i\theta_2}, \ldots, e^{i\theta_n}) \mapsto (e^{i\theta_1}, \ldots, e^{i\theta_n}, 1)$, give $(n/(n+1))^{1/2}$-Lipschitz embeddings.

Let $\mathcal{F}_n$ be the σ-algebra that is generated by the sets $A_1 \times \cdots \times A_n \times \mathbf{T} \times \mathbf{T} \times \cdots$, where $A_1, \ldots, A_n$ are Lebesgue measurable subsets of $\mathbf{T}$; then $f : \mathbf{T}^\infty \to \mathbf{C}$ is measurable with respect to $\mathcal{F}_n$ if and only if $f(\Theta)$ depends only upon the first n coordinates of $\Theta = (\theta_1, \theta_2, \ldots)$ and is a Lebesgue measurable function of these. The σ-algebras are naturally ordered by inclusion, so that

$$\mathcal{F}_1 \subset \mathcal{F}_2 \subset \mathcal{F}_3 \subset \ldots. \tag{11.1.4}$$

Haar measure on $\mathbf{T}^\infty$ arises as the weak limit of Haar measure on the $\mathbf{T}^n$ as $n \to \infty$; so that, for any continuous function $F : \mathbf{T}^\infty \to \mathbf{C}$ that depends on only finitely many coordinates,

$$\int_{\mathbf{T}^n} F(\Theta) \bigotimes_{j=1}^{n} \frac{d\theta_j}{2\pi} \to \int_{\mathbf{T}^\infty} F(\Theta) \bigotimes_{j=1}^{\infty} \frac{d\theta_j}{2\pi} \tag{11.1.5}$$

as $n \to \infty$.

We can regard $\mathbf{T}$ as the boundary of $\mathbf{D}$, and $\mathbf{T}^\infty$ as a preferred boundary of $\mathbf{D}^\infty$. The points in

$$\mathbf{D}^\infty \cap \ell^2 = \left\{ (z_j) : |z_j| < 1, j = 1, 2, \ldots; \sum_{j=1}^{\infty} |z_j|^2 < \infty \right\} \tag{11.1.6}$$

have special properties in the function theory of $\mathbf{D}^\infty$.

Let ρ_j be an absolutely continuous probability measure on $\mathbf{T}$, and let $\rho = \rho_1 \otimes \rho_2 \otimes \cdots$ be the product probability measure, which exists by Kolmogorov's extension theorem [88]. Kakutani's dichotomy theorem asserts that either:

$$(i) \quad \prod_{j=1}^{\infty} \int \left(\frac{d\rho_j}{d\theta_j} \right)^{1/2} \frac{d\theta_j}{2\pi} \tag{11.1.7}$$

converges to a positive value, and ρ is absolutely continuous with respect to $d\Theta$; or

$$(ii) \quad \prod_{j=1}^{\infty} \int \left(\frac{d\rho_j}{d\theta_j} \right)^{1/2} \frac{d\theta_j}{2\pi} \tag{11.1.8}$$

diverges to zero, and ρ is singular with respect to $d\Theta$. In case (i) when ρ is absolutely continuous we have, in terms of Exercise 1.1.8

$$\delta(d\rho, d\Theta) = \sum_{j=1}^{\infty} \delta(d\rho_j, d\theta_j/2\pi). \tag{11.1.9}$$

Exercise 11.1.2 Let $0 \le r_j < 1$ satisfy $\sum_{j=1}^{\infty} r_j^2 < \infty$, and let $\psi_j \in \mathbf{R}$; now let

$$p_j(\theta_j) = P_{r_j e^{i\psi_j}}(\theta_j) = \frac{1 - r_j^2}{1 - 2r_j \cos(\theta_j - \psi_j) + r_j^2} \tag{11.1.10}$$

be the Poisson kernels for $r_j e^{i\psi_j}$. Show that the measure

$$\rho = \bigotimes_{j=1}^{\infty} p_j(\theta_j) \frac{d\theta_j}{2\pi} \tag{11.1.11}$$

is absolutely continuous with respect to $d\Theta$. This ρ is associated with the point $(r_j e^{i\psi_j})$ which belongs to $\ell^2 \cap \mathbf{D}^{\infty}$.

11.1.3 Infinite symmetric group

Let S_n be the symmetric group on n symbols, which we identify with the permutations π_n of $\mathbf{N}$ such that $\pi_n(k) = k$ for $k = n+1, n+2, \ldots$. With $S_{\infty} = \cup_{n=1}^{\infty} S_n$ we have a natural inclusion of groups

$$S_1 \le S_2 \le \cdots \le S_{\infty}. \tag{11.1.12}$$

Since S_{∞} is countable and discrete, it has a measure $\mu(A) = \sharp(A)$ which is invariant under left translation, but no invariant probability measure by Lemma 11.1.1.

By introducing the language of noncommutative probability as in Proposition 14.2.7, one can obtain a satisfactory analytical description of the left regular representation of S_{∞}.

11.1.4 Infinite orthogonal group

Definition (*Infinite orthogonal group*). We choose some orthonormal basis $(e_j)_{j=1}^{\infty}$ for Hilbert space and form matrices with respect to this basis. Let $(O(n), c^2(n))$ be the groups of orthogonal matrices, and introduce the 1-Lipschitz map of matrices

$$U \mapsto \begin{bmatrix} U & 0 \\ 0 & I \end{bmatrix} \quad (U \in O(n)); \tag{11.1.13}$$

then the infinite orthogonal group is the inductive limit $O(\infty) = \cup_{n=1}^{\infty} O(n)$, so that

$$O(1) \le O(2) \le \cdots \le O(\infty). \tag{11.1.14}$$

Each element U of $O(\infty)$ has entries equal to δ_{jk}, with only finitely many exceptions, so $U - I$ is a finite-rank operator. We can regard $O(\infty)$ as a group of isometries of the infinite-dimensional sphere; or more precisely, as unitary operators on Hilbert space. In particular, S_∞ is a subgroup of $O(\infty)$ due to the action $\sigma : e_j \mapsto e_{\sigma(j)}$ on basis vectors. We endow $O(\infty)$ with the inductive topology; then $O(\infty)$ is not locally compact. Indeed, we let σ_j be the orthogonal transformation associated with the transposition of basis vectors $e_1 \leftrightarrow e_j$. Then one checks that the metric ρ of Proposition 2.2.3 satisfies $\rho(\sigma_j, I) \leq \sqrt{2}$ and $\rho(\sigma_j, \sigma_k) \geq 1/\sqrt{2}$ for all $j \neq k$; so $O(\infty)$ is not totally bounded.

The identity representation $\pi : O(\infty) \to U(H)$ is continuous by the definition of the inductive limit topology, and hence by Proposition 2.2.3, $O(\infty)$ has a left-invariant metric ρ such that π is continuous for the metric topology. By Lemma 11.1.1, the group $O(\infty)$ does not have a Haar probability measure.

Exercise 11.1.4a Let $g_n \in O(\infty)$ have matrix $[g_n]_{jk}$; suppose that $[g_n]_{jk} \to \delta_{jk}$ as $n \to \infty$.

(i) Show that $\langle g_n \xi, \xi \rangle_H \to \|\xi\|_H^2$ as $n \to \infty$ for all $\xi \in H$, and hence that $\|g_n \xi + \xi\|_H^2 \to 4\|\xi\|_H^2$.
(ii) Use the parallelogram law $\|g_n \xi - \xi\|_H^2 = 4\|\xi\|_H^2 - \|g_n \xi + \xi\|_H^2$ to show that $\|g_n \xi - \xi\|_H \to 0$ as $n \to \infty$.
(iii) Deduce that $g_n \to I$ as $n \to \infty$ in the inductive topology on $O(\infty)$ if and only if $[g_n]_{jk} \to \delta_{jk}$ as $n \to \infty$ for all $j, k \in \mathbf{N}$.

Exercise 11.1.4b Let K be a compact subgroup of $O(\infty)$, with Haar probability measure μ_K; let $K_n = O(n) \cap K$.

(i) Show that K_n is a closed subgroup of K.
(ii) Show that $\mu_K(K_n) > 1/2$ for all sufficiently large n.
(iii) By considering the measure of the cosets xK_n for $x \in K$, deduce that $K = K_n$ for some large n.

Definition (*Infinite unitary group*) *[87, 32]*. We define the infinite unitary group as the inductive limit $U(\infty) = \cup_{j=1}^\infty U(j)$ in a similar way to $O(\infty)$ by forming the sequence

$$U(1) \leq U(2) \leq \cdots \leq U(\infty) \tag{11.1.15}$$

via the inclusions

$$U \mapsto \begin{bmatrix} U & 0 \\ 0 & I \end{bmatrix}. \tag{11.1.16}$$

The inductive limit $U(\infty)$ is not locally compact, and does not have an invariant Haar measure by Lemma 11.1.1.

Summarizing these conclusions, we have the following theorem.

Theorem 11.1.5 *The groups $S_\infty, O(\infty)$ and $U(\infty)$ are inductive limit groups that do not have Haar probability measures; whereas the infinite torus $\mathbf{T}^\infty$ does have a Haar probability measure.*

11.2 Hua–Pickrell measure on the infinite unitary group

- The infinite unitary group inherits some of the properties of the finite unitary groups.
- The sine kernel gives a determinantal point process which is relevant to the representation theory.

In this section we review some facts about the representation theory of $U(n)$, with a view to analysing the infinite unitary group $U(\infty)$. The Lie algebra of $U(n)$ is $M_n^h(\mathbf{C})$. There are two natural maps $M_n^h(\mathbf{C}) \to U(n)$ for $n = 1, 2, \ldots, \infty$ that are significant in representation theory, namely the exponential $X \mapsto \exp(iX)$, and the Cayley transform $\varphi(X) = (X - iI)(X + iI)^{-1}$. We analyse the effect of the Cayley transform on various measures, and consider the asymptotic behaviour of these measures as the dimension $n \to \infty$. Following a similar route to Section 9.4, We consider the pseudo-Jacobi polynomials, and use them to introduce new example of a Tracy–Widom system, which is associated with a kernel involving $\sin 1/x$.

Definition (*Hua–Pickrell measure*). Let ν_n be the probability measure on $M_n^h(\mathbf{C})$

$$\nu_{n,1}(dX) = \frac{2^{n(n-1)}}{\pi^{n(n+1)/2}} \left(\prod_{j=1}^{n} \Gamma(j)\right) \det(I + X^2)^{-n}\, dX. \qquad (11.2.1)$$

Lemma 11.2.1 *The Cayley transform $\varphi : M_n^h(\mathbf{C}) \to U(n) : \varphi(X) = (X - iI)(X + iI)^{-1}$ induces Haar measure $\mu_{U(n)}$ on $U(n)$ from the Hua–Pickrell measure $\nu_{n,1}$ on $M_n^h(\mathbf{C})$.*

Proof. The proof consists of an induction on n, where the first step is the familiar calculation that $x \mapsto (x - i)/(x + i)$ induces $d\theta/2\pi$ on $\mathbf{T}$ from $dx/\pi(1 + x^2)$ on $\mathbf{R}$ as in Section 4.3. For details, see [31]. $\square$

One can compare the following probability density functions:

- Student's t-distribution on n degrees of freedom, with density

$$p_n(t) = \frac{\Gamma(n/2)}{\sqrt{(n-1)\pi}\,\Gamma((n-1)/2)}\left(1 + \frac{t^2}{n-1}\right)^{-n/2} \qquad (t \in \mathbf{R});$$

$$(11.2.2)$$

- the probability measure

$$h(x) = \mathrm{vol}_n(S^n)^{-1}\left(\frac{2}{1 + \|x\|^2}\right)^n \qquad (x \in \mathbf{R}^n) \qquad (11.2.3)$$

from 4.3, which is associated with the stereographic projection;

- the (modified) Hua–Pickrell measure on $M_n^h(\mathbf{C})$ given by

$$\nu_{n,\alpha}(dX) = C_{n,\alpha}\det(I + X^2)^{-n\alpha}\,dX; \qquad (11.2.4)$$

- the eigenvalue distribution that corresponds to the modified Hua–Pickrell measure

$$\sigma_{n,\alpha}(d\lambda) = Z_{n,\alpha}^{-1}\frac{\prod_{1 \leq j < k \leq n}(\lambda_j - \lambda_k)^2}{\prod_{j=1}^{n}(1 + \lambda_j^2)^{\alpha n}}\,d\lambda_1 \ldots d\lambda_n. \qquad (11.2.5)$$

The connection between these is as follows. The symmetric group S_n acts on $\mathbf{T}^n$ by permuting the variables, and we let $\mathcal{T} = \mathbf{T}^n/S_n$ be the quotient space for this action. Then $\mathcal{T}$ parametrizes the conjugacy classes of $U(n)$, and there is a natural map $\Lambda : U(n) \to \mathcal{T}$ which takes $V \in U(n)$ to its conjugacy class $\Lambda(V) \in \mathcal{T}$; essentially, $\Lambda(V)$ is the list of eigenvalues of V. Further, by the Weyl denominator formula Proposition 2.7.1, Λ induces Dyson's circular unitary ensemble $\sigma_n^{(2)}$ on $\mathcal{T}$ from Haar measure $\mu_{U(n)}$ on $U(n)$. So the commutative diagram of maps

$$\begin{array}{ccccccc} M_n^h(\mathbf{C}) & \to & U(n) & & X & \mapsto & (X - iI)(X + iI)^{-1} \\ \downarrow & & \downarrow & & \downarrow & & \downarrow \\ \mathbf{R}^n & \to & \mathcal{T} & & (\lambda_j) & \mapsto & ((\lambda_j - i)(\lambda_j + i)^{-1}) \end{array} \qquad (11.2.6)$$

gives a corresponding map on probability measures, as in the diagram below.

$$\begin{array}{ccc} \nu_{n,1}(dX) & \mapsto & \mu_{U(n)} \\ \downarrow & & \downarrow \\ \sigma_{n,1}(d\lambda) & \mapsto & \sigma_n^{(2)} \end{array} \qquad (11.2.7)$$

By a familiar calculation, the asymptotic form of Student's t distribution is normal as $n \to \infty$ since

$$p_n(t) \to \frac{e^{-t^2/2}}{\sqrt{2\pi}} \qquad (t \in \mathbf{R}). \qquad (11.2.8)$$

We now prove the analogous result for eigenvalues under the distribution $\sigma_{n,\alpha}$. The negative logarithm of the density of $\sigma_{n,\alpha}$ is

$$\alpha n \sum_{j=1}^{n} \log(1 + \lambda_j^2) - 2 \sum_{1 \le j < k \le n} \log |\lambda_j - \lambda_k|. \tag{11.2.9}$$

We are in the very marginal case of Theorem 4.4.1 with $\beta = 2$, so we need $\alpha > 1$ to ensure that the equilibrium measure has compact support.

Proposition 11.2.2 (i) *The scalar potential* $v_1(x) = \log(1 + x^2)$ *has equilibrium measure with density* $1/\pi(1 + x^2)$ *on* $\mathbf{R}$, *namely the Cauchy distribution or Student's t-distribution on two degrees of freedom.*

(ii) *For* $\alpha > 1$ *and* $a = (2\alpha - 1)^{1/2}/(\alpha - 1)$, *the scalar potential* $v_\alpha(x) = \alpha \log(1 + x^2)$ *has equilibrium measure supported on* $[-a, a]$ *with density*

$$\rho_\alpha(x) = \frac{(\alpha - 1)\sqrt{a^2 - x^2}}{x^2 + 1} \qquad (x \in [-a, a]). \tag{11.2.10}$$

(iii) *The scaled equilibrium measures satisfy*

$$a\rho_\alpha(au)du \to \frac{2}{\pi}\sqrt{1 - u^2} \qquad (\alpha \to \infty). \tag{11.2.11}$$

Proof. In all cases, we can express the condition that ρ_α gives the minimum as the integral equation

$$v_\alpha(x) = 2 \int_S \log |x - y| \rho_\alpha(y)\,dy + C \qquad (x \in S) \tag{11.2.12}$$

for some constant C and some set S, by Theorem 4.4.1.

(i) As in example 4.4.6, this follows from the identity

$$\log |x + i| = \int_{-\infty}^{\infty} \frac{\log |x - t|\,dt}{1 + t^2}\frac{}{\pi}, \tag{11.2.13}$$

which is the Poisson integral formula for the harmonic function $\log |z|$.

(ii) By symmetry, the equilibrium measure is supported on a single interval $[-a, a]$, and the density is strictly positive inside $(-a, a)$. Indeed, if there were a gap $[c, b]$ with $-a < c < 0 < b < a$ in the support of the measure, then by (4.4.18) we would have the identities

$$v_\alpha'(c) = 2 \int_{-a}^{c} \frac{\rho(x)dx}{c - x} + 2 \int_{b}^{a} \frac{\rho(x)dx}{c - x} \tag{11.2.14}$$

and

$$v'_\alpha(b) = 2 \int_{-a}^{c} \frac{\rho(x)dx}{b-x} + 2 \int_{b}^{a} \frac{\rho(x)dx}{b-x} \qquad (11.2.15)$$

so that $v'_\alpha(c) > v'_\alpha(b)$. However, $v'_\alpha(c) < 0 < v'_\alpha(b)$; so no such gap exists. By the general theory of singular integral equations [160, p. 173], the density of the equilibrium measure has the shape

$$\rho_\alpha(x) = \frac{1}{\pi}\sqrt{a^2 - x^2}\, r_\alpha(x) \qquad (11.2.16)$$

where

$$
\begin{aligned}
r_\alpha(x) &= \frac{1}{2\pi} \int_{-a}^{a} \frac{v'_\alpha(x) - v'_\alpha(t)}{(x-t)\sqrt{a^2 - t^2}} dt \\
&= \frac{\alpha}{\pi} \int_{-a}^{a} \frac{(1+xt)dt}{(1+x^2)(1+t^2)\sqrt{a^2 - t^2}} \\
&= \frac{\alpha}{\pi(x^2+1)} \int_{-a}^{a} \frac{dt}{(t^2+1)\sqrt{a^2 - t^2}} \\
&= \frac{\alpha}{\pi(x^2+1)} \int_{-\pi/2}^{\pi/2} \frac{d\theta}{1 + a^2 \sin^2 \theta} \\
&= \frac{\alpha}{(x^2+1)\sqrt{a^2+1}}, \qquad (11.2.17)
\end{aligned}
$$

where the final step follows by a standard contour integral. Hence

$$\rho_\alpha(x) = \frac{\alpha\sqrt{a^2 - x^2}}{\pi(1+x^2)\sqrt{a^2+1}}. \qquad (11.2.18)$$

Now we have the normalization

$$1 = \int_{-a}^{a} \rho_\alpha(x)\, dx = \frac{\alpha}{\sqrt{a^2+1}}(\sqrt{a^2+1} - 1), \qquad (11.2.19)$$

so we can recover the stated value for a.

(iii) We have

$$a\rho_\alpha(au)du = \frac{2\alpha - 1}{\alpha - 1} \frac{\sqrt{1-u^2}}{\pi(1+a^2u^2)}du, \qquad (11.2.20)$$

so we can take the limit as $\alpha \to \infty$. $\qquad\square$

In Section 9.5 we considered the Jacobi weight $(1-x)^\alpha(1+x)^\beta$ on $[-1, 1]$ and introduced the Jacobi system of orthogonal polynomials and the corresponding Jacobi ensemble. While the weight $(1+x^2)^{-n}$ on $\mathbf{R}$ has a superficial resemblance to the Jacobi weight, it decays rather slowly;

so we are able to start but not complete a similar programme as in [31, 159].

Definition (*Pseudo–Jacobi polynomials*). Let $\sigma > -1/2$ and let $n \in \mathbf{N}$. Then with the weight $(1 + x^2)^{-n-\sigma}$, we can form a sequence of monic orthogonal polynomials $1, p_1(x), \ldots, p_N(x)$ for $N < n + \sigma - 1/2$; these are called the *pseudo-Jacobi polynomials*, by analogy with Section 9.5.

One can express the p_m in terms of the hypergeometric function, and deduce that p_m satisfies the differential equation

$$-(1 + x^2)p_m''(x) + 2(\sigma + n - 1)xp_m'(x) + m(m + 1 - \sigma - 2n)p_m(x) = 0$$
(11.2.21)

as in [31]. The Möbius transformation φ, given by $\varphi(x) = (x-i)/(x+i)^{-1}$ satisfies $-\varphi(-1/x) = \varphi(x)$, which suggests that there is a natural scaling on the variables on the line associated with this inversion. One makes the change of variable $y = -1/(nx)$, and checks by calculation that the function $f_m(y) = p_m(-1/(ny))$ satisfies

$$-(n^2y^2 + 1)f_m''(y) - \frac{2}{y}(n^2y^2 + \sigma + n)f_m'(y)$$
$$+ \frac{m(m + 1 - 2\sigma - 2n)}{y^2}f_m(y) = 0. \quad (11.2.22)$$

Letting $m = n$ and letting $n \to \infty$, we obtain the limiting form of this differential equation

$$f''(y) + \frac{2}{y}f(y) + \frac{1}{y^4}f(y) = 0. \quad (11.2.23)$$

Theorem 11.2.3 (*i*) *Let f be a solution of (11.2.23). Then f gives rise to a solution of the Tracy–Widom system*

$$y^2 \frac{d}{dy}\begin{bmatrix} f \\ g \end{bmatrix} = \begin{bmatrix} 0 & 1 \\ -1 & 0 \end{bmatrix}\begin{bmatrix} f \\ g \end{bmatrix}. \quad (11.2.24)$$

(*ii*) *The general solution of (11.2.23) is*

$$f(y) = A\cos\frac{1}{y} + B\sin\frac{1}{y} \quad (11.2.25)$$

for some constants A and B.

(*iii*) *The kernel associated with $f(x) = \pi^{-1/2}\sin(1/x)$ is*

$$K(x,y) = \frac{\sin\left(\frac{1}{x} - \frac{1}{y}\right)}{\pi(y - x)} \quad (11.2.26)$$

where K is of trace class and $K \geq 0$ on $L^2(a, \infty)$ for $a > 0$.

*(iv) The Hankel operator Γ_ϕ on $L^2(0,\infty)$ with $\phi(s) = \pi^{-1}e^{-s/2}\times$
$\exp(2ie^{-s})$ satisfies*

$$\det(I - \lambda \mathbf{I}_{(1,\infty)} K \mathbf{I}_{(1,\infty)}) = \det(I - \lambda \Gamma_\phi \Gamma_\phi^*). \qquad (11.2.27)$$

*(v) The operator $K : L^2((-\infty,-1) \cup (1,\infty)) \to L^2((-\infty,-1) \cup (1,\infty))$
is unitarily equivalent to the operation of the sine kernel
$(\sin(x-y))/(\pi(x-y))$ on $L^2(-1,1)$.*

Proof. (i) and (ii) follow by direct calculation.

(iii) The kernel $K(x,y)$ that is associated with this solution is given
by

$$\frac{f(x)g(y) - f(y)g(x)}{x-y} = \frac{\sin\frac{1}{y}\cos\frac{1}{x} - \sin\frac{1}{x}\cos\frac{1}{y}}{\pi(x-y)}. \qquad (11.2.28)$$

Let $\Gamma : L^2(1,\infty) \to L^2(1,\infty)$ be the integral operator that has kernel

$$\Gamma(x,y) = \frac{e^{-i/x}e^{2i/xy}e^{i/y}}{\sqrt{\pi xy}} \qquad (x,y > 1); \qquad (11.2.29)$$

then $\Gamma\Gamma^*$ has kernel

$$
\begin{aligned}
\Gamma\Gamma^*(x,y) &= \int_1^\infty \frac{e^{-i/x}e^{(2i/w)(1/x-1/y)}e^{i/y}}{\pi xy w^2} \, dw \\
&= \frac{e^{-i(1/x-1/y)}}{\pi xy} \int_0^1 e^{2i(1/x-1/y)v} \, dv \\
&= \frac{\sin\left(\frac{1}{x}-\frac{1}{y}\right)}{\pi(y-x)} \\
&= K(x,y). \qquad (11.2.30)
\end{aligned}
$$

Clearly Γ gives a Hilbert–Schmidt operator on $L^2(a,\infty)$ for all $a > 0$,
so K is trace class; further, $K \geq 0$.

(iv) We introduce the unitary operator $U : L^2(1,\infty) \to L^2(0,\infty)$ by

$$Uf(s) = \exp(-ie^{-s})e^{s/2}f(e^s) \qquad (s > 0), \qquad (11.2.31)$$

which gives

$$
\begin{aligned}
\langle \Gamma f, g \rangle_{L^2(1,\infty)} &= \int_1^\infty \int_1^\infty \frac{e^{-i/x}e^{2i/xy}e^{i/y}}{xy} f(x)\bar{g}(y) \, dxdy \\
&= \int_0^\infty \int_0^\infty e^{-(s+t)/2}\exp(2ie^{-s-t})Uf(s)\overline{Ug(t)} \, dsdt \\
&= \langle \Gamma_\phi Uf, Ug \rangle_{L^2(0,\infty)}. \qquad (11.2.32)
\end{aligned}
$$

Hence $K = \Gamma\Gamma^* = U^*\Gamma_\phi\Gamma_\phi^* U$, so the spectra of K and $\Gamma_\phi\Gamma_\phi^*$ are equal.

(v) There is a unitary map $V : L^1(-1,1) \to L^2((-\infty) \cup (1,\infty))$: $f(x) \mapsto x^{-1}f(x^{-1})$ so that VKV^* is given by the kernel $(\sin(x-y))/(\pi(x-y))$ on $L^2(-1,1)$. □

Borodin and Olshanski proceeded to investigate the determinantal random point field that K generates [31].

For $U(\infty)$ we introduce a substitute for Haar measure. Since $U(\infty) = \lim_{n\to\infty} U(n)$, the obvious approach would be to take limits along the sequence of Haar measures $\sigma_{U(1)}, \sigma_{U(2)}, \ldots$, but this does not work since the natural inclusion $U(n) \to U(n+1)$ obviously does not induce $\mu_{U(n+1)}$ from $\mu_{U(n)}$. Further, by Proposition 3.8.1, $SU(n+1)/SU(n)$ is naturally homeomorphic to the sphere S^{2n+1} with the rotation invariant probability measure; so we get a natural measure on $\{\xi \in H : \|\xi\|_H = 1\}$ rather than on $U(\infty)$ by taking weak limits along the sequence of natural quotient spaces. In Chapter 12 we introduce the infinite dimensional sphere $S^\infty(\sqrt{\infty})$ with a suitable measure.

Meanwhile, we consider measures on the Lie algebras. The Hermitian matrices are naturally nested by inclusion

$$M_1^h(\mathbf{C}) \subset M_2^h(\mathbf{C}) \subset \cdots \subset M_\infty^h(\mathbf{C}) \tag{11.2.33}$$

$$X \mapsto \begin{bmatrix} X & 0 \\ 0 & 0 \end{bmatrix} \tag{11.2.34}$$

where $M_\infty^h(\mathbf{C}) = \cup_{j=1}^\infty M_j^h(\mathbf{C})$ has the inductive topology. The projection $P_n : M_n^h(\mathbf{C}) \to M_{n-1}^h(\mathbf{C})$ deletes the bottom row and rightmost column.

We can regard $M_\infty^h(\mathbf{C})$ as a Lie algebra for $U(\infty)$. There is a natural action of $U(\infty)$ on $M_\infty^h(\mathbf{C})$ by conjugation $U : X \mapsto UXU^*$.

Theorem 11.2.4 (*The Hua–Pickrell measure on $U(\infty)$*).

(i) *The Cayley transform $\varphi : M_n^h(\mathbf{C}) \to U(n)$ induces Haar measure $\mu_{U(n)}$ on $U(n)$ from $\nu_{n,1}$;*
(ii) *P_n induces $\nu_{n-1,1}$ on $M_{n-1}^h(\mathbf{C})$ from $\nu_{n,1}$ on $M_n^h(\mathbf{C})$;*
(iii) *the $\nu_{n,1}$ converge weakly to a probability measure ν_∞ on $M_\infty^h(\mathbf{C})$ as $n \to \infty$, where ν_∞ is invariant under conjugation by $U \in U(\infty)$.*
(iv) *Let $f : U(\infty) \to \mathbf{C}$ be a continuous function that depends on only finitely many matrix entries. Then*

$$\int_{M_\infty^h(\mathbf{C})} f(\varphi(X))\nu_\infty(dX) = \lim_{n\to\infty} \int_{U(n)} f(u)\mu_{U(n)}(du). \tag{11.2.35}$$

Proof. See [31].   □

Theorem 11.2.4 suggests the following diagram, in which the vertical arrows represent weak limits as $n \to \infty$, and the horizontal arrows represent Cayley transforms.

$$
\begin{array}{ccc}
(M_n^h(\mathbf{C}), \nu_{n,1}) & \to & (U(n), \mu_{U(n)}) \\
\downarrow & & \downarrow \\
(M_\infty^h(\mathbf{C}), \nu_\infty) & \to & (U(\infty), \mu_{U(\infty)})
\end{array}
$$

Definition (*Hua–Pickrell measure on $U(\infty)$*). The Hua–Pickrell measure on $U(\infty)$ is $\mu_{U(\infty)} = \varphi \sharp \nu_\infty$.

The Lie group $SU(n)$ has a Haar probability measure $\mu_{SU(n)}$, a maximal torus $\mathcal{T}$ of rank $n - 1$ and Weyl group S_n. One presentation of the maximal torus is the space of diagonal unitary operators, where S_n acts on $\mathcal{T}$ be permuting the diagonal entries.

Analogously, we can regard the induced Hua–Pickrell measure $\mu_{U(\infty)} = \varphi \sharp \nu_\infty$ as the natural analogue of Haar measure for $U(\infty)$, while $\mathbf{T}^\infty$ sits inside $U(\infty)$ as the space of diagonal unitary matrices, and we can view S_∞ as the Weyl group, which acts by permuting the entries of $\mathbf{T}^\infty$. The group $U(\infty)$ has natural representations on complex Hilbert space, as we shall see in Section 12.4.

In Section 12.4 we shall construct representations of S_∞ and $O(\infty)$ on a suitable Hilbert space.

Definition (*Carleman determinant*). Let S be a Hilbert–Schmidt operator on H with eigenvalues $(\lambda_j)_{j=1}^\infty$, listed according to multiplicity. Then the Carleman determinant of $I + S$ is

$$
\det{}_2(I + S) = \prod_{j=1}^\infty (1 + \lambda_j) e^{-\lambda_j}. \tag{11.2.36}
$$

Exercise 11.2.5 Let A be a Hilbert–Schmidt and self-adjoint operator on H.

(i) Prove that for each $\varepsilon > 0$, there exists $U \in U(\infty)$ such that $\|U - \exp iA\|_{c^2} < \varepsilon$.

(ii) Obtain an expression for $\det_2(\exp iA)$ in terms of the eigenvalues of A.

(iii) Let $g = \{X = iA : A \in c^2, A^* = A\}$, with $[X,Y] = XY - YX$ and $\langle X, Y \rangle = -2^{-1}\mathrm{trace}(XY)$. Find an orthonormal basis for g in the style of Exercise 2.4.2, and determine whether the proof of Proposition 2.4.1 works in this context.

11.3 Gaussian Hilbert space

- A probability measure on infinite-dimensional Hilbert space that assigns positive measure to each open set is neither unitarily invariant nor translation invariant.
- Infinite-dimensional Hilbert spaces can have Gaussian measures.
- Brownian motion can be defined by random Fourier series with Gaussian coefficients.
- There is a stochastic integral defined with respect to white noise.

Before embarking on a discussion of Gaussian measure in infinite dimensions, it is worth noting that there is no natural analogue of Lebesgue measure on infinite dimensional Hilbert space.

Proposition 11.3.1 *Let μ be a probability measure that is defined on the Borel subsets of an infinite-dimensional separable Hilbert space H, and suppose that*

(i) $\mu(B_r(x)) > 0$ for all $x \in H$ and $r > 0$.

Then neither of the following conditions can hold:

(ii) $\mu(UE) = \mu(E)$ for all open E and unitary U;
(iii) $\mu(x + E) = \mu(E)$ for all open E and $x \in H$.

Proof. Suppose that either (ii) or (iii) holds. Let $(e_j)_{j=1}^{\infty}$ be an orthonormal sequence in H, and observe that the open sets $E_j = B_{1/\sqrt{2}}(e_j)$ are disjoint. Then by (ii) or (iii), $\mu(E_j) = \mu(E_1)$ holds for all j; so by disjointness

$$1 = \mu(H) \geq \mu\left(\bigcup_{j=1}^{\infty} E_j\right) = \sum_{j=1}^{\infty} \mu(E_j) = \sum_{j=1}^{\infty} \mu(E_1). \qquad (11.3.1)$$

Hence $\mu(E_1) = 0$, contrary to (i). $\qquad\square$

Example 11.3.2 We consider various closed and bounded subsets of H. The finite-dimensional set

$$K_{n,N} = \left\{x \in H : x = \sum_{k=1}^{n} a_k e_k; \sum_{k=1}^{n} |a_k|^2 \leq N\right\} \qquad (11.3.2)$$

and Hilbert's cube

$$C = \left\{x \in H : x = \sum_{k=1}^{\infty} a_k e_k; \quad |a_k| \leq 1/k; \quad k = 1, 2, \dots\right\} \qquad (11.3.3)$$

are compact as one can show by introducing ε-nets; whereas

$$B_N(0) = \left\{ x \in H : x = \sum_{k=1}^{\infty} a_k e_k : \sum_{k=1}^{\infty} |a_k|^2 \leq N \right\} \tag{11.3.4}$$

is noncompact.

Let (Ω, d) be a Polish space with probability $\mathbf{P}$, and suppose that there exists a sequence (χ_j) of mutually independent random variables, each with $N(0,1)$ distribution. Now let W^1 be the closed real linear span in $L^2(\mathbf{P})$ generated by the χ_j.

Lemma 11.3.3 *(i) W^1 is a Hilbert space for pointwise addition of random variables with the inner product $\langle \xi, \eta \rangle = \mathbf{E}(\xi\eta)$.*

(ii) (χ_j) is a complete orthonormal basis for W^1.

(iii) Suppose that $(a_j) \in \ell^2$. Then $\xi = \sum_j a_j \chi_j$ has a Gaussian distribution with mean zero and variance $\sum_j a_j^2$.

Proof. Property (i) follows from the fact that $L^2(\mathbf{P})$ is a Hilbert space.

(ii) This is clear.

(iii) Let $\xi_n = \sum_{j=1}^n a_j \chi_j$, and observe that ξ_n has characteristic function

$$\mathbf{E} e^{it\xi_n} = e^{-t^2(a_1^2 + \cdots + a_n^2)/2}. \tag{11.3.5}$$

Now $\xi_n \to \xi$ almost surely and in $L^2(\mathbf{P})$ as $n \to \infty$, so the characteristic function of ξ is $e^{-t^2(a_1^2 + a_2^2 + \cdots)/2}$; hence ξ has a Gaussian distribution with mean zero and variance $\sum_j a_j^2$. □

Definition (*Brownian motion [112, 70]*). The Brownian motion on $[0,1]$ is the stochastic process

$$b_t = t\chi_0 + \sum_{n=-\infty}^{-1} \frac{\sqrt{2}}{2\pi n}(\sin 2\pi nt)\chi_n + \sum_{n=1}^{\infty} \frac{\sqrt{2}}{2\pi n}(1 - \cos 2\pi nt)\chi_n. \tag{11.3.6}$$

Proposition 11.3.4 *Brownian motion is a Gaussian process with stationary independent increments; so that*

(i) $b_t - b_s$ has a $N(0, t-s)$ distribution for $0 < s < t < 1$;

(ii) for $0 = t_0 < t_2 < t_2 < \cdots < t_n$, the random variables $b_{t_1} - b_{t_0}, \ldots, b_{t_n} - b_{t_{n-1}}$ are mutually independent.

(iii) Furthermore, the function $t \mapsto b_t$ belongs to $L^2[0,1]$ almost surely.

Proof. (i) We start with the indicator function $\mathbf{I}_{(0,t)}$ and generate its Fourier series with respect to the orthonormal basis

$$\{\mathbf{I}, \sqrt{2}\cos 2\pi nx, \sqrt{2}\sin 2\pi nx; n = 1, 2, \ldots\} \tag{11.3.7}$$

of $L^2[0,1]$, namely

$$\mathbf{I}_{(0,t)}(x) = t\mathbf{I} + \sum_{n=-\infty}^{-1} \frac{\sqrt{2}}{2\pi n}(\sin 2\pi nt)\sqrt{2}\cos 2\pi nx$$

$$+ \sum_{n=1}^{\infty} \frac{\sqrt{2}}{2\pi n}(1 - \cos 2\pi nt)\sqrt{2}\sin 2\pi nx. \quad (11.3.8)$$

The map $U : L^2[0,1] \to W^1$ given by changing the basis

$$\{1, \sqrt{2}\cos 2\pi nx, \sqrt{2}\sin 2\pi nx\} \leftrightarrow \{\chi_n : n \in \mathbf{Z}\} \quad (11.3.9)$$

is unitary, and satisfies $b_t = U(\mathbf{I}_{[0,t]})$.

By the preceding Lemma, $b_t - b_s = U(\mathbf{I}_{(s,t)})$ is a Gaussian random variable with mean zero and variance

$$\mathbf{E}(b_t - b_s)^2 = \int \mathbf{I}_{(s,t)}(x)\,dx = t - s. \quad (11.3.10)$$

(ii) Further, U maps the orthogonal vectors $\mathbf{I}_{(t_0,t_1)}$, $\mathbf{I}_{(t_1,t_2)}, \cdots,$ $\mathbf{I}_{(t_{n-1},t_n)}$ to uncorrelated Gaussian random variables $b_{t_1-t_0}, b_{t_2-t_1}, \cdots,$ $b_{t_n-t_{n-1}}$. A sequence of uncorrelated Gaussian random variables is mutually independent, hence the result.

(iii) By Fubini's Theorem, we have

$$\mathbf{E}\int_0^1 b_t^2\,dt = \int_0^1 \mathbf{E}b_t^2\,dt = \int_0^1 t\,dt = 1/2, \quad (11.3.11)$$

so $\int_0^1 b_t^2\,dt$ is finite almost surely. $\qquad\square$

Definition (*Brownian motion*). Brownian motion on $[0,1]$ is the stochastic process $(b_t)_{0\le t\le 1}$ where $\Omega \to \mathbf{R} : \omega \mapsto b_t(\omega)$ is the random variable at time t. Wiener measure is the probability measure on $L^2[0,1]$ that is induced by the map $(\Omega, \mathbf{P}) \to L^2[0,1] : \omega \mapsto b_t(\omega)$.

Remark. Using more sophisticated arguments, such as the theory of random Fourier series, one can show that $t \mapsto b_t(\omega)$ is almost surely continuous on $[0,1]$. Indeed b_t is almost surely Hölder continuous of order α for $0 < \alpha < 1/2$, but not differentiable. See [97, 86].

Definition. To extend Brownian motion to $(0, \infty)$, we stitch together Brownian motions on successive intervals $[n, n+1]$, as follows. Let $(b_t^{(1)})$ be an independent copy of b_t, and extend the definition of b_t by taking

$$b_t = b_1 + b_{t-1}^{(1)} \qquad (1 \le t \le 2). \quad (11.3.12)$$

Iterating, this construction, one obtains Brownian motion on $(0, \infty)$. By enlarging W^1 as necessary, we can define a unitary map $U : L^2(0, \infty) \to W^1$ such that $b_t = U(\mathbf{I}_{(0,t)})$ has stationary independent increments and

$$\mathbf{E}\, b_t b_s = \min\{t, s\} \qquad (s, t > 0). \tag{11.3.13}$$

Corollary 11.3.5 (*Joint distribution of Brownian motion*). *Let* $0 = t_0 < t_1 < t_2 < \cdots < t_n$, *and let* I_j *be an open subset of* $\mathbf{R}$ *for* $j = 1, \ldots, n$. *The Brownian motion starting from* $b_0 = y_0 = 0$ *satisfies*

$$\mathbf{P}[b_{t_1} \in I_1, b_{t_2} \in I_2, \ldots, b_{t_n} \in I_n]$$
$$= C(t_1, \ldots, t_n) \int \cdots \int_{I_1 \times \cdots \times I_n} \gamma(y_1, \ldots, y_n) dy_1 \ldots dy_n \tag{11.3.14}$$

where

$$C(t_1, \ldots, t_n) = (2\pi)^{-n/2} \prod_{j=1}^{n} (t_j - t_{j-1})^{-1/2} \tag{11.3.15}$$

and

$$\gamma(y_1, \ldots, y_n) = \exp\left(-\frac{1}{2} \sum_{j=1}^{n} \frac{(y_j - y_{j-1})^2}{t_j - t_{j-1}}\right). \tag{11.3.16}$$

Proof. By Proposition 11.3.4, the increments $b_{t_j} - b_{t_{j-1}}$ are mutually independent, and they have Gaussian distribution with mean zero and variances $t_j - t_{j-1}$. Then the formula follows immediately from the definitions. □

Proposition 11.3.6 (*Stochastic integral*). *There is a linear isometric map* $L^2(0, \infty) \to W^1$

$$\phi \mapsto \int_0^\infty \phi(t) db_t \tag{11.3.17}$$

that extends U.

Proof. We wish to define $\int \phi(s) db_s$ for $\phi \in L^2(0, \infty)$. Given $0 < T < \infty$, a continuous function $\phi_T : [0, T] \to \mathbf{R}$ and a partition $P = \{0 = t_0 < t_1 < \cdots < t_N = T\}$, we let

$$\Phi_P = \sum_{j=0}^{N-1} \phi(t_j)(b_{t_{j+1}} - b_{t_j}) \in W^1. \tag{11.3.18}$$

Evidently $\Phi_P = U(\sum_{j=0}^{N-1} \phi(t_j)\mathbf{I}_{(t_j, t_{j+1})})$ is a Gaussian random variable with mean zero and variance

$$\mathbf{E}\Phi_P^2 = \sum_{j=0}^{N-1} \phi(t_j)^2 (t_{j+1} - t_j). \tag{11.3.19}$$

Given $\varepsilon > 0$, there exists $\delta > 0$ such that

$$\left| \phi_T(t) - \sum_{j=0}^{N-1} \phi(t_j) \mathbf{I}_{(t_j, t_{j+1})}(t) \right| < \varepsilon \tag{11.3.20}$$

whenever $\max_j |t_{j+1} - t_j| < \delta$. Hence

$$\|U(\phi_T) - \Phi_P\|_{W^1}^2 \le T\varepsilon^2. \tag{11.3.21}$$

Letting $\delta \to 0+$, we obtain $U(\phi_T) \in W^1$ which we define to be $U(\phi_T) = \int_0^T \phi(t)db_t$, with

$$\mathbf{E}\left(\int_0^T \phi(t)db_t \right)^2 = \int_0^T \phi(t)^2 dt. \tag{11.3.22}$$

Furthermore, $(\int_0^N \phi(t)db_t)_{N=1}^\infty$ forms a Cauchy sequence in W^1, and converges to an element that defines $\int_0^\infty \phi(t)db_t$. □

11.4 Gaussian measures and fluctuations

- With Gaussian measures, we introduce white noise.
- We compare fluctuations of the empirical eigenvalue distribution of the generalized unitary ensemble with a fractional derivative of Brownian motion.
- We compare fluctuations of the row lengths in Young tableaux with a fractional derivative of Brownian motion.

Example 11.4.1 (*White noise*). We can regard db_t as white noise in the sense that

$$\frac{db_t}{dt} = \chi_0 + \sum_{n=-\infty}^{-1} \sqrt{2}\cos 2\pi nt\, \chi_n + \sum_{n=1}^\infty \sqrt{2}\sin 2\pi nt\, \chi_n \tag{11.4.1}$$

defines a linear functional on $C^\infty[0,1]$ almost surely, and such that

$$\int \phi(t)db_t = \phi(1)b_1 - \phi(0)b_0 - \int \phi'(t)b_t\, dt. \tag{11.4.2}$$

Thus $\phi \mapsto \int \phi(t)db_t$ gives a bounded linear operator $L^2 \to W^1$.

Next we introduce a random Fourier series with roughness midway between Brownian motion and white noise.

Example 11.4.2 (*Fractional derivative of Brownian motion*). The Gaussian random Fourier series

$$Z(\theta) = \frac{2}{\pi} \sum_{k=1}^{\infty} \frac{\sin k\theta}{\sqrt{k}} \chi_k \qquad (11.4.3)$$

does not converge in W^1 for almost all θ since $\sum_{k=2}^{\infty} 1/k$ diverges. We can regard $Z(\theta)$ as a summand in the random Fourier series $(d/dt)^{1/2}b_t$. Suppose however, that $u(\theta) = \sum_{k=0}^{\infty} a_k \cos k\theta$ belongs to Dirichlet space D_0 so that $\sum_k |k||a_k|^2$ converges; then

$$-\int_0^\pi u'(\theta)Z(\theta)\, d\theta = \sum_{k=1}^{\infty} \sqrt{k}a_k \chi_k \qquad (11.4.4)$$

converges in W. Thus Z' defines a random linear functional on D_0; that is, a bounded linear map $D_0 \to W^1$.

Using Example 11.4.2, we can express the Theorem 10.6.8 on fluctuations in a more intuitive style. We consider a potential v as in Section 4.4 with equilibrium measure as in Theorem 4.4.1.

Proposition 11.4.3 (*Fluctuations*) *[94]. Let X_n be a $n \times n$ random matrix from the generalized unitary ensemble with potential v, equilibrium measure ρ_v with support $[-1,1]$, and eigenvalues $\lambda_1, \ldots, \lambda_n$. Then*

$$\sum_{j=1}^{n} \delta_{\lambda_j}(dx) - n\rho_v(dx) \to (1/2)dZ(\cos^{-1} x) \qquad (n \to \infty) \qquad (11.4.5)$$

in the sense that

$$\mathbf{P}\Big[\sum_{j=1}^{n} f(\lambda_j) - n\int_{[-1,1]} f(x)\rho_v(dx) \le t\Big]$$

$$\to \mathbf{P}\Big[\frac{1}{2}\int_0^{2\pi} f(\cos\theta)dZ(\theta) \le t\Big] \qquad (n \to \infty) \qquad (11.4.6)$$

for all continuously differentiable functions $f : [-1,1] \to \mathbf{R}$ and all $t \in \mathbf{R}$.

Proof. With $f(\cos\theta) = \sum_{k=0}^{\infty} a_k \cos k\theta$, we have a Gaussian random variable

$$\frac{1}{2}\int_0^{2\pi} f(\cos\theta)dZ(\theta) = \frac{1}{2}\sum_{k=1}^{\infty} \sqrt{k}a_k \chi_k \qquad (11.4.7)$$

with mean zero and variance $(1/4) \sum_{k=1}^{\infty} |k| a_k^2$, hence

$$\mathbf{E} \exp \left(\frac{s}{2} \int f(\cos \theta) dZ(\theta) \right) = \exp \left(\frac{s^2}{8} \sum_{k=1}^{\infty} |k| a_k^2 \right) \qquad (s \in \mathbf{R}). \quad (11.4.8)$$

We recall from Johansson's Theorem 10.6.8 that

$$\mathbf{E} \exp \left(s \sum_{j=1}^{n} f(\lambda_j) - sn \int_{[-1,1]} f(x) \rho_v(dx) \right) \to \exp \left(\frac{s^2}{8} \sum_{k=1}^{\infty} k a_k^2 \right) (n \to \infty);$$
$$(11.4.9)$$

so by Proposition 1.8.5 the random variable $\sum_{j=1}^{n} f(\lambda_j) - n \int f(x) \rho_v(dx)$ converges in distribution to $2^{-1} \int f(\cos \theta) \, dZ(\theta)$.

Exercise 11.4.4 Let $(\chi_k, \tilde{\chi}_k; k = 1, 2, \dots)$ be mutually independent $N(0,1)$ random variables, and let

$$\zeta_\theta = \sum_{k=1}^{\infty} \frac{\cos k\theta \chi_k + \sin k\theta \tilde{\chi}_k}{\pi \sqrt{k}}. \quad (11.4.10)$$

Let $e^{i\theta_1}, \dots, e^{i\theta_n}$ be the eigenvalues of a $n \times n$ unitary matrix, chosen randomly with respect to Haar measure on $U(n)$. Use Corollary 10.6.2 to show that

$$\sum_{j=1}^{n} f(e^{i\theta_j}) - n \int_0^{2\pi} f(e^{i\theta}) \frac{d\theta}{2\pi} \to \int_0^{2\pi} f(e^{i\theta}) d\zeta_\theta \quad (11.4.11)$$

in distribution as $n \to \infty$.

The random Fourier series Z of Example 11.4.2 has another important application to fluctuations of large Young diagrams with respect to Plancherel measure. Given $\lambda \vdash N$, there exists a probability density function f_λ with graph made up of boxes of side $N^{-1/2}$ such that the graph is like a descending staircase. We introduce the rotated coordinate system $\xi = x - y$ and $\eta = y + x$ so that $N f_\lambda$ gives rise to graph $\tilde{\lambda}(\xi) \geq 0$ in the cone between the lines $\eta = \xi$ and $\eta = -\xi$ with $\tilde{\lambda}(\xi) = |\xi|$ for all sufficiently large $|\xi|$. The function $\tilde{\lambda}$ is Lipschitz continuous with a graph that consists of straight line segments with gradient ± 1.

Next we rotate the graph of the Vershik distribution f_Ω of Section 7.4, and thus obtain $x = f_\Omega(x) + 2\cos \theta$ where $f_\Omega(x) = \frac{2}{\pi}(\sin \theta - \theta \cos \theta)$, so

$$\Omega(\xi) = \frac{2}{\pi} \left(\xi \sin^{-1} \frac{\xi}{2} + \sqrt{4 - \xi^2} \right), \qquad (|\xi| \leq 2);$$
$$= |\xi|, \qquad (|\xi| \geq 2); \quad (11.4.12)$$

and let the discrepancy between $\tilde{\lambda}$ and the scaled optimum $\sqrt{\Omega}$ be

$$\tilde{\Delta}_N(\xi) = \tilde{\lambda}(\xi\sqrt{N}) - \sqrt{N}\Omega(\xi). \qquad (11.4.13)$$

Suppose that $g(\theta) = \sum_{k=0}^{\infty} a_k \cos k\theta$ is continuously differentiable, and observe that since $\tilde{\Delta}_N$ is Lipschitz continuous, we can integrate by parts and obtain

$$\int_0^{2\pi} \tilde{\Delta}_N(2\cos\theta)g'(\theta)\, d\theta = 2\int_0^{\pi} \tilde{\Delta}_N'(2\cos\theta)\sin\theta\, g(\theta)\, d\theta. \qquad (11.4.14)$$

Further, the integral

$$\int_0^{\pi} \frac{2}{\pi}\sum_{k=1}^{\infty} \frac{\chi_k}{\sqrt{k}}\sin k\theta\, g'(\theta)\, d\theta = -\sum_{k=1}^{\infty} a_k \sqrt{k}\chi_k \qquad (11.4.15)$$

gives a Gaussian random variable with mean zero and variance $\sum_{k=1}^{\infty} ka_k^2$.

Proposition 11.4.5 (*Kerov*). $\tilde{\Delta}_N(\xi)$ *converges in distribution to* $Z(\cos^{-1}(\xi/2))$ *as* $N \to \infty$.

Proof. Omitted. See [28] for further developments.   □

12

Hermite polynomials

Abstract

In this chapter we study the Hermite polynomials and the Ornstein–Uhlenbeck operator on $\mathbf{R}$, and then their extension to $\mathbf{R}^n$. Following Wiener's approach, we introduce homogeneous chaos and stochastic integrals; thus we interpret the group of rotations of $S^\infty(\sqrt{\infty})$, an invariant measure and a Laplace operator on $S^\infty(\sqrt{\infty})$. Then we construct representations of $O(\infty)$ on homogeneous chaos. We also present the logarithmic Sobolev inequality for the Ornstein–Uhlenbeck semigroup, which was an important motivating example for Chapter 6.

12.1 Tensor products of Hilbert space

Definition (*Hilbert tensor product* [92]). Let H and K be complex Hilbert spaces. Then $H \otimes K$ is the closure of the algebraic tensor product over $\mathbf{C}$, given by the set of finite sums

$$H \odot K = \left\{ \sum \xi_j \otimes \eta_j : \xi_j \in H, \eta_j \in K \right\} \qquad (12.1.1)$$

with the inner product

$$\left\langle \sum_j \xi_j \otimes \eta_j, \sum_k \alpha_k \otimes \beta_k \right\rangle = \sum_{j,k} \langle \xi_j, \alpha_k \rangle_H \langle \eta_j, \beta_k \rangle_K. \qquad (12.1.2)$$

Examples 12.1.2 (i) Clearly we have $\mathbf{C}^m \otimes \mathbf{C}^n = \mathbf{C}^{mn}$. More generally, given any complex Hilbert space, we can identify $H \otimes \mathbf{C}^n$ with the direct sum $H \oplus \cdots \oplus H$ of n copies of Hilbert space.

(ii) $\ell^2 \otimes \ell^2$ is naturally linearly isometric to the space of c^2 Hilbert–Schmidt matrices. We can identify $e_j \otimes e_k$ with the matrix unit e_{jk}.

Then

$$\langle [a_{jk}], [b_{jk}] \rangle_{c^2} = \text{trace}\big([b_{jk}]^* [a_{jk}] \big)$$

$$= \sum_{j,k} a_{jk} \bar{b}_{jk}$$

$$= \Big\langle \sum_{jk} a_{jk} e_{jk}, \sum_{jk} b_{jk} e_{jk} \Big\rangle. \qquad (12.1.3)$$

Proposition 12.1.3 (*Schur product*). *Suppose that $[a_{jk}]$ and $[b_{jk}]$ are positive semidefinite $n \times n$ matrices. Then the sum $[a_{jk} + b_{jk}]$ and Schur product $[a_{jk} b_{jk}]$ are also positive semidefinite.*

Proof. By Proposition 2.1.2(iii), there exist vectors ξ_j and η_k such that $a_{jk} = \langle \xi_j, \xi_k \rangle$ and $b_{jk} = \langle \eta_j, \eta_k \rangle$; then $a_{jk} b_{jk} = \langle \xi_j \otimes \eta_j, \xi_k \otimes \eta_k \rangle$, so we can use Proposition 2.1.2 again $\qquad\qquad\qquad\qquad\qquad\qquad\qquad \square$

Definition (*Symmetric powers*). Now we let H be a complex Hilbert space and we choose a unit vector $e_0 \in H$, traditionally called the vacuum vector, and we write $\mathbf{C} e_0 = H^{\otimes 0}$. Then we construct recursively $H^{\otimes(n+1)} = H^{\otimes n} \otimes H$, so that $H^{\otimes n} = \overline{\text{span}}\{\xi_1 \otimes \cdots \otimes \xi_n\}$, where $\overline{\text{span}}$ stands for closed linear span for the norm topology. The symmetric group S_n acts on $H^{\otimes n}$ by

$$\sigma : \xi_1 \otimes \cdots \otimes \xi_n \mapsto \xi_{\sigma(1)} \otimes \cdots \otimes \xi_{\sigma(n)}, \qquad (12.1.4)$$

and the subspace of invariant vectors $H^{s\otimes n} = \{\xi \in H^{\otimes n} : \sigma(\xi) = \xi\}$ is called the nth symmetric power of H, which is spanned by

$$\text{sym}(\xi_1 \otimes \cdots \otimes \xi_n) = \sum_{\sigma \in S_n} \xi_{\sigma(1)} \otimes \cdots \otimes \xi_{\sigma(n)}. \qquad (12.1.5)$$

Example (*Reproducing kernels*). Let H be a Hilbert function space on Ω with reproducing kernel $K(z, w)$, and let $w_1, \dots, w_n$ belong to Ω. Then $[a_{jk}] = [K(w_j, w_k)]$ is positive semidefinite, and so by Proposition 12.1.3 $[K(w_j, w_k)^m]$ is also positive semidefinite for $m = 0, 1, 2, \dots$; hence $[\exp K(w_j, w_k)]$ is positive semidefinite. In particular, this applies to $K(z, w) = \log 1/(1 - z\bar{w})$ and $\exp K(z, w) = 1/(1 - z\bar{w})$, which are the reproducing kernels for Dirichlet space and Hardy space respectively. See Section 4.2.

Definition (*Fock space* [92]). The full Fock space of H is the Hilbert space

$$\exp(H) = \Big\{ (\xi^{(n)}) \in \prod_{n=0}^{\infty} H^{\otimes n} : \sum_{n=0}^{\infty} \|\xi^{(n)}\|^2 / n! < \infty \Big\} \qquad (12.1.6)$$

with the inner product $\langle (\xi^{(j)}), (\eta^{(j)}) \rangle = \sum_{j=0}^{\infty} \langle \xi^{(j)}, \eta^{(j)} \rangle / j!$. The symmetric Fock space is the subspace of symmetric tensors

$$\Gamma(H) = \left\{ (\xi^{(n)}) \in \prod_{n=0}^{\infty} H^{s \otimes n} : \sum_{n=0}^{\infty} \|\xi^{(n)}\|^2 / n! < \infty \right\}, \quad (12.1.7)$$

and the exponential vector of $\xi \in H$ is $\exp_{\otimes}(\xi) = (e_0, \xi, \xi \otimes \xi, \dots)$ which belongs to $\Gamma(H)$ and has $\|\exp_{\otimes}(\xi)\|_{\Gamma(H)} = \exp(\|\xi\|^2 / 2)$.

Example 12.1.4 (*Left creation operators*). Let H be a separable infinite-dimensional real Hilbert space and e a unit vector in H. Each $\eta \in H$ has unique decomposition $\eta = \eta_0 e + \eta_{\perp}$, where $\eta_0 = \langle \eta, e \rangle$ and $\langle \eta_{\perp}, e \rangle = 0$. The full Fock space is $\exp(H) = \oplus_{n=0}^{\infty} H^{\otimes n}$ where $H^{\otimes 0} = \mathbf{R}e$, and write $e_0 = e \oplus 0 \oplus \cdots$ for the vacuum vector. There is a linear map $\ell : H \to B(\exp(H))$ where $\ell(h)$ raises degree by one, as in

$$\ell(h) : e_0 \mapsto h \otimes e$$
$$\ell(h) : \xi_1 \otimes \dots \xi_n \mapsto h \otimes \xi_1 \otimes \cdots \otimes \xi_n; \quad (12.1.8)$$

whereas the adjoint $\ell(h)^*$ lowers degree by one, as in

$$\ell(h)^* : \xi_1 \otimes \cdots \otimes \xi_n \mapsto \langle \xi_1, h \rangle \xi_2 \otimes \cdots \otimes \xi_n \quad (n = 2, 3, \dots), \quad (12.1.9)$$
$$\ell(h)^* \xi_1 = \langle \xi_1, h \rangle e_0, \quad \ell(h)^* e_0 = 0.$$

Clearly we have

$$\ell(h)^* \ell(k) = \langle k, h \rangle I \quad (12.1.10)$$

and

$$\ell(h)\ell(k)^* : \xi_1 \otimes \cdots \otimes \xi_n \mapsto S\xi_1 \otimes \xi_2 \otimes \cdots \otimes \xi_n \quad (12.1.11)$$

where $S : H \to H$ is the rank one operator $Sf = \langle f, k \rangle h$. Each $\ell(h)$ is a left creation operator, since the tensor factor of h appears in the left; clearly one can introduce right creation operators analogously. See also [166].

12.2 Hermite polynomials and Mehler's formula

The Hermite polynomials are:

- eigenfunctions of the Ornstein–Uhlenbeck operator on $\mathbf{R}$;
- orthogonal polynomials for the standard Gaussian measure on $\mathbf{R}$;
- a natural basis for the Wick ordering.

In this section we review some basic results that are important in that they provide an explicit construction of objects that are important in the general theory. We let $\gamma(x) = e^{-x^2/2}/\sqrt{2\pi}$ be the standard Gaussian density and form the Hilbert space $L^2(\gamma)$.

Definition (*OU operator* [54]). The Ornstein–Uhlenbeck operator L is

$$L = -\frac{d^2}{dx^2} + x\frac{d}{dx} \qquad (12.2.1)$$

and the domain includes $C_c^\infty(\mathbf{R})$.

Evidently L is symmetric and positive in the sense that

$$\int (Lf)(x)\bar{g}(x)\gamma(x)dx = \int_{-\infty}^{\infty} f(x)\overline{(Lg)}(x)\gamma(x)dx$$

$$= \int_{-\infty}^{\infty} f'(x)\bar{g}'(x)\gamma(x)\,dx. \qquad (12.2.2)$$

The Ornstein–Uhlenbeck operator satisfies the special property

$$\left(-\frac{d^2}{dx^2} + \frac{x^2}{4}\right)\left(e^{-x^2/4}\psi(x)\right) = e^{-x^2/4}\left(L + \frac{1}{2}\right)\psi(x) \qquad (12.2.3)$$

where $-d^2/dx^2 + x^2/4$ is the Hamiltonian for the linear oscillator in quantum mechanics. To carry out a spectral analysis rigorously, we begin with the eigenfunction.

Definition (*Rodrigues's formula for Hermite polynomials* [112, 142]). The Hermite polynomial of degree n is

$$H_n(x) = (-1)^n e^{x^2/2}\frac{d^n}{dx^n}e^{-x^2/2}. \qquad (12.2.4)$$

Let $P_j : L^2(\gamma) \to \operatorname{span}\{x^k; k = 0,\dots,j\}$, and for $f \in \operatorname{span}\{x^k; k = 0,\dots,j\}$ we write in Wick's notation : $f := f - P_{j-1}f$.

Lemma 12.2.1 (*i*) H_n *is a monic polynomial of degree* n.
  (*ii*) $LH_n = nH_n$.
  (*iii*) $(H_n)_{n=0}^\infty$ *is an orthogonal sequence in* $L^2(\gamma)$.
  (*iv*) $: x^n := H_n(x)$.

Proof. (i) This follows by induction.

(ii) First we show that $H'_n = nH_{n-1}$; indeed we have

$$H'_n(x) = \frac{d}{dx}\left((-1)^n e^{x^2/2}\frac{d^n}{dx^n}e^{-x^2/2}\right)$$

$$= (-1)^n xe^{x^2/2}\frac{d^n}{dx^n}e^{-x^2/2} + (-1)^n e^{x^2/2}\frac{d^n}{dx^n}\left(-xe^{-x^2/2}\right)$$

$$= (-1)^n xe^{x^2/2}\frac{d^n}{dx^n}e^{-x^2/2} + (-1)^{n-1} xe^{x^2/2}\frac{d^n}{dx^n}\left(e^{-x^2/2}\right)$$

$$+ (-1)^{n-1} ne^{x^2/2}\frac{d^{n-1}}{dx^{n-1}}\left(e^{-x^2/2}\right)$$

$$= (-1)^{n-1} ne^{x^2/2}\frac{d^{n-1}}{dx^{n-1}}\left(e^{-x^2/2}\right). \tag{12.2.5}$$

By differentiating this identity, we can deduce that

$$H''_n(x) - xH'_n(x)$$

$$= (-1)^n nxe^{x^2/2}\frac{d^{n-1}}{dx^{n-1}}\left(e^{-x^2/2}\right) + (-1)^{n-1} ne^{x^2/2}\frac{d^n}{dx^n}\left(e^{-x^2/2}\right)$$

$$+ (-1)^{n-1} nxe^{x^2/2}\frac{d^{n-1}}{dx^{n-1}}\left(e^{-x^2/2}\right) = -nH_n(x). \tag{12.2.6}$$

(iii) The operator L is symmetric in $L^2(\gamma)$, and hence eigenfunctions that correspond to distinct eigenvalues are orthogonal.

(iv) As in the Gram–Schmidt process, H_n is orthogonal to x^j in $L^2(\gamma)$ for $j = 0,\ldots,n-1$; so the result follows from (i). □

Exercise. (i) Use Rodrigues's formula to show that

$$H_{n+1}(x) = xH_n(x) - H'_n(x), \tag{12.2.7}$$

and hence obtain

$$H_0(x) = 1, \quad H_1(x) = x, \quad H_2(x) = x^2 - 1,$$
$$H_3(x) = x^3 - 3x, \quad H_4(x) = x^4 - 6x^2 + 3. \tag{12.2.8}$$

(ii) Use Rodrigues's formula to prove the three-term recurrence relation

$$xH_n(x) = H_{n+1}(x) + nH_{n-1}(x). \tag{12.2.9}$$

(iii) Let $\phi_n(x) = (2\pi)^{-1/4}(n!)^{-1/2}H_n(x)e^{-x^2/4}$. Deduce from (ii) that

$$x\phi_n(x) = \sqrt{n+1}\phi_{n+1}(x) + \sqrt{n}\phi_{n-1}(x). \tag{12.2.10}$$

(iv) Introduce a new variable ξ so that $\sqrt{2}\xi = x$, and let $\psi_n(\xi) = \sqrt{2}\phi_n(\sqrt{2}\xi)$. By repeatedly integrating by parts the identity

$$\mathcal{F}\psi_n(t) = \frac{\sqrt{2}}{(2\pi)^{3/4}2^{n/2}\sqrt{n!}}\int_{-\infty}^{\infty}e^{-i\xi t+\xi^2/2}\frac{d^n}{d\xi^n}e^{-\xi^2}d\xi, \tag{12.2.11}$$

and using Rodrigues's formula, show that

$$\mathcal{F}\psi_n(t) = \frac{\psi_n(t)}{i^n}.$$ (12.2.12)

Proposition 12.2.2 *(i)* $(H_n/\sqrt{n!})_{n=0}^\infty$ *gives a complete orthonormal basis for* $L^2(\gamma)$.

(ii) The operator L is self-adjoint and densely defined in $L^2(\gamma)$, with spectrum consisting of simple eigenvalues $\{0, 1, 2, \dots\}$ which correspond to unit eigenvectors $(H_n/\sqrt{n!})_{n=0}^\infty$.

(iii) L satisfies a spectral gap inequality with $\lambda_1 = 1$.

Proof. (i) One checks by repeated integration by parts that

$$\int_{-\infty}^\infty H_n(x)^2\gamma(x)\,dx = n!;$$ (12.2.13)

indeed,

$$\int_{-\infty}^\infty H_n(x)^2 e^{-x^2/2}\frac{dx}{\sqrt{2\pi}} = \int_{-\infty}^\infty H_n(x)(-1)^n\frac{d^n}{dx^n}e^{-x^2/2}\frac{dx}{\sqrt{2\pi}}$$

$$= \int_{-\infty}^\infty H_n^{(n)}(x)e^{-x^2/2}\frac{dx}{\sqrt{2\pi}}$$

$$= n!\int_{-\infty}^\infty e^{-x^2/2}\frac{dx}{\sqrt{2\pi}}.$$ (12.2.14)

As in Proposition 1.8.4, the linear span of the polynomials is dense in $L^2(\gamma)$, so the $(H_n/\sqrt{n!})$ gives a complete orthonormal basis.

(ii) We have found a complete orthonormal basis consisting of eigenfunctions of L, so L is self-adjoint with point spectrum.

(iii) Let $f \in C_c^\infty(\mathbf{R})$ have orthogonal expansion $f = \sum_{n=0}^\infty \langle f, H_n\rangle H_n(x)/n!$. Then

$$\int_{-\infty}^\infty |Lf(x)|^2\gamma(x)dx = \sum_{n=1}^\infty n^2\langle f, H_n\rangle^2/n!$$

$$\geq \sum_{n=1}^\infty n\langle f, H_n\rangle^2/n!$$

$$= \int_{-\infty}^\infty f(x)Lf(x)\gamma(x)dx.$$ (12.2.15)

Corollary 6.3.3 and Proposition 6.7.3 combine to give a stronger result than Proposition 12.2.2(iii), but the direct proof is much simpler. □

Corollary 12.2.3 (*Creation and annihilation*). *Let* $A : L^2(\gamma) \to L^2(\gamma)$ *be the creation operator*

$$A : \sum_{n=0}^{\infty} \frac{a_n H_n}{\sqrt{(n!)}} \mapsto \sum_{n=0}^{\infty} \frac{a_n H_{n+1}}{\sqrt{(n+1)!}}. \qquad (12.2.16)$$

Then the adjoint $A^* : L^2(\gamma) \to L^2(\gamma)$ *is the annihilation operator*

$$A^* : \sum_{n=0}^{\infty} \frac{a_n H_n}{\sqrt{(n!)}} \mapsto \sum_{n=1}^{\infty} \frac{a_n H_{n-1}}{\sqrt{(n-1)!}}, \qquad (12.2.17)$$

which satisfies $A^*A = I$, $AA^* = I - P_0$ *and* $A^*A - AA^* = P_0$, *where*

$$P_0 : A : \sum_{n=0}^{\infty} \frac{a_n H_n}{\sqrt{(n!)}} \mapsto a_0. \qquad (12.2.18)$$

Proof. Since $(H_n/\sqrt{n!})$ is a complete orthonormal basis, this is an easy calculation. $\square$

Exercise. (i) Express Corollary 12.2.3 in terms of the normalized Hermite functions, by verifying the identities

$$A\phi_n = \sqrt{n+1}\phi_{n+1} = (x/2)\phi_n(x) - \phi_n'(x), \qquad (12.2.19)$$
$$\sqrt{n}\phi_{n-1}(x) = (\phi_n'(x) + (x/2)\phi_n(x)). \qquad (12.2.20)$$

(ii) Deduce that

$$-\phi_n''(x) + \frac{x^2}{4}\phi_n(x) = (n+1/2)\phi_n(x) \qquad (n = 0, 1, 2, \dots).$$

Definition (*Mehler's kernel*). We introduce the Mehler kernel by

$$M(x, y; t) = \frac{1}{\sqrt{1 - e^{-t}}} \exp\left(-\frac{(e^{-t}x^2 - 2e^{-t/2}xy + e^{-t}y^2)}{2(1 - e^{-t})}\right) \qquad (12.2.21)$$

and the Ornstein–Uhlenbeck semigroup $(e^{-tL/2})$ by

$$e^{-tL/2}f(x) = \int_{-\infty}^{\infty} M(x, y; t)f(y)\gamma(dy) \qquad (f \in L^2(\gamma)). \qquad (12.2.22)$$

Theorem 12.2.4 (*Mehler's formula*). *For* $0 \le \theta < 1$ *and* $x, y \in \mathbf{R}$,

$$\sum_{n=0}^{\infty} \theta^n H_n(x)H_n(y) = \frac{1}{\sqrt{1 - \theta^2}} \exp\left(-\frac{((\theta x)^2 - 2\theta xy + (\theta y)^2)}{2(1 - \theta^2)}\right).$$
$$(12.2.23)$$

Proof. We follow [65] and write $\theta = e^{-t/2}$ and introduce

$$u(x,t) = \sum_{n=0}^{\infty} \frac{e^{-nt/2}}{n!} \langle f, H_n \rangle_{L^2(\gamma)} H_n(x) \qquad (12.2.24)$$

where $f \in C_c^{\infty}(\mathbf{R})$. Then $u(x,t) \to f(x)$ as $t \to 0+$, and

$$\frac{\partial}{\partial t} u(x,t) = \sum_{n=0}^{\infty} \frac{-n}{2} \frac{e^{-nt/2}}{n!} \langle f, H_n \rangle_{L^2(\gamma)} H_n(x)$$

$$= \frac{1}{2}\left(\frac{\partial^2}{\partial x^2} - x\frac{\partial}{\partial x} \right) u(x,t) \qquad (12.2.25)$$

by Lemma 12.2.1. Now we transform this to the standard heat equation by introducing $v(x,t) = u(e^{t/2}x, t)$. Then

$$\frac{\partial v}{\partial t} = e^{t/2} x \frac{\partial u}{\partial x}(e^{t/2}x, t) + \frac{\partial u}{\partial t}(e^{t/2}x, t) \qquad (12.2.26)$$

and

$$\frac{\partial^2 v}{\partial x^2} = e^t \frac{\partial^2 u}{\partial x^2}(e^{t/2}x, t), \qquad (12.2.27)$$

hence

$$e^t \frac{\partial v}{\partial t} = \frac{1}{2} \frac{\partial^2 v}{\partial x^2}. \qquad (12.2.28)$$

Now we let $\tau = 1 - e^{-t}$, so that v satisfies the standard heat equation

$$\frac{\partial v}{\partial t} = \frac{1}{2} \frac{\partial^2 v}{\partial x^2}; \qquad (12.2.29)$$

and hence the solution is

$$v(x,\tau) = \frac{1}{\sqrt{2\pi\tau}} \int_{-\infty}^{\infty} \exp\big(-(x-y)^2/(2\tau)\big) f(y)\,dy. \qquad (12.2.30)$$

Inverting the transformation, we find

$$u(x,\tau)$$
$$= v(\sqrt{1-\tau}x, \tau)$$
$$= \frac{1}{\sqrt{2\pi\tau}} \int_{-\infty}^{\infty} \exp\big(-(\sqrt{1-\tau}x - y)^2/(2\tau)\big) f(y)\,dy \qquad (12.2.31)$$
$$= \frac{1}{\sqrt{2\pi\tau}} \int_{-\infty}^{\infty} \exp\left(-\frac{\big((1-\tau)x^2 - 2\sqrt{1-\tau}xy + (1-\tau)y^2\big)}{2\tau} \right) f(y)e^{-y^2/2}\,dy$$

with $1 - \tau = e^{-t} = \theta^2$. $\qquad\qquad\square$

Exercise. Show that the density for the joint distribution of eigenvalues from the Gaussian unitary ensemble can be expressed using the determinant identity

$$\det\left[\phi_{k-1}(x_j)\right]^2_{j,k=1,\ldots,n}$$

$$= \frac{1}{(2\pi)^{n/2} \prod_{k=1}^{n-1} k!} \prod_{1 \le j < k \le n} (x_j - x_k)^2 \exp\left(-\sum_{j=1}^{n} x_j^2/2\right). \quad (12.2.32)$$

12.3 The Ornstein–Uhlenbeck semigroup

In this section we establish the main properties of e^{-tL} as an operator on $L^p(\gamma)$, as defined by the Mehler kernel.

Proposition 12.3.1 *The Ornstein–Uhlenbeck semigroup is a symmetric diffusion semigroup, and satisfies the following properties:*

(i) $e^{-tL}f \to f$ in $L^2(\gamma)$ as $t \to 0+$, for all $f \in L^2(\gamma)$;
(ii) $e^{-tL}e^{-sL} = e^{-(s+t)L}$ for $s, t > 0$;
(iii) $\|e^{-tL}\|_{B(L^2(\gamma))} = 1$;
(iv) e^{-tL} is self-adjoint on $L^2(\gamma)$;
(v) $e^{-tL}f \ge 0$ if $f \ge 0$, so the semigroup is positivity preserving;
(vi) $e^{-tL}\mathbf{I} = \mathbf{I}$;
(vii) $e^{-tL} : L^p(\gamma) \to L^p(\gamma)$ defines a bounded linear operator such that

$$\|e^{-tL}\|_{B(L^p(\gamma))} \le 1 \quad (1 \le p \le \infty);$$

(viii) $u(x, t) = e^{-tL}f(x)$ belongs to $C^\infty((0, \infty) \times \mathbf{R})$ for each $f \in L^2(\gamma)$.

Proof. Properties (i), (ii), (iii) and (iv) are clear from the expression

$$u(x, t) = \sum_{n=0}^{\infty} \frac{e^{-nt/2}}{n!} \langle f, H_n \rangle_{L^2(\gamma)} H_n(x) \quad (12.3.1)$$

which represents $e^{-tL/2}$ as a multiplier of the Hermite polynomials.

(v) Evidently the Mehler kernel is positive, so the semigroup is positivity preserving.

(vi) Since $H_0 = \mathbf{I}$, we have $e^{-tL}\mathbf{I} = \mathbf{I}$.

(vii) We can use the integral formula to define e^{-tL} on $L^p(\gamma)$ for $1 \le p \le \infty$. Properties (v) and (vi) combine to show that e^{-tL} is bounded on $L^\infty(\mathbf{R})$ with $\|e^{-tL}\|_{B(L^\infty)} = 1$. By symmetry of the Mehler kernel as in (iv), the dual operator of e^{-tL} is e^{-tL} itself, so e^{-tL} is bounded on $L^1(\gamma)$ with $\|e^{-tL}\|_{B(L^1(\gamma))} = 1$. By the Riesz–Thorin interpolation theorem [63, p. 525], e^{-tL} is bounded on $L^p(\gamma)$ for $1 \le p \le \infty$ with $\|e^{-tL}\|_{L^p(\gamma)} \le 1$.

(viii) The Mehler kernel $M(x, y; t)$ is evidently a smooth function for $(x, t) \in \mathbf{R} \times (0, \infty)$. □

Since γ is a probability density function, Hölder's inequality [71] gives a natural continuous linear inclusion $L^p(\gamma) \subset L^q(\gamma)$ for $q < p$. We now show that the operation of e^{-tL} maps $L^q(\gamma)$ into $L^p(\gamma)$ with $p > q$; that is, the semigroup improves the L^p index of each function. See [54, 75] for similar results relating to other semigroups.

Theorem 12.3.2 (*Hypercontractivity of the OU semigroup [26, 75]*). *Suppose that $2 < p < \infty$ and that $e^t > p - 1$. Then $e^{-tL/2} : L^2(\gamma) \to L^p(\gamma)$ is bounded.*

Proof. Were it not for the term in xy in the Mehler kernel, this result would be easy to prove, since $M(x, y; t)$ would factorize as a product of Gaussians. We will bound this term by using the inequality $2xy \le s^2 x^2 + y^2/s^2$ for all $s > 0$.

We choose $0 \le \alpha < \tanh(t/2)$ and a small $\beta > 0$ and so that

$$1 + p(\alpha - 1)/2 > 0, \tag{12.3.2}$$

which is possible since

$$1 + \frac{p}{2}\Big(\tanh(t/2) - 1\Big) = \frac{e^t + 1 - p}{e^t + 1} > 0. \tag{12.3.3}$$

Then we have

$$M(x, y; t) \le \frac{1}{\sqrt{1 - e^{-t}}} \exp\big((1 - \alpha)x^2/4 + (1 - \beta)y^2/4\big), \tag{12.3.4}$$

since

$$\frac{e^{-t/2}xy}{1 - e^{-t}} \le \Big(\frac{1 - \alpha}{4} + \frac{e^{-t}}{2(1 - e^{-t})}\Big)x^2 + \Big(\frac{1 - \beta}{4} + \frac{e^{-t}}{2(1 - e^{-t})}\Big)y^2 \tag{12.3.5}$$

which follows from

$$4 \le (e^{t/2}(1 - \alpha) + e^{-t/2}(1 + \alpha))(e^{t/2}(1 - \beta) + e^{-t/2}(1 + \beta)), \tag{12.3.6}$$

or

$$1 \le (\cosh(t/2) - \alpha \sinh(t/2))(\cosh(t/2) - \beta \sinh(t/2). \tag{12.3.7}$$

This holds for $\alpha < \tanh(t/2)$ and sufficiently small $\beta > 0$.

Then from the integral formula we have the pointwise bound

$$|e^{-tL/2}f(x)| \leq \frac{e^{(1-\alpha)x^2/4}}{\sqrt{1-e^{-t}}} \int_{-\infty}^{\infty} |f(y)||e^{(1-\beta)y^2/4}\gamma(dy)$$

$$\leq \frac{e^{(1-\alpha)x^2/4}}{\sqrt{1-e^{-t}}} \|f\|_{L^2(\gamma)} \|e^{(1-\beta)y^2/4}\|_{L^2(\gamma)}. \quad (12.3.8)$$

We evaluate the integrals here, obtaining

$$\|e^{(1-\beta)y^2/4}\|_{L^2(\gamma)}^2 = \int_{-\infty}^{\infty} e^{(1-\beta)y^2/2 - y^2/2} \frac{dy}{\sqrt{2\pi}} = \frac{1}{\sqrt{\beta}}, \quad (12.3.9)$$

and

$$\|e^{(1-\alpha)x^2/4}\|_{L^p(\gamma)}^p = \int_{-\infty}^{\infty} e^{p(1-\alpha)x^2/4 - x^2/2} \frac{dx}{\sqrt{2\pi}} = \frac{1}{\sqrt{1 + p(\alpha-1)/2}}.$$
$$(12.3.10)$$

Consequently we have

$$\|e^{-tL/2}\|_{L^2 \to L^p} \leq \beta^{-1/4}(1-e^{-t})^{-1/2}\big(1 + p(\alpha-1)/2\big)^{-1/(2p)}; \quad (12.3.11)$$

so $e^{-tL/2}$ is bounded $L^2(\gamma) \to L^p(\gamma)$ when $e^t + 1 > p$, hence the result. $\qquad \square$

Gross [75] showed that under a mild technical hypothesis, hypercontractivity of e^{-tA} on $L^2(\rho)$ is equivalent to the logarithmic Sobolev inequality for the corresponding Dirichlet form $\int f A f \rho dx$. In particular, his result shows that Theorem 12.3.2 follows from Corollary 6.3.3. We now give one implication from this theorem. The idea is to choose $p(t)$ such that $e^{-tL} : L^2(\rho) \to L^{p(t)}(\rho)$ is a bounded linear operator with bound $M(t)$.

Theorem 12.3.3 (*Gross [75]*). *Suppose that ρ is a probability density function on $\mathbf{R}^n$ and that e^{-tA} ($t \geq 0$) is a semigroup of continuous linear operators on $L^2(\rho)$ that satisfies $(i)-(viii)$ of Proposition 12.3.1. Suppose further that there exist differentiable functions $p : [0,\infty) \to [2,\infty)$ and $M : [0,\infty) \to (0,\infty)$ such that $p(0) = 2$, $M(0) = 1$, that $p'(0) > 0$ and that $e^{-tA} : L^2(\rho) \to L^{p(t)}(\rho)$ is bounded with*

$$\|e^{-tA}f\|_{L^{p(t)}(\rho)} \leq M(t)\|f\|_{L^2(\rho)} \qquad (f \in L^2(\rho)). \quad (12.3.12)$$

Then ρ satisfies the logarithmic Sobolev inequality

$$\int f(x)^2 \log(f(x)^2/\|f\|_{L^2(\rho)}^2)\rho(x)dx$$

$$\leq \frac{4}{p'(0)}\int f(x)Af(x)\rho(x)dx + \frac{4M'(0)}{p'(0)}\|f\|_{L^2(\rho)}^2 \quad (12.3.13)$$

for all nonnegative f in the domain of the quadratic form $\langle A, \rangle$.

Proof. By (viii) the function $u(x,t) = e^{-tA}f(x)$ is non-negative and differentiable on $\mathbf{R}^n \times (0,\infty)$; further, by an approximation argument discussed in [54] we can assume that the time derivative exists at $t = 0+$. By hypothesis, we have

$$\log \int \left(e^{-tA}f(x)\right)^{p(t)}\rho(x)\,dx \leq p(t)\log M(t) + p(t)\log\|f\|_{L^2(\rho)}$$

$$(12.3.14)$$

with equality at $t = 0$; so when we compare the derivatives at $t = 0+$ of the left- and right-hand sides of this inequality, we obtain

$$\frac{1}{\|f\|_{L^2(\rho)}^2}\left(\int \left(p'(0)f(x)^2\log f(x) - 2f(x)Af(x)\right)\rho(x)dx\right)$$

$$\leq p'(0)\log M(t) + p(0)\frac{M'(0)}{M(0)} + p'(0)\log\|f\|_{L^2(\rho)}. \quad (12.3.15)$$

We multiply this through by $2\|f\|_{L^2(\rho)}^2/p'(0)$ to simplify, and obtain the stated result. $\qquad\square$

Remark. Nelson [121] showed that for the Ornstein–Uhlenbeck semi-group, one can take $M(t) = 1$ for all $t \geq 0$.

12.4 Hermite polynomials in higher dimensions

The Hermite polynomials are:

- eigenfunctions of the Ornstein–Uhlenbeck operator on $\mathbf{R}^\infty$;
- spherical polynomials on the sphere of infinite dimension;
- a natural basis for the Wick ordering.

In view of Proposition 12.2.2, we can regard L as the number operator $H_n \mapsto nH_n$ in $L^2(\gamma)$. In this section, we extend this interpretation to higher dimensions. So long as we work with functions that depend upon only finitely many variables at a time, we can safely introduce

$$x = (x_1, x_2, \dots) \in \mathbf{R}^\infty, \qquad p = (p_1, p_2, \dots) \in \mathbf{Z}_+^\infty, \quad (12.4.1)$$

where $p_j = 0$ for all but finitely many j. Then let $|p| = \sum_{j=1}^{\infty} p_j$ and $p! = \prod_{j=1}^{\infty} p_j!$, and $H_p(x) = \prod_{j=1}^{\infty} H_{p_j}(x_j)$ where all but finitely many of the factors are equal to one, so $H_p(x)$ is a polynomial of total degree $|p|$. We can write $H_p(x) =: \prod_{j=1}^{\infty} x_j^{p_j}$:.

In $\mathbf{R}^N$ for each m, we can form the orthogonal projection

$$\pi_m : L^2(\Gamma_N) \to \text{span}\Big\{ H_{p_1}(x_1) \ldots H_{p_N}(x_N) : \sum_{j=1}^{N} p_j = m \Big\} \quad (12.4.2)$$

and find that $\pi_m(x^{p_1} \ldots x_N^{p_N}) = H_{p_1}(x_1) \ldots H_{p_N}(x_N)$. By using the Ornstein–Uhlenbeck operator, we can suppress Γ_N and carry out a similar construction in infinite-dimensional Hilbert space. This is often referred to as the Wick ordering of polynomials.

Proposition 12.4.1 *Let the Ornstein–Uhlenbeck operator on $\mathbf{R}^\infty$ be*

$$L = \sum_{j=1}^{\infty} \Big(-\frac{\partial^2}{\partial x_j^2} + x_j \frac{\partial}{\partial x_j} \Big) \quad (12.4.3)$$

where the domain includes functions $f(x_1, \ldots, x_N)$ such that $f \in C_c^\infty(\mathbf{R}^N; \mathbf{R})$.

(i) Then

$$Z_m = \overline{\text{span}}\{ H_p : |p| = m \} \quad (12.4.4)$$

is an eigenspace corresponding to eigenvalue m.
(ii) L satisfies a spectral gap inequality with $\lambda_1 = 1$.
(iii) There is a natural representation π of S_∞ on Z_m, given by the permutation of variables $\sigma : (x_j) \mapsto (x_{\sigma(j)})$, which commutes with the operation of L.

Proof. (i) This follows directly from Lemma 12.2.1(i).

(ii) Suppose that $f \in C_c^\infty(\mathbf{R}^N; \mathbf{R})$. Then we have $f(x) = \sum_p \langle f, H_p \rangle H_p / p!$ where the Hermite polynomials depend upon only finitely many coordinates, so the obvious extension of Proposition 12.2.2 gives

$$\int_{\mathbf{R}^N} |Lf(x)|^2 \gamma_N(dx) \geq \int_{\mathbf{R}^N} \sum_{j=1}^{N} \Big| \frac{\partial f}{\partial x_j} \Big|^2 \gamma_N(dx). \quad (12.4.5)$$

Once again, Corollary 6.3.3 and Proposition 6.7.3 imply a stronger result than Proposition 12.4.1(ii).

(iii) The elements of S_∞ permute finitely many variables, and leave the others fixed, so the representation $\pi_\sigma : H_p(x) \mapsto H_p(\sigma x)$ is well defined. The variables appear symmetrically in the definition of L, so $\pi_\sigma LH_p(x) = LH_p(\sigma x)$.   □

Lemma 12.4.2 *The following identities hold:*

$$(i) \quad \exp\left(\lambda x - \lambda^2/2\right) = \sum_{n=0}^{\infty} \frac{\lambda^n}{n!} H_n(x) \quad (x \in \mathbf{R}, \lambda \in \mathbf{C}); \quad (12.4.6)$$

$$(ii) \quad H_m(\langle x, y \rangle) = \sum_{|p|=m} \binom{m}{p} \prod_j H_{p_j}(x_j) \prod_j y_j^{p_j}. \quad (12.4.7)$$

Proof. (i) By Taylor's theorem we have

$$\sum_{n=0}^{\infty} \frac{\lambda^n}{n!} H_n(x) = \sum_{n=0}^{\infty} \frac{(-1)^n \lambda^n}{n!} e^{x^2/2} \frac{d^n}{dx^n} e^{-x^2/2}$$

$$= \sum_{n=0}^{\infty} \frac{\lambda^n}{n!} \left(\frac{d^n}{dt^n}\right)_{t=0} e^{x^2/2-(x-t)^2/2}$$

$$= \sum_{n=0}^{\infty} \frac{\lambda^n}{n!} \left(\frac{d^n}{dt^n}\right)_{t=0} e^{tx-t^2/2}$$

$$= e^{\lambda x - \lambda^2/2}. \quad (12.4.8)$$

(ii) For each j, by (i) we have

$$\exp\left(y_j x_j - y_j^2/2\right) = \sum_{p_j=0}^{\infty} \frac{y_j^{p_j}}{p_j!} H_{p_j}(x_j) \quad (12.4.9)$$

and hence we can take the product of j to obtain

$$\exp\left(\langle x, y \rangle - \|y\|^2/2\right) = \prod_{p=(p_j)} \sum_{p_j=0}^{\infty} \frac{y_j^{p_j}}{p_j!} H_{p_j}(x_j)$$

$$= \sum_m \sum_{p:|p|=m} \frac{1}{m!} \frac{m!}{\prod p_j!} H_{p_j}(x_j) y_j^{p_j}. \quad (12.4.10)$$

The terms that have degree m are

$$\frac{1}{m!} H_n(\langle x, y \rangle) = \frac{1}{m!} \sum_{p:|p|=m} \binom{m}{p_1 \ldots} \prod_j H_{p_j}(x_j) \prod_j y_j^{p_j}. \quad (12.4.11)$$

□

Pursuing the analogy with the preceding section, we wish to regard the H_p as orthogonal with respect to a Gaussian measure on $\mathbf{R}^\infty$. Hence we introduce a sequence of mutually independent random variables

$\chi = (\chi_j)_{j=1}^{\infty}$ on $(\Omega, \mathbf{P})$, where χ_j has a $N(0,1)$ distribution, and then we let $H_p(\chi) = \prod_{j=1}^{\infty} H_{p_j}(\chi_j)$.

Definition (*Homogeneous chaos*). The space $Z_m = \overline{\text{span}}\{H_p(\chi) : |p| = m\}$ is homogeneous chaos of degree m. The domain of the self-adjoint operator L is

$$D(L) = \Big\{ f = \sum_{m=0}^{\infty} f_m : f_m \in Z_m; \sum_{m=0}^{\infty} m^2 \|f_m\|^2 < \infty \Big\}. \quad (12.4.12)$$

Theorem 12.4.3 *(i)* $(H_p/p!)_p$ *gives an orthonormal sequence in* $L^2(\Omega, \mathbf{P})$.

(ii) For each $(y_j) \in \ell^2$ *and* $m \in \mathbf{Z}_+$, $H_m(\sum_{j=1}^{\infty} y_j \chi_j)$ *belongs to* Z_m.

(iii) There is a natural representation π *of* $O(\infty)$ *on* $L^2(\Omega, \mathbf{P})$ *given by*

$$[u_{jk}] : (\chi_j)_{j=1}^{\infty} \mapsto \Big(\sum_{j=1}^{\infty} u_{kj} \chi_j \Big)_{k=1}^{\infty} \quad (12.4.13)$$

and $\pi_u H_p(\chi) = H_p(u\chi)$, *and* Z_m *is a submodule for this representation.*

(iv) The operator L *commutes with the action of* $O(\infty)$ *on* $D(L)$, *so that* $\pi_u L f = L(\pi_u f)$ *for all* $f \in D(L)$ *and* $u \in O(\infty)$.

Proof. (i) When $p \neq q$, there exists an index j such that $p_j \neq q_j$ and $\mathbf{E} H_p(\chi) H_q(\chi)$ includes a factor

$$\mathbf{E} H_{p_j}(\chi_j) H_{q_j}(\chi_j) = \int_{-\infty}^{\infty} H_{p_j}(x) H_{q_j}(x) \gamma(x)\, dx = 0. \quad (12.4.14)$$

Further, by independence of the x_j, we have

$$\mathbf{E} H_p(\chi)^2 / p! = \prod_{j=1}^{\infty} \frac{1}{p_j!} \int_{-\infty}^{\infty} H_{p_j}(x)^2 \gamma(x) dx = 1. \quad (12.4.15)$$

(ii) This follows immediately from Lemma 12.4.2(ii).

(iii) For any matrix $[u_{jk}] \in O(\infty)$, each column involves only finitely many nonzero terms, so the sums $\sum_{j=1}^{\infty} u_{kj} \chi_j$ are actually finite. The random variables $(\sum_{j=1}^{\infty} u_{kj} \chi_j)$ are Gaussian by Lemma 11.3.3 and mutually independent since

$$\mathbf{E}\Big\{ \Big(\sum_{j=1}^{\infty} u_{kj} \chi_j\Big)\Big(\sum_{j=1}^{\infty} u_{\ell j} \chi_j\Big) \Big\} = \sum_{j=1}^{\infty} u_{kj} u_{\ell j} = \delta_{k\ell} \quad (12.4.16)$$

by orthogonality. Further, by (ii), each $H_m(\sum_j u_{kj} \chi_j)$ belongs to Z_m. By forming products, we see that Z_m is invariant under the operation of $O(\infty)$.

(iv) For each $f_m \in Z_m$, we have $\pi_u f \in Z_m$ where Z_m is an eigenspace of L by Proposition 12.4.1(i), so $L(\pi_u f_m) = m\pi_u f_m = \pi_u(Lf_m)$.   □

Remark. In this section, we have followed McKean fairly closely, giving more explicit statements of some of the results. McKean suggests that the spaces Z_m should give a complete list of irreducible submodules of $L^2(\Omega, \mathbf{P})$.

Malliavin [112] gives a detailed discussion of the subspaces of L^2 spanned by Hermite polynomials. In [14], the authors extend these ideas to the context of free probability.

Proposition 12.4.4 *The infinite unitary group $U(\infty)$ has a natural representation on $\mathbf{C} \otimes Z_m$ given by $\pi_u H_p(z) = H_p(uz)$ such that $H_m(\chi_1)$ is a cyclic vector.*

Proof. The basic identities in Lemma 12.4.2 for Hermite polynomials hold for complex variables since we only used identities for convergent power series in the proofs. Hence

$$H_m\left(\sum_{j=1}^{N} a_j e^{i\theta_j} \chi_j\right) = \sum_{p:|p|=m} \binom{m}{p_1 \ldots p_N} \prod_{j=1}^{N} H_{p_j}(x_j) \prod_j a_j^{p_j} e^{ip_j\theta_j},$$

$$(12.4.17)$$

so we use orthogonality of characters to pick off the term

$$\binom{m}{p_1 \ldots p_N} \prod_{j=1}^{N} H_{p_j}(x_j) \prod_j a_j^{p_j}$$

$$= \int_{\mathbf{T}^N} H_m\left(\sum_{j=1}^{n} a_j e^{i\theta_j} \chi_j\right) \exp\left(-i\sum_{j=1}^{N} p_j\theta_j\right) \frac{d\theta_1}{2\pi} \cdots \frac{d\theta_N}{2\pi}.$$

$$(12.4.18)$$

For each real sequence $(a_j)_{j=1}^{N}$ such that $\sum_{j=1}^{N} a_j^2 = 1$, and each real sequence $(\theta_j)_{j=1}^{N}$ there exists $u \in U(N)$ that has first row $(e^{i\theta_j} a_j)_{j=1}^{N}$. Hence the right-hand side of (12.4.17) lies in the closed linear span of $\pi_u H_m(\chi_1)$. Since the products $\prod_j H_{p_j}(\chi_j)$ generate Z_m as a complex vector space, we deduce that $H_m(\chi_1)$ is a cyclic vector.   □

The sphere $S^{n-1}(R)$ has constant Ricci curvature tensor $(n - 2)R^{-2}I_{n-1}$. This suggests that when $R = \sqrt{n}$ and $n \to \infty$, we should obtain interesting results on some limiting object. In Theorem 1.6.2 we saw that the surface area measure on $S^{n-1}(\sqrt{n})$ induces measures on $\mathbf{R}^m$ that converges weakly to the standard Gaussian. By Theorem 12.4.3(ii) the measures are invariant under the action of $O(\infty)$. Further, the next result shows that the Laplace operator on $S^{n-1}(\sqrt{n})$ converges to the Ornstein–Uhlenbeck operator on $\mathbf{R}^\infty$.

Proposition 12.4.5 *Let* $\pi_{m,n} : \mathbf{R}^n \to \mathbf{R}^m$ *be the projection onto the first* m *coordinates, so* $\pi_{m,n}(\xi_1,\dots,\xi_n) = (\xi_1,\dots,\xi_m)$, *let* $g : \mathbf{R}^m \to \mathbf{R}$ *be a smooth and compactly supported function and let* $f_n(x) = g(\sqrt{n}\pi_{m,n}(x)/\|x\|)$. *Then*

$$-\sum_{j=1}^\infty \frac{\partial^2}{\partial x_j^2} f_n(x) \to \sum_{j=1}^m \Big(-\frac{\partial^2}{\partial \xi_j^2} + \xi_j \frac{\partial}{\partial \xi_j}\Big) g \qquad (n \to \infty). \quad (12.4.19)$$

Proof. McKean [115] sketches a proof of this result which involves transforming the Laplacian into polar coordinates and then finding the limit as $n \to \infty$; the details are wearisome to write out. We provide a similar proof which involves a little less computation.

When $\xi = (\xi_1,\dots,\xi_n)$ is a typical vector in $\mathbf{R}^n$, the coordinates can vary independently, but the point $\sqrt{n}\xi/\|\xi\|$ lies on the sphere of radius $\sqrt{n}$, so the coordinates are dependent. If we let $n \to \infty$, then the first m coordinates, become asymptotically independent again; this subtle effect leads to the stated partial differential operator.

Evidently f_n does not depend upon $r = \|x\|$. Then we convert the Laplace operator on Euclidean space into into polar coordinates, and take

$$-\sum_{j=1}^n \frac{\partial^2}{\partial x_j^2} = -\frac{\partial^2}{\partial r^2} - \frac{(n-1)}{r}\frac{\partial}{\partial r} + \frac{1}{r^2}\Delta_\sigma \quad (12.4.20)$$

where Δ_σ is the Laplace operator on $S^{n-1}(r)$; likewise $\nabla f_n = r^{-2}\langle \nabla f_n, x\rangle x + \nabla_\sigma f_n$, where ∇_σ is the gradient on $S^{n-1}(r)$. The gradient $\nabla_\sigma f_n(x)$ lies in the hyperplane that is tangent to the sphere at x; further, $\langle \nabla f_n, x\rangle = 0$ since f_n does not depend upon $\|x\|$. By the divergence theorem, we have

$$\int_{S^{n-1}(\sqrt{n})} f_n \Delta_\sigma f_n \hat{\sigma}_{n-1,\sqrt{n}}(dx) = \int_{S^{n-1}(\sqrt{n})} \|\nabla_\sigma f_n\|^2 \hat{\sigma}_{n-1,\sqrt{n}}(dx),$$

$$(12.4.21)$$

where

$$\|\nabla_\sigma f_n\|^2 = \sum_{j=1}^n \Big(\frac{\partial f_n}{\partial x_j}\Big)^2 - \frac{1}{r^2}\Big(\sum_{j=1}^n x_j \frac{\partial f_n}{\partial x_j}\Big)^2$$

$$= \sum_{j=1}^m \Big(\frac{\partial g}{\partial x_j}\Big)^2 - \frac{1}{n}\Big(\sum_{j=1}^m x_j \frac{\partial g}{\partial x_j}\Big)^2, \quad (12.4.22)$$

on $S^{n-1}(\sqrt{n})$ since f_n depends upon the first m coordinates; the final term goes to 0 as $n \to \infty$.

As in Theorem 1.6.2, the coordinate map $\pi_{m,n} : S^{n-1}(\sqrt{n}) \to \mathbf{R}^m$ induces $\gamma_{m,n} = \pi_{m,n} \sharp \hat{\sigma}_{n-1,n}$ from the normalized surface area on $S^{n-1}(\sqrt{n})$, where $\gamma_{m,n} \to \gamma_m$ weakly as $n \to \infty$. Hence we have

$$\int_{S^{n-1}(\sqrt{n})} \sum_{j=1}^m \left(\frac{\partial g}{\partial x_j}\right)^2 \hat{\sigma}_{n-1,\sqrt{n}}(dx) \to \int_{\mathbf{R}^m} \sum_{j=1}^m \left(\frac{\partial g}{\partial x_j}\right)^2$$

$$\times \exp\left(-\sum_{j=1}^m x_j^2/2\right) \frac{dx_1 \dots dx_m}{(2\pi)^{m/2}} \qquad (12.4.23)$$

as $n \to \infty$. Finally, we integrate by parts to obtain

$$\int_{\mathbf{R}^m} \sum_{j=1}^m \left(\frac{\partial g}{\partial x_j}\right)^2 \exp\left(-\sum_{j=1}^m x_j^2/2\right) \frac{dx_1 \dots dx_m}{(2\pi)^{m/2}} \qquad (12.4.24)$$

$$= \int_{\mathbf{R}^m} \sum_{j=1}^m \left(-\frac{\partial^2 g}{\partial x_j^2} + x_j \frac{\partial g}{\partial x_j}\right) g(x) \exp\left(-\sum_{j=1}^m x_j^2/2\right) \frac{dx_1 \dots dx_m}{(2\pi)^{m/2}} \qquad (12.4.25)$$

and thence the stated result by polarization. □

In Chapter 3 we used the Gaussian concentration of measure to deduce concentration of measure on spheres. In Chapter 6, we saw how the Gaussian density $N(0, I)$ on $\mathbf{R}^m$ satisfies a logarithmic Sobolev inequality with constant $\alpha = 1$. We now show how this arises as the limiting case of Sobolev inequalities on the spheres $S^{n-1}(\sqrt{n})$ as $n \to \infty$.

Proposition 12.4.6 (*Sobolev inequality on the sphere*). *Let σ_n be the rotation invariant probability measure on S^n. Then*

$$\left(\int_{S^n} |F(\omega)|^{2n/(n-2)} \sigma_n(d\omega)\right)^{(n-2)/n} \leq \frac{4}{n(n-2)} \int_{S^n} \|\nabla F(\omega)\|^2 \sigma_n(d\omega)$$

$$+ \int_{S^n} |F(\omega)|^2 \sigma_n(d\omega) \qquad (12.4.26)$$

holds for all $F \in L^2(S^n)$ such that $\nabla F \in L^2(S^n)$.

Proof. See Bechner [12]. □

Suppose that $f_n : S^{n-1}(\sqrt{n}) \to \mathbf{R}$ depends on the first m coordinates, and hence is given by $g_m : \mathbf{R}^m \to \mathbf{R}$ as in Proposition 12.4.5. Then the terms in the spherical Sobolev inequality converge as follows: the rescaled gradient term satisfies

$$\frac{1}{n} \int_{S^{n-1}(\sqrt{n})} \|\nabla f_n\|^2 d\hat{\sigma}_{n-1,n} \to \int_{\mathbf{R}^m} \|\nabla g_m\|^2 \gamma_m(dx), \quad (12.4.27)$$

as $n \to \infty$, and

$$\left(\int_{S^{n-1}(\sqrt{n})} |f_n|^{2(n-1)/(n-3)} d\hat{\sigma}_{n-1,n} \right)^{(n-3)/(n-1)} - \int_{S^{n-1}(\sqrt{n})} f_n^2 d\hat{\sigma}_{n-1,n}$$

$$= \frac{2}{n-1} \int_{S^{n-1}(\sqrt{n})} f_n(\omega)^2 \log\left(f_n(\omega)^2 \Big/ \int f_n^2 d\hat{\sigma}_{n-1,n} \right) \hat{\sigma}_{n-1,n}(d\omega)$$

$$+ O(1/n^2). \quad (12.4.28)$$

Hence the limiting form of the Sobolev inequality is

$$\int_{\mathbf{R}^m} g_m(x)^2 \log\left(g_m(x)^2 \Big/ \int g_m^2 d\gamma_m \right) \gamma_m(dx) \le 2 \int_{\mathbf{R}^m} \|\nabla g_m\|^2 \gamma_m(dx).$$

$$(12.4.29)$$

In the preceding section, we have given meaning to the following heuristic statements.

- $O(\infty)$ acts as the group of rotations of $S^\infty(\sqrt{\infty})$.
- The Ornstein–Uhlenbeck operator gives the rotation–invariant Laplace operator on $S^\infty(\sqrt{\infty})$.
- Gaussian measure is the rotation–invariant measure on $S^\infty(\sqrt{\infty})$.
- The logarithmic Sobolev inequality for the Gaussian measure is the limiting case of the Sobolev inequality on the spheres.

13
From the Ornstein–Uhlenbeck process to the Burgers equation

Abstract

In Chapter 12 the Ornstein–Uhlenbeck operator L was defined as a self-adjoint differential operator, and we obtained the Hermite polynomials as the eigenfunctions in $L^2(\gamma)$. In Chapter 13, we introduce the Ornstein–Uhlenbeck stochastic process via the corresponding stochastic differential equation, and we show that the Mehler kernel gives the transition densities. We then extend these ideas to stochastic processes with values in the real symmetric $n \times n$ matrices, obtaining the matrix Ornstein–Uhlenbeck process. More generally, we prove an existence theorem for matrix stochastic differential equations in L^2 with Lipschitz coefficients. Following Dyson's approach, we obtain corresponding stochastic differential equations for the eigenvalues and then derive the Burgers equation for the equilibrium density as $n \to \infty$.

13.1 The Ornstein–Uhlenbeck process

- The Ornstein–Uhlenbeck process is introduced by a simple stochastic differential equation.
- The Ornstein–Uhlenbeck process is the unique Gaussian process that is stationary and Markovian.
- The transition densities are given by the Mehler kernel.

Let b_t be Brownian motion, and let $\alpha > 0$. Suppose that z_t represents the velocity of a particle at time t, where the particle is subject to a resistive force $-\alpha z_t$ proportional to z_t and subject to random impulses. The suggests the stochastic differential equation for the Ornstein–Uhlenbeck process:

$$dz_t = -\alpha z_t dt + db_t, \qquad (13.1.1)$$

to be interpreted as shorthand for the integral equation

$$z_t = z_0 - \int_0^t \alpha z_s \, ds + b_t - b_0 \qquad (13.1.2)$$

in $L^2(dt \, d\mathbf{P})$. See [70, 73].

Lemma 13.1.1 *Suppose that z_0 is independent of $(b_t)_{t>0}$ and has a $N(0, 1/(2\alpha))$ distribution. Then any solution of the Ornstein–Uhlenbeck equation satisfies*

$$\mathbf{E}z_t = 0,$$
$$\mathbf{E}(z_t z_s) = e^{-\alpha(t-s)} \mathbf{E}(z_s^2) \qquad (t \geq s \geq 0). \qquad (13.1.3)$$

Proof. We have

$$z_t = z_s - \int_s^t \alpha z_u \, du + b_t - b_s \qquad (t > s > 0) \qquad (13.1.4)$$

where $b_t - b_s$ is independent of z_s, hence

$$\mathbf{E}(z_t z_s) = \mathbf{E}z_s^2 - \int_s^t \alpha \mathbf{E}(z_u z_s) \, du; \qquad (13.1.5)$$

hence

$$\mathbf{E}(z_t z_s) = e^{-\alpha(t-s)} \mathbf{E}z_s^2. \qquad (13.1.6)$$

Furthermore, the expectation satisfies

$$\mathbf{E}z_t = \mathbf{E}z_0 - \alpha \int_0^t \mathbf{E}z_u \, du, \qquad (13.1.7)$$

and $\mathbf{E}z_0 = 0$, hence $\mathbf{E}z_s = 0$. $\qquad \square$

Proposition 13.1.2 *There exists a unique Gaussian process that satisfies the OU equation, and which has the following properties.*

(i) (z_t) *has Markov's property,*

$$\mathbf{E}(f(z_t) \mid z_s, z_{s_j}, \ldots, z_{s_1}) = \mathbf{E}(f(z_t) \mid z_s) \qquad (13.1.8)$$

for all $0 < s_1 < \cdots < s_j < s < t$ and $f \in C_b(\mathbf{R})$.

(ii) (z_t) *is a strongly stationary Gaussian process; so that $z_{t_0}, z_{t_1}, \ldots, z_{t_n}$ have joint Gaussian distribution equal to that of $z_{t_0+t}, z_{t_1+t}, \ldots, z_{t_n+t}$ whenever $0 < t$ and $0 \leq t_0 < t_1 < \cdots < t_n$.*

Proof. (ii) By successive substitutions, we have

$$z_t = e^{-\alpha t} z_0 + b_t - \alpha \int_0^t e^{-\alpha(t-s)} b_s \, ds$$

$$= e^{-\alpha t} z_0 + \int_0^t e^{-\alpha(t-s)} db_s, \qquad (13.1.9)$$

so

$$\mathbf{E}(z_t^2) = \mathbf{E}z_0^2 + \int_0^t e^{-2(t-s)} \, ds = \frac{1}{2\alpha}. \qquad (13.1.10)$$

Hence (z_t) is a Gaussian random variable with mean zero and variance $1/2\alpha$. By the Lemma, $\mathbf{E}z_t = 0$, and

$$\mathbf{E}z_s z_t = \mathbf{E}z_{s+h} z_{t+h} = \frac{e^{-\alpha|t-s|}}{2\alpha}. \qquad (13.1.11)$$

Since (z_t) is a Gaussian process, the correlations determine the joint distribution, and this suffices to prove strong stationarity [73].

(i) The Markov property is clear from (13.1.11): knowledge of z_0 does not change the conditional distribution of z_t given z_s.

Any solution to the OU equation has Markov's property and is covariance stationary by the Lemma. For a Gaussian process, the auto-covariance $\mathbf{E}z_t z_s$ uniquely determines the joint distribution. See [73]. □

Definition (*OU(α) process*). The process Z_t is the Ornstein–Uhlenbeck process with drift coefficient α. (When $\alpha = 1/2$ the process Z_t has $Z_0 \sim N(0,1)$ and the standard Gaussian γ_1 as its stationary distribution.) See also [73, 112].

Theorem 13.1.3 *Suppose that f is a bounded and continuous function. Then the transition densities of (z_t) are given by the Mehler kernel; so that*

$$\mathbf{E}\big(f(z_t) \mid z_0 = x\big) = \sqrt{\frac{\alpha}{\pi(1 - e^{-2\alpha t})}}$$

$$\times \int_{-\infty}^{\infty} \exp\Big(-\frac{\alpha(e^{-2\alpha t} x^2 - 2e^{-\alpha t} xy + e^{-2\alpha t} y^2)}{1 - e^{-2\alpha t}}\Big)$$

$$\times \exp(-\alpha y^2) f(y) \, dy. \qquad (13.1.12)$$

Proof. Since (z_t) is a stationary Markov process, the joint distribution is completely specified by the distribution of z_0 and the conditional distribution of z_t given z_0. We observe that when $z_0 = x$, the integral equation reduces to

$$z_t = e^{-\alpha t} x + \int_0^t e^{-\alpha(t-s)} db_s, \qquad (13.1.13)$$

so that the conditional distribution of z_t is

$$z_t \sim N\left(e^{-\alpha t} x, \frac{1 - e^{-2\alpha t}}{2\alpha}\right). \qquad (13.1.14)$$

Hence the formula for the Gaussian probability density function gives

$$\mathbf{E}\big(f(z_t) \mid z_0 = x\big) = \sqrt{\frac{\alpha}{\pi(1 - e^{-2\alpha t})}}$$
$$\times \int_{-\infty}^{\infty} \exp\left(\frac{-\alpha z^2}{1 - e^{-2\alpha t}}\right) f(z + xe^{-\alpha t}) \, dz \quad (13.1.15)$$

and we obtain the stated formula by substituting $z = z + e^{-\alpha t} x$. See also [65]. $\qquad \square$

Corollary 13.1.4 (*Backward equation*). *Let* $u(x,t) = \mathbf{E}(f(z_t) \mid z_0 = x)$ *where* $f \in C_b(\mathbf{R}; \mathbf{R})$. *Then* $u : \mathbf{R} \times (0, \infty) \to \mathbf{R}$ *is twice continuously differentiable and satisfies*

$$\frac{\partial u}{\partial t} = \frac{1}{2} \frac{\partial^2 u}{\partial x^2} - \alpha x \frac{\partial u}{\partial x}. \qquad (13.1.16)$$

In particular, when $\alpha = 1/2$, *the differential equation is* $\partial u/\partial t = -Lu/2$, *where* L *is the Ornstein–Uhlenbeck operator of Chapter 6.*

Proof. See the proof of Theorem 12.2.4. $\qquad \square$

Let b_t^j for $j = 1, 2$ be mutually independent copies of Brownian motion, and let

$$\zeta_t = \frac{1}{\sqrt{2}} (b_t^1 + i b_t^2) \qquad (t > 0)$$
$$= \frac{1}{\sqrt{2}} (b_{-t}^1 - i b_{-t}^2) \qquad (t < 0). \qquad (13.1.17)$$

Proposition 13.1.5 *The Ornstein–Uhlenbeck process has spectral representation*

$$z_t = \frac{1}{\sqrt{2\pi\alpha}} \int_{-\infty}^{\infty} \frac{e^{-i\alpha t u}}{\sqrt{1 + u^2}} d\zeta_u \qquad (t > 0). \qquad (13.1.18)$$

Proof. We have $dz_u \overline{dz_u} = du$ and $dz_u \overline{dz_{-u}} = 0$, so that

$$
\begin{aligned}
\mathbf{E}z_t \bar{z}_s &= \frac{1}{2\pi\alpha} \int_{-\infty}^{\infty} \frac{e^{-i\alpha(t-s)u} du}{1+u^2} \\
&= \frac{1}{2\alpha} e^{-\alpha|t-s|} \qquad (s,t \geq 0)
\end{aligned}
\tag{13.1.19}
$$

holds by a familiar identity from Fourier analysis. Further, $\zeta_{-t} = \bar{\zeta}_t$, so that z_t is a real Gaussian process. Hence $(z_t)_{t>0}$ is the Ornstein–Uhlenbeck process. See also [73].   □

13.2 The logarithmic Sobolev inequality for the Ornstein–Uhlenbeck generator

- The Ornstein–Uhlenbeck process is associated with the logarithmic Sobolev inequality for the Gaussian measure.
- The gradient flow reduces to a simple form of the Fokker–Planck equation.

Let X be a random variable with probability density function q_0, and let b_t be Brownian motion independent of X. The stochastic differential equation

$$
dz_t = -\alpha z_t dt + db_t, \tag{13.2.1}
$$
$$
z_0 = X
$$

has solutions

$$
z_t = e^{-\alpha t} X + \int_0^t e^{-\alpha(t-s)} db_s; \tag{13.2.2}
$$

so that

$$
z_t \sim e^{-\alpha t} X + \frac{(1 - e^{-2\alpha t})^{1/2}}{(2\alpha)^{1/2}} Z, \tag{13.2.3}
$$

where Z is a $N(0,1)$ Gaussian random variable that is independent of X.

Lemma 13.2.1 *The probability density function of z_t is*

$$
q_t(x) = e^{-\alpha x^2} \int_{\mathbf{R}} M_t(x,y) q_0(y)\, dy \tag{13.2.4}
$$

where $M_t(x,y)$ is the Mehler kernel.

Proof. The random variable $e^{-\alpha t}X$ has probability density function $e^{\alpha t}q_0(e^{\alpha t}y)$, and the probability density function of z_t is the convolution of the probability density function of $e^{-\alpha t}X$ with the probability density function of $(1 - e^{-2\alpha t})^{1/2}Z/(2\alpha)^{1/2}$. Hence by some simple manipulations we have

$$q_t(x) = \left(\frac{\alpha}{\pi(1 - e^{-2\alpha t})}\right)^{1/2} \int_{\mathbf{R}} \exp\left(-\frac{\alpha(x - y)^2}{1 - e^{-2\alpha t}}\right) e^{\alpha t}q_0(e^{\alpha t}y)dy$$

$$= \left(\frac{\alpha}{\pi(1 - e^{-2\alpha t})}\right)^{1/2} e^{-\alpha x^2}$$

$$\times \int \exp\left(-\frac{\alpha(e^{-2\alpha t}x^2 - 2e^{-\alpha t}xy + e^{-2\alpha t}y^2)}{1 - e^{-2\alpha t}}\right)q_0(y)\,dy, \quad (13.2.5)$$

where the final expression involves the Mehler kernel. $\square$

We now compare the rate of change of relative entropy with respect to the stationary distribution with the relative information. This is an instance of Theorem 6.2.2.

Proposition 13.2.2 (*Barron [10]*). *Let q_∞ be the probability density function of the $N(0, 1/(2\alpha))$ distribution, and let q_t be the probability density function of z_t. Then*

$$\frac{\partial}{\partial t}\mathrm{Ent}(q_t \mid q_\infty) = \frac{-1}{2}I(q_t \mid q_\infty) \qquad (13.2.6)$$

and $q_t \to q_\infty$ in L^1 as $t \to \infty$.

Proof. By Corollary 13.1.4, we have

$$\frac{\partial}{\partial t}M_t(x, y) = \frac{1}{2}\frac{\partial^2}{\partial x^2}M_t(x, y) - \alpha x\frac{\partial}{\partial x}M_t(x, y) \qquad (13.2.7)$$

and so by an elementary calculation we have

$$\frac{\partial}{\partial t}e^{-\alpha x^2}M_t(x, y) = \left(\frac{1}{2}\frac{\partial^2}{\partial x^2} + \alpha x\frac{\partial}{\partial x} + \alpha\right)e^{-\alpha x^2}M_t(x, y); \quad (13.2.8)$$

hence by Lemma 13.2.1, we deduce that q_t satisfies the forward equation

$$\frac{\partial}{\partial t}q_t(x) = \left(\frac{1}{2}\frac{\partial^2}{\partial x^2} + \alpha x\frac{\partial}{\partial x} + \alpha\right)q_t(x). \qquad (13.2.9)$$

In view of Chapter 6, a more illuminating version of this Fokker–Planck partial differential equation is

$$\frac{\partial}{\partial t}q_t(x) = \frac{1}{2}\frac{\partial}{\partial x}\left(q_t\frac{\partial}{\partial x}\log\frac{q_t}{q_\infty}\right), \qquad (13.2.10)$$

which we recognise as the gradient flow for $2^{-1}\text{Ent}(q_t \mid q_\infty)$. Now we can calculate

$$\frac{\partial}{\partial t} \int q_t(x)dx = 0 \qquad (13.2.11)$$

and

$$\begin{aligned}
\frac{\partial}{\partial t}\text{Ent}(q_t \mid q_\infty) &= \frac{\partial}{\partial t} \int_{\mathbf{R}} q_t(x) \log \frac{q_t(x)}{q_\infty(x)} dx \\
&= \int_{\mathbf{R}} \left(\frac{\partial q_t}{\partial t} \log \frac{q_t(x)}{q_\infty(x)} + \frac{\partial q_t}{\partial t} \right) dx \\
&= \frac{1}{2} \int_{\mathbf{R}} \log \frac{q_t(x)}{q_\infty(x)} \frac{\partial}{\partial x} \left(q_t(x) \frac{\partial}{\partial x} \log \frac{q_t(x)}{q_\infty(x)} \right) dx \\
&= \frac{-1}{2} \int_{\mathbf{R}} q_t(x) \left(\frac{\partial}{\partial x} \log \frac{q_t(x)}{q_\infty(x)} \right)^2 dx,
\end{aligned}$$

so by integration by parts

$$\frac{\partial}{\partial t}\text{Ent}(q_t \mid q_\infty) = \frac{-1}{2} I(q_t \mid q_\infty). \qquad (13.2.12)$$

By the logarithmic Sobolev inequality Corollary 6.3.3, we have

$$\text{Ent}(q_t \mid q_\infty) \leq \frac{1}{4\alpha} I(q_t \mid q_\infty) \qquad (13.2.13)$$

which gives the differential inequality

$$\frac{\partial}{\partial t}\text{Ent}(q_t \mid q_\infty) \leq -2\alpha\text{Ent}(q_t \mid q_\infty), \qquad (13.2.14)$$

hence by Gronwall's lemma

$$\text{Ent}(q_t \mid q_\infty) \leq e^{-2\alpha t}\text{Ent}(q \mid q_\infty). \qquad (13.2.15)$$

Hence q_t converges to q_∞ in entropy, and so the result follows by Csiszár's inequality Proposition 3.1.8. □

13.3 The matrix Ornstein–Uhlenbeck process

- There is a natural extension of the Ornstein–Uhlenbeck process to $n \times n$ symmetric matrix-valued processes.
- The eigenvalues satisfy a singular stochastic differential equation with a simple limiting form as $n \to \infty$.

The results of the previous sections generalize easily to the matrix-valued setting; for illustration, we consider real symmetric matrices. Given a processes Y_t in $M_n^s(\mathbf{R})$ with $\mathbf{E}Y_t = 0$, the auto-covariance is

$$\langle Y_t, Y_s \rangle = \frac{1}{n}\mathbf{E}\operatorname{trace}(Y_t Y_s). \qquad (13.3.1)$$

Definition (*Matrix Brownian motion*). Let $b_{jk}^{(t)}$ for $1 \leq j \leq k \leq n$ be mutually independent copies of Brownian motion, let $b_{jk} = b_{kj}$ and let $[B_t]_{jk} = \frac{1}{\sqrt{2n}}b_{jk}^{(t)}$ for $1 \leq j < k \leq n$, $[B_t]_{jj} = \frac{1}{\sqrt{n}}b_{jj}^{(t)}$, and $[B_t]_{jk} = [B_t]_{kj}$ so that

$$B_t = \frac{1}{n}\sum_{1 \leq j \leq k \leq n} E_{jk}b_{jk}^{(t)} \qquad (13.3.2)$$

with respect to the standard orthonormal basis of $M_n^s(\mathbf{R})$, as in Section 2.2. Note that the normalizations are consistent with those that we used in the definition of the Gaussian orthogonal ensemble; in particular, B_1 has the same distribution as the GOE, and the process has stationary independent increments with

$$\langle B_t - B_s, B_t - B_s \rangle = \frac{1}{2}\left(1 + \frac{1}{n}\right)(t - s). \qquad (13.3.3)$$

When $n = 1$, we have the standard Brownian motion of Section 11.4.

Definition (*Matrix OU process*). Let $z_{jk}^{(t)}$ for $1 \leq j, k \leq n$ be mutually independent copies of the $OU(\alpha)$ process as in the previous section, then let

$$[Z_t]_{jk} = \frac{1}{n}\sum_{1 \leq j \leq k \leq n} z_{jk}^{(t)} E_{jk}. \qquad (13.3.4)$$

Then the process $[Z_t]$ is the matrix Ornstein–Uhlenbeck process.

Proposition 13.3.1 *For $\alpha > 0$, there exists a stationary Gaussian process (Z_t) in $M_n^s(\mathbf{R})$ that satisfies*

$$dZ_t = -\alpha Z_t dt + dB_t \qquad (13.3.5)$$

and is such that

$$\langle Z_t, Z_s \rangle = \frac{n+1}{4n\alpha}e^{-\alpha|t-s|} \qquad (s, t \geq 0). \qquad (13.3.6)$$

Proof. We have by orthogonality

$$\langle Z_t, Z_s \rangle = \frac{1}{n^2} \sum_{1 \leq j \leq k \leq n} \mathbf{E} z_{jk}^{(t)} z_{jk}^{(s)}$$

$$= \frac{n+1}{4n\alpha} e^{-\alpha|t-s|}. \tag{13.3.7}$$

$\square$

Stochastic differential equations for eigenvalues: motivation

The corresponding stochastic differential equation for the eigenvalues is more complicated, see [14]. Suppose that $(\lambda_j^{(t)})_{j=1}^n$ are the eigenvalues of Z_t, in some order. Then formal manipulations show that

$$d\lambda_j^{(t)} = \left(-\alpha\lambda_j^{(t)} + \frac{1}{n} \sum_{k:k \neq j} \frac{1}{\lambda_j^{(t)} - \lambda_k^{(t)}}\right) dt + \frac{1}{\sqrt{n}} db_j^{(t)} \tag{13.3.8}$$

where $b_j^{(t)}$ are mutually independent copies of Brownian motion. See [112].

This stochastic differential equation is difficult to interpret in its own right, since there are singularities where the eigenvalues coincide or cross. Fortunately, Proposition 13.3.1 shows us that there exists a solution to this family of equations, for reasonable interpretations. We are mainly interested in the empirical eigenvalue distribution $\mu_n^{(t)} = \frac{1}{n} \sum_{j=1}^n \delta_{\lambda_j^{(t)}}$ which is clearly independent of the ordering of the eigenvalues, and satisfies

$$\int f(x) \mu_n^{(t)}(dx) = \frac{1}{n} \text{trace} f(Z_t). \tag{13.3.9}$$

We can carry out computations at three different levels, corresponding to the random objects

$$Z_t \in M_n^s(\mathbf{R}) \quad \mapsto \quad \lambda^{(t)} \in \mathbf{R}^n \quad \mapsto \quad \mu_n^{(t)} \in Prob(\mathbf{R}).$$

So by working at the level of matrices, we can hope to avoid the singularities that appear in the stochastic differential equation for the eigenvalues. In many applications, one takes a further expectation so as to have a non-random probability measure that describes the eigenvalue distribution.

Definition (*IDS* [37]). Given an empirical probability measure μ_n, the integrated density of states is that probability measure ρ_n that satisfies

$$\int f(x)\rho_n(dx) = \mathbf{E}\int f(x)\mu_n(dx) \qquad (f \in C_b(\mathbf{R})). \qquad (13.3.10)$$

More generally, given a differentiable scalar potential function $u : \mathbf{R} \to \mathbf{R}$, the formal system of stochastic differential equations

$$d\lambda_j = \left(-u'(\lambda_j) + \frac{1}{n}\sum_{k:k\neq j}\frac{1}{\lambda_k - \lambda_j}\right)dt + \frac{1}{\sqrt{n}}db_t \qquad (13.3.11)$$

with $\lambda_j = \lambda_j^{(t)}$ gives rise to the diffusion equation for the empirical distribution

$$\frac{d}{dt}\mathbf{E}\frac{1}{n}\sum_{j=1}^{n}f(\lambda_j) = \frac{-1}{n}\mathbf{E}\sum_{j=1}^{n}u'(\lambda_j)f'(\lambda_j)$$

$$+\mathbf{E}\frac{1}{2n^2}\sum_{k:k\neq j}\frac{f'(\lambda_j)-f'(\lambda_k)}{\lambda_j-\lambda_k}+\mathbf{E}\frac{1}{2n^2}\sum_{j=1}^{n}f''(\lambda_j),$$

$$(13.3.12)$$

and hence to an equation for the integrated density of states

$$\frac{d}{dt}\int f(x)\rho_n^{(t)}(x)dx = -\int u'(x)f(x)\rho_n^{(t)}(x)dx$$

$$+\frac{1}{2}\iint\frac{f'(x)-f'(y)}{x-y}\rho_n^{(t)}(x)\rho_n^{(t)}(y)dxdy$$

$$+\frac{1}{2n}\int f''(x)\rho_n^{(t)}(x)dx. \qquad (13.3.13)$$

This argument is not rigorous since the standard hypotheses of Itô's formula are violated by the function $1/(x-y)$. We overcome these difficulties by proving the existence of a matrix process such that the eigenvalues of Z_t satisfy equations such as (13.3.11).

13.4 Solutions for matrix stochastic differential equations

- There is a simple existence and uniqueness theorem for stochastic differential equations for symmetric $n \times n$ matrices with similar form to the matrix OU equation.
- The corresponding eigenvalues satisfy a stochastic differential equation which reduces to the Burgers equation as $n \to \infty$.

Let $\|\,.\,\|$ be a matricial norm on $M_n^s(\mathbf{R})$, such as the operator norm or $c^p(n)$, and let $v : (M_n^s(\mathbf{R}), \|\,.\,\|) \to (M_n^s(\mathbf{R}), \|\,.\,\|)$ a Lipschitz function with constant κ. We consider matrix stochastic processes to be elements of the Banach space

$$E = \left\{ Y : \Omega \times [0,\tau] \to M_n^s(\mathbf{R}) \quad \text{measurable} : \int_0^\tau \mathbf{E}\|Y_t\|^q \, dt < \infty \right\} \tag{13.4.1}$$

with the norm

$$\|Y\|_E = \left(\int_0^\tau \mathbf{E}\|Y_t\|^q dt \right)^{1/q} \tag{13.4.2}$$

where $1 < q < \infty$ and $\tau > 0$. In particular, E is a Hilbert space when we choose $q = 2$ and $\|\,.\,\|$ to be the norm of $c^2(n)$.

Theorem 13.4.1 *Suppose that $q^{-1/q}\tau\kappa < 1$. Then for any $X_0 : \Omega \to M_n^s(\mathbf{R})$ such that $\mathbf{E}\|X_0\|^q < C_0$, there exists a unique solution in E for*

$$dX_t = -v(X_t)dt + dB_t,$$
$$X_0 = X_0, \tag{13.4.3}$$

and there exists $C(n,k,v,q,\tau,C_0) < \infty$ such that $\mathbf{E}\|X_t\|^q \leq C(n,k,v,q,\tau,C_0)$.

Proof. We require to solve the integral equation

$$X_t = -\int_0^t v(X_s)\, ds + X_0 + B_t \quad (0 \leq t \leq \tau) \tag{13.4.4}$$

by the iteration scheme $X_t^0 = X_0$ and

$$X_t^{m+1} = X_0 + B_t - \int_0^t v(X_s^m)\, ds. \tag{13.4.5}$$

Now $X_0, B \in E$, and we can introduce $\Phi : E \to E$ by

$$\Phi(Y)_t = X_0 + B_t - \int_0^t v(Y_s)\, ds. \tag{13.4.6}$$

By Banach's fixed point theorem [150], there exists a unique fixed point for Φ, which gives a solution to the stochastic differential equation. Indeed,

$$\Phi(Y)_t - \Phi(Z)_t = -\int_0^t \big(v(Y_s) - v(Z_s)\big)\, ds \tag{13.4.7}$$

and so by Hölder's inequality [71]

$$\|\Phi(Y)_t - \Phi(Z)_t\|^q \le t^{q-1} \int_0^t \|v(Y_s) - v(Z_s)\|^q \, ds \qquad (13.4.8)$$

so

$$\int_0^\tau \|\Phi(Y)_t - \Phi(Z)_t\|^q \, dt \le \frac{\tau^q}{q} \kappa^q \int_0^\tau \|Y_s - Z_s\|^q \, ds \qquad (13.4.9)$$

and hence

$$\|\Phi(Y) - \Phi(Z)\|_E \le q^{-1/q} \tau \kappa \|Y - Z\|_E \qquad (13.4.10)$$

so Φ is uniformly a strict contraction on E. Given the solution of (13.4.4), one can use Gronwall's Lemma to bound $\mathbf{E}\|X_t\|^q$. □

We use the orthonormal basis for $(M_n^s(\mathbf{R}), c^2(n))$ that was introduced in Section 2.1.

Lemma 13.4.2 *For the above process, the drift is*

$$\frac{1}{t}\mathbf{E}(Z_t - Z_0) \to -v(Z_0) \qquad (13.4.11)$$

and the quadratic variation is

$$\frac{1}{t}\mathbf{E}\Big((Z_t - Z_0) \otimes (Z_t - Z_0)\Big) \to \frac{1}{n^2} \sum_{1 \le j \le k \le n} E_{jk} \otimes E_{jk} \qquad (t \to 0+).$$
$$(13.4.12)$$

Proof. We introduce the maximal functions $\zeta_t = \sup_{0 \le u \le t} \|Z_t - Z_0\|_{c^2(n)}$ and $\beta_t = \sup_{0 \le u \le t} \|B_t\|_{c^2(n)}$. Then by the reflection principle for Brownian motion [73], or the submartingale maximal theorem [139], there exists a universal constant such that

$$\mathbf{E}\beta_t^2 \le C\mathbf{E}\|B_t\|_{c^2(n)}^2 \le Ct. \qquad (13.4.13)$$

We write the stochastic integral equation as

$$Z_t - Z_0 = -v(Z_0)t + \int_0^t \big(v(z_0) - v(Z_u)\big) \, du + B_t \qquad (13.4.14)$$

so

$$\zeta_t \le t\|v(z_0)\|_{c^2(n)} + t\kappa\zeta_t + \beta_t; \qquad (13.4.15)$$

and hence

$$\zeta_t^2 \le \frac{2t^2\|v(z_0)\|^2 + 2\beta_t^2}{(1 - t\kappa)^2}, \qquad (13.4.16)$$

hence $\mathbf{E}\zeta_t^2 = O(t)$ as $t \to 0+$, and so $\mathbf{E}(v(z_u) - v(Z_0)) = O(\sqrt{u})$ as $u \to 0 +$. Now we have

$$\frac{1}{t}\mathbf{E}(Z_t - Z_0) = -v(Z_0) + \frac{1}{t}\int_0^t \left(\mathbf{E}v(Z_u) - v(Z_0)\right) du$$

$$\to -v(Z_0) \quad (t \to 0+). \qquad (13.4.17)$$

For the second order terms, the integral equation gives

$$(Z_t - Z_0) \otimes (Z_t - Z_0) = \int_0^t v(Z_u)\, du \otimes \int_0^t v(Z_u)\, du$$

$$- \int_0^t v(Z_u)\, du \otimes B_t$$

$$+ B_t \otimes \int_0^t v(Z_u)\, du + B_t \otimes B_t, \quad (13.4.18)$$

and we split up the mixed terms as

$$- \int_0^t v(Z_u)\, du \otimes B_t = \int_0^t \left(v(Z_0) - v(Z_u)\right) du \otimes B_t - tv(Z_0) \otimes B_t$$

$$(13.4.19)$$

where $\mathbf{E}tv(Z_0) \otimes B_t = 0$ and

$$\mathbf{E}\left(\left\|\int_0^t \left(v(Z_0) - v(Z_u)\right) du\right\|_{c^2(n)} \left\|B_t\right\|_{c^2(n)}\right) \leq t\kappa \mathbf{E}(\zeta_t \beta_t)$$

$$\leq t\kappa \left(\mathbf{E}\zeta_t^2\right)^{1/2}\left(\mathbf{E}\beta_t^2\right)^{1/2}$$

$$= O(t^2). \qquad (13.4.20)$$

Finally we have the main contribution to the quadratic variation

$$\mathbf{E}\frac{1}{t}B_t \otimes B_t = \frac{1}{n^2}\sum_{1 \leq j \leq k \leq n} E_{jk} \otimes E_{jk}. \qquad (13.4.21)$$

$\square$

Theorem 13.4.3 (*Itô's formula*). *Let* $F : M_n^s(\mathbf{R}) \to \mathbf{R}$ *have continuous and bounded derivatives of orders* $0, 1, 2$. *Then*

$$\left(\frac{d}{dt}\right)_{t=0} \mathbf{E}F(Z_t) = -\operatorname{trace}_n\left(\nabla F(Z_0)v(Z_0)\right) + \frac{1}{2n^2}\Delta F(Z_0). \quad (13.4.22)$$

Proof. By the mean value theorem, there exists $\bar{Z}_t$ on the line segment joining Z_0 to Z_t such that

$$F(Z_t) = F(Z_0) + \langle \nabla F(Z_0), (Z_t - Z_0)\rangle_{c^2(n)}$$

$$+ \frac{1}{2}\langle \operatorname{Hess}F(\bar{Z}_t), (Z_t - Z_0) \otimes (Z_t - Z_0)\rangle. \quad (13.4.23)$$

Now $\mathrm{Hess}F(\bar{Z}_t)$ is bounded and by continuity

$$\mathrm{Hess}F(\bar{Z}_t) \to \mathrm{Hess}F(Z_0) \qquad (13.4.24)$$

almost as surely as $t \to 0+$. Hence we have

$$\frac{\mathbf{E}F(Z_t) - \mathbf{E}F(Z_0)}{t} = \frac{1}{t}\langle \nabla F(Z_0), \mathbf{E}(Z_t - Z_0)\rangle_{c^2(n)}$$
$$+ \mathbf{E}\frac{1}{2t}\langle \mathrm{Hess}F(\bar{Z}_t), (Z_t - Z_0) \otimes (Z_t - Z_0)\rangle$$
$$(13.4.25)$$

and the limit of the right-hand side as $t \to 0+$ is

$$-\langle \nabla F(Z_0), v(Z_0)\rangle_{c^2(n)} + \frac{1}{2n^2}\Big\langle \mathrm{Hess}F(Z_0), \sum_{1\leq j\leq k\leq n} E_{jk} \otimes E_{jk}\Big\rangle_{c^2(n)}$$

$$= -\mathrm{trace}_n\big(\nabla F(Z_0)v(Z_0)\big) + \frac{1}{2n^2}\Delta F(Z_0), \qquad (13.4.26)$$

by definition of the operator Δ. $\qquad\qquad\square$

In Section 13.2, we solved the stochastic differential equation $dZ_t = -\alpha Z_t dt + dB_t$ by introducing the matrix OU process; here we perturb the potential $\alpha x^2/2$ by adding a bounded term g, and solve the resulting stochastic differential equation by using Theorem 13.4.1.

Theorem 13.4.4 *Let* $u(x) = \frac{\alpha}{2}x^2 + g(x)$, *where* $g \in C^3(\mathbf{R})$ *has* $g', g''' \in L^2(\mathbf{R})$ *and* $\alpha > 0$. *Then for each* $X_0 : \Omega \to M_n^s(\mathbf{R})$ *such that* $\mathbf{E}\|X_0\|^2 < \infty$, *the stochastic differential equation*

$$dX_t = -\big(\alpha X_t + g'(X_t)\big)\, dt + dB_t \qquad (13.4.27)$$

has a solution X_t *such that* $X_t : \Omega \to M_n^s(\mathbf{R})$ *has* $\mathbf{E}\|X_t\|^2 < \infty$. *Further, there exists* $Z < \infty$ *such that*

$$\nu_n(dX) = Z^{-1}\exp\big(-2n^2\mathrm{trace}_n\, u(X)\big)dX \qquad (13.4.28)$$

is a stationary probability measure on $M_n^s(\mathbf{R})$ *for this flow.*

Proof. By Proposition 2.2.8, the map $X \mapsto u'(X)$ is Lipschitz, so by Theorem 13.4.1 the stochastic differential equation has a unique solution on $[0,\tau]$ for sufficiently small $\tau > 0$, where τ depends only upon u.

We now show that for any smooth and compactly supported function $f : M_n^s(\mathbf{R}) \to \mathbf{R}$, the expression

$$\int \mathbf{E}(f(X_t) \mid X_0)\nu_n(dX_0) \qquad (13.4.29)$$

is constant as t varies over $[0, \infty)$. By Itô's formula

$$\left(\frac{d}{dt}\right)_{t=0} \mathbf{E}\big(f(X_t) \mid X_0\big) = \mathbf{E}\Big(\big(-\mathrm{trace}_n \left(\nabla f(X_0) u'(X_0)\right)$$
$$+ \frac{1}{2n^2} \Delta f(Z_0)\big) \mid X_0\Big) \qquad (13.4.30)$$

and by the divergence theorem

$$\frac{1}{2n^2 Z} \int_{M_n^s(\mathbf{R})} \Delta f(X) \exp\big(-2n^2 \mathrm{trace}_n\, u(X)\big)\, dX$$
$$= \frac{1}{Z} \int_{M_n^s(\mathbf{R})} \mathrm{trace}_n \left(\nabla f(X) u'(X)\right) \exp\big(-2n^2 \mathrm{trace}_n\, u(X)\big)\, dX$$

$$(13.4.31)$$

where Z is the normalizing constant. Hence

$$\frac{d}{dt} \int_{M_n^s(\mathbf{R})} \mathbf{E}\big(f(X_t) \mid X_0\big) \nu(dX_0) \qquad (13.4.32)$$

vanishes at $t = 0$. The coefficients of the stochastic differential equation do not depend upon time, so this suffices to prove stationarity. $\qquad \square$

When $\alpha = 1/2, n = 1$ and $u(x) = x^2/4$, Theorem 13.4.4 reduces to Lemma 13.2.1. For $n > 1$, the eigenvalues interact, giving an extra term in Itô's formula.

Corollary 13.4.5 *Let $f \in C_b^2(\mathbf{R})$ and let $F : M_n^s(\mathbf{R}) \to \mathbf{R}$ be $F(X) =$ $\mathrm{trace}_n f(X)$. Let Z_t be the solution of the stochastic differential equation*

$$dZ_t = -u'(Z_t)dt + dB_t \qquad (13.4.33)$$

with u as in Theorem 13.4.4. Suppose that the eigenvalues of Z_0 are $\lambda_1, \ldots, \lambda_n$ and that they are distinct. Then

$$\left(\frac{d}{dt}\right)_{t=0} \mathbf{E}F(Z_t) = \frac{-1}{n} \sum_{j=1}^{n} u'(\lambda_j) f'(\lambda_j)$$
$$+ \frac{1}{2n^2} \sum_{j \neq k} \frac{f'(\lambda_j) - f'(\lambda_k)}{\lambda_j - \lambda_k} + \frac{1}{2n^2} \sum_{j=1}^{n} f''(\lambda_j).$$

$$(13.4.34)$$

Proof. By Proposition 2.2.4, we have

$$\langle u'(Z_t), \nabla F(Z_t) \rangle_{c^2(n)} = \frac{1}{n} \sum_{j=1}^{n} u'(\lambda_j^{(t)}) f'(\lambda_j^{(t)}). \qquad (13.4.35)$$

We choose an orthonormal basis (ξ_j) consisting of eigenvectors of Z_t with corresponding eigenvalues λ_j, then introduce matrix units $e_{jk} = \xi_j \otimes \xi_k$ and the corresponding orthonormal basis $\{E_{jk} : 1 \le j \le k \le n\}$ of $(M_n^s(\mathbf{R}), c^2(n))$. From the Rayleigh–Schrödinger formula we obtain

$$\langle \text{Hess } F, E_{jj} \otimes E_{jj} \rangle = f''(\lambda_j) \qquad (j = 1, \dots, n), \qquad (13.4.36)$$

$$\langle \text{Hess } F, E_{jk} \otimes E_{jk} \rangle = 2\frac{f'(\lambda_j) - f'(\lambda_k)}{\lambda_j - \lambda_k} \qquad (1 \le j < k \le n), \qquad (13.4.37)$$

since $\langle E_{jk}\xi_j, \xi_k \rangle = \langle E_{jk}\xi_k, \xi_j \rangle = \sqrt{n/2}$. Hence

$$\left\langle \text{Hess } F, \sum_{1 \le j \le k \le n} E_{jk} \otimes E_{jk} \right\rangle = \sum_{j=1}^{n} f''(\lambda_j) + 2 \sum_{1 \le j < k \le n} \frac{f'(\lambda_j) - f'(\lambda_k)}{\lambda_j - \lambda_k}.$$
$$(13.4.38)$$

When we substitute these terms into Theorem 13.4.3, we obtain the stated result. □

Biane and Speicher [14, 15] showed that the stochastic differential equation for the empirical distribution leads to a quasi differential equation for the integrated density of states. In the special case of $u(x) = x^2/2$, this is the complex Burgers equation.

Proposition 13.4.6 *Let Z_t be as in Corollary 13.4.5, and let $\mu_n^{(t)}$ be the empirical distribution of its eigenvalues. Suppose that $(\rho_t)_{t>0}$ are probability density functions on $\mathbf{R}$ such that $\mu_n^{(t)}$ converges weakly to $\rho_t(x)dx$ in probability as $n \to \infty$. Then $(\rho_t)_{t>0}$ gives a weak solution of*

$$\frac{\partial \rho_t}{\partial t} = \frac{\partial}{\partial x}\Big(\rho_t(x)\big(u'(x) - \pi \mathcal{H}\rho_t(x)\big)\Big), \qquad (13.4.39)$$

namely the gradient flow for the functional

$$F(\rho) = \int_{\mathbf{R}} u(x)\rho(x)\,dx + \frac{1}{2}\iint_{\mathbf{R}^2} \log\frac{1}{|x-y|}\rho(x)\rho(y)\,dx dy. \qquad (13.4.40)$$

Proof. Note that $\mu_n^{(t)} \otimes \mu_n^{(t)}$ converges weakly to $\rho_t \otimes \rho_t$ in probability as $n \to \infty$, and that

$$\mathbf{E}\,\text{trace}_n f(Z_t) \to \int_{\mathbf{R}} f(x)\rho_t(x)\,dx \qquad (n \to \infty) \qquad (13.4.41)$$

for all $f \in C_b(\mathbf{R}; \mathbf{R})$. Suppose that $f \in C^1(\mathbf{R})$ has compact support; then $(f(x) - f(y))/(x - y)$ extends to define a bounded and continuous function on $\mathbf{R}^2$, and $u'(x)f'(x)$ is a bounded and continuous function. The limit of the equation

$$\frac{d}{dt}\mathbf{E}\frac{1}{n}\sum_{j=1}^{n}f(\lambda_j^{(t)}) = -\mathbf{E}\frac{1}{n}\sum_{j=1}^{n}u'(\lambda_j^{(t)})f'(\lambda_j^{(t)})$$

$$+\mathbf{E}\frac{1}{2n^2}\sum_{j,k:j\neq k}\frac{f'(\lambda_j^{(t)}) - f'(\lambda_k^{(t)})}{\lambda_j^{(t)} - \lambda_k^{(t)}}$$

$$+\mathbf{E}\frac{1}{2n^2}\sum_{j=1}^{n}f''(\lambda_j^{(t)}) \tag{13.4.42}$$

as $n \to \infty$ is

$$\frac{\partial}{\partial t}\int f(x)\rho_t(x)\,dx = -\int f'(x)u'(x)\rho_t(x)\,dx$$

$$+\frac{1}{2}\iint \frac{f'(x) - f'(y)}{x - y}\rho_t(x)\rho_t(y)\,dxdy$$

$$= -\int f'(x)\Big(u'(x) - \text{p.v.}\int\frac{\rho_t(y)dy}{x - y}\Big)\rho_t(x)\,dx$$

$$= \int f(x)\frac{\partial}{\partial x}\Big(\rho_t(x)\big(u'(x) - \pi\mathcal{H}\rho_t(x)\big)\Big)dx, \tag{13.4.43}$$

so ρ_t is a weak solution of the stated equation. Further, the weak derivative of F is

$$\frac{\delta F}{\delta\rho} = u(x) - \int \log|x - y|\rho(y)\,dy \tag{13.4.44}$$

so the gradient flow is

$$\frac{\partial\rho_t}{\partial t} = \frac{\partial}{\partial x}\Big(\rho_t(x)\big(u'(x) - \pi\mathcal{H}\rho_t(x)\big)\Big). \tag{13.4.45}$$

Recall that if u is a potential that has equilibrium density ρ, then $u' = \pi\mathcal{H}\rho$ by Theorem 4.4.1. □

13.5 The Burgers equation

The complex Burgers equation arises:

- from the free Fokker–Planck equation;
- as a limiting case of the Riccati equation for heat flow.
- The complex Burgers equation plays a rôle analogous to the Ornstein–Uhlenbeck equation.

In Section 6.5 we saw how $\rho = \Im G$ satisfies the gradient flow for free entropy when G satisfies the complex Burgers equation. In this section we consider the other interpretations.

We previously considered the score function ϕ'/ϕ of a probability density function, and now we consider how this evolves under the heat equation. Let $f(x, t)$ be a positive solution to

$$\frac{\partial f}{\partial t} = \frac{\partial^2 f}{\partial x^2} \tag{13.5.1}$$

and observe that $u(x, t) = -\frac{\partial f}{\partial x}/f$ satisfies the Riccati equation

$$\frac{\partial u}{\partial t} + 2u\frac{\partial u}{\partial x} - \frac{\partial^2 u}{\partial x^2} = 0. \tag{13.5.2}$$

By rescaling u and x, we can write this as

$$\frac{\partial w}{\partial t} + 2w\frac{\partial w}{\partial x} - \varepsilon^2\frac{\partial^2 w}{\partial x^2} = 0, \tag{13.5.3}$$

for some scaling parameter $\varepsilon > 0$. Thus the diffusive term can be made to appear as a perturbation to the Burgers equation

$$\frac{\partial w}{\partial t} + 2w\frac{\partial w}{\partial x} = 0. \tag{13.5.4}$$

We now present some solutions of the Burgers equation involving the distributions of Section 4.5. In [166, 83] there is a discussion of how μ_t arises in free probability.

Proposition 13.5.1 *Let μ_t be the probability measure that has $S(0, \sqrt{t})$ distribution, and let*

$$G(z, t) = \frac{2}{z + \sqrt{z^2 - t}} \qquad (t > 0, z \in \mathbf{C} \setminus [-\sqrt{t}, \sqrt{t}]) \tag{13.5.5}$$

be its Cauchy transform. Then G satisfies the complex Burgers equation

$$\frac{\partial G}{\partial t} + \frac{G}{4}\frac{\partial G}{\partial z} = 0. \tag{13.5.6}$$

Proof. The formula for $G(z, t)$ was computed in Corollary 4.5.7. One can check by direct calculation that G satisfies the differential equation. $\square$

Proposition 13.5.2 *Let* $z = e^w$ *and* $\lambda = 1 - e^{-\tau}$ *in the Cauchy transform*

$$G_\lambda(z) = \frac{z + 1 - \lambda - \sqrt{(z - 1 - \lambda)^2 - 4\lambda}}{2z} \qquad (13.5.7)$$

of the Marchenko–Pastur distribution as in (5.5.1). Then G satisfies the equation

$$\frac{\partial G}{\partial \tau} + \frac{G}{1 - G} \frac{\partial G}{\partial w} = 0. \qquad (13.5.8)$$

Proof. This follows by direct calculation. $\square$

14

Noncommutative probability spaces

Abstract

In the axioms of Chapter 1, the basic object in classical probability theory is a probability measure on a compact metric space Ω, or equivalently a positive normalized linear functional on $C(\Omega; \mathbf{C})$. In noncommutative probability, the basic object is a state on a C^*-algebra which is generally noncommutative, but which can always be realized as a norm-closed subalgebra of $B(H)$, where H is a Hilbert space. Of particular interest are tracial probability spaces, which include the type II_1 von Neumann factors. In Section 14.3, we consider an important example of a noncommutative distribution, namely the semicircular distribution, which arises naturally in random matrix theory.

14.1 Noncommutative probability spaces

Noncommutative probability spaces provide a unified framework in which we can describe:

- classical probability spaces;
- states on C^* algebras;
- random matrices;
- free probability theory.

In this section we state the axioms and then look at some basic examples.

Definition (*-probability spaces [166, 16]). (A, φ) is a *-probability space when A is a unital $*$ algebra over $\mathbf{C}$ and $\varphi : A \to \mathbf{C}$ is a state; so that

(i) $a \mapsto a^*$ is conjugate linear;
(ii) $(ab)^* = b^* a^*$;

411

(iii) $a^{**} = a$;

(iv) $\varphi(1) = 1$, $\varphi(sa + tb) = s\varphi(a) + t\varphi(b)$, $\varphi(a^*) = \overline{\varphi(a)}$, $\varphi(a^*a) \geq 0$.

Further, when $\| \cdot \|$ is a norm on A such that A is a Banach space, we often impose the conditions:

(v) $\|ab\| \leq \|a\| \|b\|$, and $\|a^*\| = \|a\|$;

(vi) $\|a^*a\| = \|a\|^2$, the C^* norm condition.

When conditions (i)–(vi) all hold, we have a C^* probability space. (In Section 14.2, we shall introduce further axioms (vii), (viii) and (ix) to define a tracial probability space.)

Example 14.1.1 (*Classical probability spaces*). For a Polish space Ω, let A be $C_b(\Omega; \mathbf{C})$ with the supremum norm and pointwise multiplication and let $\mu \in Prob(\Omega)$. Then (A, ϕ) becomes a C^* probability space with $\phi(f) = \int f d\mu$. The algebra here is commutative, so classical examples of probability spaces can be regarded as commutative C^* probability spaces.

Example 14.1.2 (*Matrices*). Let $A = M_n(\mathbf{C})$ and let $\xi \in \mathbf{C}^n$ have $\|\xi\| = 1$. Then (A, ϕ) is a C^* probability space with $\phi(X) = \langle X\xi, \xi \rangle$. Likewise, (A, τ) is a C^* probability space with $\tau(X) = \text{trace}(X)/n$.

Example 14.1.3 (*Discrete group algebras*). Let G be a discrete countable group with identity element e, and for each $g \in g$, let $\delta_g : G \to \mathbf{C}$ be $\delta_g(h) = 1$ for $h = g$ and $\delta_g(h) = 0$ for $h \in G \setminus \{g\}$. Then we form the Hilbert space

$$\ell^2(G) = \left\{ \sum_{g \in G} \alpha_g \delta_g : \sum_{g \in G} |\alpha_g|^2 < \infty \right\} \tag{14.1.1}$$

with the inner product

$$\left\langle \sum_{g \in G} \alpha_g \delta_g, \sum_{h \in G} \beta_h \delta_h \right\rangle = \sum_{g \in G} \alpha_g \bar{\beta}_g. \tag{14.1.2}$$

Similarly we form

$$\ell^1(G) = \left\{ \alpha = \sum_{g \in G} \alpha_g \delta_g : \|\alpha\|_{\ell^1} = \sum_{g \in G} |\alpha_g| < \infty \right\} \tag{14.1.3}$$

which forms a unital $*$-algebra for the convolution product

$$\alpha * \beta = \sum_{k \in G} (\alpha * \beta)_k \delta_k = \sum_{g, h \in G} \alpha_g \beta_h \delta_{gh}, \tag{14.1.4}$$

for the involution

$$\alpha^* = \left(\sum_{g \in G} \alpha_g \delta_g\right)^* = \sum_{g \in G} \bar{\alpha}_g \delta_{g^{-1}}. \qquad (14.1.5)$$

Now each $h \in G$ gives a unitary operator λ_h on $\ell^2(G)$ by $\lambda_h :$ $\sum_{g \in G} \alpha_g \delta_g \mapsto \sum_{g \in G} \alpha_g \delta_{hg}$, and the representation $h \mapsto \lambda_h$ extends to a $*$ representation

$$\lambda : \ell^1(G) \to B(\ell^2(G)) : \sum_{g \in G} \alpha_g \delta_g \mapsto \sum_{g \in G} \alpha_g \lambda_g. \qquad (14.1.6)$$

Example 14.1.4 (*Free algebras*).

(1°) Let E be the free unital $*$algebra over $\mathbf{C}$ that is generated by I, ℓ, ℓ^*, subject to $\ell^* \ell = I$. Then

$$E = \mathrm{span}\{\ell^m (\ell^*)^n : m, n = 0, 1, 2, \dots\} \qquad (14.1.7)$$

and the listed terms give a basis. To check that E is closed under multiplication, we consider the product of $\ell^k (\ell^*)^j$ and $\ell^m (\ell^*)^n$ in two cases: if $j > m$, then $\ell^k (\ell^*)^j \ell^m (\ell^*)^n = \ell^k (\ell^*)^{n+j-m}$; whereas if $j \leq m$, then $\ell^k (\ell^*)^j \ell^m (\ell^*)^n = \ell^{k+m-j} (\ell^*)^n$. Furthermore, $\varphi : E \to \mathbf{C}$, as defined by

$$\varphi : \sum_{m,n} a_{mn} \ell^m (\ell^*)^n \mapsto a_{00}, \qquad (14.1.8)$$

gives a positive functional. Indeed, we have

$$\varphi\left(\left(\sum_{m,n} a_{mn} \ell^m (\ell^*)^n\right)^* \left(\sum_{m,n} a_{mn} \ell^m (\ell^*)^n\right)\right) = \sum_m |a_{m0}|^2$$
$$(14.1.9)$$

since $\varphi(\ell^a (\ell^*)^b \ell^c (\ell^*)^d) = 0$ unless $a = d = 0$ and $b = c$. Hence (E, φ) is a noncommutative probability space.

(2°) More generally, we let E_K be the free unital algebra that is generated by

$$I, \ell_1, \dots, \ell_K, \ell_1^*, \dots, \ell_K^*, \qquad (14.1.10)$$

subject to $\ell_j^* \ell_k = \delta_{jk} I$; then we let $\varphi : E_K \to \mathbf{C}$ be the linear functional

$$\varphi : \sum_{\alpha(1)\dots\alpha(m);\beta(1)\dots\beta(n)} w_{\alpha(1)\dots\alpha(m);\beta(1)\dots\beta(n)} \ell_{\alpha(1)} \cdots \ell_{\alpha(m)} \ell_{\beta(1)}^* \cdots \ell_{\beta(n)}^*$$
$$\mapsto w_{\emptyset\emptyset}. \qquad (14.1.11)$$

Let H be a separable Hilbert space with orthonormal basis $(e_j)_{j=1}^\infty$, and let $\exp(H)$ be the full Fock space as in Section 12.1 and let $\ell(e_j)$ be the left creation operator associated with e_j as in Example 12.1.4.

Proposition 14.1.4 *There exists a unique $*$-algebra homomorphism*

$$\Phi : E_K \to \mathrm{alg}\{I, \ell(e_j), \ell(e_k)^*; j, k = 1, \dots, K\}, \qquad (14.1.12)$$

such that $\Phi(1) = I$ *and* $\Phi(\ell_j) = \ell(e_j)$.

Proof. The creation operators satisfy

$$\ell(e_j)^* \ell(e_k) = \delta_{jk} I \qquad (j, k = 1, \dots, K), \qquad (14.1.13)$$

so the existence and uniqueness of Φ follow from the universal properties of free algebras. $\square$

14.2 Tracial probability spaces

- Finite factors give the main examples of tracial probability spaces.
- There exists a unique hyperfinite II_1 factor on H.
- The factors constructed from the left-regular representations of free discrete groups are generally not hyperfinite.

Let H be separable, infinite-dimensional complex Hilbert space and let $B(H)$ be the space of bounded linear operators on H with the operator norm. The natural $*$ operation on a bounded linear operator on Hilbert space is the adjoint $x \mapsto x^*$. Each $\xi \in H$ such that $\|\xi\|_H = 1$ gives rise to a state $x \mapsto \langle x\xi, \xi \rangle$ on $B(H)$. See [60, 120].

Theorem 14.2.1 (*Gelfand–Naimark*). *Let A be a unital C^*-algebra that satisfies (i), (ii), (iii), (v) and (vi) of Section 14.1. Then A is $*$-isomorphic to an algebra of bounded operators on Hilbert space.*

With A safely embedded as a unital $*$ subalgebra of $B(H)$, we can formulate some definitions which we shall use in later sections.

Definition (*Positivity*). Let A be a unital C^* algebra. A self-adjoint element $a \in A$ is positive if any one of the three equivalent conditions are satisfied:

(1) $a = b^* b$ for some $b \in A$;
(2) $\langle a\xi, \xi \rangle_H \geq 0$ for all $\xi \in H$, where A is a C^*-subalgebra of $B(H)$;
(3) $\lambda 1 + a$ is invertible in A for all $\lambda > 0$.

Definition (*Von Neumann Algebra*). Suppose that A is a unital C^* subalgebra of $B(H)$ such that A is closed for the weak operator topology. Then A is a von Neumann algebra.

In a von Neumann algebra, one can form the least upper bound of a bounded family of positive elements. Let $A^+ = \{a \in A : \langle a\xi, \xi \rangle \geq 0 \ \forall \xi \in H\}$. Let $a_j \in A^+$ be a net indexed by a partially ordered set J such that $\|a_j\| \leq K$ for all $j \in J$ and $j \leq k$ implies $a_j \leq a_k$ as operators on H. Then there exists a least upper bound $a_\infty \in A^+$ such that $a_j \to a_\infty$ weakly as j increases through J.

Definition (*Normal*). Let A and B be von Neumann algebras, and $\varphi : A \to B$ be a linear map such that $\varphi(A^+) \subseteq B^+$, so φ preserves positivity. Say that φ is normal if $\varphi(a_\infty)$ is the least upper bound of $\varphi(a_j)$.

Let $S(A) = \{\varphi \in A^* : \varphi \ \text{ satisfies } \ (iv)\}$ be the set of states on A; then $S(A)$ is a convex subset of the dual space A^*, and $S(A)$ is compact for the $\sigma(A^*, A)$ topology.

On a unital C^* algebra there is an abundance of states, and

$$\|x\|^2 = \sup\{\phi(x^*x) : \phi \in S(A)\} \qquad (x \in A). \qquad (14.2.1)$$

Theorem 14.2.2 (*Gelfand, Naimark, Segal*). *Let (A, φ) be a C^* probability space, so (i)–(vi) all hold. Then there exists a Hilbert space H and a $*$ representation π of A on H, such that*

$$\varphi(a) = \langle \pi(a)v, v \rangle_H \qquad (a \in A) \qquad (14.2.2)$$

for some $v \in H$. Further, if A is separable, then H may be chosen to be separable.

Proof. The crucial observation is that for any $a \in A$, there exists $c \in A$ such that $\|a\|^2 I - a^*a = c^*c$; so $\|a\|^2 \geq \varphi(a^*a)$ and $b^*b\|a\|^2 - b^*a^*ab = b^*c^*cb$ for all $b \in A$. The state φ gives an inner product on A by $\langle a, b \rangle = \varphi(b^*a)$; so we let H be the completion of A for the associated norm, let $j : A \to H$ the natural inclusion map. Then we let $v = j(1) \in H$ and define $\pi(a) : b \mapsto ab$. Now since $j(A)$ is a dense linear subspace of H, we have

$$\|\pi(a)\|^2 = \sup\{\phi(b^*a^*ab) : \phi(b^*b) \leq 1\}$$
$$\leq \|a^*a\| = \|a\|^2, \qquad (14.2.3)$$

so $\pi(a)$ extends to a bounded linear operator on H and $\pi : A \to B(H)$ is a bounded $*$-representation. Further, $\varphi(a) = \varphi(1a1) = \langle \pi(a)v, v \rangle$. $\square$

Definition (*Tracial probability space* [166, 16]). Let (A, τ) be a C^* probability space such that (i)–(vi) all hold, together with

(vii) $\tau(ab) = \tau(ba)$ for all $a, b \in A$, so τ is a tracial state;
(viii) $\tau(a^*a) = 0 \Rightarrow a = 0$, so τ is faithful.

Then (A, τ) is a tracial C^* probability space. When A is a von Neumann algebra, we often impose the further condition:

(ix) τ is normal.

Some authors use the general term 'noncommutative probability space' for the tracial probability spaces that satisfy (i)–(ix).

Now we exhibit examples of tracial probability spaces (M, τ), where M is a finite von Neumann factor with its unique normal tracial state τ. Our purpose is to express some statements in the language of noncommutative probability; the proofs are omitted, as they are clearly presented in standard books on von Neumann algebra theory such as [60, 120]. We give examples of tracial probability spaces that can be approximated by finite matrix models. We recall some basic definitions.

Definition (*Finite factors*). Let M be a von Neumann algebra. Then M is finite if there exists a normal, faithful and tracial state $\tau : M \to \mathbf{C}$. Further, M is a factor if its centre consists of scalar operators that is,

$$\{b \in M : ba = ab \quad \text{for all} \quad a \in M\} = \{zI : z \in \mathbf{C}\}. \quad (14.2.4)$$

Proposition 14.2.4 *A finite factor M has a unique normal tracial state τ, hence (M, τ) satisfies (i)–(ix).*

Example 14.2.5 Let M be a finite-dimensional factor. Then M is $*$-isomorphic to $M_n(\mathbf{C})$ for some $n < \infty$, and the unique tracial state is τ_n where

$$\tau_n(a) = \frac{1}{n} \sum_{j=1}^n \langle ae_j, e_j \rangle \qquad (a \in M) \quad (14.2.5)$$

where (e_j) is some orthonormal basis of $\mathbf{C}^n$.

Let (M, τ) be a von Neumann factor and consider the set D of dimensions of orthogonal projections

$$D = \{\tau(p) : p \in M; \ p^*p = p, \ p^2 = p\}. \quad (14.2.6)$$

Then either D is discrete, and (M, τ) is $*$ isomorphic to $M_n(\mathbf{C})$ with the standard trace for some $n \in \mathbf{N}$; or D is continuous, and M is said to be a II_1 factor.

Definition (*Hyperfinite*). A factor (R, τ) is called hyperfinite if there exists a sequence of factors (M_k) $(k = 1, 2, \dots)$ such that:

(i) M_k is *isomorphic to $M_{n_k}(\mathbf{C})$ for some $n_k < \infty$;
(ii) $M_1 \subseteq M_2 \subseteq \cdots \subseteq R$;
(iii) $\cup_{k=1}^{\infty} M_k$ is dense in R for the strong operator topology.

Theorem 14.2.6 (*Murray–von Neumann [118, Theorem XIV]*). *On H, there exists a unique hyperfinite II_1 factor R.*

By considering discrete groups, we can construct finite factors which are hyperfinite, and some which are not hyperfinite.

Definition (*Group von Neumann algebra [60]*). Let $\lambda(G)$ be the closure of $\lambda(\ell^1(G))$ in $B(\ell^2(G))$ for the weak operator topology. Let $\tau : \lambda(G) \to \mathbf{C}$ be $\tau(a) = \langle a\delta_e, \delta_e \rangle_{\ell^2(G)}$.

Proposition 14.2.7 *The left regular representation of the group S_∞ generates the hyperfinite II_1 factor.*

Proof. Here τ gives the tracial state; the tracial property reduces to the fact that $gh = e$ if and only if $hg = e$. For each $g \in S(\infty) \setminus \{e\}$, the conjugacy class $\{hgh^{-1} : h \in S(\infty)\}$ is infinite by an obvious extension of the results of Section 7.2. Hence the left regular representation of S_∞ is factorial by [60].

Since S_∞ is the union of an increasing sequence of finite subgroups S_n, it is easy to verify that the representation is hyperfinite. See [118, Lemma 5.2.2]. □

Let $g_1, \dots, g_n$ be free generators of $G = \mathbf{F}_n$, so that G has distinct elements e and $g_{\beta(1)}^{\alpha(1)} g_{\beta(2)}^{\alpha(2)} \cdots g_{\beta(m)}^{\alpha(m)}$ with $\alpha(j) \in \mathbf{Z} \setminus \{0\}$ and $\beta(k) \in \{1, \dots, n\}$ such that $\beta(1) \neq \beta(2), \beta(2) \neq \beta(3), \dots, \beta(m-1) \neq \beta(m)$.

Proposition 14.2.8 (*Nonhyperfinite factors*). *For $n > 1$ and $G = \mathbf{F}_n$, the von Neumann algebra $\lambda(G)$ is a finite factor with canonical trace τ which is not hyperfinite.*

Proof. See [118, Theorem XVI]. □

Open Problem Which of the factors $\lambda(\mathbf{F}_n)$ with $n \in \{2, 3, \dots\} \cup \{\infty\}$ are isomorphic as von Neumann algebras?

This question is suggested by the Appendix to [118] and motivated much of free probability theory. Kadison drew attention to the problem in his Baton Rouge problem list, and so the question is sometimes attributed there.

14.3 The semicircular distribution

• Wigner's semicircular distribution plays a rôle in free probability analogous to that of the Gaussian distribution in classical probability.

Definition (*Distribution* [166]). Let a be an element of a $*$ probability space (A, ϕ). Then the distribution of a consists of the sequence $m_n = \phi(a^n)$ of moments of a.

Proposition 14.3.1 *Let (A, ϕ) be a C^* probability space. Given $a \in A$ such that $a = a^*$, there exists a unique $\mu \in Prob[-\|a\|, \|a\|]$ such that*

$$\phi(a^n) = \int_{[-\|a\|, \|a\|]} x^n \, \mu(dx) \qquad (n = 0, 1, 2, \dots). \qquad (14.3.1)$$

The Cauchy transform of μ satisfies $G(z) = \phi((zI - a)^{-1})$ for $z \in \mathbf{C} \setminus [-\|a\|, \|a\|]$.

Proof. The commutative case of the Gelfand–Naimark theorem, which is relatively easy to prove [120], leads to the conclusion that the unital C^* subalgebra of A generated by a is a subalgebra of $B(H)$ for some Hilbert space H. The spectrum of a as an element of $B(H)$ is contained in $[-\|a\|, \|a\|]$, and for any $f \in C[-\|a\|, \|a\|]$ we can form $f(a)$ by functional calculus. Further, $f \mapsto \phi(f(a))$ gives a state on $C[-\|a\|, \|a\|]$, so is given by a unique measure by Theorem 1.1.1. By Proposition 1.8.6, the moments determine μ uniquely. In particular, when we take $z \in \mathbf{C} \setminus [-\|a\|, \|a\|]$ and introduce the continuous function $f(x) = (z - x)^{-1}$, we have the Cauchy transform of μ, namely $\phi((zI - a)^{-1}) = \int (z - x)^{-1} \mu(dx)$. $\qquad \square$

Definition (*Semicircular distribution*). Let X be a self-adjoint element of a $*$probability space (A, φ). We say that $X \in A$ has a semicircular $S(a, r)$ distribution when

$$\phi(X^n) = \frac{2}{\pi r^2} \int_{a-r}^{a+r} x^n \sqrt{r^2 - (x - a)^2} \, dx. \qquad (14.3.2)$$

We show how this distribution arises in three apparently diverse contexts.

Definition (*Finite shift*). For each N, let $\ell_N^2 = \mathbf{C}^N$ with the usual inner product and orthonormal basis $(e_j)_{j=1}^N$. Let

$$S_N : \sum_{j=1}^N a_j e_j \to \sum_{j=1}^{N-1} a_j e_{j+1}, \qquad (14.3.3)$$

be the finite shift which has adjoint

$$S_N^* : \sum_{j=1}^{N} a_j e_j \to \sum_{j=2}^{N} a_j e_{j-1}; \qquad (14.3.4)$$

so $S_N^N = 0 = (S_N^*)^N$.

Proposition 14.3.2 *The spectral measure of $S_N^* + S_N$ consists of the average of N unit point masses on $[-2, 2]$, namely*

$$\frac{1}{N} \sum_{k=1}^{N} \delta_{2\cos(k\pi/(N+1))}, \qquad (14.3.5)$$

and converges weakly to the Chebyshev distribution on $[-2, 2]$ as $N \to \infty$, while

$$\langle (S_N^* + S_N)^n e_1, e_1 \rangle \to \frac{1}{2\pi} \int_{-2}^{2} x^n \sqrt{4 - x^2} \, dx \qquad (N \to \infty). \quad (14.3.6)$$

Proof. See [56]. The operator $S_N + S_N^*$ is represented by the tridiagonal $N \times N$ matrix

$$\begin{bmatrix} 0 & 1 & 0 & 0 & \cdots & 0 \\ 1 & 0 & 1 & 0 & \cdots & 0 \\ 0 & 1 & 0 & 1 & \cdots & 0 \\ 0 & 0 & 1 & 0 & \cdots & 0 \\ \vdots & \vdots & \vdots & \vdots & \ddots & 1 \\ 0 & 0 & 0 & 0 & 1 & 0 \end{bmatrix} \qquad (14.3.7)$$

and the eigenvalues are

$$2\cos\frac{k\pi}{N+1} \qquad (k = 1, 2, \ldots, N) \qquad (14.3.8)$$

with corresponding unit eigenvectors

$$v_k = \sqrt{\frac{2}{N}} \text{col}\left[\sin\frac{jk\pi}{N+1}\right]_{j=1}^{N}. \qquad (14.3.9)$$

To see this, we use the relation

$$\sin(j-1)\theta + \sin(j+1)\theta = 2\sin j\theta \cos\theta \qquad (14.3.10)$$

and observe that $\sin(N+1)\theta = 0$ provided $\theta = k\pi/(N+1)$ for some $k = 1, \ldots, N$.

By the standard functional calculus for real symmetric matrices, we can form $f(S_N + S_N^*)$ for any continuous function $f : [-2, 2] \to \mathbf{R}$. We have

$$\frac{1}{N}\text{trace}f(S_N + S_N^*) \to \int_0^1 f(2\cos\pi\theta)\, d\theta$$

$$= \frac{1}{\pi}\int_{-2}^2 \frac{f(x)\, dx}{\sqrt{4 - x^2}} \qquad (N \to \infty). \quad (14.3.11)$$

In terms of the orthonormal basis of eigenvectors, we have

$$e_1 = \sqrt{\frac{2}{N}}\sum_{k=1}^N \sin\frac{\pi k}{N+1}v_k$$

so that

$$\langle (S_N^* + S_N)^n e_1, e_1 \rangle = \frac{2^{n+1}}{N}\sum_{k=1}^N \cos^n\frac{\pi k}{N+1}\sin^2\frac{\pi k}{N+1}$$

$$\to 2^{n+1}\int_0^1 \cos^n\pi x \sin^2\pi x\, dx \quad (14.3.12)$$

as $N \to \infty$. For odd n, this gives zero; whereas for even n we have with $t = 2\cos\pi x$ the semicircle law

$$\langle (S_N^* + S_N)^n e_1, e_1 \rangle \to \frac{1}{2\pi}\int_{-2}^2 t^n\sqrt{4 - t^2}\, dt \qquad (N \to \infty). \quad (14.3.13)$$

$\square$

We now exhibit an element of a C^* probability space that has such a distribution.

Theorem 14.3.3 *Let $\exp(\mathbf{C})$ be the full Fock space on $\mathbf{C}$ as in (12.1.6), and let $S = (\ell + \ell^*)/2$, where ℓ is the left creation operator. Then the distribution of S is the $S(0,1)$ law, and the von Neumann algebra generated by S is canonically isomorphic to $L^\infty([-1,1], \sigma_{0,1})$ via the map $S \mapsto x$.*

Proof. To prove Theorem 14.3.3, we digress into the theory of random walks, where we recover the Cauchy transform $G(z)$ of the semicircular distribution from a suitable probability generating function. See [73]. Suppose that $X_1, X_2, \ldots$ are mutually independent random variables, each distributed as X, where $\mathbf{P}[X = 1] = \mathbf{P}[X = -1] = 1/2$. Then $S_m = X_1 + \cdots + X_m$ gives the position after m steps of the random walk on the integers that starts from zero and has j^{th} step X_j either one to the right or one to the left. Given S_m, the future progress of the walk does not depend upon the path up until S_m, so the walk has no memory of its previous history; this is Markov's property. Let $p(m) = \mathbf{P}[S_m = 0]$ be the probability that the walk is at zero after m steps; let B_m be the event that the walk returns to zero for the first time at step m, with $f(m) = \mathbf{P}(B_m)$. $\qquad\square$

Lemma 14.3.4 *The corresponding generating functions*

$$P(z) = \sum_{m=0}^{\infty} p(m)z^m$$

and

$$F(z) = \sum_{m=1}^{\infty} f(m)z^m \qquad (|z| \le 1) \tag{14.3.14}$$

satisfy (1) $P(z) = (1 - z^2)^{-1/2}$ and (2) $F(z) = 1 - (1 - z^2)^{1/2}$.

Proof. (1) Clearly the walk cannot be at the origin after an odd number of steps, so only even powers appear in the probability generating functions. After $2n$ steps, the walk is at the origin if and only if the walk has taken n steps to the left and n steps to the right in some order, so by definition of the binomial coefficients

$$p(2n) = \frac{1}{2^n}\binom{2n}{n}. \tag{14.3.15}$$

Hence

$$P(z) = \sum_{n=0}^{\infty}\binom{2n}{n}\frac{z^{2n}}{2^n} = (1 - z^2)^{-1/2}. \tag{14.3.16}$$

(2) The event $[S_{2n} = 0]$ is partitioned into disjoint events $[S_{2n} = 0] \cap B_{2k}$ for $k = 1, \ldots, n$, so

$$\mathbf{P}[S_{2n} = 0] = \sum_{k=1}^{n} \mathbf{P}[S_{2n} = 0 \mid B_{2k}]\mathbf{P}(B_{2k}), \tag{14.3.17}$$

where $\mathbf{P}[S_{2n} = 0 \mid B_{2k}] = \mathbf{P}[S_{2n-2k} = 0]$ since the walk has no memory and the steps are homogeneous in time. Hence p satisfies the recurrence relation

$$p(2n) = \sum_{k=1}^{n} p(2n - 2k)f(2k) \qquad (n = 1, 2, \ldots) \tag{14.3.18}$$

so by multiplying by z^{2n} and summing over n, we obtain $P(z) = 1 + P(z)F(z)$, hence the result by (1). $\qquad\square$

Lemma 14.3.5 (*Catalan numbers*). *Let E_{2k} be the event that the random walk is nonnegative up until time $2k$ and is zero at time $2k$; namely*

$$E_{2k} = [X_1 + \cdots + X_j \geq 0 \quad (j = 1, \ldots, 2k) \quad and \quad X_1 + \cdots + X_{2k} = 0].$$
$$(14.3.19)$$

Then

$$\mathbf{P}(E_{2k}) = 2f(2k+2) = \frac{1}{2^{2k}(k+1)}\binom{2k}{k}. \qquad (14.3.20)$$

Proof. We introduce a preliminary step X_0, chosen to be distributed as X and independent of the other X_j, and introduce

$$B_{2k+1}^+ = [X_0 = 1, X_0 + \cdots + X_j > 0, j = 1, \ldots, 2k; X_0 + \cdots + X_{2k+1} = 0]$$
$$(14.3.21)$$

which is the event that a random walk starts rightward and returns to the origin for the first time after $(2k + 2)$ steps, so has probability $f(2k + 2)/2$. But this event equals the event

$$[X_0 = 1, \quad X_1 + \cdots + X_j \geq 0 \quad (j = 1, \ldots, 2k);$$
$$X_1 + \cdots + X_{2k} = 0, \quad X_{2k+1} = -1]; \qquad (14.3.22)$$

hence by independence

$$(1/2)f(2k+2) = (1/4)\mathbf{P}[X_1 + \cdots + X_j \geq 0 \quad (j = 1, \ldots, 2k);$$
$$X_1 + \cdots + X_{2k} = 0], \qquad (14.3.23)$$

so $\mathbf{P}(E_{2k}) = 2f(2k + 2)$. Now we observe that

$$\frac{2F(z)}{z^2} = \sum_{k=0}^{\infty}(-1)^k 2\binom{1/2}{k+1}z^{2k} = \sum_{k=0}^{\infty}\binom{2k}{k}\frac{z^{2k}}{2^{2k}(k+1)}, \qquad (14.3.24)$$

where the required probability is the coefficient of z^{2k}.   □

Conclusion of the proof of Theorem 14.3.3 We convert the random walk of m steps into a summand of the expansion of S^m. We introduce a product of m factors, where the jth factor from the left is ℓ when $X_j = 1$ and ℓ^* when $X_j = -1$; in either case we write $\ell^{(X_j)}$. For instance, the sequence of steps $(1, -1, 1, 1, -1, -1)$ gives rise to $\ell\ell^*\ell\ell\ell^*\ell^*$. Generally, the product $\ell^{(X_1)}\ldots\ell^{(X_m)}$ that corresponds to $(X_1, X_2, \ldots, X_m)$ satisfies

$$\langle \ell^{(X_1)}\ldots\ell^{(X_m)}e_0, e_0\rangle = 1 \qquad (14.3.25)$$

when $X_1 + X_2 + \cdots + X_j \geq 0$ for $j = 1, 2, \ldots, m-1$ and $X_1 + X_2 + \cdots + X_m = 0$; that is the walk starts by moving right, is always nonnegative, and is at the origin after m steps. Otherwise

$$\langle \ell^{(X_1)} \ldots \ell^{(X_m)} e_0, e_0 \rangle = 0. \tag{14.3.26}$$

By Lemma 14.3.5 we have

$$\left\langle \left(\frac{\ell + \ell^*}{2} \right)^{2k} e_0, e_0 \right\rangle = \frac{1}{2^{2k}(k+1)} \binom{2k}{k}. \tag{14.3.27}$$

Here S is self-adjoint as viewed as an operator on $\exp \mathbf{C}$; hence S has spectrum σ where σ is a compact subset of $\mathbf{R}$, and the von Neumann algebra generated by I and S is abelian and consists of $\{f(S)\}$ where f belongs to $L^\infty(\sigma)$. Now φ is a faithful tracial state with $\varphi(f) = \langle f(S)e_0, e_0 \rangle$, and by comparison with Proposition 4.5.5 we have

$$\varphi((zI - S)^{-1}) = \sum_{k=0}^\infty \frac{\varphi(S^k)}{z^{k+1}}$$

$$= \sum_{k=0}^\infty z^{-2k-1} \frac{1}{2^{2k}(k+1)} \binom{2k}{k}$$

$$= \frac{2}{\pi} \int_{-1}^1 \frac{\sqrt{1-t^2}}{z-t} dt, \quad (z \in \mathbf{C} \setminus [-1,1]) \tag{14.3.28}$$

hence

$$\varphi(S^n) = \frac{2}{\pi} \int_{-1}^1 s^n \sqrt{1-s^2}\, ds, \tag{14.3.29}$$

hence $\sigma = [-1,1]$ and S has a $S(0,1)$ distribution. □

Example 14.3.6 For $(\Omega, \mathbf{P})$ a probability space, let A_n be the space of strongly measurable functions $X : \Omega \to M_n(\mathbf{C})$ such that $\mathbf{E}\|X\|^k < \infty$ for all $k \in \mathbf{N}$, and let $\phi_n(X) = \mathbf{E}\mathrm{trace}_n(X)$ for $X \in A_n$. Then (A_n, ϕ_n) satisfies axioms (i), (ii), (iii), (iv), (vii) and (viii). Suppose that $Y_n \in A_n$ is a random matrix from the Gaussian unitary ensemble as in Theorem 2.5.3 and Example 4.5.4. Then we have $\phi_n(Y_n^k) = \int y^k \rho_n(dy)$, where ρ_n is the integrated density of states, and so

$$\phi_n(Y_n^k) \to \frac{2}{\pi} \int_{-1}^1 y^k \sqrt{1-y^2}\, dy \quad (n \to \infty). \tag{14.3.30}$$

References

[1] M.J. Ablowitz, A.S. Fokas, *Complex Variables: Introduction and Applications*, second edition, (Cambridge University Press, 2003).

[2] M.J. Ablowitz and Segur, Exact linearization of a Painlevé transcendent, *Phys. Rev. Lett.* **38** (1977), 1103–1106.

[3] C.J. Adkins, *Equilibrium Thermodynamics*, 3rd Edition, (Cambridge University Press, 1983).

[4] A.C. Aitken, *Determinants and Matrices*, 9th Edition, (Oliver and Boyd, Edinburgh, 1956).

[5] D. Aldous and P. Diaconis, Longest increasing subsequences: from patience sorting to the Baik–Deift Johansson Theorem, *Bull. Amer. Math. Soc. (N.S.)* **36** (1999), 413–432.

[6] G.E. Andrews, *The Theory of Partitions*, (Cambridge University Press, 1984).

[7] J. Baik, P. Deift and K. Johansson, On the distribution of the length of the second row of a Young diagram under Plancherel measure, *Geom. Funct. Anal.* **10** (2000), 702–731.

[8] J. Baik, P. Deift, K. Johansson, On the distribution of the length of the longest increasing subsequence of random permutations, *J. Amer. Math. Soc.* **12** (1999), 1119–1178.

[9] D. Bakry and M. Emery, Diffusions hypercontractives, *Séminaire de Probabilités XIX*, J. Azéma and M. Yor (Edrs), Lecture Notes in Mathematics 1123 (Springer, Berlin 1985)

[10] A.R. Barron, Entropy and the central limit theorem, *Ann. Probab.* **14** (1986), 336–342.

[11] E.L. Basor, Toeplitz determinants, Fisher–Hartwig symbols and random matrices, pp. 309–336 in *Recent Perspectives in Random Matrix Theory and Number Theory*, edrs. F. Mezzadri and N.C. Snaith, (Cambridge University Press, 2005).

[12] W. Beckner, Sharp Sobolev inequalities on the sphere and the Moser–Trudinger inequality, *Ann. of Math. (2)* **138** (1993), 213–242.

[13] P. Biane and F. Lehner, Computation of some examples of Brown's spectral measure in free probability, *Colloq. Math.* **90** (2001), 181–211.

[14] P. Biane and R. Speicher, Stochastic calculus with respect to free Brownian motion and analysis on Wigner space, *Probab. Theory Related Fields* **112** (1998), 373–409.

[15] P. Biane and R. Speicher, Free diffusions, free entropy and free Fisher information, *Ann. Inst. H. Poincaré Probab. Statisit.* **37** (2001), 581–606.

[16] P. Biane, and D. Voiculescu, A free probability analogue of the Wasserstein metric on the trace-state space, *Geom. Funct. Anal.* **11** (2001), 1125–1138.

[17] G. Blower, Almost sure weak convergence for the generalized orthogonal ensemble, *J. Statist. Phys.* **105** (2001), 309–355.

[18] G. Blower, Almost sure weak convergence and concentration for the circular ensembles of Dyson, *Stochastics: Stochastic Reports* **75** (2003), 425–433.

[19] G. Blower, Displacement convexity for the generalized orthogonal ensemble, *J. Statist. Phys.* **116** (2004), 1359–1387.

[20] G. Blower, Transportation of measure, Young diagrams and random matrices, *Bernoulli* **10** (2004), 755–782.

[21] G. Blower, Operators associated with soft and hard spectral edges for unitary ensembles, *J. Math. Anal. Appl.* **337** (2008), 239–265.

[22] G. Blower, Integrable operators and the squares of Hankel operators, *J. Math. Anal. Appl.* **140** (2008), 943–953.

[23] G. Blower, Hankel operators that commute with second order differential operators', *J. Math. Anal. Appl.* **342** (2008), 601–614.

[24] G. Blower and A. McCafferty, Discrete Tracy–Widom operators, *Proc. Edin. Math. Soc.* **52** (2009).

[25] S.G. Bobkov and F. Götze, Exponential integrability and transportation cost related to logarithmic Sobolev inequalities. *J. Funct. Anal.* **163** (1999), 1–28.

[26] S. Bobkov, I. Gentil and M. Ledoux, Hypercontractivity of Hamilton–Jacobi equations, *J. Math. Pures Appl. (9)* **80** (2001), 669–696.

[27] S.G. Bobkov and M. Ledoux, From Brunn–Minkowski to Brascamp–Lieb and to logarithmic Sobolev inequalities, *Geom. Funct. Anal.* **10** (2000), 1028–1052.

[28] L.V. Bogachev and Z. Su, Gaussian fluctuations of Young diagrams under the Plancherel measure, *Proc. R. Soc. Lond. Ser. A Math. Phys. Eng. Sci.* **463** (2007), 1069–1080.

[29] F. Bolley, *Applications du transport optimal à des problèmes de limites de champ moyen*, Thesis, Ecole Normale Supérieure De Lyon, December 2005.

[30] A. Borodin, A. Okounkov, G. Olshanski, Asymptotics of Plancherel measures for symmetric groups, *J. Amer. Math. Soc.* **13** (2000), 481–515.

[31] A. Borodin and G. Olshanski, Infinite random matrices and ergodic measures, *Comm. Math. Phys.* **223** (2001), 87–123.

[32] A. Borodin and G. Olshanski, Harmonic analysis on the infinite-dimensional unitary group and determinantal point processes, *Ann. of Math. (2)* **161** (2005), 1319–1422.

[33] A. Borodin and G. Olshanski, Asymptotics of Plancherel type random partitions, *arXiv:math/0610240v2*

[34] A. Borodin and G. Olshanski, z-measures on partitions, Robinson–Schensted–Knuth correspondence and $\beta = 2$ matrix ensembles, pp. 71–94 in *Random Matrix Models and Their Applications*, edrs P. M. Bleher and A.I. Its, MSRI Publication (Cambridge University Press, 2001.)

[35] V.H. Boros and V.H. Moll, *Irresistible Integrals: Symbolics, Analysis and Experiments in the Evaluation of Integrals*, (Cambridge University Press, 2004).

[36] R. Bott, The geometry and representation theory of compact Lie groups, pp. 65–90, *Representation theory of Lie groups*, edr. G.L. Luke, LMS Lecture Note Series 34, (Cambridge University Press, 1979).

[37] A. Boutet de Monvel, L. Pastur and M. Shcherbina, On the statistical mechanics approach in the random matrix theory: integrated density of states, *J. Statist. Phys.* **79** (1995), 585–611.

[38] L. De Branges, The Riemann Hypothesis for Hilbert spaces of entire functions, *Bull. Amer. Math. Soc. (N.S.)* **15** (1986), 1–17.

[39] E. Brézin, C. Itzykson, G. Parisi, J.B. Zuber, Planar diagrams, *Comm. Math. Phys.* **59** (1978), 35–51.

[40] D. Burago, Y. Burago and S. Ivanov, *A Course in Metric Geometry*, (American Mathematical Society, Rhode Island, 2001).

[41] T. Cabanal–Duvillard and A. Guionnet, Discussions around Voiculescu's free entropies, *Adv. Math.* **174** (2003), 167–226.

[42] E.A. Carlen, E.H. Lieb and M. Loss, A sharp analog of Young's inequality on S^N and related entropy inequalities, *J. Geom. Anal.* **14** (2004), 487–520.

[43] E.A. Carlen, and M.C. Carvalho, Strict entropy production bounds and stability of the rate of convergence to equilibrium for the Boltzmann equation, *J. Statist. Phys.* **67** (1992), 575–608.

[44] E.A. Carlen and W. Gangbo, Constrained steepest descent in the 2-Wasserstein metric, *Ann. of Math. (2)* **157** (2003), 807–846.

[45] E. Carlen and M. Loss, Competing symmetries, the logarithmic HLS inequality and Onofri's inequality on S^n, *Geom. Funct. Anal.* **2** (1992), 90–104.

[46] E.A. Carlen, M.C. Carvahlo, R. Esposito, J.L. Lebowitz, and R. Marra, Free energy minimizers for a two-species model with segregation and liquid–vapour transition, *Nonlinearity* **16** (2003), 1075–1105.

[47] I. Chavel, *Riemannian Geometry: A Modern Introduction*, (Cambridge University Press, 1993).

[48] J. Cheeger and D.G. Ebin, *Comparison Theorems in Riemannian Geometry*, (North–Holland, New York, 1975).

[49] A. Connes, Classification of injective factors, cases II_1, II_∞ and III_λ $\lambda \neq 1$, *Ann. of Math (2)* **104** (1976), 73–116.

[50] B. Conrey, Notes on eigenvalue distributions for the classical compact groups, pp. 111–145, *Recent Perspectives in Random Matrix Theory and Number Theory*, F. Mezzadri and N.C. Snaith (edrs.), LMS lecture Note Series 322, (Cambridge University Press, 2005).

[51] D. Cordero-Erausquin, R.J. McCann and M. Schmuckenschläger, A Riemannian interpolation inequality à la Borell, Brascamp and Lieb, *Invent. Math.* **146** (2001), 219–257.

[52] O. Costin and J.L. Lebowitz, Gaussian fluctuations in random matrices, *Physics Review Letters* **75** (1995), 69–72.

[53] K.R. Davidson and S.J. Szarek, Local operator theory, random matrices and Banach spaces, *Handbook of the Geometry of Banach Spaces,* J. Lindenstrauss (edr.) Vol. I, 317–366, (North-Holland, Amsterdam, 2001).

[54] E.B. Davies, *Heat Kernels and Spectral Theory*, (Cambridge University Press, 1989).

[55] P.A. Deift, A.R. Its, Alexander and X. Zhou, A Riemann–Hilbert approach to asymptotic problems arising in the theory of random matrix models, and also in the theory of integrable statistical mechanics, *Ann. of Math. (2)* **146** (1997), 149–235.

[56] P. Deift, *Orthogonal Polynomials and Random matrices: A Riemann–Hilbert approach*, (American Mathematical Society, Rhode Island 1998).

[57] P. Deift, T. Kriecherbauer, K.T.-R. McLaughlin, New results on the equilibrium measure for logarithmic potentials in the presence of an external field, *J. Approx. Theory* **95** (1998), 388–475.

[58] J.D. Deuschel and D.W. Stroock, Hypercontractivity and spectral gap of symmetric diffusions with applications to the stochastic Ising models, *J. Funct. Anal.* **92** (1990), 30–48.

[59] J.-D. Deuschel and D.W. Stroock, *Large Deviations*, (Academic Press, Boston, 1989).

[60] J. Dixmier, *Von Neumann Algebras*, (North-Holland, 1981).

[61] H. Djellout, A. Guillin, and L. Wu, Transportation cost-information inequalities and applications to random dynamical systems and diffusions, *Ann. Probab.* **32** (2004), 2702–2732.

[62] P.G. Drazin and R.S. Johnson, *Solitons: an introduction,* (Cambridge University Press, 1989).

[63] N. Dunford and J.T. Schwartz, *Linear Operators Part 1: General Theory*, (John Wiley and Sons, New York, 1957).

[64] F.J. Dyson, Statistical theory of the energy levels of complex systems I, *J. Mathematical Physics* **3** 140–156 (1962).

[65] W. Feller, *An Introduction to Probability Theory and Its Applications*, Volume II, Second Edition, (John Wiley and Sons, New York, 1971).

[66] P.J. Forrester, Log gases and random matrices, http://www.ms.unimelb.edu.au/~matpjf/matpjf.html.

[67] P.J. Forrester, Hard and soft spacing distributions for random matric ensembles with orthogonal and symplectic symmetry, *Nonlinearity* **19** (2006), 2989–3002.

[68] W. Fulton, *Young Tableaux*, London Mathematical Society Student Texts, (Cambridge University Press, 1997).

[69] S. Gallot, D. Hulin, J. Lafontaine, *Riemannian Geometry*, Second edition, (Springer–Verlag, Berlin, 1987).

[70] T.C. Gard, *Introduction to Stochastic Differential Equations*, (Marcel Dekker, New York, 1998).

[71] D.J.H. Garling, *Inequalities: A Journey into Linear Analysis*, (Cambridge University Press, 2007).

[72] J.B. Garnett, *Bounded Analytic Functions*, (Academic Press, London 1981).

[73] G.R. Grimmett and D.R. Stirzaker, *Probability and Random Processes*, (Oxford Science Publications, Oxford, 1982).

[74] M. Gromov and V.D. Milman, A topological application of the isoperimetric inequality, *Amer. J. Math.* **105** (1983), 843–854.

[75] L. Gross, Logarithmic Sobolev inequalities and contractivity properties of semigroups, Lecture Notes in Mathematics **1563** (Springer, Berlin 1993).

[76] A. Guionnet and O. Zeitouni, Concentration of the spectral measure for large matrices, *Electron. Comm. Probab.* **5** (2000), 119–136.

[77] U. Haagerup and S. Thorbjørnsen, Random matrices and K-theory for exact C^*-algebras. *Doc. Math.* **4** (1999), 341–450.

[78] G.H. Hardy, *Divergent Series*, (Clarendon Press, Oxford, 1956).

[79] G.H. Hardy and S. Ramanujan, Asymptotic distribution of integers of various types, *Proc. London Math. Soc. (2)* **16** (1917), 112–137.

[80] L.L. Helms, *Introduction to Potential Theory*, (John Wiley, New York, 1969).

[81] J.W. Helton, *Analytic Functions, Matrices and Electrical Engineering*, AMS regional conference series, (American Mathematical Society, 1987).

[82] F. Hiai, M. Mizuno, and D. Petz, Free relative entropy for measures and a corresponding perturbation theory, *J. Math. Soc. Japan* **54** (2002), 679–718.

[83] F. Hiai and D. Petz, *The Semicircle Law, Free Random Variables and Entropy*, (American Mathematical Society, Rhode Island, 2000).

[84] F. Hiai and D. Petz, A free analogue of the transportation cost inequality on the circle, *Banach Centre Publications,* **73** pp. 199–206, (Warsaw, 2006).

[85] F. Hiai, D. Petz, and Y. Ueda, Free transportation cost inequalities via random matrix approximation, *Probab. Theory Related Fields,* **130** (2004), 199–221.

[86] T. Hida, *Brownian Motion*, (Springer–Verlag, New York, 1980).

[87] T. Husain, *Introduction to Topological Groups*, (W.B. Saunders, Philadelphia and London, 1966).

[88] K. Itô, *Introduction to Probability Theory*, (Cambridge University Press, 1984).

[89] C. Itzyson and J.-M. Drouffe, *Statistical Field Theory* Vol. 2, (Cambridge University Press, 1989).

[90] N. Jacobson, *Basic Algebra I*, (W.H. Freeman and Company, San Francisco, 1974).

[91] G. James and M. Liebeck, *Representations and Characters of Groups*, second edition, (Cambridge University Press, 2001).

[92] S. Janson, *Gaussian Hilbert Spaces*, (Cambridge University Press, 1997).

[93] S. Jitomirskaya, Metal-insulator transition for the almost Mathieu operator, *Ann. of Math. (2)* **150** (1999), 1159–1175.

[94] K. Johansson, On fluctuations of eigenvalues of random hermitian matrices, *Duke Math. J.* **91** (1998), 151–204.

[95] K. Johansson, The longest increasing subsequence in a random permutation and a unitary random matrix model, *Math. Res. Lett.* **5** (1998), 63–82.

[96] K. Johansson, Discrete orthogonal polynomial ensembles and the Plancherel measure, *Ann. Math. (2)* **153** (2001), 259–296.

[97] J.-P. Kahane, *Some Random Series of Functions*, second edition, (Cambridge University Press, 1985).

[98] N.M. Katz and P. Sarnak, *Random Matrices, Frobenius Eigenvalues and Monodromy*, (American Mathematical Society, Providence, 1999).

[99] S.V. Kerov and A.M. Vershik, Characters and factor representations of the infinite symmetric group, *Soviet Math. Dokl.* **23** (1981), 389–392.

[100] J.F.C. Kingman, Subadditive ergodic theory (with discussion), *Ann. Probability* **1** (1973), 883–909.

[101] J.F.C. Kingman, *Poisson Processes*, (Oxford Science Publications, Clarendon Press, Oxford, 1993).

[102] P. Koosis, *Introduction to H_p Spaces*, (Cambridge University Press, 1980).

[103] P.D. Lax, *Functional Analysis*, (Wiley–Interscience, 2002).

[104] M.R. Leadbetter, G. Lindgren, H. Rootzén, *Extremes and related properties of random sequences and processes*, (Springer–Verlag, New York, 1982).

[105] M. Ledoux, A (one-dimensional) free Brunn–Minkowski inequality, *C. R. Math. Acad. Sci. Paris* **340** (2005), 301–304.

[106] D.E. Littlewood, *The Theory of Group Characters and Matrix Representations of Groups*, (Clarendon Press, Oxford, 1940).

[107] B.F. Logan and L.A. Shepp, A variational problem for random Young tableaux, *Adv. Math.* **26** (1977), 206–222.

[108] J. Lott and C. Villani, Weak curvature conditions and functional inequalities, *J. Funct. Anal.* **245** (2007), 311–333.

[109] I.G. MacDonald, *Symmetric Functions and Hall Polynomials*, Second Edition, (Oxford University Press, New York, 1998).

[110] I.G. MacDonald, Algebraic structure of Lie groups, pp. 91–150 in *Representation theory of Lie groups*, edr. G.L. Luke, (Cambridge University Press, 1979).

[111] W. Magnus and S. Winkler, *Hill's Equation*, (Dover Publications, New York, 1979).

[112] P. Malliavin, *Stochastic Analysis*, (Springer–Verlag, Berlin, 1997).

[113] A. McCafferty, *Operators and Special Functions in Random Matrix Theory*, (PhD Thesis, Lancaster University, 2008).

[114] R.J. McCann, A convexity principle for interacting gases, *Adv. Math.* **128** (1997), 153–179.

[115] H.P. McKean, Geometry of differential space, *Ann. Probability* **1** (1973), 197–206.

[116] M.L. Mehta, *Random Matrices,* second edition, (Academic Press, San Diego, 1991).

[117] V.D. Milman and G. Schechtman, *Asymptotic Theory of Finite-Dimensional Normed Spaces,* with an Appendix by M. Gromov on *Isoperimetric Inequalities in Riemannian Geometry,* Springer Lecture Notes in Mathematics **1200**, (Springer, Berlin, 1986).

[118] F.J. Murray and J. von Neumann, On rings of operators IV: Appendix, *Ann. of Math. (2)* **44** (1943), 716–808.

[119] N.I. Muskhelishvili, *Singular Integral Equations: Boundary Problems of Function Theory and Their Application to Mathematical Physics,* (P. Noordhoff N.V., Groningen, 1966).

[120] M.A. Naimark, *Normed Algebras,* (Wolters–Noordhoff, Groningen, 1972).

[121] E. Nelson, The free Markoff field, *J. Funct. Anal.* **12** (1973), 211–227.

[122] N. Nikolski, *Operators, Functions and Systems: An Easy Reading Volume 2: Model Operators and Systems,* (American Mathematical Society, Providence, 2002).

[123] I. Olkin, The 70th anniversary of the distribution of random matrices: a survey, *Linear algebra and its applications* **354** (2002), 231–243.

[124] B. Osgood, R. Phillips, and P. Sarnak, Extremals of determinants of Laplacians, *J. Funct. Anal.* **80** (1988), 148–211.

[125] L. Pastur and M. Shcherbina, Universality of the local eigenvalue statistics for a class of unitary invariant random matrix ensembles, *J. Statist. Phys.* **86** (1997), 109–147.

[126] L. Pastur and M. Shcherbina, Bulk universality and related properties of Hermitian matrix models, ArXiv: 0705:1050.

[127] V.V. Peller, *Hankel Operators and Their Applications,* (Springer, New York, 2003).

[128] P. Petersen, *Riemannian Geometry,* second edition, (Springer, Berlin, 2006).

[129] G. Pisier, *The Volume of Convex Bodies and the Geometry of Banach Spaces,* (Cambridge University Press, 1989).

[130] L.S. Pontryagin, *Topological Groups,* second edition, (Gordon and Breach, New York, 1966).

[131] T. Ransford, *Potential Theory in the Complex Plane,* London Mathematical Society student texts **28**, (Cambridge University Press, 1995).

[132] M. Reed and B. Simon, *Methods of Modern Mathematical Physics: Volume II,* (Academic Press, New York, 1975).

[133] C. Remling, Schrödinger operators and de Branges spaces, *J. Funct. Anal.* **196** (2002), 323–394.

[134] W. Rossmann, *Lie Groups: An Introduction Through Linear Groups,* (Oxford Science Publications, 2002).

[135] O.S. Rothaus, Hypercontractivity and the Bakry–Emery criterion for compact Lie groups, *J. Funct. Anal.* **65** (1986), 358–367.

[136] M. Rudelson, Lower estimates for the singular values of random matrices, *C.R. Math. Acad. Sci. Paris* **342** (2006), 247–252.

[137] E.B. Saff, and V. Totik, *Logarithmic Potentials With External Fields*, (Springer, Berlin, 1997).

[138] B.E. Sagan, *The symmetric group, representations, combinatorial algorithms and symmetric functions*, (Springer, New York, 2001).

[139] R.L. Schilling, *Measures, Integrals and Martingales*, (Cambridge University Press, 2005).

[140] J.W. Silverstein, The smallest eigenvalue of a large Wishart matrix, *Ann. Probab.* **13** (1985), 1364–1368.

[141] G.F. Simmons, *Introduction to Topology and Modern Analysis*, (McGraw–Hill, Auckland, 1963).

[142] G.F. Simmons, *Differential Equations with Applications and Historical Notes*, (Tata McGraw–Hill, New Delhi, 1972).

[143] B. Simon, *Trace Ideals and Their Applications*, (Cambridge University Press, 1979).

[144] B. Simon, *Representations of Finite and Compact Groups*, (American Mathematical Society, 1996).

[145] B. Simon, *Orthogonal Polynomials on the Unit Circle, Part 1: Classical Theory*, (American Mathematical Society, 2005).

[146] I.N. Sneddon, *Fourier Transforms*, (Dover, New York, 1995).

[147] A.G. Soshnikov, Gaussian limit for determinantal random point fields. *Ann. Probab.* **30** (2002), 171–187.

[148] A.G. Soshnikov. Determinantal random point fields (2000), *Russian Math. Surveys* **55** (2000), 923–975.

[149] K.-T. Sturm, Convex functional of probability measures and nonlinear diffusions on manifolds, *J. Math. Pures Appl. (9)* **84** (2005), 149–168.

[150] W.A. Sutherland, *Introduction to Metric and Topological Spaces*, (Clarendon Press, Oxford, 1975).

[151] S.J. Szarek, Spaces with large distance to ℓ_∞^n and random matrices, *Amer. J. Math.* **112** (1990), 899–942.

[152] S.J. Szarek and D. Voiculescu, Shannon's entropy power inequality via restricted Minkowski sums. *Geometric aspects of functional analysis*, 257–262, Lecture Notes in Math., 1745, (Springer, Berlin, 2000).

[153] G. Szegő, *Orthogonal Polynomials*, American Mathematical Society Colloquium Publications, XXIII, (American Mathematical Society, New York, 1959).

[154] E.C. Titchmarsh, *The Theory of Functions*, (Oxford University Press, 1939).

[155] V. Totik, Asymptotics for Christoffel functions with varying weights, *Adv. Apply. Math.* **25** (2000), 322–351.

[156] C.A. Tracy and H. Widom, Correlation functions, cluster functions, and spacing distributions for random matrices, *J. Statist. Phys.* **92** (1998), 89–835.

[157] C.A. Tracy and H. Widom, Level spacing distributions and the Airy kernel, *Comm. Math. Phys.* **159** (1994), 151–174.

[158] C.A. Tracy and H. Widom, Level spacing distributions and the Bessel kernel, *Comm. Math. Phys.* **161** (1994), 289–309.

[159] C.A. Tracy and H. Widom, Fredholm determinants, differential equations and matrix models, *Comm. Math. Phys.* **163** (1994), 33–72.

[160] F.G. Tricomi, *Integral Equations*, (Interscience Publishers, New York, 1957).

[161] A.M. Vershik and S.V. Kerov, Asymptotic behavior of the Plancherel measure of the symmetric group and the limit form of Young tableaux, *Dokl. Akad. Nauk. SSSR* **233**, 1024–1027.

[162] C. Villani, *Topics in Optimal Transportation*, (American Mathematical Society, Rhode Island, 2003).

[163] D. Voiculescu, The analogues of entropy and of Fisher's information measure in free probability theory I, *Comm. Math. Phys.* **155** (1993), 71–92.

[164] D.V. Voiculescu, The analogues of entropy and of Fisher's information measure in free probability theory II, *Invent. Math.* **118** (1994), 411–440.

[165] D.V. Voiculescu, The analogues of entropy and of Fisher's information measure in free probability V: Noncommutative Hilbert transforms, *Invent. Math.* **132** (1998), 189–227.

[166] D.V. Voiculescu, K.J. Dykema and A. Nica, *Free Random Variables*, (American Mathematical Society, Rhode Island, 1992).

[167] G.N. Watson, *A Treatise on the Theory of Bessel Functions*, second edition, (Cambridge University Press, 1962).

[168] A. Wainwright, *A Pictorial Guide to the Lakeland Fells 4: The Southern Fells*, Kendal: Westmorland Gazette 1960.

[169] E.T. Whittaker and G.N. Watson, *A Course of Modern Analysis*, fourth edition (Cambridge University Press, 1963).

[170] J. Wishart, The generalised product moment distribution in samples from a normal multivariate population, *Biometrika* **20A** (1928), 32–52.

[171] N. Young, *An Introduction to Hilbert Space*, (Cambridge University Press, 1988).

[172] F. Morgan, Manifolds with density and Perelman's proof of the Poincaré conjecture, *The American Mathematical Monthly*, **116** (2009), 134–142.

[173] I.S. Gradshteyn and M. Ryzhik, *Table of Integrals, Series, and Products*, Fifth edition (Academic Press, San Diego, 1994).

[174] M. Ledoux and I. Popescu, Mass transportation proofs of free functional inequalities, and free Poincaré inequalities, *J. Funct. Anal.*

[175] E.A. Carlen, Superadditivity of Fisher's information and logarithmic Sobolev inequalities, *J. Funct. Anal.* **101** (1991), 194–211.

[176] W. Rudin, *Real and Complex Analysis*, Second edition (Tata McGraw-Hill, New Delhi, 1974).

Index

actions 16
action, adjoint 59
action, group 16
adjoint action 59
aerofoil Equations 155
affine connection 63
Airy's equation 291
Airy function, asymptotic expansion 300
Airy kernel 274, 300
algebra, group von Neumann 417
algebra, Lie 57
algebra, Von Neumann 415
almost sure weak convergence 180
almost sure weak convergence of
    empirical distribution 190
Andréief's identity 51
annihilation and creation operators 379
arclength 177

backward equation 395
Baik–Deift–Johansson Theorem 342
Bakry–Emery LSI 207
Banach's fixed point theorem 402
band-limited function 295
Barron's theorem 203, 397
Bessel function 291
Bessel function, modified
    (MacDonald) 291
Bessel function of integral order 246
Bessel kernel 306
Biane–Voiculescu Theorem 221
Bobkov–Götze Theorem 105
Bochner's theorem 267
Bochner–Weitzenbock formula 224
Bogachev–Su Theorem 343
Boltzmann's entropy 84
Borel measure 7
Borel–Cantelli Lemma, first 9
Borodin–Okounkov–Olshanski Theorem
    341

Boutet de Monvel, Pastur, Shcherbina
    Theorem 182
Brascamp–Lieb inequality 91
Brownian motion 366
Brownian motion, joint distribution 366
Brownian motion, matrix 399
Brunn–Minkowski Theorem 22
Burgers equation 213, 407

C^* probability space 411
canonical system 276
Carlen's theorem on marginals 210
Carlen and Loss's Theorem 146
Carleson's interpolation theorem 314
Cartan's criterion 60
Cartan's theorem (on energy
    spaces) 140
Cartan–Killing classification 78
Catalan numbers 158, 422
Cauchy transform 38
chaos, homogeneous 387
characteristic function 33
Chebyshev's distribution 141
Chebyshev's polynomials 141, 349
Christoffel–Darboux formula 271
circular orthogonal ensemble 81
circular symplectic ensemble 83
circular unitary ensemble 79
classical groups 43, 72
classical probability spaces 412
compact metric group 10
compact metric space 5
complement principle 335
complex Burgers equation 214, 407
concentration inequality 94
concentration inequality for compact
    Lie groups 128
concentration of measure 96
concentration of measure for generalized
    ensembles 109

concentration of measure for GOE 108
concentration of measure on spheres 124
configurations 254
confluent hypergeometric equation 291, 293
conjugate function 136
conjugation 43
constrained variational problem 166
continuity equation 198
continuity, Hölder 39
convergence, almost sure weak 180
convergence of distribution sequences 38
convergence of IDS 180
convergence, Operator 48
convergence in energy 140
convexity 86
convexity, uniform 106
convolution 33
correlation function 261
correlation measure 255
cost, transportation 99
Costin–Lebowitz central limit theorem 324
creation and annihilation operators 379
Csiszár's inequality 92
cumulants 36
curvature 63, 223

De Branges's axioms 275
de Bruijn's identity 51
density of states 180, 185, 401
derivations 53
determinantal random point field 257
dimensions of classical ensembles 44
dimension formula 44
Dirichlet kernel 294
Dirichlet space 137
discrete Bessel kernel 328
discrete Tracy–Widom system 327
displacement convexity 163, 169
distribution sequence 37
derivative, weak 197
dual space 5
Duhamel's Formula 55
Dyson's circular ensembles 78
Dyson's universality principle 320

electrostatic energy 147
empirical eigenvalue distribution 47
endomorphism 59
energy, convergence in 140
energy, Logarithmic 141
energy norm 140
energy space 140
enlargement 32
ensemble, orthogonal circular 81
ensemble, unitary circular 79

ensemble, symplectic circular 83
ensemble, generalized orthogonal 68
ensemble, generalized symplectic 68
ensemble, generalized unitary 68
entropy 84
entropy, Boltzmann's 84
entropy, free on circle 142
entropy power, Shannon's 211
entropy, relative 87
entropy, relative free 143
entropy, Shannon 84
ε-net 5
equidistribution theorem, Weyl's 14
equilibrium measure on circle 177
equilibrium measure on conductor 141
equilibrium measure on the real line 147
    integral formula 149
exponential bases 313
Eulerian integrals 165

factor 416
factors, finite 416
factors, nonhyperfinite 417
Fatou's Theorem 135
finite factors 416
finite Hilbert transform 155
finite shift operator 418
Fisher's measure of information 210
fluctuations 321
Fock space 374
Fokker–Planck equation 202
Fourier transform 34, 279
flow 197
Fredholm's formula 263
free algebras 413
free entropy 197, 171
free entropy on circle 142
free information 215
free LSI 216
free LSI (Ledoux) 218
free Relative entropy 200
free Relative information 215
free transportation inequality 200
Frobenius's coordinates 235
Frobenius's formula 235
functional calculus 55

Gaudin's Lemma 272
Gauss–Legendre multiplication formula 133
Gaussian measure 25
Gaussian orthogonal ensemble 26
Gaussian Hilbert space 365
Gelfand–Naimark Theorem 414
generalized orthogonal ensemble 68
generalized symplectic ensemble 68
generalized unitary ensemble 68

generator 57
Gessel's formula 246
Gibbs measure 223
Glivenko's theorem 34
GNS theorem 414
gradient flow 202
Green's function 140
Gromov–Lévy Theorem 125
Gross's LSI 208
Gross's theorem on hypercontractivity 383
group algebra 53, 227
group, compact metric 10
groups classical 42, 72
group, Lie 56
group representation 66, 227
group von Neumann algebra 416
group, Weyl 72
Gumbel distribution 114

Haar measure 11
Hamilton–Jacobi equation 206
Hammersley's theorem 245
Hankel determinant 348
Hankel integral operator 283
Hankel matrix 310
Hankel transform 279
hard edge 274, 304
Hardy–Ramanujan formula 249
Hardy space 136
harmonic functions 136
Hausdorff distance 32
Heine's Theorem 270
Helmholtz equation 297
Helson–Szegö Theorem 174
Hermite function 279
Hermite polynomials 297
Hermite polynomials on $\mathbf{R}^n$ 384
Hessian for matrices 53
Hilbert–Schmidt norm 47
Hilbert space, Gaussian 365
Hilbert tensor product 373
Hilbert transform 40
Hilbert transform, finite 155
Hilbert transform on circle 136
Hölder continuity 39
Hölder's inequality 20
homogeneous chaos 387
hook length formula 234
Hua's lemma 116
Hua–Pickrell measure 357
hypercontractivity 382
hyperfinite factor 416
hypergeometric equation 291, 293

ideals in Lie algebra 60
IDS integrated density of states 180

IDS, convergence of 182
inclusion-exclusion principle 250
induced measure 9
inductive limits 352
information, relative 211
information, Fisher's 196, 210
information, relative Fisher 211
information, free 215
information, relative free 215
inner regularity 17
integrable operators 282
interpolating sequence 313
invariant measure 10, 16, 25
inverse temperature β 88
involutions 43
irreducible representation 228
isoperimetric inequality 24, 96
Itô's formula for matrix stochastic differential equation 404

Jacobi's ensemble 284, 304
Jacobi's identity 60
Jacobi's polynomials 305
Jensen's inequality 87
Johansson's fluctuation theorem 350

Kac's theorem 34
Kantorovich's Theorem 100
Killing form 60
Killing–Cartan classification 61
Kronecker's Theorem 15
Kronecker's Theorem on Hankel operators 312
Kuratowski 17

λ-tableaux 232
Laguerre functions 284, 311
Laplacian on matrices 53
Laplacian on $S(\sqrt{\infty})^{\infty}$ 389
Lebedev–Milin Theorem 138
Ledoux's Theorem (free LSI) 218
Legendre transform 94
Lévy's Theorem 28
Lévy–Gromov Theorem 125
Lidskii's Lemma 48
Lidskii's Lemma for unitary matrices 129
Lie algebra 57
Lie group 43, 56
linear Lie group 56
linear statistic 176
Lipschitz function 31
Logan and Shepp's variational formula 238
logarithmic moment generating function 94
logarithmic energy 132

logarithmic Sobolev inequality LSI 203
LSI for generalized ensembles 208
LSI for eigenvalue distributions 209
LSI, free 216
Lusin's Lemma 17

MacDonald function 291
Marchenko–Pastur distribution 193
marginal densities 31, 89, 210
Markov's property 393
Maschke's theorem 228
Mathieu's equation 328
matricial norms 46
maximal torus 72
Mazur's Theorem 6
mean field convergence 183
measure 7
Mehler's formula 379
Mehler's kernel 379
Mehler–Heine formula 309
Mercer's theorem 263
metrics 4, 47
Möbius function 250
modules 227
molecular field equation 185
moments 34
moments, Gaussian 35
moments, semicircular 418
moments; Stieltjes's example 35
Monge's transportation problem 99
monogenic 14
multiplicity 254

net, ε-net 5
Nicolson's approximation 332
normal 414
norms, matricial 46
norms Schatten 46

operator convergence 48
optimal transportation 101
orthogonal circular ensemble 79
OU process 392
Ornstein–Uhlenbeck process 393, 394
OU process, matrix 398
Ornstein–Uhlenbeck semigroup 379
Ornstein–Uhlenbeck operator 376
orthogonal circular ensemble 79
orthogonal group, infinite 355
orthogonal polynomials 270
Ω distribution, 237

Painlevé equations 303, 315
parabolic cylinder function 297
partition function 249
Paley–Wiener theorem 295
Pfaffian 50

Plancherel measure of symmetric group
234
Plancherel–Rotach asymptotic formula
299
Plemelj's formula 40
Poincaré's inequality 222
Poincaré–McKean theorem 29
Poisson kernel 135
Poisson process 243
Poissonization 345
Polish spaces 4
positive semidefinite 45
positivity 414
potential energy 148
Prékopa–Leindler inequality 22
principal value integral p.v. 39
probability space 7
prolate spheroidal functions 295
pseudo–Jacobi functions 361

quartic potential 159
quaternions 42

Rain's formula 245
random point field 253
random walks and generating
functions 420
rational approximation of Hankel
operators 310
Rayleigh–Schrödinger formula 54
relative Fisher information 211
relative entropy 87
relative free entropy
relative free information 215
reproducing kernels 135
Ricci curvature 64, 223
Riesz basis 314
Riesz Representation Theorem 7
Riesz's theorem on Hilbert transform 40
Robinson–Schensted–Knuth theorem
230
Rodrigues's formula 376
roots 74
root system 77
Rudelson's theorem 122

Schatten norms 46
Schur product 45, 374
Schur polynomial 232
score function 212
semicircular distribution 157
semigroup, Ornstein–Uhlenbeck 379
semigroup, symmetric diffusion 381
semisimple 60
Shannon entropy 84
Shannon's Entropy Power 211
Schatten norm 46

shift, finite 418
Silverstein's Lemma 119
sine kernel 274, 293
singular numbers 46
singular numbers of Hankel operator
    310
Sobolev space 139
Sobolev inequality on sphere 390
soft edge 299
Sonine 291
Specht module 232
spectral bulk 293
spectral gap 221
spectrum of Hankel operator 310
sphere, surface area 12
spheroidal function, prolate 295
states 415
Stieltjes's example on moments 35
Stirling numbers 252
stochastic integral 368
Student's t-distribution 358
subadditive ergodic theorem 245
subadditivity of relative entropy 89
sum of sets 22
support of equilibrium measure 148, 150
symmetric powers 374
symmetric diffusion semigroup 381
symmetric group, infinite 352
symmetric tensor powers 375
symplectic group 43
symplectic circular ensemble 83
Szegő's limit theorem 175
Szegő's limit theorem, strong 346

Talagrand's transportation theorem 108
Tauberian theorem for Borel
    summability 345
temperature, inverse β 88
tensor product, Hilbert 373
tensorization of LSI 205
three-term recurrence relation 270
tightness 9
Toeplitz determinant 247
torus, maximal 72
torus, infinite 353
total boundedness 5
trace, on matrices 46
trace on von Neumann algebra 416
tracial probability space 414

Tracy–Widom distribution 315
Tracy–Widom system 284
Tracy–Widom system, discrete 327
Tracy–Widom theorem on soft edges
    319
transportation cost 100
transportation inequality 103
Tricomi's Lemma 156
Tychonov theorem 6

Ulam's problem 229
uniform convexity 106
unitary circular ensemble 79
unitary group, infinite 356
unitary representation 51
universality 320

Vandermonde's determinant 49
variation norm 7
Vershik–Kerov theorem 248
Vershik's Ω distribution 240
Von Neumann algebra 415
von Neumann–Schatten norm 46
von Neumann algebra, group 417

Wallis's product 29
Wasserstein metric 100
Wasserstein metric dual form 101
Wasserstein metric in noncommutative
    case
weak convergence, characterization of
    102
weak convergence, almost sure 189
weak derivative 197
Weber's parabolic cylinder function 297
Weyl's equidistribution theorem 14
Weyl denominator 76
Weyl group 72, 78
Weyl integration formula 72
white noise 369
Whittaker's function 293
Wick ordering 376
Wigner's semicircle law 157
Wishart matrices 193
Wishart eigenvalue distribution 193
Wishart eigenvalue convergence 193

Young diagram 229
Young tableaux 230